Java
核心技术 卷II
高级特性（原书第11版）

Core Java Volume II—Advanced Features
(Eleventh Edition)

[美] 凯·S. 霍斯特曼（Cay S. Horstmann） 著

陈昊鹏 译

图书在版编目（CIP）数据

Java 核心技术 卷II 高级特性（原书第 11 版）/（美）凯·S. 霍斯特曼（Cay S. Horstmann）著；陈昊鹏译 . —北京：机械工业出版社，2019.12（2023.3 重印）

（Java 核心技术系列）

书名原文：Core Java, Volume II—Advanced Features (Eleventh Edition)

ISBN 978-7-111-64343-2

I. J… II. ①凯… ②陈… III. JAVA 语言 – 程序设计 IV. TP312.8

中国版本图书馆 CIP 数据核字（2019）第 268688 号

北京市版权局著作权合同登记 图字：01-2019-2815 号。

Authorized translation from the English language edition, entitled *Core Java, Volume II—Advanced Features (Eleventh Edition)*, ISBN: 978-0-13-516631-4, by Cay S. Horstmann, published by Pearson Education, Inc., Copyright © 2019 Pearson Education Inc., Portions Copyright © 1996-2013 Oracle and/or its affiliates.

All rights reserved. No part of this book may be reproduced or transmitted in any form or by any means, electronic or mechanical, including photocopying, recording or by any information storage retrieval system, without permission from Pearson Education, Inc.

Chinese simplified language edition published by China Machine Press, Copyright © 2020.

本书中文简体字版由 Pearson Education（培生教育出版集团）授权机械工业出版社在中国大陆地区（不包括香港、澳门特别行政区及台湾地区）独家出版发行。未经出版者书面许可，不得以任何方式抄袭、复制或节录本书中的任何部分。

本书封底贴有 Pearson Education（培生教育出版集团）激光防伪标签，无标签者不得销售。

Java 核心技术 卷II 高级特性（原书第 11 版）

出版发行：机械工业出版社（北京市西城区百万庄大街 22 号　邮政编码：100037）

责任编辑：关　敏　　　　　　　　　　　责任校对：殷　虹

印　　刷：北京联兴盛业印刷股份有限公司　版　　次：2023 年 3 月第 1 版第 8 次印刷

开　　本：186mm×240mm　1/16　　　　　印　　张：43.5

书　　号：ISBN 978-7-111-64343-2　　　　定　　价：149.00 元

客服电话：(010) 88361066　68326294

版权所有・侵权必究
封底无防伪标均为盗版

译 者 序

《Java核心技术 卷II 高级特性》（原书第11版）就要出版了！本书在上一版的基础上针对Java 11进行了全面修订，除了之前版本中包含的高级UI特性、企业编程、网络、安全等主题之外，还新纳入了模块系统等内容，以全面反映Java 11的最新特性。从某种程度上说，这本书的版本演进反映了Java语言本身的演进过程。虽然在人工智能和大数据的发展风起云涌的大环境下，以Python为代表的新兴语言不断涌现并迅速被广为接受，但Java在企业级Web应用方面的优势还是非常明显，其完备的语言特性和丰富的生态圈使得它在编程语言排行榜上一直位居前列，深受广大程序员的青睐和追捧。可以预见，Java在未来仍然会有广阔的市场空间，Java程序员的数量会持续增长，Java的生态圈也会不断地扩展壮大。

本书在写作上保留了以往的风格，并针对新内容进行了扩展，读者既可以将其作为学习Java语言的进阶教材，学习利用Java语言编写企业级Web应用的高级特性，又可以将其当作Java语言的API手册，在编程时作为查询API细节的工具书。本书提供的所有样例代码几乎都进行了更新，以反映Java 11中新的变化。

本书中文版是在之前版本的基础上完成的。在翻译本书的过程中，译者不但对新增加的内容进行了翻译，还对之前版本中存在的错误和不符合中文表达习惯的地方进行了修正，力求以准确和流畅的语言重现英文版的内容和韵味，希望读者在阅读本书时能够感受到Java的魅力和作者的风格。

Java技术在物联网时代还会因其卓越的跨平台能力继续大放异彩，让我们一起拥抱Java，拥抱新一代互联网技术，让世界变得更智慧！

陈昊鹏

前 言

致读者

本书是按照 Java SE 11 进行更新的。卷 I 主要介绍了 Java 语言的一些关键特性，而本卷主要介绍编程人员进行专业软件开发时需要了解的高级主题。因此，与卷 I 和之前的版本一样，我们仍将本书定位于用 Java 技术进行实际项目开发的编程人员。

编写任何一本书籍都难免会有一些错误或不准确的地方。我们非常乐意听到读者的意见。当然，我们更希望对相同问题的报告只出现一次。为此，我们创建了一个 FAQ、bug 修正以及应急方案的网站 http://horstmann.com/corejava。你可以在 bug 报告网页的末尾处（鼓励读者阅读以前的报告）添加 bug 报告，以此来发布 bug 和问题并给出建议，以便我们提高本书将来版本的质量。

内容提要

本书中的章节大部分是相互独立的。你可以研究自己最感兴趣的主题，并可以按照任意顺序阅读这些章节。

在**第 1 章**中，你将学习 Java 的流库，它带来了现代风格的数据处理机制，即只需指定想要的结果，而无须详细描述应该如何获得该结果。这使得流库可以专注于优化的计算策略，对于优化并发计算来说，这显得特别有利。

第 2 章的主题是输入/输出处理。在 Java 中，所有 I/O 都是通过输入/输出流来处理的。这些流（不要与第 1 章的那些流混淆了）使你可以按照统一的方式来处理与各种数据源之间的通信，例如文件、网络连接或内存块。我们对各种读入器和写出器类进行了详细的讨论，它们使得对 Unicode 的处理变得很容易。我们还展示了如何使用对象序列化机制使保存和加载对象变得容易而方便，以及对象序列化机制背后的原理。然后，我们讨论了正则表达式以及如何操作文件与路径。该章通篇都包含了最新的 Java 版本中引入的广受欢迎的改进和优化。

第 3 章介绍 XML，展示怎样解析 XML 文件、怎样生成 XML 以及怎样使用 XSL 转换。在一个实用示例中，我们将展示怎样在 XML 中指定 Swing 窗体的布局。我们还讨论了 XPath API，它使得"在 XML 的干草堆中寻找绣花针"变得更加容易。

第 4 章介绍网络 API。Java 使复杂的网络编程工作变得很容易实现。我们将介绍怎样连接到服务器，怎样实现你自己的服务器，以及怎样创建 HTTP 连接。该章还讨论了新的 HTTP 客户端。

第 5 章介绍数据库编程，重点讲解 JDBC，即 Java 数据库连接 API，这是用于将 Java 程序与关系数据库进行连接的 API。我们将介绍怎样通过使用 JDBC API 的核心子集，编写能够处理实际的数据库日常操作事务的实用程序。（如果要完整介绍 JDBC API 的功能，可能需要编写一本像本书一样厚的书才行。）最后我们简要介绍了层次数据库，探讨了 JNDI（Java 命名及目录接口）以及 LDAP（轻量级目录访问协议）。

Java 对于处理日期和时间的类库做出过两次设计，而在 Java 8 中做出的第三次设计则极富魅力。在第 6 章，你将学习如何使用新的日期和时间库来处理日历和时区的复杂性。

第 7 章讨论一个我们认为其重要性将会不断提升的特性——国际化。Java 编程语言是少数几种一开始就被设计为可以处理 Unicode 的语言之一，不过 Java 平台的国际化支持则走得更远。因此，你可以对 Java 应用程序进行国际化，使其不仅可以跨平台，而且还可以跨国界。例如，我们会展示怎样编写一个使用英语、德语和汉语的退休金计算器。

第 8 章讨论三种处理代码的技术。脚本机制和编译器 API 允许程序去调用使用诸如 JavaScript 或 Groovy 之类的脚本语言编写的代码，并且允许程序去编译 Java 代码。可以使用注解向 Java 程序中添加任意信息（有时称为元数据）。我们将展示注解处理器怎样在源码级别或者在类文件级别上收集注解，以及怎样运用注解来影响运行时的类行为。注解只有在工具的支持下才有用，因此，我们希望这些讨论能够帮助你根据需要选择有用的注解处理工具。

第 9 章讲解从 Java 9 开始引入的 Java 平台模块系统，以促进 Java 平台和核心类库的有序演化。这个模块系统提供了对包的封装和用于描述模块需求的机制。你将学习模块的属性，以便决定是否要在自己的应用程序中使用它们。即使你决定不使用，也需要了解这些新规则，这样你才能和 Java 平台以及其他模块化的类库交互。

第 10 章继续介绍 Java 安全模型。Java 平台一开始就是基于安全而设计的，该章会带你深入内部，查看这种设计是怎样实现的。我们将展示怎样编写用于特殊应用的类加载器和安全管理器。然后介绍允许使用消息、代码签名、授权和认证以及加密等重要特性的安全 API。最后，我们用一个使用 AES 和 RSA 加密算法的示例进行总结。

第 11 章讨论没有纳入卷 I 的所有 Swing 知识，尤其是重要但很复杂的树形构件和表格构件。我们还会介绍 Java 2D API，你可以用它来创建实际的图形和特殊的效果。当然，如今已经没有多少程序员需要编写 Swing 用户界面了，因此我们会将注意力放到在服务器端生成图像的实用特性上。

第 12 章介绍本地方法，这个功能支持你调用为微软 Windows API 这样的特殊机制而编写的各种方法。很显然，这种特性具有争议：使用本地方法，那么 Java 平台的跨平台特性将会随之消失。毫无疑问，每个为特定平台编写 Java 应用程序的专业开发人员都需要了解这些技术。有时，当你与不支持 Java 平台的设备或服务进行交互时，为了你的目标平台，你可能需要求助于操作系统 API。我们将通过展示如何从某个 Java 程序访问 Windows 注册表 API 来阐明这一点。

所有章节都按照最新版本的 Java 进行了修订，过时的材料都删除了，Java 9、10 和 11 的新 API 都详细地进行了讨论。

约定

我们使用等宽字体表示计算机代码，这种格式在众多的计算机书籍中极为常见。各种图标的含义如下：

> **注释**：需要引起注意的地方。

> **提示**：有用的提示。

> **警告**：关于缺陷或危险情况的警告信息。

> **C++ 注释**：本书中有许多这类提示，用于解释 Java 程序设计语言和 C++ 语言之间的不同。如果你对这部分不感兴趣，可以跳过。

Java 平台配备有大量的编程类库或者应用程序编程接口（API）。当第一次使用某个 API 时，我们在每一节的末尾都添加了简短的描述。这些描述可能有点不太规范，但是比官方在线 API 文档更具指导性。接口的名字以斜体显示，就像许多官方文档一样。类、接口或方法名后面的数字是 JDK 的版本，表示在该版本中才引入了相应的特性。

> **API** 应用程序编程接口

本书示例代码以程序清单的形式列举出来，例如：

> **程序清单 1-1** ScriptTest.java

可以从网站 http://horstmann.com/corejava 下载示例代码。

致 谢

写一本书需要投入大量的精力，升级一本书也并不像想象的那样轻松，尤其是 Java 技术一直在持续不断地更新。出版一本书会让很多人耗费很多心血，在此衷心地感谢《Java 核心技术》小组的每一位成员。

Pearson 公司的许多人提供了非常有价值的帮助，却甘愿做幕后英雄。在此，我希望大家都能够知道我对他们努力的感恩。与以往一样，我要真诚地感谢我的编辑 Greg Doench，从本书的写作到出版他一直在给予我们指导，同时感谢那些不知其姓名的为本书做出贡献的幕后人士。非常感谢 Julie Nahil 在图书制作方面给予的支持，感谢 Dmitry Kirsanov 和 Alina Kirsanova 完成手稿的编辑与排版工作。我还要感谢早期版本中我的合作者 Gary Cornell，他已经转向了其他行业。

感谢早期版本的许多读者，他们指出了许多令人尴尬的错误并给出了许多具有建设性的修改意见。我还要特别感谢本书优秀的审校小组，他们仔细地审阅我的手稿，使本书减少了许多错误。

这一版及以前版本是由以下人员评审的：Chuck Allison（特约编辑，*C/C++ Users Journal*）、Lance Anderson（Oracle）、Alec Beaton（PointBase, Inc.）、Cliff Berg（iSavvix Corporation）、Joshua Bloch、David Brown、Corky Cartwright、Frank Cohen（PushToTest）、Chris Crane（devXsolution）、Nicholas J. De Lillo 博士（曼哈顿学院）、Rakesh Dhoopar（Oracle）、Robert Evans（资深教师，约翰·霍普金斯大学应用物理实验室）、David Geary（Sabreware）、Jim Gish（Oracle）、Brian Goetz（Oracle）、Angela Gordon、Dan Gordon、Rob Gordon、John Gray（Hartford 大学）、Cameron Gregory（olabs.com）、Steve Haines、Marty Hall（约翰·霍普金斯大学应用物理实验室）、Vincent Hardy、Dan Harkey（圣何塞州立大学）、William Higgins（IBM）、Vladimir Ivanovic（PointBase）、Jerry Jackson（ChannelPoint Software）、Tim Kimmet（Preview Systems）、Chris Laffra、Charlie Lai、Angelika Langer、Doug Langston、Hang Lau（McGill 大学）、Mark Lawrence、Doug Lea（SUNY Oswego）、Gregory Longshore、Bob Lynch（Lynch Associates）、Philip Milne（顾问）、Mark Morrissey（俄勒冈研究生院）、Mahesh Neelakanta（佛罗里达大西洋大学）、Hao Pham、Paul Philion、Blake Ragsdell、Ylber Ramadani（Ryerson 大学）、Stuart Reges（亚利桑那大学）、Simon Ritter、Rich Rosen（Interactive Data Corporation）、Peter Sanders（法国尼斯 ESSI 大学）、Paul Sanghera 博士（圣何塞州立大学和布鲁克斯学院）、Paul Sevinc（Teamup AG）、Yoshiki Shabata、Devang Shah、Richard Slywczak（NASA/Glenn 研究中心）、Bradley A. Smith、Steven Stelting、Christopher Taylor、Luke Taylor（Valtech）、George Thiruvathukal、Kim Topley（*Core JFC*，*Second Edition* 的作者）、Janet Traub、Paul Tyma（顾问）、Christian Ullenboom、Peter van der Linden、Burt Walsh、Joe Wang（Oracle）和 Dan Xu（Oracle）。

Cay Horstmann
2018 年 12 月于加州旧金山

目　录

译者序
前言
致谢

第 1 章　Java 8 的流库 ………… 1
1.1 从迭代到流的操作 ………… 1
1.2 流的创建 ………… 3
1.3 filter、map 和 flatMap 方法 ………… 8
1.4 抽取子流和组合流 ………… 9
1.5 其他的流转换 ………… 10
1.6 简单约简 ………… 11
1.7 Optional 类型 ………… 13
1.7.1 获取 Optional 值 ………… 13
1.7.2 消费 Optional 值 ………… 13
1.7.3 管道化 Optional 值 ………… 14
1.7.4 不适合使用 Optional 值的方式 ………… 15
1.7.5 创建 Optional 值 ………… 16
1.7.6 用 flatMap 构建 Optional 值的函数 ………… 16
1.7.7 将 Optional 转换为流 ………… 17
1.8 收集结果 ………… 19
1.9 收集到映射表中 ………… 24
1.10 群组和分区 ………… 27
1.11 下游收集器 ………… 28
1.12 约简操作 ………… 32
1.13 基本类型流 ………… 34
1.14 并行流 ………… 39

第 2 章　输入与输出 ………… 43
2.1 输入 / 输出流 ………… 43
2.1.1 读写字节 ………… 43
2.1.2 完整的流家族 ………… 46
2.1.3 组合输入 / 输出流过滤器 ………… 50
2.1.4 文本输入与输出 ………… 53
2.1.5 如何写出文本输出 ………… 53
2.1.6 如何读入文本输入 ………… 55
2.1.7 以文本格式存储对象 ………… 56
2.1.8 字符编码方式 ………… 59
2.2 读写二进制数据 ………… 61
2.2.1 DataInput 和 DataOutput 接口 ………… 61
2.2.2 随机访问文件 ………… 63
2.2.3 ZIP 文档 ………… 67
2.3 对象输入 / 输出流与序列化 ………… 70
2.3.1 保存和加载序列化对象 ………… 70
2.3.2 理解对象序列化的文件格式 ………… 74
2.3.3 修改默认的序列化机制 ………… 79
2.3.4 序列化单例和类型安全的枚举 ………… 81
2.3.5 版本管理 ………… 82
2.3.6 为克隆使用序列化 ………… 84
2.4 操作文件 ………… 86
2.4.1 Path ………… 86
2.4.2 读写文件 ………… 89
2.4.3 创建文件和目录 ………… 90
2.4.4 复制、移动和删除文件 ………… 91
2.4.5 获取文件信息 ………… 92
2.4.6 访问目录中的项 ………… 94
2.4.7 使用目录流 ………… 95
2.4.8 ZIP 文件系统 ………… 98
2.5 内存映射文件 ………… 99

2.5.1	内存映射文件的性能	99
2.5.2	缓冲区数据结构	105
2.6	文件加锁机制	107
2.7	正则表达式	109
2.7.1	正则表达式语法	109
2.7.2	匹配字符串	112
2.7.3	找出多个匹配	115
2.7.4	用分隔符来分割	117
2.7.5	替换匹配	117

第 3 章 XML 120
- 3.1 XML 概述 120
- 3.2 XML 文档的结构 122
- 3.3 解析 XML 文档 124
- 3.4 验证 XML 文档 133
  - 3.4.1 文档类型定义 134
  - 3.4.2 XML Schema 140
  - 3.4.3 一个实践示例 142
- 3.5 使用 XPath 来定位信息 148
- 3.6 使用命名空间 152
- 3.7 流机制解析器 154
  - 3.7.1 使用 SAX 解析器 154
  - 3.7.2 使用 StAX 解析器 159
- 3.8 生成 XML 文档 162
  - 3.8.1 不带命名空间的文档 162
  - 3.8.2 带命名空间的文档 163
  - 3.8.3 写出文档 163
  - 3.8.4 使用 StAX 写出 XML 文档 165
  - 3.8.5 示例：生成 SVG 文件 170
- 3.9 XSL 转换 171

第 4 章 网络 180
- 4.1 连接到服务器 180
  - 4.1.1 使用 telnet 180
  - 4.1.2 用 Java 连接到服务器 182
  - 4.1.3 套接字超时 184
  - 4.1.4 因特网地址 185
- 4.2 实现服务器 186
  - 4.2.1 服务器套接字 186
  - 4.2.2 为多个客户端服务 189
  - 4.2.3 半关闭 192
  - 4.2.4 可中断套接字 193
- 4.3 获取 Web 数据 199
  - 4.3.1 URL 和 URI 199
  - 4.3.2 使用 URLConnection 获取信息 201
  - 4.3.3 提交表单数据 207
- 4.4 HTTP 客户端 215
- 4.5 发送 E-mail 221

第 5 章 数据库编程 225
- 5.1 JDBC 的设计 225
  - 5.1.1 JDBC 驱动程序类型 226
  - 5.1.2 JDBC 的典型用法 227
- 5.2 结构化查询语言 227
- 5.3 JDBC 配置 232
  - 5.3.1 数据库 URL 232
  - 5.3.2 驱动程序 JAR 文件 233
  - 5.3.3 启动数据库 233
  - 5.3.4 注册驱动器类 234
  - 5.3.5 连接到数据库 234
- 5.4 使用 JDBC 语句 237
  - 5.4.1 执行 SQL 语句 237
  - 5.4.2 管理连接、语句和结果集 240
  - 5.4.3 分析 SQL 异常 240
  - 5.4.4 组装数据库 242
- 5.5 执行查询操作 246
  - 5.5.1 预备语句 246
  - 5.5.2 读写 LOB 252
  - 5.5.3 SQL 转义 253
  - 5.5.4 多结果集 254
  - 5.5.5 获取自动生成的键 255
- 5.6 可滚动和可更新的结果集 256

5.6.1	可滚动的结果集	256
5.6.2	可更新的结果集	258
5.7	行集	261
5.7.1	构建行集	262
5.7.2	被缓存的行集	262
5.8	元数据	265
5.9	事务	274
5.9.1	用 JDBC 对事务编程	274
5.9.2	保存点	275
5.9.3	批量更新	275
5.9.4	高级 SQL 类型	277
5.10	Web 与企业应用中的连接管理	278

第 6 章 日期和时间 API 280

6.1	时间线	280
6.2	本地日期	284
6.3	日期调整器	288
6.4	本地时间	289
6.5	时区时间	290
6.6	格式化和解析	294
6.7	与遗留代码的互操作	298

第 7 章 国际化 300

7.1	locale	300
7.1.1	为什么需要 locale	300
7.1.2	指定 locale	301
7.1.3	默认 locale	303
7.1.4	显示名字	304
7.2	数字格式	305
7.2.1	格式化数字值	306
7.2.2	货币	310
7.3	日期和时间	311
7.4	排序和规范化	318
7.5	消息格式化	323
7.5.1	格式化数字和日期	324
7.5.2	选择格式	325
7.6	文本输入和输出	327
7.6.1	文本文件	327
7.6.2	行结束符	327
7.6.3	控制台	328
7.6.4	日志文件	328
7.6.5	UTF-8 字节顺序标志	329
7.6.6	源文件的字符编码	329
7.7	资源包	330
7.7.1	定位资源包	330
7.7.2	属性文件	331
7.7.3	包类	332
7.8	一个完整的例子	333

第 8 章 脚本、编译与注解处理 348

8.1	Java 平台的脚本机制	348
8.1.1	获取脚本引擎	348
8.1.2	脚本计算与绑定	349
8.1.3	重定向输入和输出	351
8.1.4	调用脚本的函数和方法	352
8.1.5	编译脚本	353
8.1.6	示例：用脚本处理 GUI 事件	354
8.2	编译器 API	358
8.2.1	调用编译器	358
8.2.2	发起编译任务	359
8.2.3	捕获诊断消息	359
8.2.4	从内存中读取源文件	360
8.2.5	将字节码写出到内存中	360
8.2.6	示例：动态 Java 代码生成	362
8.3	使用注解	367
8.3.1	注解简介	368
8.3.2	示例：注解事件处理器	369
8.4	注解语法	373
8.4.1	注解接口	373
8.4.2	注解	375
8.4.3	注解各类声明	376
8.4.4	注解类型用法	377

	8.4.5	注解 this ························ 378
8.5	标准注解 ······························ 379
	8.5.1	用于编译的注解 ················· 380
	8.5.2	用于管理资源的注解 ·········· 381
	8.5.3	元注解 ···························· 381
8.6	源码级注解处理 ······················· 383
	8.6.1	注解处理器 ······················ 384
	8.6.2	语言模型 API ··················· 384
	8.6.3	使用注解来生成源码 ·········· 385
8.7	字节码工程 ···························· 388
	8.7.1	修改类文件 ······················ 388
	8.7.2	在加载时修改字节码 ·········· 393

第 9 章 Java 平台模块系统 ······ 395

9.1	模块的概念 ···························· 395
9.2	对模块命名 ···························· 396
9.3	模块化的"Hello, World!" 程序 ································· 397
9.4	对模块的需求 ························ 398
9.5	导出包 ·································· 400
9.6	模块化的 JAR ························ 403
9.7	模块和反射式访问 ················· 404
9.8	自动模块 ······························· 406
9.9	不具名模块 ···························· 408
9.10	用于迁移的命令行标识 ··········· 409
9.11	传递的需求和静态的需求 ········ 410
9.12	限定导出和开放 ····················· 411
9.13	服务加载 ······························· 412
9.14	操作模块的工具 ····················· 414

第 10 章 安全 ······························ 417

10.1	类加载器 ······························· 417
	10.1.1	类加载过程 ··················· 418
	10.1.2	类加载器的层次结构 ········ 419
	10.1.3	将类加载器用作命名空间 ··············· 420
	10.1.4	编写你自己的类加载器 ······ 421
	10.1.5	字节码校验 ··················· 426

10.2	安全管理器与访问权限 ··········· 429
	10.2.1	权限检查 ······················ 429
	10.2.2	Java 平台安全性 ·············· 431
	10.2.3	安全策略文件 ················· 434
	10.2.4	定制权限 ······················ 439
	10.2.5	实现权限类 ··················· 440
10.3	用户认证 ······························· 446
	10.3.1	JAAS 框架 ····················· 446
	10.3.2	JAAS 登录模块 ················ 451
10.4	数字签名 ······························· 459
	10.4.1	消息摘要 ······················ 460
	10.4.2	消息签名 ······················ 463
	10.4.3	校验签名 ······················ 465
	10.4.4	认证问题 ······················ 467
	10.4.5	证书签名 ······················ 469
	10.4.6	证书请求 ······················ 469
	10.4.7	代码签名 ······················ 470
10.5	加密 ····································· 472
	10.5.1	对称密码 ······················ 473
	10.5.2	密钥生成 ······················ 474
	10.5.3	密码流 ·························· 478
	10.5.4	公共密钥密码 ················· 479

第 11 章 高级 Swing 和图形化编程 ······························ 483

11.1	表格 ····································· 483
	11.1.1	一个简单表格 ················· 483
	11.1.2	表格模型 ······················ 486
	11.1.3	对行和列的操作 ·············· 489
	11.1.4	单元格的绘制和编辑 ········· 503
11.2	树 ··· 513
	11.2.1	简单的树 ······················ 514
	11.2.2	节点枚举 ······················ 526
	11.2.3	绘制节点 ······················ 528
	11.2.4	监听树事件 ··················· 530
	11.2.5	定制树模型 ··················· 536
11.3	高级 AWT ······························ 544

11.3.1 绘图操作流程 …………………… 544
11.3.2 形状 ……………………………… 546
11.3.3 区域 ……………………………… 560
11.3.4 笔画 ……………………………… 561
11.3.5 着色 ……………………………… 567
11.3.6 坐标变换 ………………………… 569
11.3.7 剪切 ……………………………… 574
11.3.8 透明与组合 ……………………… 575
11.4 像素图 ………………………………… 583
11.4.1 图像的读取器和写入器 ………… 583
11.4.2 图像处理 ………………………… 591
11.5 打印 …………………………………… 604
11.5.1 图形打印 ………………………… 604
11.5.2 打印多页文件 …………………… 612
11.5.3 打印服务程序 …………………… 620
11.5.4 流打印服务程序 ………………… 622
11.5.5 打印属性 ………………………… 625

第 12 章 本地方法 ………………………… 632
12.1 从 Java 程序中调用 C 函数 ………… 633
12.2 数值参数与返回值 …………………… 637
12.3 字符串参数 …………………………… 639
12.4 访问域 ………………………………… 644
12.4.1 访问实例域 ……………………… 644
12.4.2 访问静态域 ……………………… 648
12.5 编码签名 ……………………………… 648
12.6 调用 Java 方法 ……………………… 650
12.6.1 实例方法 ………………………… 650
12.6.2 静态方法 ………………………… 653
12.6.3 构造器 …………………………… 654
12.6.4 另一种方法调用 ………………… 654
12.7 访问数组元素 ………………………… 656
12.8 错误处理 ……………………………… 659
12.9 使用调用 API ………………………… 663
12.10 完整的示例：访问 Windows
       注册表 ……………………………… 668
12.10.1 Windows 注册表概述 ……… 668
12.10.2 访问注册表的 Java 平台
         接口 …………………………… 669
12.10.3 以本地方法实现注册表访问
         函数 …………………………… 670

第 1 章　Java 8 的流库

- ▲ 从迭代到流的操作
- ▲ 流的创建
- ▲ filter、map 和 flatMap 方法
- ▲ 抽取子流和组合流
- ▲ 其他的流转换
- ▲ 简单约简
- ▲ Optional 类型
- ▲ 收集结果
- ▲ 收集到映射表中
- ▲ 群组和分区
- ▲ 下游收集器
- ▲ 约简操作
- ▲ 基本类型流
- ▲ 并行流

与集合相比，流提供了一种可以让我们在更高的概念级别上指定计算任务的数据视图。通过使用流，我们可以说明想要完成什么任务，而不是说明如何去实现它。我们将操作的调度留给具体实现去解决。例如，假设我们想要计算某个属性的平均值，那么我们就可以指定数据源和该属性，然后，流库就可以对计算进行优化，例如，使用多线程来计算总和与个数，并将结果合并。

在本章中，你将会学习如何使用 Java 的流库，它是在 Java 8 中引入的，用来以"做什么而非怎么做"的方式处理集合。

1.1　从迭代到流的操作

在处理集合时，我们通常会迭代遍历它的元素，并在每个元素上执行某项操作。例如，假设我们想要对某本书中的所有长单词进行计数。首先，将所有单词放到一个列表中：

```
var contents = new String(Files.readAllBytes(
   Paths.get("alice.txt")), StandardCharsets.UTF_8); // Read file into string
List<String> words = List.of(contents.split("\\PL+"));
   // Split into words; nonletters are delimiters
```

现在，我们可以迭代它了：

```
int count = 0;
for (String w : words) {
   if (w.length() > 12) count++;
}
```

在使用流时，相同的操作看起来像下面这样：

```
long count = words.stream()
   .filter(w -> w.length() > 12)
   .count();
```

现在我们不必扫描整个代码去查找过滤和计数操作，方法名就可以直接告诉我们其代码意欲何为。而且，循环需要非常详细地指定操作的顺序，而流却能够以其想要的任何方式来调度这些操作，只要结果是正确的即可。

仅将 stream 修改为 parallelStream 就可以让流库以并行方式来执行过滤和计数。

```
long count = words.parallelStream()
    .filter(w -> w.length() > 12)
    .count();
```

流遵循了"做什么而非怎么做"的原则。在流的示例中，我们描述了需要做什么：获取长单词，并对它们计数。我们没有指定该操作应该以什么顺序或者在哪个线程中执行。相比之下，本节开头处的循环要确切地指定计算应该如何工作，因此也就丧失了进行优化的机会。

流表面上看起来和集合很类似，都可以让我们转换和获取数据。但是，它们之间存在着显著的差异：

1. 流并不存储其元素。这些元素可能存储在底层的集合中，或者是按需生成的。
2. 流的操作不会修改其数据源。例如，filter 方法不会从流中移除元素，而是会生成一个新的流，其中不包含被过滤掉的元素。
3. 流的操作是尽可能惰性执行的。这意味着直至需要其结果时，操作才会执行。例如，如果我们只想查找前 5 个长单词而不是所有长单词，那么 filter 方法就会在匹配到第 5 个单词后停止过滤。因此，我们甚至可以操作无限流。

我们再来看看这个示例。stream 和 parallelStream 方法会产生一个用于 words 列表的流。filter 方法会返回另一个流，其中只包含长度大于 12 的单词。count 方法会将这个流化简为一个结果。

这个工作流是操作流时的典型流程。我们建立了一个包含三个阶段的操作管道：

1. 创建一个流。
2. 指定将初始流转换为其他流的中间操作，可能包含多个步骤。
3. 应用终止操作，从而产生结果。这个操作会强制执行之前的惰性操作。从此之后，这个流就再也不能用了。

在程序清单 1-1 的示例中，流是用 stream 或 parallelStream 方法创建的。filter 方法对其进行转换，而 count 方法是终止操作。

程序清单 1-1 streams/CountLongWords.java

```
1  package streams;
2
3  /**
4   * @version 1.01 2018-05-01
5   * @author Cay Horstmann
6   */
7
8  import java.io.*;
9  import java.nio.charset.*;
10 import java.nio.file.*;
```

```
11  import java.util.*;
12
13  public class CountLongWords
14  {
15     public static void main(String[] args) throws IOException
16     {
17        var contents = new String(Files.readAllBytes(
18           Paths.get("../gutenberg/alice30.txt")), StandardCharsets.UTF_8);
19        List<String> words = List.of(contents.split("\\PL+"));
20
21        long count = 0;
22        for (String w : words)
23        {
24           if (w.length() > 12) count++;
25        }
26        System.out.println(count);
27
28        count = words.stream().filter(w -> w.length() > 12).count();
29        System.out.println(count);
30
31        count = words.parallelStream().filter(w -> w.length() > 12).count();
32        System.out.println(count);
33     }
34  }
```

在下一节中，你将会看到如何创建流。后续的三节将讨论流的转换。再后面的五节将讨论终止操作。

API *java.util.stream.Stream<T>* 8

- Stream<T> filter(Predicate<? super T> p)
  产生一个流，其中包含当前流中满足 p 的所有元素。
- long count()
  产生当前流中元素的数量。这是一个终止操作。

API *java.util.Collection<E>* 1.2

- default Stream<E> stream()
- default Stream<E> parallelStream()
  产生当前集合中所有元素的顺序流或并行流。

1.2 流的创建

你已经看到了可以用 Collection 接口的 stream 方法将任何集合转换为一个流。如果你有一个数组，那么可以使用静态的 Stream.of 方法。

```
Stream<String> words = Stream.of(contents.split("\\PL+"));
   // split returns a String[] array
```

of 方法具有可变长参数，因此我们可以构建具有任意数量引元的流：

```java
Stream<String> song = Stream.of("gently", "down", "the", "stream");
```

使用 Array.stream(array, from, to) 可以用数组中的一部分元素来创建一个流。

为了创建不包含任何元素的流，可以使用静态的 Stream.empty 方法：

```java
Stream<String> silence = Stream.empty();
    // Generic type <String> is inferred; same as Stream.<String>empty()
```

Stream 接口有两个用于创建无限流的静态方法。generate 方法会接受一个不包含任何引元的函数（或者从技术上讲，是一个 Supplier<T> 接口的对象）。无论何时，只要需要一个流类型的值，该函数就会被调用以产生一个这样的值。我们可以像下面这样获得一个常量值的流：

```java
Stream<String> echos = Stream.generate(() -> "Echo");
```

或者像下面这样获取一个随机数的流：

```java
Stream<Double> randoms = Stream.generate(Math::random);
```

如果要产生像 0 1 2 3 … 这样的序列，可以使用 iterate 方法。它会接受一个"种子"值，以及一个函数（从技术上讲，是一个 UnaryOperation<T>），并且会反复地将该函数应用到之前的结果上。例如，

```java
Stream<BigInteger> integers
    = Stream.iterate(BigInteger.ZERO, n -> n.add(BigInteger.ONE));
```

该序列中的第一个元素是种子 BigInteger.ZERO，第二个元素是 f(seed)，即 1（作为大整数），下一个元素是 f(f(seed))，即 2，后续以此类推。

如果要产生一个有限序列，则需要添加一个谓词来描述迭代应该如何结束：

```java
var limit = new BigInteger("10000000");
Stream<BigInteger> integers
    = Stream.iterate(BigInteger.ZERO,
        n -> n.compareTo(limit) < 0,
        n -> n.add(BigInteger.ONE));
```

只要该谓词拒绝了某个迭代生成的值，这个流即结束。

最后，Stream.ofNullable 方法会用一个对象来创建一个非常短的流。如果该对象为 null，那么这个流的长度就为 0；否则，这个流的长度为 1，即只包含该对象。这个方法与 flatMap 相结合时最有用，可以查看 1.7.7 节中的示例。

> 📖 **注释**：Java API 中有大量方法都可以产生流。例如，Pattern 类有一个 splitAsStream 方法，它会按照某个正则表达式来分割一个 CharSequence 对象。可以使用下面的语句来将一个字符串分割为一个个的单词：
>
> ```java
> Stream<String> words = Pattern.compile("\\PL+").splitAsStream(contents);
> ```
>
> Scanner.tokens 方法会产生一个扫描器的符号流。另一种从字符串中获取单词流的方式是：

```
Stream<String> words = new Scanner(contents).tokens();
```

静态的 Files.lines 方法会返回一个包含了文件中所有行的 Stream：

```
try (Stream<String> lines = Files.lines(path)) {
    Process lines
}
```

> **注释**：如果我们持有的 Iterable 对象不是集合，那么可以通过下面的调用将其转换为一个流：
>
> ```
> StreamSupport.stream(iterable.spliterator(), false);
> ```
>
> 如果我们持有的是 Iterator 对象，并且希望得到一个由它的结果构成的流，那么可以使用下面的语句：
>
> ```
> StreamSupport.stream(Spliterators.spliteratorUnknownSize(
>     iterator, Spliterator.ORDERED), false);
> ```

> **警告**：至关重要的是，在执行流的操作时，我们并没有修改流背后的集合。记住，流并没有收集其数据，数据一直存储在单独的集合中。如果修改了该集合，那么流操作的结果就会变成未定义的。JDK 文档称这种要求为**不干涉性**。
>
> 准确地讲，因为中间的流操作是惰性的，所以在终止操作得以执行时，集合有可能已经发生了变化。例如，尽管我们不推荐下面这段代码，但是它仍旧可以工作：
>
> ```
> List<String> wordList = . . .;
> Stream<String> words = wordList.stream();
> wordList.add("END");
> long n = words.distinct().count();
> ```
>
> 但是下面的代码是错误的：
>
> ```
> Stream<String> words = wordList.stream();
> words.forEach(s -> if (s.length() < 12) wordList.remove(s));
>     // ERROR--interference
> ```

程序清单 1-2 中的示例程序展示了创建流的各种方式。

程序清单 1-2　streams/CreatingStreams.java

```
1  package streams;
2
3  /**
4   * @version 1.01 2018-05-01
5   * @author Cay Horstmann
6   */
7
8  import java.io.IOException;
9  import java.math.BigInteger;
10 import java.nio.charset.StandardCharsets;
11 import java.nio.file.*;
12 import java.util.*;
```

```java
import java.util.regex.Pattern;
import java.util.stream.*;

public class CreatingStreams
{
   public static <T> void show(String title, Stream<T> stream)
   {
      final int SIZE = 10;
      List<T> firstElements = stream
         .limit(SIZE + 1)
         .collect(Collectors.toList());
      System.out.print(title + ": ");
      for (int i = 0; i < firstElements.size(); i++)
      {
         if (i > 0) System.out.print(", ");
         if (i < SIZE) System.out.print(firstElements.get(i));
         else System.out.print("...");
      }
      System.out.println();
   }

   public static void main(String[] args) throws IOException
   {
      Path path = Paths.get("../gutenberg/alice30.txt");
      var contents = new String(Files.readAllBytes(path), StandardCharsets.UTF_8);

      Stream<String> words = Stream.of(contents.split("\\PL+"));
      show("words", words);
      Stream<String> song = Stream.of("gently", "down", "the", "stream");
      show("song", song);
      Stream<String> silence = Stream.empty();
      show("silence", silence);

      Stream<String> echos = Stream.generate(() -> "Echo");
      show("echos", echos);

      Stream<Double> randoms = Stream.generate(Math::random);
      show("randoms", randoms);

      Stream<BigInteger> integers = Stream.iterate(BigInteger.ONE,
         n -> n.add(BigInteger.ONE));
      show("integers", integers);

      Stream<String> wordsAnotherWay = Pattern.compile("\\PL+").splitAsStream(contents);
      show("wordsAnotherWay", wordsAnotherWay);

      try (Stream<String> lines = Files.lines(path, StandardCharsets.UTF_8))
      {
         show("lines", lines);
      }

      Iterable<Path> iterable = FileSystems.getDefault().getRootDirectories();
      Stream<Path> rootDirectories = StreamSupport.stream(iterable.spliterator(), false);
      show("rootDirectories", rootDirectories);
```

```
 67
 68         Iterator<Path> iterator = Paths.get("/usr/share/dict/words").iterator();
 69         Stream<Path> pathComponents = StreamSupport.stream(Spliterators.spliteratorUnknownSize(
 70             iterator, Spliterator.ORDERED), false);
 71         show("pathComponents", pathComponents);
 72     }
 73 }
```

api **java.util.stream.Stream** 8

- static <T> Stream<T> of(T... values)
  产生一个元素为给定值的流。
- static <T> Stream<T> empty()
  产生一个不包含任何元素的流。
- static <T> Stream<T> generate(Supplier<T> s)
  产生一个无限流，它的值是通过反复调用函数 s 而构建的。
- static <T> Stream<T> iterate(T seed, UnaryOperator<T> f)
- static <T> Stream<T> iterate(T seed, Predicate<? super T> hasNext, UnaryOperator<T> f)
  产生一个无限流，它的元素包含 seed、在 seed 上调用 f 产生的值、在前一个元素上调用 f 产生的值，等等。第一个方法会产生一个无限流，而第二个方法的流会在碰到第一个不满足 hasNext 谓词的元素时终止。
- static <T> Stream<T> ofNullable(T t) 9
  如果 t 为 null，返回一个空流，否则返回包含 t 的流。

api **java.util.Spliterators** 8

- static <T> Spliterator<T> spliteratorUnknownSize(Iterator<? extends T> iterator, int characteristics)
  用给定的特性（一种包含诸如 Spliterator.ORDERED 之类的常量的位模式）将一个迭代器转换为一个具有未知尺寸的可分割的迭代器。

api **java.util.Arrays** 1.2

- static <T> Stream<T> stream(T[] array, int startInclusive, int endExclusive) 8
  产生一个流，它的元素是由数组中指定范围内的元素构成的。

api **java.util.regex.Pattern** 1.4

- Stream<String> splitAsStream(CharSequence input) 8
  产生一个流，它的元素是输入中由该模式界定的部分。

api **java.nio.file.Files** 7

- static Stream<String> lines(Path path) 8
- static Stream<String> lines(Path path, Charset cs) 8

产生一个流，它的元素是指定文件中的行，该文件的字符集为 UTF-8，或者为指定的字符集。

API java.util.stream.StreamSupport 8

- static <T> Stream<T> stream(Spliterator<T> spliterator, boolean parallel)

  产生一个流，它包含了由给定的可分割迭代器产生的值。

API java.lang.Iterable 5

- Spliterator<T> spliterator() 8

  为这个 Iterable 产生一个可分割的迭代器。默认实现不分割也不报告尺寸。

API java.util.Scanner 5

- public Stream<String> tokens() 9

  产生一个字符串流，该字符串是调用这个扫描器的 next 方法时返回的。

API java.util.function.Supplier<T> 8

- T get()

  提供一个值。

1.3 filter、map 和 flatMap 方法

流的转换会产生一个新的流，它的元素派生自另一个流中的元素。我们已经看到了 filter 转换会产生一个新流，它的元素与某种条件相匹配。下面，我们将一个字符串流转换为只包含长单词的另一个流：

```
List<String> words = . . .;
Stream<String> longWords = words.stream().filter(w -> w.length() > 12);
```

filter 的引元是 Predicate<T>，即从 T 到 boolean 的函数。

通常，我们想要按照某种方式来转换流中的值，此时，可以使用 map 方法并传递执行该转换的函数。例如，我们可以像下面这样将所有单词都转换为小写：

```
Stream<String> lowercaseWords = words.stream().map(String::toLowerCase);
```

这里，我们使用的是带有方法引用的 map，但是，通常我们可以使用 lambda 表达式来代替：

```
Stream<String> firstLetters = words.stream().map(s -> s.substring(0, 1));
```

上面的语句所产生的流中包含了所有单词的首字母。

在使用 map 时，会有一个函数应用到每个元素上，并且其结果是包含了应用该函数后所产生的所有结果的流。现在，假设我们有一个函数，它返回的不是一个值，而是一个包含众多值的流。下面的示例展示的方法会将字符串转换为字符串流，即一个个的编码点：

```
public static Stream<String> codePoints(String s)
{
    var result = new ArrayList<String>();
```

```
    int i = 0;
    while (i < s.length())
    {
        int j = s.offsetByCodePoints(i, 1);
        result.add(s.substring(i, j));
        i = j;
    }
    return result.stream();
}
```

这个方法可以正确地处理需要用两个 char 值来表示的 Unicode 字符,因为本来就应该这样处理。但是,我们不用再次纠结其细节。

例如,codePoints("boat") 的返回值是流 ["b", "o", "a", "t"]。

假设我们将 codePoints 方法映射到一个字符串流上:

```
Stream<Stream<String>> result = words.stream().map(w -> codePoints(w));
```

那么会得到一个包含流的流,就像 [...["y","o","u","r"],["b","o","a","t"],...]。为了将其摊平为单个流 [..."y","o","u","r","b","o","a","t",...],可以使用 flatMap 方法而不是 map 方法:

```
Stream<String> flatResult = words.stream().flatMap(w -> codePoints(w));
    // Calls codePoints on each word and flattens the results
```

> **注释**:在流之外的类中你也会发现 flatMap 方法,因为它是计算机科学中的一种通用概念。假设我们有一个泛型 G(例如 Stream),以及将某种类型 T 转换为 G<U> 的函数 f 和将类型 U 转换为 G<V> 的函数 g。我们可以通过使用 flatMap 来组合它们,即首先应用 f,然后应用 g。这是单子论的关键概念。但是不必担心,我们无须了解任何有关单子的知识就可以使用 flatMap。

API java.util.stream.Stream 8

- Stream<T> filter(Predicate<? super T> predicate)
  产生一个流,它包含当前流中所有满足谓词条件的元素。
- <R> Stream<R> map(Function<? super T,? extends R> mapper)
  产生一个流,它包含将 mapper 应用于当前流中所有元素所产生的结果。
- <R> Stream<R> flatMap(Function<? super T,? extends Stream<? extends R>> mapper)
  产生一个流,它是通过将 mapper 应用于当前流中所有元素所产生的结果连接到一起而获得的。(注意,这里的每个结果都是一个流。)

1.4 抽取子流和组合流

调用 *stream*.limit(n) 会返回一个新的流,它在 n 个元素之后结束(如果原来的流比 n 短,那么就会在该流结束时结束)。这个方法对于裁剪无限流的尺寸特别有用。例如,

```
Stream<Double> randoms = Stream.generate(Math::random).limit(100);
```

会产生一个包含 100 个随机数的流。

调用 *stream*.skip(n) 正好相反：它会丢弃前 n 个元素。这个方法对于本书的读操作示例很方便，因为按照 split 方法的工作方式，第一个元素是没什么用的空字符串。我们可以通过调用 skip 来跳过它：

```
Stream<String> words = Stream.of(contents.split("\\PL+")).skip(1);
```

stream.takeWhile(*predicate*) 调用会在谓词为真时获取流中的所有元素，然后停止。

例如，假设我们使用上一节的 codePoints 方法将字符串分割为字符，然后收集所有的数字元素。takeWhile 方法可以实现此目标：

```
Stream<String> initialDigits = codePoints(str).takeWhile(
    s -> "0123456789".contains(s));
```

dropWhile 方法的做法正好相反，它会在条件为真时丢弃元素，并产生一个由第一个使该条件为假的字符开始的所有元素构成的流：

```
Stream<String> withoutInitialWhiteSpace = codePoints(str).dropWhile(
    s -> s.trim().length() == 0);
```

我们可以用 Stream 类的静态 concat 方法将两个流连接起来：

```
Stream<String> combined = Stream.concat(
    codePoints("Hello"), codePoints("World"));
    // Yields the stream ["H", "e", "l", "l", "o", "W", "o", "r", "l", "d"]
```

当然，第一个流不应该是无限的，否则第二个流永远都不会有机会处理。

API **java.util.stream.Stream** 8

- Stream<T> limit(long maxSize)

  产生一个流，其中包含了当前流中最初的 maxSize 个元素。

- Stream<T> skip(long n)

  产生一个流，它的元素是当前流中除了前 n 个元素之外的所有元素。

- Stream<T> takeWhile(Predicate<? super T> predicate) 9

  产生一个流，它的元素是当前流中所有满足谓词条件的元素。

- Stream<T> dropWhile(Predicate<? super T> predicate) 9

  产生一个流，它的元素是当前流中排除不满足谓词条件的元素之外的所有元素。

- static <T> Stream<T> concat(Stream<? extends T> a, Stream<? extends T> b)

  产生一个流，它的元素是 a 的元素后面跟着 b 的元素。

1.5 其他的流转换

distinct 方法会返回一个流，它的元素是从原有流中产生的，即原来的元素按照同样的顺序剔除重复元素后产生的。这些重复元素并不一定是毗邻的。

```
Stream<String> uniqueWords
    = Stream.of("merrily", "merrily", "merrily", "gently").distinct();
    // Only one "merrily" is retained
```

对于流的排序，有多种 sorted 方法的变体可用。其中一种用于操作 Comparable 元素的流，而另一种可以接受一个 Comparator。下面，我们对字符串排序，使得最长的字符串排在最前面：

```
Stream<String> longestFirst
    = words.stream().sorted(Comparator.comparing(String::length).reversed());
```

与所有的流转换一样，sorted 方法会产生一个新的流，它的元素是原有流中按照顺序排列的元素。

当然，我们在对集合排序时可以不使用流。但是，当排序处理是流管道的一部分时，sorted 方法就会显得很有用。

最后，peek 方法会产生另一个流，它的元素与原来流中的元素相同，但是在每次获取一个元素时，都会调用一个函数。这对于调试来说很方便：

```
Object[] powers = Stream.iterate(1.0, p -> p * 2)
    .peek(e -> System.out.println("Fetching " + e))
    .limit(20).toArray();
```

当实际访问一个元素时，就会打印出来一条消息。通过这种方式，你可以验证 iterate 返回的无限流是被惰性处理的。

> 提示：当我们使用调试器来调试流的计算程序时，可以针对各个流转换操作中的某一个，在它所调用的方法中设置断点。对于大多数 IDE，我们都可以在 lambda 表达式中设置断点。如果只想了解在流管道的某个特定点上会发生什么，那么可以添加下面的代码，并在第二行上设置断点：
> ```
> .peek(x -> {
>     return; })
> ```

API *java.util.stream.Stream* 8

- Stream<T> distinct()
  产生一个流，包含当前流中所有不同的元素。
- Stream<T> sorted()
- Stream<T> sorted(Comparator<? super T> comparator)
  产生一个流，它的元素是当前流中的所有元素按照顺序排列的。第一个方法要求元素是实现了 Comparable 的类的实例。
- Stream<T> peek(Consumer<? super T> action)
  产生一个流，它与当前流中的元素相同，在获取其中每个元素时，会将其传递给 action。

1.6 简单约简

现在你已经看到了如何创建和转换流，我们终于可以讨论最重要的内容了，即从流数据中获得答案。我们在本节所讨论的方法被称为约简。约简是一种终结操作（terminal operation),它们会将流约简为可以在程序中使用的非流值。

你已经看到过一种简单约简:count 方法会返回流中元素的数量。

其他的简单约简还有 max 和 min,它们分别返回最大值和最小值。这里要稍作解释,这些方法返回的是一个类型 Optional<T> 的值,它要么在其中包装了答案,要么表示没有任何值(因为流碰巧为空)。在过去,碰到这种情况返回 null 是很常见的,但是这样做会导致在未做完备测试的程序中产生空指针异常。Optional 类型是一种表示缺少返回值的更好的方式。我们将在下一节中详细讨论 Optional 类型。下面展示了如何获得流中的最大值:

```
Optional<String> largest = words.max(String::compareToIgnoreCase);
System.out.println("largest: " + largest.orElse(""));
```

findFirst 返回的是非空集合中的第一个值。它通常在与 filter 组合使用时很有用。例如,下面展示了如何找到第一个以字母 Q 开头的单词,前提是存在这样的单词:

```
Optional<String> startsWithQ
    = words.filter(s -> s.startsWith("Q")).findFirst();
```

如果不强调使用第一个匹配,而是使用任意的匹配都可以,那么就可以使用 findAny 方法。这个方法在并行处理流时很有效,因为流可以报告任何它找到的匹配而不是被限制为必须报告第一个匹配。

```
Optional<String> startsWithQ
    = words.parallel().filter(s -> s.startsWith("Q")).findAny();
```

如果只想知道是否存在匹配,那么可以使用 anyMatch。这个方法会接受一个断言引元,因此不需要使用 filter。

```
boolean aWordStartsWithQ
    = words.parallel().anyMatch(s -> s.startsWith("Q"));
```

还有 allMatch 和 noneMatch 方法,它们分别在所有元素和没有任何元素匹配谓词的情况下返回 true。这些方法也可以通过并行运行而获益。

API java.util.stream.Stream 8

- Optional<T> max(Comparator<? super T> comparator)
- Optional<T> min(Comparator<? super T> comparator)
  分别产生这个流的最大元素和最小元素,使用由给定比较器定义的排序规则,如果这个流为空,会产生一个空的 Optional 对象。这些操作都是终结操作。
- Optional<T> findFirst()
- Optional<T> findAny()
  分别产生这个流的第一个和任意一个元素,如果这个流为空,会产生一个空的 Optional 对象。这些操作都是终结操作。
- boolean anyMatch(Predicate<? super T> predicate)
- boolean allMatch(Predicate<? super T> predicate)
- boolean noneMatch(Predicate<? super T> predicate)

分别在这个流中任意元素、所有元素和没有任何元素匹配给定谓词时返回 true。这些操作都是终结操作。

1.7 Optional 类型

Optional<T> 对象是一种包装器对象，要么包装了类型 T 的对象，要么没有包装任何对象。对于第一种情况，我们称这种值是存在的。Optional<T> 类型被当作一种更安全的方式，用来替代类型 T 的引用，这种引用要么引用某个对象，要么为 null。但是，它只有在正确使用的情况下才会更安全，接下来的三个小节我们将讨论如何正确使用。

1.7.1 获取 Optional 值

有效地使用 Optional 的关键是要使用这样的方法：它在值不存在的情况下会产生一个可替代物，而只有在值存在的情况下才会使用这个值。

本小节我们先来看看第一条策略。通常，在没有任何匹配时，我们会希望使用某种默认值，可能是空字符串：

```
String result = optionalString.orElse("");
    // The wrapped string, or "" if none
```

你还可以调用代码来计算默认值：

```
String result = optionalString.orElseGet(() -> System.getProperty("myapp.default"));
    // The function is only called when needed
```

或者可以在没有任何值时抛出异常：

```
String result = optionalString.orElseThrow(IllegalStateException::new);
    // Supply a method that yields an exception object
```

API java.util.Optional 8

- T orElse(T other)

  产生这个 Optional 的值，或者在该 Optional 为空时，产生 other。

- T orElseGet(Supplier<? extends T> other)

  产生这个 Optional 的值，或者在该 Optional 为空时，产生调用 other 的结果。

- <X extends Throwable> T orElseThrow(Supplier<? extends X> exceptionSupplier)

  产生这个 Optional 的值，或者在该 Optional 为空时，抛出调用 exceptionSupplier 的结果。

1.7.2 消费 Optional 值

在上一小节，我们看到了如何在不存在任何值的情况下产生相应的替代物。另一条使用可选值的策略是只有在其存在的情况下才消费该值。

ifPresent 方法会接受一个函数。如果可选值存在，那么它会被传递给该函数。否则，不会发生任何事情。

```
optionalValue.ifPresent(v -> Process v);
```

例如，如果在该值存在的情况下想要将其添加到某个集中，那么就可以调用

```
optionalValue.ifPresent(v -> results.add(v));
```

或者直接调用

```
optionalValue.ifPresent(results::add);
```

如果想要在可选值存在时执行一种动作，在可选值不存在时执行另一种动作，可以使用 ifPresentOrElse：

```
optionalValue.ifPresentOrElse(
    v -> System.out.println("Found " + v),
    () -> logger.warning("No match"));
```

API java.util.Optional 8

- void ifPresent(Consumer<? super T> action)

  如果该 Optional 不为空，就将它的值传递给 action。

- void ifPresentOrElse(Consumer<? super T> action, Runnable emptyAction) 9

  如果该 Optional 不为空，就将它的值传递给 action，否则调用 emptyAction。

1.7.3 管道化 Optional 值

在上一节中，你看到了如何从 Optional 对象获取值。另一种有用的策略是保持 Optional 完整，使用 map 方法来转换 Optional 内部的值：

```
Optional<String> transformed = optionalString.map(String::toUpperCase);
```

如果 optionalString 为空，那么 transformed 也为空。

下面是另一个例子，我们将一个结果添加到列表中，如果它存在的话：

```
optionalValue.map(results::add);
```

如果 optionalValue 为空，则什么也不会发生。

> **注释**：这个 map 方法与 1.3 节中描述的 Stream 接口的 map 方法类似。你可以直接将可选值想象成尺寸为 0 或 1 的流。结果的尺寸也是 0 或 1，并且在后一种情况中，函数会应用于其上。

类似地，可以使用 filter 方法来只处理那些在转换它之前或之后满足某种特定属性的 Optional 值。如果不满足该属性，那么管道会产生空的结果：

```
Optional<String> transformed = optionalString
    .filter(s -> s.length() >= 8)
    .map(String::toUpperCase);
```

你也可以用 or 方法将空 Optional 替换为一个可替代的 Optional。这个可替代值将以惰性方式计算。

```
Optional<String> result = optionalString.or(() -> // Supply an Optional
    alternatives.stream().findFirst());
```

如果 optionalString 的值存在，那么 result 为 optionalString。如果值不存在，那么就会计算 lambda 表达式，并使用计算出来的结果。

API **java.util.Optional** 8

- <U> Optional<U> map(Function<? super T,? extends U> mapper)

  产生一个 Optional，如果当前的 Optional 的值存在，那么所产生的 Optional 的值是通过将给定的函数应用于当前的 Optional 的值而得到的；否则，产生一个空的 Optional。

- Optional<T> filter(Predicate<? super T> predicate)

  产生一个 Optional，如果当前的 Optional 的值满足给定的谓词条件，那么所产生的 Optional 的值就是当前 Optional 的值；否则，产生一个空 Optional。

- Optional<T> or(Supplier<? extends Optional<? extends T>> supplier) 9

  如果当前 Optional 不为空，则产生当前的 Optional；否则由 supplier 产生一个 Optional。

1.7.4 不适合使用 Optional 值的方式

如果没有正确地使用 Optional 值，那么相比以往得到"某物或 null"的方式，你并没有得到任何好处。

get 方法会在 Optional 值存在的情况下获得其中包装的元素，或者在不存在的情况下抛出一个 NoSuchElementException 异常。因此，

```
Optional<T> optionalValue = ...;
optionalValue.get().someMethod()
```

并不比下面的方式更安全：

```
T value = ...;
value.someMethod();
```

isPresent 方法会报告某个 Optional<T> 对象是否具有值。但是

```
if (optionalValue.isPresent()) optionalValue.get().someMethod();
```

并不比下面的方式更容易处理：

```
if (value != null) value.someMethod();
```

> **注释**：Java 10 为 get 方法引入了一个骇人听闻的同义词，称为 optionalValue.orElseThrow()，这个名字明确表示该方法会在 optionalValue 为空时抛出一个 NoSuchElementException。这样命名是希望程序员只有在非常明确地知道 Optional 永远都不会为空时才去调用该方法。

下面是一些有关 Optional 类型正确用法的提示：

- Optional 类型的变量永远都不应该为 null。
- 不要使用 Optional 类型的域。因为其代价是额外多出来一个对象。在类的内部，使用 null 表示缺失的域更易于操作。

- 不要在集合中放置 Optional 对象，并且不要将它们用作 map 的键。应该直接收集其中的值。

API java.util.Optional 8

- T get()
- T orElseThrow() 10
  产生这个 Optional 的值，或者在该 Optional 为空时，抛出一个 NoSuchElementException 异常。
- boolean isPresent()
  如果该 Optional 不为空，则返回 true。

1.7.5 创建 Optional 值

到目前为止，我们已经讨论了如何使用其他人创建的 Optional 对象。如果想要编写方法来创建 Optional 对象，那么有多个方法可以用于此目的的，包括 Optional.of(result) 和 Optional.empty()。例如，

```
public static Optional<Double> inverse(Double x)
{
    return x == 0 ? Optional.empty() : Optional.of(1 / x);
}
```

ofNullable 方法被用来作为可能出现的 null 值和可选值之间的桥梁。Optional.ofNullable(obj) 会在 obj 不为 null 的情况下返回 Optional.of(obj)，否则会返回 Optional.empty()。

API java.util.Optional 8

- static <T> Optional<T> of(T value)
- static <T> Optional<T> ofNullable(T value)
  产生一个具有给定值的 Optional。如果 value 为 null，那么第一个方法会抛出一个 NullPointerException 异常，而第二个方法会产生一个空 Optional。
- static <T> Optional<T> empty()
  产生一个空 Optional。

1.7.6 用 flatMap 构建 Optional 值的函数

假设你有一个可以产生 Optional<T> 对象的方法 f，并且目标类型 T 具有一个可以产生 Optional<U> 对象的方法 g。如果它们都是普通的方法，那么你可以通过调用 s.f().g() 来将它们组合起来。但是这种组合无法工作，因为 s.f() 的类型为 Optional<T>，而不是 T。因此，需要调用

```
Optional<U> result = s.f().flatMap(T::g);
```

如果 s.f() 的值存在，那么 g 就可以应用到它上面。否则，就会返回一个空 Optional<U>。

很明显，如果有更多可以产生 Optional 值的方法或 lambda 表达式，那么就可以重复此过程。你可以直接将对 flatMap 的调用链接起来，从而构建由这些步骤构成的管道，只有所有步

骤都成功，该管道才会成功。

例如，考虑前一节中安全的 inverse 方法。假设我们还有一个安全的平方根：

```
public static Optional<Double> squareRoot(Double x)
{
    return x < 0 ? Optional.empty() : Optional.of(Math.sqrt(x));
}
```

那么你可以像下面这样计算倒数的平方根：

```
Optional<Double> result = inverse(x).flatMap(MyMath::squareRoot);
```

或者，你可以选择下面的方式：

```
Optional<Double> result
    = Optional.of(-4.0).flatMap(Demo::inverse).flatMap(Demo::squareRoot);
```

无论是 inverse 方法还是 squareRoot 方法返回 Optional.empty()，整个结果都会为空。

> **注释**：你已经在 Stream 接口中看到过 flatMap 方法（参见 1.3 节），当时这个方法被用来将产生流的两个方法组合起来，其实现方式是摊平由流构成的流。如果将可选值解释为具有 0 个或 1 个元素，那么 Optional.flatMap 方法与其操作方式一样。

API java.util.Optional 8

- <U> Optional<U> flatMap(Function<? super T,? extends Optional<? extends U>> mapper)

  如果 Optional 存在，产生将 mapper 应用于当前 Optional 值所产生的结果，或者在当前 Optional 为空时，返回一个空 Optional。

1.7.7 将 Optional 转换为流

stream 方法会将一个 Optional<T> 对象转换为一个具有 0 个或 1 个元素的 Stream<T> 对象。这种做法看起来很自然，但是我们为什么希望这么做呢？

这会使返回 Optional 结果的方法变得很有用。假设我们有一个用户 ID 流和下面的方法：

```
Optional<User> lookup(String id)
```

怎样才能在获取用户流时，跳过那些无效的 ID 呢？

当然，我们可以过滤掉无效 ID，然后将 get 方法应用于剩余的 ID：

```
Stream<String> ids = . . .;
Stream<User> users = ids.map(Users::lookup)
    .filter(Optional::isPresent)
    .map(Optional::get);
```

但是这样就需要使用我们之前警告过要慎用的 isPresent 和 get 方法。下面的调用显得更优雅：

```
Stream<User> users = ids.map(Users::lookup)
    .flatMap(Optional::stream);
```

每一个对 stream 的调用都会返回一个具有 0 个或 1 个元素的流。flagMap 方法将这些方法组合在一起，这意味着不存在的用户会直接被丢弃。

> 📝 **注释**：本节我们研究了一些令人愉快的场景，在其中我们拥有可以返回 Optional 值的方法。当前，许多方法都会在没有任何有效结果的情况下返回 null。假设 Users.classicLookup(id) 会返回一个 User 对象或者 null，而不是 Optional<User>，我们当然可以过滤掉 null 值：
>
> ```
> Stream<User> users = ids.map(Users::classicLookup)
>     .filter(Objects::nonNull);
> ```
>
> 但是如果更喜欢 flatMap 的方式，那么我们可以使用下面的代码：
>
> ```
> Stream<User> users = ids.flatMap(
>     id -> Stream.ofNullable(Users.classicLookup(id)));
> ```
>
> 或者是下面的代码：
>
> ```
> Stream<User> users = ids.map(Users::classicLookup)
>     .flatMap(Stream::ofNullable);
> ```
>
> Stream.ofNullable(obj) 这个调用在 obj 为 null 时，会产生一个空的流，否则会产生一个只包含 obj 的流。

程序清单 1-3 中的示例程序演示了 Optional API 的使用方式。

程序清单 1-3 optional/OptionalTest.java

```java
  1  package optional;
  2
  3  /**
  4   * @version 1.01 2018-05-01
  5   * @author Cay Horstmann
  6   */
  7
  8  import java.io.*;
  9  import java.nio.charset.*;
 10  import java.nio.file.*;
 11  import java.util.*;
 12
 13  public class OptionalTest
 14  {
 15     public static void main(String[] args) throws IOException
 16     {
 17        var contents = new String(Files.readAllBytes(
 18           Paths.get("../gutenberg/alice30.txt")), StandardCharsets.UTF_8);
 19        List<String> wordList = List.of(contents.split("\\PL+"));
 20
 21        Optional<String> optionalValue = wordList.stream()
 22           .filter(s -> s.contains("fred"))
 23           .findFirst();
 24        System.out.println(optionalValue.orElse("No word") + " contains fred");
 25
 26        Optional<String> optionalString = Optional.empty();
 27        String result = optionalString.orElse("N/A");
```

```java
28        System.out.println("result: " + result);
29        result = optionalString.orElseGet(() -> Locale.getDefault().getDisplayName());
30        System.out.println("result: " + result);
31        try
32        {
33           result = optionalString.orElseThrow(IllegalStateException::new);
34           System.out.println("result: " + result);
35        }
36        catch (Throwable t)
37        {
38           t.printStackTrace();
39        }
40
41        optionalValue = wordList.stream()
42           .filter(s -> s.contains("red"))
43           .findFirst();
44        optionalValue.ifPresent(s -> System.out.println(s + " contains red"));
45
46        var results = new HashSet<String>();
47        optionalValue.ifPresent(results::add);
48        Optional<Boolean> added = optionalValue.map(results::add);
49        System.out.println(added);
50
51        System.out.println(inverse(4.0).flatMap(OptionalTest::squareRoot));
52        System.out.println(inverse(-1.0).flatMap(OptionalTest::squareRoot));
53        System.out.println(inverse(0.0).flatMap(OptionalTest::squareRoot));
54        Optional<Double> result2 = Optional.of(-4.0)
55           .flatMap(OptionalTest::inverse).flatMap(OptionalTest::squareRoot);
56        System.out.println(result2);
57     }
58
59     public static Optional<Double> inverse(Double x)
60     {
61        return x == 0 ? Optional.empty() : Optional.of(1 / x);
62     }
63
64     public static Optional<Double> squareRoot(Double x)
65     {
66        return x < 0 ? Optional.empty() : Optional.of(Math.sqrt(x));
67     }
68  }
```

API `java.util.Optional` 8

- `<U> Optional<U> flatMap(Function<? super T,Optional<U>> mapper)` 9

产生将 mapper 应用于当前 Optional 值所产生的结果，或者在当前 Optional 为空时，返回一个空 Optional。

1.8 收集结果

当处理完流之后，通常会想要查看其结果。此时可以调用 iterator 方法，它会产生用来访问元素的旧式风格的迭代器。

或者，可以调用 forEach 方法，将某个函数应用于每个元素：

```
stream.forEach(System.out::println);
```

在并行流上，forEach 方法会以任意顺序遍历各个元素。如果想要按照流中的顺序来处理它们，可以调用 forEachOrdered 方法。当然，这个方法会丧失并行处理的部分甚至全部优势。

但是，更常见的情况是，我们想要将结果收集到数据结构中。此时，可以调用 toArray，获得由流的元素构成的数组。

因为无法在运行时创建泛型数组，所以表达式 stream.toArray() 会返回一个 Object[] 数组。如果想要让数组具有正确的类型，可以将其传递到数组构造器中：

```
String[] result = stream.toArray(String[]::new);
   // stream.toArray() has type Object[]
```

针对将流中的元素收集到另一个目标中，有一个便捷方法 collect 可用，它会接受一个 Collector 接口的实例。收集器是一种收集众多元素并产生单一结果的对象，Collectors 类提供了大量用于生成常见收集器的工厂方法。要想将流的元素收集到一个列表中，应该使用 Collectors.toList() 方法产生的收集器：

```
List<String> result = stream.collect(Collectors.toList());
```

类似地，下面的代码展示了如何将流的元素收集到一个集中：

```
Set<String> result = stream.collect(Collectors.toSet());
```

如果想要控制获得的集的种类，那么可以使用下面的调用：

```
TreeSet<String> result = stream.collect(Collectors.toCollection(TreeSet::new));
```

假设想要通过连接操作来收集流中的所有字符串。我们可以调用

```
String result = stream.collect(Collectors.joining());
```

如果想要在元素之间增加分隔符，可以将分隔符传递给 joining 方法：

```
String result = stream.collect(Collectors.joining(", "));
```

如果流中包含除字符串以外的其他对象，那么我们需要先将其转换为字符串，就像下面这样：

```
String result = stream.map(Object::toString).collect(Collectors.joining(", "));
```

如果想要将流的结果约简为总和、数量、平均值、最大值或最小值，可以使用 summarizing(Int|Long|Double) 方法中的某一个。这些方法会接受一个将流对象映射为数值的函数，产生类型为 (Int|Long|Double)SummaryStatistics 的结果，同时计算总和、数量、平均值、最大值和最小值。

```
IntSummaryStatistics summary = stream.collect(
   Collectors.summarizingInt(String::length));
double averageWordLength = summary.getAverage();
double maxWordLength = summary.getMax();
```

程序清单 1-4 中的示例程序展示了如何从流中收集元素。

程序清单 1-4 collecting/CollectingResults.java

```java
package collecting;

/**
 * @version 1.01 2018-05-01
 * @author Cay Horstmann
 */

import java.io.*;
import java.nio.charset.*;
import java.nio.file.*;
import java.util.*;
import java.util.stream.*;

public class CollectingResults
{
   public static Stream<String> noVowels() throws IOException
   {
      var contents = new String(Files.readAllBytes(
         Paths.get("../gutenberg/alice30.txt")),
         StandardCharsets.UTF_8);
      List<String> wordList = List.of(contents.split("\\PL+"));
      Stream<String> words = wordList.stream();
      return words.map(s -> s.replaceAll("[aeiouAEIOU]", ""));
   }

   public static <T> void show(String label, Set<T> set)
   {
      System.out.print(label + ": " + set.getClass().getName());
      System.out.println("["
         + set.stream().limit(10).map(Object::toString).collect(Collectors.joining(", "))
         + "]");
   }

   public static void main(String[] args) throws IOException
   {
      Iterator<Integer> iter = Stream.iterate(0, n -> n + 1).limit(10).iterator();
      while (iter.hasNext())
         System.out.println(iter.next());

      Object[] numbers = Stream.iterate(0, n -> n + 1).limit(10).toArray();
      System.out.println("Object array:" + numbers);
         // Note it's an Object[] array

      try
      {
         var number = (Integer) numbers[0]; // OK
         System.out.println("number: " + number);
         System.out.println("The following statement throws an exception:");
         var numbers2 = (Integer[]) numbers; // Throws exception
      }
```

```
51            catch (ClassCastException ex)
52            {
53               System.out.println(ex);
54            }
55
56            Integer[] numbers3 = Stream.iterate(0, n -> n + 1)
57               .limit(10)
58               .toArray(Integer[]::new);
59            System.out.println("Integer array: " + numbers3);
60               // Note it's an Integer[] array
61
62            Set<String> noVowelSet = noVowels().collect(Collectors.toSet());
63            show("noVowelSet", noVowelSet);
64
65            TreeSet<String> noVowelTreeSet = noVowels().collect(
66               Collectors.toCollection(TreeSet::new));
67            show("noVowelTreeSet", noVowelTreeSet);
68
69            String result = noVowels().limit(10).collect(Collectors.joining());
70            System.out.println("Joining: " + result);
71            result = noVowels().limit(10)
72               .collect(Collectors.joining(", "));
73            System.out.println("Joining with commas: " + result);
74
75            IntSummaryStatistics summary = noVowels().collect(
76               Collectors.summarizingInt(String::length));
77            double averageWordLength = summary.getAverage();
78            double maxWordLength = summary.getMax();
79            System.out.println("Average word length: " + averageWordLength);
80            System.out.println("Max word length: " + maxWordLength);
81            System.out.println("forEach:");
82            noVowels().limit(10).forEach(System.out::println);
83         }
84      }
```

API *java.util.stream.BaseStream* 8

- Iterator<T> iterator()

  产生一个用于获取当前流中各个元素的迭代器。这是一种终结操作。

API *java.util.stream.Stream* 8

- void forEach(Consumer<? super T> action)

  在流的每个元素上调用 action。这是一种终结操作。

- Object[] toArray()

- <A> A[] toArray(IntFunction<A[]> generator)

  产生一个对象数组，或者在将引用 A[]::new 传递给构造器时，返回一个 A 类型的数组。这些操作都是终结操作。

- <R,A> R collect(Collector<? super T,A,R> collector)

使用给定的收集器来收集当前流中的元素。Collectors 类有用于多种收集器的工厂方法。

API java.util.stream.Collectors 8

- static <T> Collector<T,?,List<T>> toList()
- static <T> Collector<T,?,List<T>> toUnmodifiableList() 10
- static <T> Collector<T,?,Set<T>> toSet()
- static <T> Collector<T,?,Set<T>> toUnmodifiableSet() 10

  产生一个将元素收集到列表或集合中的收集器。

- static <T,C extends Collection<T>> Collector<T,?,C> toCollection(Supplier<C> collectionFactory)

  产生一个将元素收集到任意集合中的收集器。可以传递一个诸如 TreeSet::new 的构造器引用。

- static Collector<CharSequence,?,String> joining()
- static Collector<CharSequence,?,String> joining(CharSequence delimiter)
- static Collector<CharSequence,?,String> joining(CharSequence delimiter, CharSequence prefix, CharSequence suffix)

  产生一个连接字符串的收集器。分隔符会置于字符串之间,而第一个字符串之前可以有前缀,最后一个字符串之后可以有后缀。如果没有指定,那么它们都为空。

- static <T> Collector<T,?,IntSummaryStatistics> summarizingInt(ToIntFunction<? super T> mapper)
- static <T> Collector<T,?,LongSummaryStatistics> summarizingLong(ToLongFunction<? super T> mapper)
- static <T> Collector<T,?,DoubleSummaryStatistics> summarizingDouble(ToDoubleFunction<? super T> mapper)

  产生能够生成 (Int|Long|Double)SummaryStatistics 对象的收集器,通过它们可以获得将 mapper 应用于每个元素后所产生的结果的数量、总和、平均值、最大值和最小值。

API IntSummaryStatistics 8
LongSummaryStatistics 8
DoubleSummaryStatistics 8

- long getCount()

  产生汇总后的元素的个数。

- (int|long|double) getSum()
- double getAverage()

  产生汇总后的元素的总和或平均值,或者在没有任何元素时返回 0。

- (int|long|double) getMax()
- (int|long|double) getMin()

  产生汇总后的元素的最大值和最小值,或者在没有任何元素时,产生 (Integer| Long| Double).(MAX|MIN)_VALUE。

1.9 收集到映射表中

假设我们有一个 Stream<Person>，并且想要将其元素收集到一个映射表中，这样后续就可以通过它们的 ID 来查找人员了。Collectors.toMap 方法有两个函数引元，它们用来产生映射表的键和值。例如，

```
Map<Integer, String> idToName = people.collect(
    Collectors.toMap(Person::getId, Person::getName));
```

通常情况下，值应该是实际的元素，因此第二个函数可以使用 Function.identity()。

```
Map<Integer, Person> idToPerson = people.collect(
    Collectors.toMap(Person::getId, Function.identity()));
```

如果有多个元素具有相同的键，就会存在冲突，收集器将会抛出一个 IllegalStateException 异常。可以通过提供第 3 个函数引元来覆盖这种行为，该函数会针对给定的已有值和新值来解决冲突并确定键对应的值。这个函数应该返回已有值、新值或它们的组合。

在下面的代码中，我们构建了一个映射表，存储了所有可用 locale 中的语言，其中每种语言在默认 locale 中的名字（例如 "German"）为键，而其本地化的名字（例如 "Deutsch"）为值：

```
Stream<Locale> locales = Stream.of(Locale.getAvailableLocales());
Map<String, String> languageNames = locales.collect(
    Collectors.toMap(
        Locale::getDisplayLanguage,
        loc -> loc.getDisplayLanguage(loc),
        (existingValue, newValue) -> existingValue));
```

我们不关心同一种语言是否可能会出现两次（例如，德国和瑞士都使用德语），因此我们只记录第一项。

> **注释**：在本章中，我们使用 Locale 类作为感兴趣的数据集的数据源。请参阅第 7 章以了解有关 locale 的更多信息。

现在，假设我们想要了解给定国家的所有语言，这样我们就需要一个 Map<String, Set<String>>。例如，"Switzerland" 的值是集 [French, German, Italian]。首先，我们为每种语言都存储一个单例集。无论何时，只要找到了给定国家的新语言，我们就会对已有集和新集进行并操作。

```
Map<String, Set<String>> countryLanguageSets = locales.collect(
    Collectors.toMap(
        Locale::getDisplayCountry,
        l -> Collections.singleton(l.getDisplayLanguage()),
        (a, b) -> { // Union of a and b
            var union = new HashSet<String>(a);
            union.addAll(b);
            return union; }));
```

在下一节中，你将会看到一种更简单的获取这种映射表的方式。

如果想要得到 TreeMap，那么可以将构造器作为第 4 个引元来提供。你必须提供一种合并函数。下面是本节一开始所列举的示例之一，现在它会产生一个 TreeMap：

```
Map<Integer, Person> idToPerson = people.collect(
    Collectors.toMap(
        Person::getId,
        Function.identity(),
        (existingValue, newValue) -> { throw new IllegalStateException(); },
        TreeMap::new));
```

> 注释：对于每一个 toMap 方法，都有一个等价的可以产生并发映射表的 toConcurrentMap 方法。单个并发映射表可以用于并行集合处理。当使用并行流时，共享的映射表比合并映射表更高效。注意，元素不再是按照流中的顺序收集的，但是通常这不会有什么问题。

程序清单 1-5 给出了将流的结果收集到映射表中的示例。

程序清单 1-5   collecting/CollectingIntoMaps.java

```java
 1  package collecting;
 2
 3  /**
 4   * @version 1.00 2016-05-10
 5   * @author Cay Horstmann
 6   */
 7
 8  import java.io.*;
 9  import java.util.*;
10  import java.util.function.*;
11  import java.util.stream.*;
12
13  public class CollectingIntoMaps
14  {
15
16      public static class Person
17      {
18          private int id;
19          private String name;
20
21          public Person(int id, String name)
22          {
23              this.id = id;
24              this.name = name;
25          }
26
27          public int getId()
28          {
29              return id;
30          }
31
32          public String getName()
33          {
34              return name;
```

```java
35        }
36
37        public String toString()
38        {
39           return getClass().getName() + "[id=" + id + ",name=" + name + "]";
40        }
41     }
42
43     public static Stream<Person> people()
44     {
45        return Stream.of(new Person(1001, "Peter"), new Person(1002, "Paul"),
46           new Person(1003, "Mary"));
47     }
48
49     public static void main(String[] args) throws IOException
50     {
51        Map<Integer, String> idToName = people().collect(
52           Collectors.toMap(Person::getId, Person::getName));
53        System.out.println("idToName: " + idToName);
54
55        Map<Integer, Person> idToPerson = people().collect(
56           Collectors.toMap(Person::getId, Function.identity()));
57        System.out.println("idToPerson: " + idToPerson.getClass().getName()
58           + idToPerson);
59
60        idToPerson = people().collect(
61           Collectors.toMap(Person::getId, Function.identity(),
62              (existingValue, newValue) -> { throw new IllegalStateException(); },
63              TreeMap::new));
64        System.out.println("idToPerson: " + idToPerson.getClass().getName()
65           + idToPerson);
66
67        Stream<Locale> locales = Stream.of(Locale.getAvailableLocales());
68        Map<String, String> languageNames = locales.collect(
69           Collectors.toMap(
70              Locale::getDisplayLanguage,
71              l -> l.getDisplayLanguage(l),
72              (existingValue, newValue) -> existingValue));
73        System.out.println("languageNames: " + languageNames);
74
75        locales = Stream.of(Locale.getAvailableLocales());
76        Map<String, Set<String>> countryLanguageSets = locales.collect(
77           Collectors.toMap(
78              Locale::getDisplayCountry,
79              l -> Set.of(l.getDisplayLanguage()),
80              (a, b) ->
81              { // union of a and b
82                 Set<String> union = new HashSet<>(a);
83                 union.addAll(b);
84                 return union;
85              }));
86        System.out.println("countryLanguageSets: " + countryLanguageSets);
87     }
88  }
```

API `java.util.stream.Collectors` 8

- static <T,K,U> Collector<T,?,Map<K,U>> toMap(Function<? super T,? extends K> keyMapper, Function<? super T,? extends U> valueMapper)
- static <T,K,U> Collector<T,?,Map<K,U>> toMap(Function<? super T,? extends K> keyMapper, Function<? super T,? extends U> valueMapper, BinaryOperator<U> mergeFunction)
- static <T,K,U,M extends Map<K,U>> Collector<T,?,M> toMap(Function<? super T,? extends K> keyMapper, Function<? super T,? extends U> valueMapper, BinaryOperator<U> mergeFunction, Supplier<M> mapSupplier)
- static <T,K,U> Collector<T,?,Map<K,U>> toUnmodifiableMap(Function<? super T,? extends K> keyMapper, Function<? super T,? extends U> valueMapper) 10
- static <T,K,U> Collector<T,?,Map<K,U>> toUnmodifiableMap(Function<? super T,? extends K> keyMapper, Function<? super T,? extends U> valueMapper, BinaryOperator<U> mergeFunction) 10
- static <T,K,U> Collector<T,?,ConcurrentMap<K,U>> toConcurrentMap(Function<? super T,? extends K> keyMapper, Function<? super T,? extends U> valueMapper)

产生一个收集器，它会产生一个映射表、不可修改的映射表或并发映射表。keyMapper 和 valueMapper 函数会应用于每个收集到的元素上，从而在所产生的映射表中生成一个键/值项。默认情况下，当两个元素产生相同的键时，会抛出一个 IllegalStateException 异常。你可以提供一个 mergeFunction 来合并具有相同键的值。默认情况下，其结果是一个 HashMap 或 ConcurrentHashMap。你可以提供一个 mapSupplier，它会产生所期望的映射表实例。

1.10 群组和分区

在上一节中，你看到了如何收集给定国家的所有语言，但是其处理显得有些冗长。你必须为每个映射表的值都生成单例集，然后指定如何将现有值与新值合并。将具有相同特性的值群聚成组是非常常见的，并且 groupingBy 方法直接就支持它。

我们来看看通过国家聚成组 Locale 的问题。首先，构建该映射表：

```
Map<String, List<Locale>> countryToLocales = locales.collect(
    Collectors.groupingBy(Locale::getCountry));
```

函数 Locale::getCountry 是群组的分类函数，你现在可以查找给定国家代码对应的所有地点了，例如：

```
List<Locale> swissLocales = countryToLocales.get("CH");
    // Yields locales de_CH, fr_CH, it_CH and maybe more
```

> **注释**：快速复习一下 locale：每个 locale 都有一个语言代码（例如英语的 en）和一个国家代码（例如美国的 US）。locale en_US 描述的是美国英语，而 en_IE 是爱尔兰英语。某些国家有多个 locale。例如，ga_IE 是爱尔兰的盖尔语，而前面的示例也展示了我的 JDK 知道瑞士至少有三个 locale。

当分类函数是断言函数（即返回 boolean 值的函数）时，流的元素可以分为两个列表：该函数返回 true 的元素和其他的元素。在这种情况下，使用 partitioningBy 比使用 groupingBy 更高效。例如，在下面的代码中，我们将所有 locale 分成了使用英语和使用所有其他语言的两类：

```
Map<Boolean, List<Locale>> englishAndOtherLocales = locales.collect(
    Collectors.partitioningBy(l -> l.getLanguage().equals("en")));
List<Locale> englishLocales = englishAndOtherLocales.get(true);
```

> **注释**：如果调用 groupingByConcurrent 方法，就会在使用并行流时获得一个被并行组装的并行映射表。这与 toConcurrentMap 方法完全类似。

API java.util.stream.Collectors 8

- static <T,K> Collector<T,?,Map<K,List<T>>> groupingBy(Function<? super T,? extends K> classifier)
- static <T,K> Collector<T,?,ConcurrentMap<K,List<T>>> groupingByConcurrent(Function<? super T,? extends K> classifier)

  产生一个收集器，它会产生一个映射表或并发映射表，其键是将 classifier 应用于所有收集到的元素上所产生的结果，而值是由具有相同键的元素构成的一个个列表。

- static <T> Collector<T,?,Map<Boolean,List<T>>> partitioningBy(Predicate<? super T> predicate)

  产生一个收集器，它会产生一个映射表，其键是 true/false，而值是由满足/不满足断言的元素构成的列表。

1.11 下游收集器

groupingBy 方法会产生一个映射表，它的每个值都是一个列表。如果想要以某种方式来处理这些列表，就需要提供一个"下游收集器"。例如，如果想要获得集而不是列表，那么可以使用上一节中看到的 Collectors.toSet 收集器：

```
Map<String, Set<Locale>> countryToLocaleSet = locales.collect(
    groupingBy(Locale::getCountry, toSet()));
```

> **注释**：在本节的这个示例以及后续示例中，我们认为静态导入 java.util.stream.Collectors.* 会使表达式更容易阅读。

Java 提供了多种可以将收集到的元素约简为数字的收集器：

- counting 会产生收集到的元素的个数。例如：

```
Map<String, Long> countryToLocaleCounts = locales.collect(
    groupingBy(Locale::getCountry, counting()));
```

  可以对每个国家有多少个 locale 进行计数。

- summing(Int|Long|Double) 会接受一个函数作为引元，将该函数应用到下游元素中，并产生它们的和。例如：

```
Map<String, Integer> stateToCityPopulation = cities.collect(
    groupingBy(City::getState, summingInt(City::getPopulation)));
```

可以计算城市流中每个州的人口总和。

- maxBy 和 minBy 会接受一个比较器，并分别产生下游元素中的最大值和最小值。例如：

```
Map<String, Optional<City>> stateToLargestCity = cities.collect(
    groupingBy(City::getState,
        maxBy(Comparator.comparing(City::getPopulation))));
```

可以产生每个州中最大的城市。

collectingAndThen 收集器在收集器后面添加了一个最终处理步骤。例如，如果我们想要知道有多少不同的结果，那么就可以将它们收集到一个集中，然后计算其尺寸：

```
Map<Character, Integer> stringCountsByStartingLetter = strings.collect(
    groupingBy(s -> s.charAt(0),
        collectingAndThen(toSet(), Set::size)));
```

mapping 收集器的做法正好相反，它会将一个函数应用于收集到的每个元素，并将结果传递给下游收集器。

```
Map<Character, Set<Integer>> stringLengthsByStartingLetter = strings.collect(
    groupingBy(s -> s.charAt(0),
        mapping(String::length, toSet())));
```

这里，我们按照首字符对字符串进行了分组。在每个组内部，我们会计算字符串的长度，然后将这些长度收集到一个集中。

mapping 方法还针对上一节中的问题，即把某国所有的语言收集到一个集中，产生了一种更佳的解决方案。

```
Map<String, Set<String>> countryToLanguages = locales.collect(
    groupingBy(Locale::getDisplayCountry,
        mapping(Locale::getDisplayLanguage,
            toSet())));
```

还有一个 flatMapping 方法，可以与返回流的函数一起使用。

如果群组和映射函数的返回值为 int、long 或 double，那么可以将元素收集到汇总统计对象中，就像 1.8 节中所讨论的一样。例如，

```
Map<String, IntSummaryStatistics> stateToCityPopulationSummary = cities.collect(
    groupingBy(City::getState,
        summarizingInt(City::getPopulation)));
```

然后，可以从每个组的汇总统计对象中获取这些函数值的总和、数量、平均值、最小值和最大值。

filtering 收集器会将一个过滤器应用到每个组上，例如：

```
Map<String, Set<City>> largeCitiesByState
    = cities.collect(
        groupingBy(City::getState,
            filtering(c -> c.getPopulation() > 500000,
                toSet()))); // States without large cities have empty sets
```

> **注释**：还有 3 个版本的 reducing 方法，它们都应用了通用的约简操作，正如 1.12 节中所描述的一样。

将收集器组合起来是一种很强大的方式，但是它也可能会导致产生非常复杂的表达式。最佳用法是与 groupingBy 和 partitioningBy 一起处理"下游的"映射表中的值。否则，应该直接在流上应用诸如 map、reduce、count、max 或 min 这样的方法。

程序清单 1-6 中的示例程序演示了下游收集器。

程序清单 1-6　collecting/DownstreamCollectors.java

```java
package collecting;

/**
 * @version 1.00 2016-05-10
 * @author Cay Horstmann
 */

import static java.util.stream.Collectors.*;

import java.io.*;
import java.nio.file.*;
import java.util.*;
import java.util.stream.*;

public class DownstreamCollectors
{
   public static class City
   {
      private String name;
      private String state;
      private int population;

      public City(String name, String state, int population)
      {
         this.name = name;
         this.state = state;
         this.population = population;
      }

      public String getName()
      {
         return name;
      }

      public String getState()
      {
         return state;
      }

      public int getPopulation()
```

```java
42      {
43          return population;
44      }
45   }
46
47   public static Stream<City> readCities(String filename) throws IOException
48   {
49       return Files.lines(Paths.get(filename))
50          .map(l -> l.split(", "))
51          .map(a -> new City(a[0], a[1], Integer.parseInt(a[2])));
52   }
53
54   public static void main(String[] args) throws IOException
55   {
56       Stream<Locale> locales = Stream.of(Locale.getAvailableLocales());
57       locales = Stream.of(Locale.getAvailableLocales());
58       Map<String, Set<Locale>> countryToLocaleSet = locales.collect(groupingBy(
59          Locale::getCountry, toSet()));
60       System.out.println("countryToLocaleSet: " + countryToLocaleSet);
61
62       locales = Stream.of(Locale.getAvailableLocales());
63       Map<String, Long> countryToLocaleCounts = locales.collect(groupingBy(
64          Locale::getCountry, counting()));
65       System.out.println("countryToLocaleCounts: " + countryToLocaleCounts);
66
67       Stream<City> cities = readCities("cities.txt");
68       Map<String, Integer> stateToCityPopulation = cities.collect(groupingBy(
69          City::getState, summingInt(City::getPopulation)));
70       System.out.println("stateToCityPopulation: " + stateToCityPopulation);
71
72       cities = readCities("cities.txt");
73       Map<String, Optional<String>> stateToLongestCityName = cities
74          .collect(groupingBy(City::getState,
75              mapping(City::getName, maxBy(Comparator.comparing(String::length)))));
76       System.out.println("stateToLongestCityName: " + stateToLongestCityName);
77
78       locales = Stream.of(Locale.getAvailableLocales());
79       Map<String, Set<String>> countryToLanguages = locales.collect(groupingBy(
80          Locale::getDisplayCountry, mapping(Locale::getDisplayLanguage, toSet())));
81       System.out.println("countryToLanguages: " + countryToLanguages);
82
83       cities = readCities("cities.txt");
84       Map<String, IntSummaryStatistics> stateToCityPopulationSummary = cities
85          .collect(groupingBy(City::getState, summarizingInt(City::getPopulation)));
86       System.out.println(stateToCityPopulationSummary.get("NY"));
87
88       cities = readCities("cities.txt");
89       Map<String, String> stateToCityNames = cities.collect(groupingBy(
90          City::getState,
91          reducing("", City::getName, (s, t) -> s.length() == 0 ? t : s + ", " + t)));
92
93       cities = readCities("cities.txt");
94       stateToCityNames = cities.collect(groupingBy(City::getState,
95          mapping(City::getName, joining(", "))));
```

```
96          System.out.println("stateToCityNames: " + stateToCityNames);
97       }
98    }
```

API java.util.stream.Collectors 8

- public static <T,K,A,D> Collector<T,?,Map<K,D>> groupingBy(Function<? super T,? extends K> classifier, Collector<? super T,A,D> downstream)

  产生一个收集器，该收集器会产生一个映射表，其中的键是将 classifier 应用到所有收集到的元素上之后产生的结果，而值是使用下游收集器收集具有相同的键的元素所产生的结果。

- static <T> Collector<T,?,Long> counting()

  产生一个可以对收集到的元素进行计数的收集器。

- static <T> Collector<T,?,Integer> summingInt(ToIntFunction<? super T> mapper)
- static <T> Collector<T,?,Long> summingLong(ToLongFunction<? super T> mapper)
- static <T> Collector<T,?,Double> summingDouble(ToDoubleFunction<? super T> mapper)

  产生一个收集器，对将 mapper 应用到收集到的元素上之后产生的结果计算总和。

- static <T> Collector<T,?,Optional<T>> maxBy(Comparator<? super T> comparator)
- static <T> Collector<T,?,Optional<T>> minBy(Comparator<? super T> comparator)

  产生一个收集器，使用 comparator 指定的排序方法，计算收集到的元素中的最大值和最小值。

- static <T,A,R,RR> Collector<T,A,RR> collectingAndThen(Collector<T,A,R> downstream, Function<R,RR> finisher)

  产生一个收集器，它会将元素发送到下游收集器中，然后将 finisher 函数应用到其结果上。

- static <T,U,A,R> Collector<T,?,R> mapping(Function<? super T,? extends U> mapper, Collector<? super U,A,R> downstream)

  产生一个收集器，它会在每个元素上调用 mapper，并将结果发送到下游收集器中。

- static <T,U,A,R> Collector<T,?,R> flatMapping(Function<? super T,? extends Stream<? extends U>> mapper, Collector<? super U,A,R> downstream)

  产生一个收集器，它会在每个元素上调用 mapper，并将结果中的元素发送到下游收集器中。

- static <T,A,R> Collector<T,?,R> filtering(Predicate<? super T> predicate, Collector<? super T,A,R> downstream)

  产生一个收集器，它会将满足谓词逻辑的元素发送到下游收集器中。

1.12 约简操作

reduce 方法是一种用于从流中计算某个值的通用机制，其最简单的形式将接受一个二元函数，并从前两个元素开始持续应用它。如果该函数是求和函数，那么就很容易解释这种机制：

```
List<Integer> values = . . .;
Optional<Integer> sum = values.stream().reduce((x, y) -> x + y);
```

在上面的情况中，reduce 方法会计算 $v_0+v_1+v_2+\cdots$，其中 v_i 是流中的元素。如果流为空，那么该方法会返回一个 Optional，因为没有任何有效的结果。

> **注释**：在上面的情况中，可以写成 reduce(Integer::sum) 而不是 reduce((x, y) -> x+y)。

更一般地，我们可以使用任何约简操作将部分结果 x 与下一个值 y 组合起来以产生新的部分结果。

下面是另一种看待约简的方式。给定约简操作 op，该约简会产生 v_0 op v_1 op v_2 op...，其中我们将函数调用 $op(v_i, v_{i+1})$ 写作 v_i op v_{i+1}。有很多种在实践中很有用的可结合操作，例如求和、乘积、字符串连接、求最大值和最小值、求集的并与交等。

如果要用并行流来约简，那么这项约简操作必须是可结合的，即组合元素时使用的顺序不会产生任何影响。在数学标记法中，(x op y) op z 必须等于 x op (y op z)。减法是一个不可结合操作的例子，例如，(6 − 3) − 2 ≠ 6 − (3 − 2)。

通常，会有一个幺元值 e 使得 e op x = x，可以使用这个元素作为计算的起点。例如，0 是加法的幺元值。由此，我们可以使用第 2 种形式的 reduce：

```
List<Integer> values = . . .;
Integer sum = values.stream().reduce(0, (x, y) -> x + y);
    // Computes 0 + v₀ + v₁ + v₂ + . . .
```

如果流为空，则会返回幺元值，你就再也不需要处理 Optional 类了。

现在，假设你有一个对象流，并且想要对某些属性求和，例如字符串流中所有字符串的长度，那么你就不能使用简单形式的 reduce，而是需要 (T,T)->T 这样的函数，即引元和结果的类型相同的函数。但是在这种情况下，你有两种类型：流的元素具有 String 类型，而累积结果是整数。有一种形式的 reduce 可以处理这种情况。

首先，你需要提供一个"累积器"函数 (total, word) -> total + word.length()。这个函数会被反复调用，产生累积的总和。但是，当计算被并行化时，会有多个这种类型的计算，你需要将它们的结果合并。因此，你需要提供第二个函数来执行此处理。完整的调用如下：

```
int result = words.reduce(0,
    (total, word) -> total + word.length(),
    (total1, total2) -> total1 + total2);
```

> **注释**：在实践中，你可能并不会频繁地用到 reduce 方法。通常，映射为数字流并使用其方法来计算总和、最大值和最小值会更容易。（我们将在 1.13 节中讨论数字流。）在这个特定示例中，你可以调用 words.mapToInt(String::length).sum()，因为它不涉及装箱操作，所以更简单也更高效。

> **注释**：有时 reduce 会显得不够通用。例如，假设我们想要收集 BitSet 中的结果。如果收集操作是并行的，那么就不能直接将元素放到单个 BitSet 中，因为 BitSet 对象不是线程安全的。因此，我们不能使用 reduce，因为每个部分都需要以其自己的空集开始，并且 reduce 只能让我们提供一个幺元值。此时，应该使用 collect，它会接受单个引元：
> 1. 一个提供者，它会创建目标对象的新实例，例如散列集的构造器。
> 2. 一个累积器，它会将一个元素添加到该目标上，例如 add 方法。
> 3. 一个组合器，它会将两个对象合并成一个，例如 addAll。
>
> 下面的代码展示了 collect 方法是如何操作位集的：
> ```
> BitSet result = stream.collect(BitSet::new, BitSet::set, BitSet::or);
> ```

API java.util.Stream 8

- Optional<T> reduce(BinaryOperator<T> accumulator)
- T reduce(T identity, BinaryOperator<T> accumulator)
- <U> U reduce(U identity, BiFunction<U,? super T,U> accumulator, BinaryOperator<U> combiner)

  用给定的 accumulator 函数产生流中元素的累积总和。如果提供了幺元，那么第一个被累积的元素就是该幺元。如果提供了组合器，那么它可以用来将分别累积的各个部分整合成总和。

- <R> R collect(Supplier<R> supplier, BiConsumer<R,? super T> accumulator, BiConsumer<R,R> combiner)

  将元素收集到类型 R 的结果中。在每个部分上，都会调用 supplier 来提供初始结果，调用 accumulator 来交替地将元素添加到结果中，并调用 combiner 来整合两个结果。

1.13 基本类型流

到目前为止，我们都是将整数收集到 Stream<Integer> 中，尽管很明显，但是将每个整数都包装到包装器对象中却是很低效的。对其他基本类型来说，情况也是一样，这些基本类型是 double、float、long、short、char、byte 和 boolean。流库中具有专门的类型 IntStream、LongStream 和 DoubleStream，用来直接存储基本类型值，而无须使用包装器。如果想要存储 short、char、byte 和 boolean，可以使用 IntStream；而对于 float，可以使用 DoubleStream。

为了创建 IntStream，需要调用 IntStream.of 和 Arrays.stream 方法：

```
IntStream stream = IntStream.of(1, 1, 2, 3, 5);
stream = Arrays.stream(values, from, to); // values is an int[] array
```

与对象流一样，我们还可以使用静态的 generate 和 iterate 方法。此外，IntStream 和 LongStream 有静态方法 range 和 rangeClosed，可以生成步长为 1 的整数范围：

```
IntStream zeroToNinetyNine = IntStream.range(0, 100); // Upper bound is excluded
IntStream zeroToHundred = IntStream.rangeClosed(0, 100); // Upper bound is included
```

CharSequence 接口拥有 codePoints 和 chars 方法，可以生成由字符的 Unicode 码或由 UTF-16 编码机制的码元构成的 IntStream。

```
String sentence = "\uD835\uDD46 is the set of octonions.";
    // \uD835\uDD46 is the UTF-16 encoding of the letter 𝕆, unicode U+1D546

IntStream codes = sentence.codePoints();
    // The stream with hex values 1D546 20 69 73 20 . . .
```

当你有一个对象流时，可以用 mapToInt、mapToLong 或 mapToDouble 将其转换为基本类型流。例如，如果你有一个字符串流，并想将其长度处理为整数，那么就可以在 IntStream 中实现此目的：

```
Stream<String> words = . . .;
IntStream lengths = words.mapToInt(String::length);
```

为了将基本类型流转换为对象流，需要使用 boxed 方法：

```
Stream<Integer> integers = IntStream.range(0, 100).boxed();
```

通常，基本类型流上的方法与对象流上的方法类似。下面是主要的差异：

- toArray 方法会返回基本类型数组。
- 产生可选结果的方法会返回一个 OptionalInt、OptionalLong 或 OptionalDouble。这些类与 Optional 类类似，但是具有 getAsInt、getAsLong 和 getAsDouble 方法，而不是 get 方法。
- 具有分别返回总和、平均值、最大值和最小值的 sum、average、max 和 min 方法。对象流没有定义这些方法。
- summaryStatistics 方法会产生一个类型为 IntSummaryStatistics、LongSummaryStatistics 或 DoubleSummaryStatistics 的对象，它们可以同时报告流的总和、数量、平均值、最大值和最小值。

> 注释：Random 类具有 ints、longs 和 doubles 方法，它们会返回由随机数构成的基本类型流。如果需要的是并行流中的随机数，那么需要使用 SplittableRandom 类。

程序清单 1-7 给出了基本类型流的 API 的示例。

程序清单 1-7　streams/PrimitiveTypeStreams.java

```
1  package streams;
2
3  /**
4   * @version 1.01 2018-05-01
5   * @author Cay Horstmann
6   */
7
8  import java.io.IOException;
9  import java.nio.charset.StandardCharsets;
10 import java.nio.file.Files;
11 import java.nio.file.Path;
12 import java.nio.file.Paths;
13 import java.util.stream.Collectors;
```

```java
14 import java.util.stream.IntStream;
15 import java.util.stream.Stream;
16
17 public class PrimitiveTypeStreams
18 {
19    public static void show(String title, IntStream stream)
20    {
21       final int SIZE = 10;
22       int[] firstElements = stream.limit(SIZE + 1).toArray();
23       System.out.print(title + ": ");
24       for (int i = 0; i < firstElements.length; i++)
25       {
26          if (i > 0) System.out.print(", ");
27          if (i < SIZE) System.out.print(firstElements[i]);
28          else System.out.print("...");
29       }
30       System.out.println();
31    }
32
33    public static void main(String[] args) throws IOException
34    {
35       IntStream is1 = IntStream.generate(() -> (int) (Math.random() * 100));
36       show("is1", is1);
37       IntStream is2 = IntStream.range(5, 10);
38       show("is2", is2);
39       IntStream is3 = IntStream.rangeClosed(5, 10);
40       show("is3", is3);
41
42       Path path = Paths.get("../gutenberg/alice30.txt");
43       var contents = new String(Files.readAllBytes(path), StandardCharsets.UTF_8);
44
45       Stream<String> words = Stream.of(contents.split("\\PL+"));
46       IntStream is4 = words.mapToInt(String::length);
47       show("is4", is4);
48       var sentence = "\uD835\uDD46 is the set of octonions.";
49       System.out.println(sentence);
50       IntStream codes = sentence.codePoints();
51       System.out.println(codes.mapToObj(c -> String.format("%X ", c)).collect(
52          Collectors.joining()));
53
54       Stream<Integer> integers = IntStream.range(0, 100).boxed();
55       IntStream is5 = integers.mapToInt(Integer::intValue);
56       show("is5", is5);
57    }
58 }
```

API *java.util.stream.IntStream* 8

- static IntStream range(int startInclusive, int endExclusive)
- static IntStream rangeClosed(int startInclusive, int endInclusive)

  产生一个由给定范围内的整数构成的 IntStream。

- static IntStream of(int... values)

产生一个由给定元素构成的 IntStream。

- int[] toArray()

  产生一个由当前流中的元素构成的数组。

- int sum()
- OptionalDouble average()
- OptionalInt max()
- OptionalInt min()
- IntSummaryStatistics summaryStatistics()

  产生当前流中元素的总和、平均值、最大值和最小值，或者产生一个可以从中获取所有这四个值的对象。

- Stream<Integer> boxed()

  产生用于当前流中的元素的包装器对象流。

API *java.util.stream.LongStream* 8

- static LongStream range(long startInclusive, long endExclusive)
- static LongStream rangeClosed(long startInclusive, long endInclusive)

  用给定范围内的整数产生一个 LongStream。

- static LongStream of(long... values)

  用给定元素产生一个 LongStream。

- long[] toArray()

  用当前流中的元素产生一个数组。

- long sum()
- OptionalDouble average()
- OptionalLong max()
- OptionalLong min()
- LongSummaryStatistics summaryStatistics()

  产生当前流中元素的总和、平均值、最大值和最小值，或者产生一个可以从中获取所有这四个值的对象。

- Stream<Long> boxed()

  产生用于当前流中的元素的包装器对象流。

API *java.util.stream.DoubleStream* 8

- static DoubleStream of(double... values)

  用给定元素产生一个 DoubleStream。

- double[] toArray()

  用当前流中的元素产生一个数组。

- double sum()

- OptionalDouble average()
- OptionalDouble max()
- OptionalDouble min()
- DoubleSummaryStatistics summaryStatistics()

  产生当前流中元素的总和、平均值、最大值和最小值，或者产生一个可以从中获取所有这四个值的对象。

- Stream<Double> boxed()

  产生用于当前流中的元素的包装器对象流。

API *java.lang.CharSequence* 1.0

- IntStream codePoints() 8

  产生由当前字符串的所有 Unicode 码点构成的流。

API *java.util.Random* 1.0

- IntStream ints()
- IntStream ints(int randomNumberOrigin, int randomNumberBound) 8
- IntStream ints(long streamSize) 8
- IntStream ints(long streamSize, int randomNumberOrigin, int randomNumberBound) 8
- LongStream longs() 8
- LongStream longs(long randomNumberOrigin, long randomNumberBound) 8
- LongStream longs(long streamSize) 8
- LongStream longs(long streamSize, long randomNumberOrigin, long randomNumberBound) 8
- DoubleStream doubles() 8
- DoubleStream doubles(double randomNumberOrigin, double randomNumberBound) 8
- DoubleStream doubles(long streamSize) 8
- DoubleStream doubles(long streamSize, double randomNumberOrigin, double randomNumberBound) 8

  产生随机数流。如果提供了 streamSize，这个流就是具有给定数量元素的有限流。当提供了边界时，其元素将位于 randomNumberOrigin（包含）和 randomNumberBound（不包含）的区间内。

API *java.util.Optional(Int|Long|Double)* 8

- static Optional(Int|Long|Double) of((int|long|double) value)

  用所提供的基本类型值产生一个可选对象。

- (int|long|double) getAs(Int|Long|Double)()

  产生当前可选对象的值，或者在其为空时抛出一个 NoSuchElementException 异常。

- (int|long|double) orElse((int|long|double) other)
- (int|long|double) orElseGet((Int|Long|Double)Supplier other)

产生当前可选对象的值，或者在这个对象为空时产生可替代的值。
- void ifPresent((Int|Long|Double)Consumer consumer)
  如果当前可选对象不为空，则将其值传递给 consumer。

API java.util.(Int|Long|Double)SummaryStatistics 8

- long getCount()
- (int|long|double) getSum()
- double getAverage()
- (int|long|double) getMax()
- (int|long|double) getMin()

产生收集到的元素的数量、总和、平均值、最大值和最小值。

1.14 并行流

流使并行处理块操作变得很容易。这个过程几乎是自动的，但是需要遵守一些规则。首先，必须有一个并行流。可以用 Collection.parallelStream() 方法从任何集合中获取一个并行流：

```
Stream<String> parallelWords = words.parallelStream();
```

而且，parallel 方法可以将任意的顺序流转换为并行流。

```
Stream<String> parallelWords = Stream.of(wordArray).parallel();
```

只要在终结方法执行时流处于并行模式，所有的中间流操作就都将被并行化。

当流操作并行运行时，其目标是让其返回结果与顺序执行时返回的结果相同。重要的是，这些操作是无状态的，并且可以以任意顺序执行。

下面的示例是一项你无法完成的任务。假设你想要对字符串流中的所有短单词计数：

```
var shortWords = new int[12];
words.parallelStream().forEach(
    s -> { if (s.length() < 12) shortWords[s.length()]++; });
    // ERROR--race condition!
System.out.println(Arrays.toString(shortWords));
```

这是一种非常糟糕的代码。传递给 forEach 的函数会在多个并发线程中运行，每个都会更新共享的数组。正如我们在卷 I 第 12 章中所解释的，这是一种经典的竞争情况。如果多次运行这个程序，你很可能就会发现每次运行都产生不同的计数值，而且每个都是错的。

你的职责是确保传递给并行流操作的任何函数都可以安全地并行执行，达到这个目的的最佳方式是远离易变状态。在本例中，如果用长度将字符串分组，然后分别对它们进行计数，那么就可以安全地并行化这项计算。

```
Map<Integer, Long> shortWordCounts
    = words.parallelStream()
        .filter(s -> s.length() < 12)
        .collect(groupingBy(
            String::length,
            counting()));
```

默认情况下，从有序集合（数组和列表）、范围、生成器和迭代器产生的流，或者通过调用 Stream.sorted 产生的流，都是有序的。它们的结果是按照原来元素的顺序累积的，因此是完全可预知的。如果运行相同的操作两次，将会得到完全相同的结果。

排序并不排斥高效的并行处理。例如，当计算 stream.map(fun) 时，流可以被划分为 n 部分，它们会被并行地处理。然后，结果将会按照顺序重新组装起来。

当放弃排序需求时，有些操作可以被更有效地并行化。通过在流上调用 Stream.unordered 方法，就可以明确表示我们对排序不感兴趣。Stream.distinct 就是从这种方式中获益的一种操作。在有序的流中，distinct 会保留所有相同元素中的第一个，这对并行化是一种阻碍，因为处理每个部分的线程在其之前的所有部分都被处理完之前，并不知道应该丢弃哪些元素。如果可以接受保留唯一元素中任意一个的做法，那么所有部分就可以并行地处理（使用共享的集合来跟踪重复元素）。

还可以通过放弃排序要求来提高 limit 方法的速度。如果只想从流中取出任意 n 个元素，而并不在意到底要获取哪些，那么可以调用：

```
Stream<String> sample = words.parallelStream().unordered().limit(n);
```

正如 1.9 节所讨论的，合并映射表的代价很高昂。正是这个原因，Collectors.groupingByConcurrent 方法使用了共享的并发映射表。为了从并行化中获益，映射表中值的顺序不会与流中的顺序相同。

```
Map<Integer, List<String>> result = words.parallelStream().collect(
    Collectors.groupingByConcurrent(String::length));
    // Values aren't collected in stream order
```

当然，如果使用独立于排序的下游收集器，那么就不必在意了，例如：

```
Map<Integer, Long> wordCounts
    = words.parallelStream()
        .collect(
            groupingByConcurrent(
                String::length,
                counting()));
```

不要指望通过将所有的流都转换为并行流就能够加速操作，要牢记下面几条：
- 并行化会导致大量的开销，只有面对非常大的数据集才划算。
- 只有在底层的数据源可以被有效地分割为多个部分时，将流并行化才有意义。
- 并行流使用的线程池可能会因诸如文件 I/O 或网络访问这样的操作被阻塞而饿死。

只有面对海量的内存数据和运算密集处理，并行流才会工作最佳。

> ✓ 提示：在 Java 9 之前，对 Files.lines 方法返回的流进行并行化是没有意义的。因为数据是不可分割的，所以我们只能在读取文件的后半部分之前读取前半部分。现在，该方法使用的是内存映射文件，因此可以有效地进行分割。如果想要处理一个大型文件的各个行，并行化这个流可能会提高性能。

> **注释**：默认情况下，并行流使用的是 ForkJoinPool.commonPool 返回的全局 fork-join 池。只有在操作不会阻塞并且我们不会将这个池与其他任务共享的情况下，这种方式才不会有什么问题。有一种解决方法是使用另一个不同的池，即把操作放置到定制的池的 submit 方法中：
>
> ```
> ForkJoinPool customPool = ....;
> result = customPool.submit(() ->
>     stream.parallel().map(...).collect(...)).get();
> ```
>
> 或者，使用异步方式：
>
> ```
> CompletableFuture.supplyAsync(() ->
>     stream.parallel().map(...).collect(...),
>     customPool).thenAccept(result -> ...);
> ```

> **注释**：如果想要并行化基于随机数的流计算，那么请不要以从 Random.ints、Random.longs 或 Random.doubles 方法中获得的流为起点，因为这些流不可分割。应该使用 SplittableRandom 类的 ints、longs 或 doubles。

程序清单 1-8 中的示例程序展示了如何操作并行流。

程序清单 1-8　parallel/ParallelStreams.java

```
 1  package parallel;
 2
 3  /**
 4   * @version 1.01 2018-05-01
 5   * @author Cay Horstmann
 6   */
 7
 8  import static java.util.stream.Collectors.*;
 9
10  import java.io.*;
11  import java.nio.charset.*;
12  import java.nio.file.*;
13  import java.util.*;
14  import java.util.stream.*;
15
16  public class ParallelStreams
17  {
18      public static void main(String[] args) throws IOException
19      {
20          var contents = new String(Files.readAllBytes(
21              Paths.get("../gutenberg/alice30.txt")), StandardCharsets.UTF_8);
22          List<String> wordList = List.of(contents.split("\\PL+"));
23
24          // Very bad code ahead
25          var shortWords = new int[10];
26          wordList.parallelStream().forEach(s ->
27              {
28                  if (s.length() < 10) shortWords[s.length()]++;
```

```
29        });
30     System.out.println(Arrays.toString(shortWords));
31
32     // Try again--the result will likely be different (and also wrong)
33     Arrays.fill(shortWords, 0);
34     wordList.parallelStream().forEach(s ->
35        {
36           if (s.length() < 10) shortWords[s.length()]++;
37        });
38     System.out.println(Arrays.toString(shortWords));
39
40     // Remedy: Group and count
41     Map<Integer, Long> shortWordCounts = wordList.parallelStream()
42        .filter(s -> s.length() < 10)
43        .collect(groupingBy(String::length, counting()));
44
45     System.out.println(shortWordCounts);
46
47     // Downstream order not deterministic
48     Map<Integer, List<String>> result = wordList.parallelStream().collect(
49        Collectors.groupingByConcurrent(String::length));
50
51     System.out.println(result.get(14));
52
53     result = wordList.parallelStream().collect(
54        Collectors.groupingByConcurrent(String::length));
55
56     System.out.println(result.get(14));
57
58     Map<Integer, Long> wordCounts = wordList.parallelStream().collect(
59        groupingByConcurrent(String::length, counting()));
60
61     System.out.println(wordCounts);
62  }
63 }
```

API *java.util.stream.BaseStream<T,S extends BaseStream<T,S>>* 8

- S parallel()
  产生一个与当前流中元素相同的并行流。
- S unordered()
  产生一个与当前流中元素相同的无序流。

API *java.util.Collection<E>* 1.2

- Stream<E> parallelStream() 8
  用当前集合中的元素产生一个并行流。

在本章中，你学习到了如何运用 Java 8 的流库。下一章将讨论另一个重要的主题：输入与输出。

第 2 章　输入与输出

- ▲ 输入 / 输出流
- ▲ 读写二进制数据
- ▲ 对象输入 / 输出流与序列化
- ▲ 操作文件
- ▲ 内存映射文件
- ▲ 文件锁机制
- ▲ 正则表达式

本章将介绍 Java 中用于输入和输出的各种应用编程接口（Application Programming Interface，API）。你将要学习如何访问文件与目录，以及如何以二进制格式和文本格式来读写数据。本章还要向你展示对象序列化机制，它可以使存储对象像存储文本和数值数据一样容易。然后，我们将介绍如何使用文件和目录。最后，本章将讨论正则表达式，尽管这部分内容实际上与输入和输出并不相关，但是我们确实也找不到更合适的地方来处理这个话题。很明显，Java 设计团队在这个问题的处理上和我们一样，因为正则表达式 API 的规格说明隶属于"新 I/O"特性的规格说明。

2.1　输入 / 输出流

在 Java API 中，可以从其中读入一个字节序列的对象称作输入流，而可以向其中写入一个字节序列的对象称作输出流。这些字节序列的来源地和目的地可以是文件，而且通常都是文件，但是也可以是网络连接，甚至是内存块。抽象类 InputStream 和 OutputStream 构成了输入 / 输出（I/O）类层次结构的基础。

> **注释**：这些输入 / 输出流与在前一章中看到的流没有任何关系。为了清楚起见，只要是讨论用于输入和输出的流，我们都将使用术语输入流、输出流或输入 / 输出流。

因为面向字节的流不便于处理以 Unicode 形式存储的信息（回忆一下，Unicode 中每个字符都使用了多个字节来表示），所以从抽象类 Reader 和 Writer 中继承出来了一个专门用于处理 Unicode 字符的单独的类层次结构。这些类拥有的读入和写出操作都是基于两字节的 Char 值的（即 Unicode 码元），而不是基于 byte 值的。

2.1.1　读写字节

InputStream 类有一个抽象方法：

```
abstract int read()
```

这个方法将读入一个字节，并返回读入的字节，或者在遇到输入源结尾时返回 -1。在设计具体的输入流类时，必须覆盖这个方法以提供适用的功能，例如，在 FileInputStream 类中，这个方法将从某个文件中读入一个字节，而 System.in（它是 InputStream 的一个子类的预定义对象）却是从"标准输入"中读入信息，即从控制台或重定向的文件中读入信息。

InputStream 类还有若干个非抽象的方法，它们可以读入一个字节数组，或者跳过大量的字节。从 Java 9 开始，有了一个非常有用的可以读取流中所有字节的方法：

```
byte[] bytes = in.readAllBytes();
```

还有多个用来读取给定数量字节的方法，可以参见 API 说明。

这些方法都要调用抽象的 read 方法，因此，各个子类都只需覆盖这一个方法。

与此类似，OutputStream 类定义了下面的抽象方法：

```
abstract void write(int b)
```

它可以向某个输出位置写出一个字节。

如果我们有一个字节数组，那么就可以一次性地写出它们：

```
byte[] values = . . .;
out.write(values);
```

transferTo 方法可以将所有字节从一个输入流传递到一个输出流：

```
in.transferTo(out);
```

read 和 write 方法在执行时都将阻塞，直至字节确实被读入或写出。这就意味着如果流不能被立即访问（通常是因为网络连接忙），那么当前的线程将被阻塞。这使得在这两个方法等待指定的流变为可用的这段时间里，其他的线程就有机会去执行有用的工作。

available 方法使我们可以去检查当前可读入的字节数量，这意味着像下面这样的代码片段不可能被阻塞：

```
int bytesAvailable = in.available();
if (bytesAvailable > 0)
{
   var data = new byte[bytesAvailable];
   in.read(data);
}
```

当你完成对输入 / 输出流的读写时，应该通过调用 close 方法来关闭它，这个调用会释放掉十分有限的操作系统资源。如果一个应用程序打开了过多的输入 / 输出流而没有关闭，那么系统资源将被耗尽。关闭一个输出流的同时还会冲刷用于该输出流的缓冲区：所有被临时置于缓冲区中，以便用更大的包的形式传递的字节在关闭输出流时都将被送出。特别是，如果不关闭文件，那么写出字节的最后一个包可能永远也得不到传递。当然，我们还可以用 flush 方法来人为地冲刷这些输出。

即使某个输入 / 输出流类提供了使用原生的 read 和 write 功能的某些具体方法，应用系统的程序员还是很少使用它们，因为大家感兴趣的数据可能包含数字、字符串和对象，而不是原生字节。

我们可以使用众多的构建于基本的 InputStream 和 OutputStream 类之上的某个输入 / 输出类，而不只是直接使用字节。

API `java.io.InputStream` 1.0

- `abstract int read()`
  从数据中读入一个字节,并返回该字节。这个 read 方法在碰到输入流的结尾时返回 -1。
- `int read(byte[] b)`
  读入一个字节数组,并返回实际读入的字节数,或者在碰到输入流的结尾时返回 -1。这个 read 方法最多读入 b.length 个字节。
- `int read(byte[] b, int off, int len)`
- `int readNBytes(byte[] b, int off, int len)` 9
  如果未阻塞(read),则读入由 len 指定数量的字节,或者阻塞至所有的值都被读入(readNBytes)。读入的值将置于 b 中从 off 开始的位置。返回实际读入的字节数,或者在碰到输入流的结尾时返回 -1。
- `byte[] readAllBytes()` 9
  产生一个数组,包含可以从当前流中读入的所有字节。
- `long transferTo(OutputStream out)` 9
  将当前输入流中的所有字节传送到给定的输出流,返回传递的字节数。这两个流都不应该处于关闭状态。
- `long skip(long n)`
  在输入流中跳过 n 个字节,返回实际跳过的字节数(如果碰到输入流的结尾,则可能小于 n)。
- `int available()`
  返回在不阻塞的情况下可获取的字节数(回忆一下,阻塞意味着当前线程将失去它对资源的占用)。
- `void close()`
  关闭这个输入流。
- `void mark(int readlimit)`
  在输入流的当前位置打一个标记(并非所有的流都支持这个特性)。如果从输入流中已经读入的字节多于 readlimit 个,则这个流允许忽略这个标记。
- `void reset()`
  返回到最后一个标记,随后对 read 的调用将重新读入这些字节。如果当前没有任何标记,则这个流不被重置。
- `boolean markSupported()`
  如果这个流支持打标记,则返回 true。

API `java.io.OutputStream` 1.0

- `abstract void write(int n)`
  写出一个字节的数据。
- `void write(byte[] b)`

- void write(byte[] b, int off, int len)
  写出所有字节或者某个范围的字节到数组 b 中。
- void close()
  冲刷并关闭输出流。
- void flush()
  冲刷输出流，也就是将所有缓冲的数据发送到目的地。

2.1.2 完整的流家族

与 C 语言只有单一类型 FILE* 包打天下不同，Java 拥有一个流家族，包含各种输入 / 输出流类型，其数量超过 60 个！请参见图 2-1 和图 2-2。

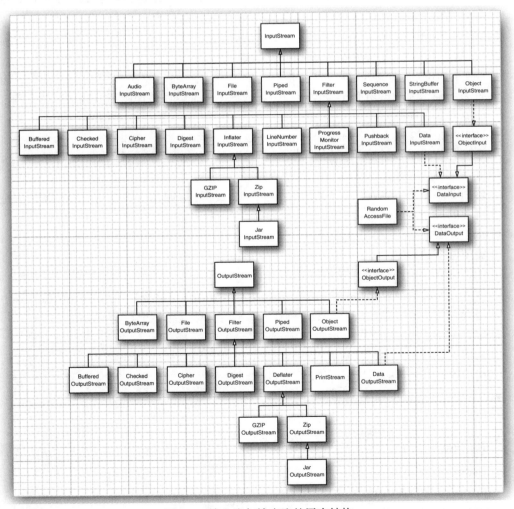

图 2-1 输入流与输出流的层次结构

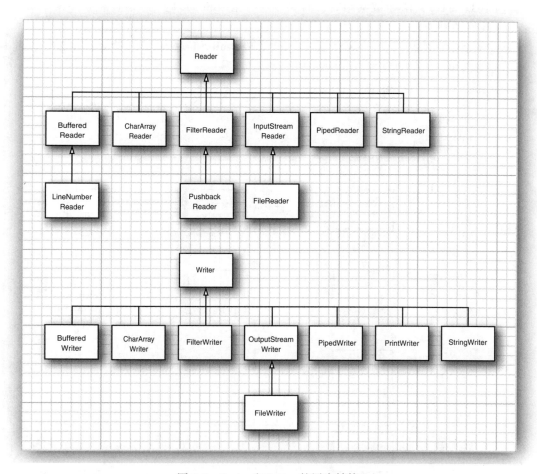

图 2-2 Reader 和 Writer 的层次结构

让我们把输入/输出流家族中的成员按照它们的使用方法来进行划分，这样就形成了处理字节和字符的两个单独的层次结构。正如所见，InputStream 和 OutputStream 类可以读写单个字节或字节数组，这些类构成了图 2-1 所示的层次结构的基础。要想读写字符串和数字，就需要功能更强大的子类，例如，DataInputStream 和 DataOutputStream 可以以二进制格式读写所有的基本 Java 类型。最后，还包含了多个很有用的输入/输出流，例如，ZipInputStream 和 ZipOutputStream 可以以我们常见的 ZIP 压缩格式读写文件。

另一方面，对于 Unicode 文本，可以使用抽象类 Reader 和 Writer 的子类（请参见图 2-2）。Reader 和 Writer 类的基本方法与 InputStream 和 OutputStream 中的方法类似。

```
abstract int read()
abstract void write(int c)
```

read 方法将返回一个 Unicode 码元（一个在 0～65535 之间的整数），或者在碰到文件结

尾时返回 -1。write 方法在被调用时，需要传递一个 Unicode 码元（请查看卷 I 第 3 章有关 Unicode 码元的讨论）。

还有 4 个附加的接口：Closeable、Flushable、Readable 和 Appendable（请查看图 2-3）。前两个接口非常简单，它们分别拥有下面的方法：

void close() throws IOException

和

void flush()

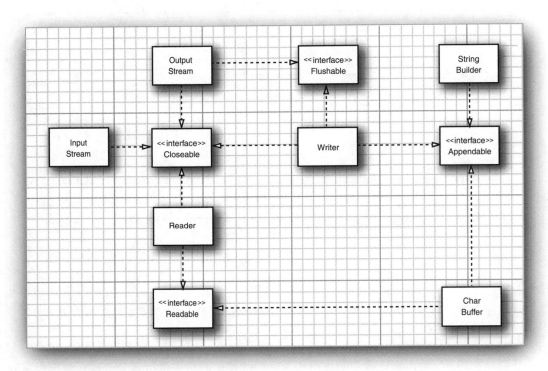

图 2-3　Closeable、Flushable、Readable 和 Appendable 接口

InputStream、OutputStream、Reader 和 Writer 都实现了 Closeable 接口。

> **注释**：java.io.Closeable 接口扩展了 java.lang.AutoCloseable 接口。因此，对任何 Closeable 进行操作时，都可以使用 try-with-resource 语句[⊖]。为什么要有两个接口呢？因为 Closeable 接口的 close 方法只抛出 IOException，而 AutoCloseable.close 方法可以抛出任何异常。

而 OutputStream 和 Writer 还实现了 Flushable 接口。

⊖ try-with-resource 语句是指声明了一个或多个资源的 try 语句。——译者注

Readable 接口只有一个方法：

```
int read(CharBuffer cb)
```

CharBuffer 类拥有按顺序和随机地进行读写访问的方法，它表示一个内存中的缓冲区或者一个内存映像的文件（请参见 2.5.2 节以了解细节）。

Appendable 接口有两个用于添加单个字符和字符序列的方法：

```
Appendable append(char c)
Appendable append(CharSequence s)
```

CharSequence 接口描述了一个 char 值序列的基本属性，String、CharBuffer、StringBuilder 和 StringBuffer 都实现了它。

在流类的家族中，只有 Writer 实现了 Appendable。

API *java.io.Closeable* 5.0

- void close()
  关闭这个 Closeable，这个方法可能会抛出 IOException。

API *java.io.Flushable* 5.0

- void flush()
  冲刷这个 Flushable。

API *java.lang.Readable* 5.0

- int read(CharBuffer cb)
  尝试着向 cb 读入其可持有数量的 char 值。返回读入的 char 值的数量，或者当从这个 Readable 中无法再获得更多的值时返回 -1。

API *java.lang.Appendable* 5.0

- Appendable append(char c)
- Appendable append(CharSequence cs)
  向这个 Appendable 中追加给定的码元或者给定的序列中的所有码元，返回 this。

API *java.lang.CharSequence* 1.4

- char charAt(int index)
  返回给定索引处的码元。
- int length()
  返回在这个序列中的码元的数量。
- CharSequence subSequence(int startIndex, int endIndex)
  返回由存储在 startIndex 到 endIndex-1 处的所有码元构成的 CharSequence。
- String toString()
  返回这个序列中所有码元构成的字符串。

2.1.3 组合输入/输出流过滤器

FileInputStream 和 FileOutputStream 可以提供附着在一个磁盘文件上的输入流和输出流，而你只需向其构造器提供文件名或文件的完整路径名。例如：

```
var fin = new FileInputStream("employee.dat");
```

这行代码可以查看用户目录下名为 "employee.dat" 的文件。

> **提示**：所有在 java.io 中的类都将相对路径名解释为以用户工作目录开始，你可以通过调用 System.getProperty("user.dir") 来获得这个信息。

> **警告**：由于反斜杠字符在 Java 字符串中是转义字符，因此要确保在 Windows 风格的路径名中使用 \\（例如，C:\\Windows\\win.ini）。在 Windows 中，还可以使用单斜杠字符（C:/Windows/win.ini），因为大部分 Windows 文件处理的系统调用都会将斜杠解释成文件分隔符。但是，并不推荐这样做，因为 Windows 系统函数的行为会因与时俱进而发生变化。因此，对于可移植的程序来说，应该使用程序所运行平台的文件分隔符，我们可以通过常量字符串 java.io.File.separator 获得它。

与抽象类 InputStream 和 OutputStream 一样，这些类只支持在字节级别上的读写。也就是说，我们只能从 fin 对象中读入字节和字节数组。

```
byte b = (byte) fin.read();
```

正如下节中看到的，如果我们只有 DataInputStream，那么我们就只能读入数值类型：

```
DataInputStream din = ...;
double x = din.readDouble();
```

但是正如 FileInput Stream 没有任何读入数值类型的方法一样，DataInputStream 也没有任何从文件中获取数据的方法。

Java 使用了一种灵巧的机制来分离这两种职责。某些输入流（例如 FileInputStream 和由 URL 类的 openStream 方法返回的输入流）可以从文件和其他更外部的位置上获取字节，而其他的输入流（例如 DataInputStream）可以将字节组装到更有用的数据类型中。Java 程序员必须对二者进行组合。例如，为了从文件中读入数字，首先需要创建一个 FileInputStream，然后将其传递给 DataInputStream 的构造器：

```
var fin = new FileInputStream("employee.dat");
var din = new DataInputStream(fin);
double x = din.readDouble();
```

如果再次查看图 2-1，你就会看到 FilterInputStream 和 FilterOutputStream 类。这些文件的子类用于向处理字节的输入/输出流添加额外的功能。

你可以通过嵌套过滤器来添加多重功能。例如，输入流在默认情况下是不被缓冲区缓存的，也就是说，每个对 read 的调用都会请求操作系统再分发一个字节。相比之下，请求一个数据块并将其置于缓冲区中会显得更加高效。如果我们想使用缓冲机制和用于文件的数据输入方法，那么就需要使用下面这种相当复杂的构造器序列：

```
var din = new DataInputStream(
   new BufferedInputStream(
      new FileInputStream("employee.dat")));
```

注意，我们把 DataInputStream 置于构造器链的最后，这是因为我们希望使用 DataInputStream 的方法，并且希望它们能够使用带缓冲机制的 read 方法。

有时当多个输入流链接在一起时，你需要跟踪各个中介输入流（intermediate input stream）。例如，当读入输入时，你经常需要预览下一个字节，以了解它是否是你想要的值。Java 提供了用于此目的的 PushbackInputStream：

```
var pbin = new PushbackInputStream(
   new BufferedInputStream(
      new FileInputStream("employee.dat")));
```

现在你可以预读下一个字节：

```
int b = pbin.read();
```

并且在它并非你所期望的值时将其推回流中。

```
if (b != '<') pbin.unread(b);
```

但是读入和推回是可应用于可回推（pushback）输入流的仅有的方法。如果你希望能够预先浏览并且还可以读入数字，那么就需要一个既是可回推输入流，又是一个数据输入流的引用。

```
var din = new DataInputStream(
   pbin = new PushbackInputStream(
      new BufferedInputStream(
         new FileInputStream("employee.dat"))));
```

当然，在其他编程语言的输入/输出流类库中，诸如缓冲机制和预览等细节都是自动处理的。因此，相比较而言，Java 就有一点麻烦，它必须将多个流过滤器组合起来。但是，这种混合并匹配过滤器类以构建真正有用的输入/输出流序列的能力，将带来极大的灵活性，例如，你可以从一个 ZIP 压缩文件中通过使用下面的输入流序列来读入数字（请参见图 2-4）：

```
var zin = new ZipInputStream(new FileInputStream("employee.zip"));
var din = new DataInputStream(zin);
```

（请查看 2.3.3 节以了解更多有关 Java 处理 ZIP 文件功能的知识。）

API java.io.FileInputStream 1.0

- FileInputStream(String name)
- FileInputStream(File file)

  使用由 name 字符串或 file 对象指定路径名的文件创建一个新的文件输入流（File 类在本章结尾处描述）。非绝对的路径名将按照相对于 VM 启动时所设置的工作目录来解析。

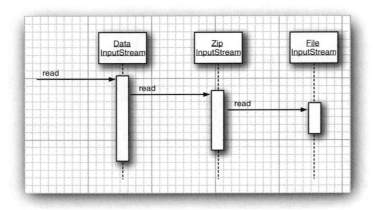

图 2-4　过滤器流序列

API java.io.FileOutputStream 1.0

- FileOutputStream(String name)
- FileOutputStream(String name, boolean append)
- FileOutputStream(File file)
- FileOutputStream(File file, boolean append)

  使用由 name 字符串或 file 对象指定路径名的文件创建一个新的文件输出流（File 类在本章结尾处描述）。如果 append 参数为 true，那么数据将被添加到文件尾，而具有相同名字的已有文件不会被删除；否则，这个方法会删除所有具有相同名字的已有文件。

API java.io.BufferedInputStream 1.0

- BufferedInputStream(InputStream in)

  创建一个带缓冲区的输入流。带缓冲区的输入流在从流中读入字符时，不会每次都访问设备。当缓冲区为空时，会向缓冲区中读入一个新的数据块。

API java.io.BufferedOutputStream 1.0

- BufferedOutputStream(OutputStream out)

  创建一个带缓冲区的输出流。带缓冲区的输出流在收集要写出的字符时，不会每次都访问设备。当缓冲区填满或当流被冲刷时，数据就被写出。

API java.io.PushbackInputStream 1.0

- PushbackInputStream(InputStream in)
- PushbackInputStream(InputStream in, int size)

  构建一个可以预览一个字节或者具有指定尺寸的回推缓冲区的输入流。

- void unread(int b)

  回推一个字节，它可以在下次调用 read 时被再次获取。

2.1.4 文本输入与输出

在保存数据时，可以选择二进制格式或文本格式。例如，整数 1234 存储成二进制数时，会被写为由字节 00 00 04 D2 构成的序列（十六进制表示法），而存储成文本格式时，则被存成了字符串 "1234"。尽管二进制格式的 I/O 高速且高效，但是不适合人类阅读。我们首先讨论文本格式的 I/O，然后在 2.2 节中讨论二进制格式的 I/O。

在存储文本字符串时，需要考虑字符编码（character encoding）方式。在 Java 内部使用的 UTF-16 编码方式中，字符串"José"编码为 00 4A 00 6F 00 73 00 E9（十六进制）。但是，许多程序都希望文本文件按照其他的编码方式编码。在 UTF-8 这种在互联网上最常用的编码方式中，这个字符串将写出为 4A 6F 73 C3 A9，其中并没有用于前 3 个字母的任何 0 字节，而字符 é 占用了两个字节。

OutputStreamWriter 类将使用选定的字符编码方式，把 Unicode 码元的输出流转换为字节流。而 InputStreamReader 类将包含字节（用某种字符编码方式表示的字符）的输入流转换为可以产生 Unicode 码元的读入器。

例如，下面的代码就展示了如何让输入读入器从控制台读入键盘敲击信息，并将其转换为 Unicode：

```
var in = new InputStreamReader(System.in);
```

这个输入流读入器会假定使用主机系统所使用的默认字符编码方式。在桌面操作系统中，它可能是像 Windows 1252 或 MacRoman 这样古老的字符编码方式。你应该总是在 InputStreamReader 的构造器中选择一种具体的编码方式。例如，

```
var in = new InputStreamReader(new FileInputStream("data.txt"), StandardCharsets.UTF_8);
```

请查看 2.1.8 节以了解字符编码方式的更多信息。

Reader 和 Writer 类都只有读入和写出单个字符的基础方法。在使用流时，可以使用处理字符串和数字的子类。

2.1.5 如何写出文本输出

对于文本输出，可以使用 PrintWriter。这个类拥有以文本格式打印字符串和数字的方法。为了打印文件，需要用文件名和字符编码方式构建一个 PrintStream 对象：

```
var out = new PrintWriter("employee.txt", StandardCharsets.UTF_8);
```

为了输出到打印写出器，需要使用与使用 System.out 时相同的 print、println 和 printf 方法。你可以用这些方法来打印数字（int、short、long、float、double）、字符、boolean 值、字符串和对象。

例如，考虑下面的代码：

```
String name = "Harry Hacker";
double salary = 75000;
out.print(name);
out.print(' ');
out.println(salary);
```

它将把字符

```
Harry Hacker 75000.0
```

输出到写出器 out，之后这些字符将会被转换成字节并最终写入 employee.txt 中。

println 方法在行中添加了对目标系统来说恰当的行结束符（Windows 系统是 "\r\n"，UNIX 系统是 "\n"），也就是通过调用 System.getProperty("line.separator") 而获得的字符串。

如果写出器设置为自动冲刷模式，那么只要 println 被调用，缓冲区中的所有字符都会被发送到它们的目的地（打印写出器总是带缓冲区的）。默认情况下，自动冲刷机制是禁用的，你可以通过使用 PrintWriter(Writer writer, boolean autoFlush) 来启用或禁用自动冲刷机制：

```
var out = new PrintWriter(
    new OutputStreamWriter(
        new FileOutputStream("employee.txt"), StandardCharsets.UTF_8),
    true); // autoflush
```

print 方法不抛出异常，你可以调用 checkError 方法来查看输出流是否出现了某些错误。

> **注释**：Java 的老手们可能会很想知道 PrintStream 类和 System.out 到底怎么了。在 Java 1.0 中，PrintStream 类只是通过将高字节丢弃的方式把所有 Unicode 字符截断成 ASCII 字符。（那时，Unicode 仍旧是 16 位编码方式。）很明显，这并非一种干净利落和可移植的方式，这个问题在 Java 1.1 中通过引入读入器和写出器得到了修正。为了与已有的代码兼容，System.in、System.out 和 System.err 仍旧是输入/输出流而不是读入器和写出器。但是现在 PrintStream 类在内部采用与 PrintWriter 相同的方式将 Unicode 字符转换成了默认的主机编码方式。当你在使用 print 和 println 方法时，PrintStream 类型的对象的行为看起来确实很像打印写出器，但是与打印写出器不同的是，它们允许用 write(int) 和 write(byte[]) 方法输出原生字节。

API java.io.PrintWriter 1.1

- PrintWriter(Writer out)
- PrintWriter(Writer writer)

  创建一个向给定的写出器写出的新的 PrintWriter。

- PrintWriter(String filename, String encoding)
- PrintWriter(File file, String encoding)

  创建一个使用给定的编码方式向给定的文件写出的新的 PrintWriter。

- void print(Object obj)

  通过打印从 toString 产生的字符串来打印一个对象。

- void print(String s)

  打印一个包含 Unicode 码元的字符串。

- void println(String s)

  打印一个字符串，后面紧跟一个行终止符。如果这个流处于自动冲刷模式，那么就会冲刷这个流。

- void print(char[] s)

  打印在给定的字符串中的所有 Unicode 码元。

- void print(char c)

  打印一个 Unicode 码元。

- void print(int i)
- void print(long l)
- void print(float f)
- void print(double d)
- void print(boolean b)

  以文本格式打印给定的值。

- void printf(String format, Object... args)

  按照格式字符串指定的方式打印给定的值。请查看卷 I 第 3 章以了解格式化字符串的相关规范。

- boolean checkError()

  如果产生格式化或输出错误，则返回 true。一旦这个流碰到了错误，它就受到了污染，并且所有对 checkError 的调用都将返回 true。

2.1.6 如何读入文本输入

最简单的处理任意文本的方式就是使用在卷 I 中我们广泛使用的 Scanner 类。我们可以从任何输入流中构建 Scanner 对象。

或者，我们也可以将短小的文本文件像下面这样读入到一个字符串中：

```
var content = (Files.read String path, charset);
```

但是，如果想要将这个文件一行行地读入，那么可以调用：

```
List<String> lines = Files.readAllLines(path, charset);
```

如果文件太大，那么可以将行惰性处理为一个 Stream<String> 对象：

```
try (Stream<String> lines = Files.lines(path, charset))
{
    ...
}
```

还可以使用扫描器来读入符号（token），即由分隔符分隔的字符串，默认的分隔符是空白字符。可以将分隔符修改为任意的正则表达式。例如，下面的代码

```
Scanner in = ...;
in.useDelimiter("\\PL+");
```

将接受任何非 Unicode 字母作为分隔符。之后，这个扫描器将只接受 Unicode 字母。

调用 next 方法可以产生下一个符号：

```
while (in.hasNext())
{
   String word = in.next();
   ...
}
```

或者,可以像下面这样获取一个包含所有符号的流:

```
Stream<String> words = in.tokens();
```

在早期的 Java 版本中,处理文本输入的唯一方式就是通过 BufferedReader 类。它的 readLine 方法会产生一行文本,或者在无法获得更多的输入时返回 null。典型的输入循环看起来像下面这样:

```
InputStream inputStream = . . .;
try (var in = new BufferedReader(new InputStreamReader(inputStream, charset)))
{
   String line;
   while ((line = in.readLine()) != null)
   {
      do something with line
   }
}
```

如今,BufferedReader 类又有了一个 lines 方法,可以产生一个 Stream<String> 对象。但是,与 Scanner 不同,BufferedReader 没有用于任何读入数字的方法。

2.1.7 以文本格式存储对象

在本节,我们将带你领略一个示例程序,它将一个 Employee 记录数组存储成了一个文本文件,其中每条记录都保存成单独的一行,而实例字段彼此之间使用分隔符分离开,这里我们使用竖线(|)作为分隔符(冒号(:)是另一种流行的选择,有趣的是,每个人都会使用不同的分隔符)。因此,我们这里是在假设不会发生在要存储的字符串中存在 | 的情况。

下面是一个记录集的样本:

```
Harry Hacker|35500|1989-10-01
Carl Cracker|75000|1987-12-15
Tony Tester|38000|1990-03-15
```

写出记录相当简单,因为是要写出到一个文本文件中,所以我们使用 PrintWriter 类。我们直接写出所有的字段,每个字段后面跟着一个 |,而最后一个字段的后面跟着一个换行符。这项工作是在下面这个我们添加到 Employee 类中的 writeEmployee 方法里完成的:

```
public static void writeEmployee(PrintWriter out, Employee e)
{
   out.println(e.getName() + "|" + e.getSalary() + "|" + e.getHireDay());
}
```

为了读入记录,我们每次读入一行,然后分离所有的字段。我们使用一个扫描器来读入每一行,然后用 String.split 方法将这一行断开成一组标记。

```java
public static Employee readEmployee(Scanner in)
{
   String line = in.nextLine();
   String[] tokens = line.split("\\|");
   String name = tokens[0];
   double salary = Double.parseDouble(tokens[1]);
   LocalDate hireDate = LocalDate.parse(tokens[2]);
   int year = hireDate.getYear();
   int month = hireDate.getMonthValue();
   int day = hireDate.getDayOfMonth();
   return new Employee(name, salary, year, month, day);
}
```

split 方法的参数是一个描述分隔符的正则表达式，我们在本章的末尾将详细讨论正则表达式。碰巧的是，竖线在正则表达式中具有特殊的含义，因此需要用 \ 字符来表示转义，而这个 \ 又需要用另一个 \ 来转义，这样就产生了 "\\|" 表达式。

完整的程序如程序清单 2-1 所示。静态方法

```java
void writeData(Employee[] e, PrintWriter out)
```

首先写出该数组的长度，然后写出每条记录。静态方法

```java
Employee[] readData(BufferedReader in)
```

首先读入该数组的长度，然后读入每条记录。这显得稍微有点棘手：

```java
int n = in.nextInt();
in.nextLine(); // consume newline
var employees = new Employee[n];
for (int i = 0; i < n; i++)
{
   employees[i] = new Employee();
   employees[i].readData(in);
}
```

对 nextInt 的调用读入的是数组长度，但不包括行尾的换行字符，我们必须处理掉这个换行符，这样，在调用 nextLine 方法后，readData 方法就可以获得下一行输入了。

程序清单 2-1　textFile/TextFileTest.java

```java
 1  package textFile;
 2
 3  import java.io.*;
 4  import java.nio.charset.*;
 5  import java.time.*;
 6  import java.util.*;
 7
 8  /**
 9   * @version 1.15 2018-03-17
10   * @author Cay Horstmann
11   */
12  public class TextFileTest
13  {
14     public static void main(String[] args) throws IOException
15     {
```

```java
16        var staff = new Employee[3];
17
18        staff[0] = new Employee("Carl Cracker", 75000, 1987, 12, 15);
19        staff[1] = new Employee("Harry Hacker", 50000, 1989, 10, 1);
20        staff[2] = new Employee("Tony Tester", 40000, 1990, 3, 15);
21
22        // save all employee records to the file employee.dat
23        try (var out = new PrintWriter("employee.dat", StandardCharsets.UTF_8))
24        {
25           writeData(staff, out);
26        }
27
28        // retrieve all records into a new array
29        try (var in = new Scanner(
30              new FileInputStream("employee.dat"), "UTF-8"))
31        {
32           Employee[] newStaff = readData(in);
33
34           // print the newly read employee records
35           for (Employee e : newStaff)
36              System.out.println(e);
37        }
38     }
39
40     /**
41      * Writes all employees in an array to a print writer
42      * @param employees an array of employees
43      * @param out a print writer
44      */
45     private static void writeData(Employee[] employees, PrintWriter out)
46           throws IOException
47     {
48        // write number of employees
49        out.println(employees.length);
50
51        for (Employee e : employees)
52           writeEmployee(out, e);
53     }
54
55     /**
56      * Reads an array of employees from a scanner
57      * @param in the scanner
58      * @return the array of employees
59      */
60     private static Employee[] readData(Scanner in)
61     {
62        // retrieve the array size
63        int n = in.nextInt();
64        in.nextLine(); // consume newline
65
66        var employees = new Employee[n];
67        for (int i = 0; i < n; i++)
68        {
69           employees[i] = readEmployee(in);
```

```
70        }
71        return employees;
72     }
73
74     /**
75      * Writes employee data to a print writer
76      * @param out the print writer
77      */
78     public static void writeEmployee(PrintWriter out, Employee e)
79     {
80        out.println(e.getName() + "|" + e.getSalary() + "|" + e.getHireDay());
81     }
82
83     /**
84      * Reads employee data from a buffered reader
85      * @param in the scanner
86      */
87     public static Employee readEmployee(Scanner in)
88     {
89        String line = in.nextLine();
90        String[] tokens = line.split("\\|");
91        String name = tokens[0];
92        double salary = Double.parseDouble(tokens[1]);
93        LocalDate hireDate = LocalDate.parse(tokens[2]);
94        int year = hireDate.getYear();
95        int month = hireDate.getMonthValue();
96        int day = hireDate.getDayOfMonth();
97        return new Employee(name, salary, year, month, day);
98     }
99  }
```

2.1.8 字符编码方式

输入和输出流都是用于字节序列的,但是在许多情况下,我们希望操作的是文本,即字符序列。于是,字符如何编码成字节就成了问题。

Java 针对字符使用的是 Unicode 标准。每个字符或 "编码点" 都具有一个 21 位的整数。有多种不同的字符编码方式,也就是说,将这些 21 位数字包装成字节的方法有多种。

最常见的编码方式是 UTF-8,它会将每个 Unicode 编码点编码为 1 到 4 个字节的序列(请参阅表 2-1)。UTF-8 的好处是传统的包含了英语中用到的所有字符的 ASCII 字符集中的每个字符都只会占用一个字节。

表 2-1 UTF-8 编码方式

字符范围	编码方式
0...7F	$0a_6a_5a_4a_3a_2a_1a_0$
80...7FF	$110a_{10}a_9a_8a_7a_6\ 10a_5a_4a_3a_2a_1a_0$
800...FFFF	$1110a_{15}a_{14}a_{13}a_{12}\ 10a_{11}a_{10}a_9a_8a_7a_6\ 10a_5a_4a_3a_2a_1a_0$
10000...10FFFF	$11110a_{20}a_{19}a_{18}\ 10a_{17}a_{16}a_{15}a_{14}a_{13}a_{12}\ 10a_{11}a_{10}a_9a_8a_7a_6\ 10a_5a_4a_3a_2a_1a_0$

另一种常见的编码方式是 UTF-16，它会将每个 Unicode 编码点编码为 1 个或 2 个 16 位值（请参阅表 2-2）。这是一种在 Java 字符串中使用的编码方式。实际上，有两种形式的 UTF-16，被称为"高位优先"和"低位优先"。考虑一下 16 位值 0x2122。在高位优先格式中，高位字节会先出现：0x21 后面跟着 0x22。但是在低位优先格式中，是另外一种排列方式：0x22 0x21。为了表示使用的是哪一种格式，文件可以以"字节顺序标记"开头，这个标记为 16 位数值 0xFEFF。读入器可以使用这个值来确定字节顺序，然后丢弃它。

表 2-2 UTF-16 编码方式

字符范围	编码方式
0...FFFF	$a_{15}a_{14}a_{13}a_{12}a_{11}a_{10}a_9a_8\ a_7a_6a_5a_4a_3a_2a_1a_0$
10000...10FFFF	$110110b_{19}b_{18}\ b_{17}b_{16}a_{15}a_{14}a_{13}a_{12}a_{11}a_{10}\ 110111a_9a_8\ a_7a_6a_5a_4a_3a_2a_1a_0$ 其中 $b_{19}b_{18}b_{17}b_{16} = a_{20}a_{19}a_{18}a_{17}a_{16} - 1$

> **警告**：有些程序，包括 Microsoft Notepad（微软记事本）在内，都在 UTF-8 编码的文件开头处添加了一个字节顺序标记。很明显，这并不需要，因为在 UTF-8 中，并不存在字节顺序的问题。但是 Unicode 标准允许这样做，甚至认为这是一种好的做法，因为这样做可以使编码机制不留疑惑。遗憾的是，Java 并没有这么做，有关这个问题的缺陷报告最终是以"will not fix"（不做修正）关闭的。对你来说，最好的做法是将输入中发现的所有先导的 \uFEFF 都剥离掉。

除了 UTF 编码方式，还有一些编码方式，它们各自都覆盖了适用于特定用户人群的字符范围。例如，ISO 8859-1 是一种单字节编码，它包含了西欧各种语言中用到的带有重音符号的字符，而 Shift-JIS 是一种用于日文字符的可变长编码。类似这些的大量编码方式至今仍被广泛使用。

不存在任何可靠的方式可以自动地探测出字节流中所使用的字符编码方式。某些 API 方法让我们使用"默认字符集"，即计算机的操作系统首选的字符编码方式。这种字符编码方式与我们的字节源中所使用的编码方式相同吗？字节源中的字节可能来自世界上的其他国家或地区，因此，你应该总是明确指定编码方式。例如，在编写网页时，应该检查 Content-Type 头信息。

> **注释**：平台使用的编码方式可以由静态方法 Charset.defaultCharset 返回。静态方法 Charset.availableCharsets 会返回所有可用的 Charset 实例，返回结果是一个从字符集的规范名称到 Charset 对象的映射表。

> **警告**：Oracle 的 Java 实现有一个用于覆盖平台默认值的系统属性 file.encoding。但是它并非官方支持的属性，并且 Java 库的 Oracle 实现的所有部分并非都以一致的方式处理该属性，因此，你不应该设置它。

StandardCharsets 类具有类型为 Charset 的静态变量，用于表示每种 Java 虚拟机都必须支持的字符编码方式：

```
StandardCharsets.UTF_8
StandardCharsets.UTF_16
StandardCharsets.UTF_16BE
StandardCharsets.UTF_16LE
StandardCharsets.ISO_8859_1
StandardCharsets.US_ASCII
```

为了获得另一种编码方式的 Charset，可以使用静态的 forName 方法：

```
Charset shiftJIS = Charset.forName("Shift-JIS");
```

在读入或写出文本时，应该使用 Charset 对象。例如，我们可以像下面这样将一个字节数组转换为字符串：

```
var str = new String(bytes, StandardCharsets.UTF_8);
```

> **提示**：在 Java 10 中，java.io 包中的所有方法都允许我们用一个 Charset 对象或字符串来指定字符编码方式。应该选择的是 StandardCharsets 常量，这样就不会在编译时捕获到任何拼写错误了。

> **警告**：在不指定任何编码方式时，有些方法（例如 String(byte[]) 构造器）会使用默认的平台编码方式，而其他方法（例如 Files.readAllLines）会使用 UTF-8。

2.2 读写二进制数据

文本格式对于测试和调试而言会显得很方便，因为它是人类可阅读的，但是它并不像以二进制格式传递数据那样高效。在下面的各小节中，你将会学习如何用二进制数据来完成输入和输出。

2.2.1 DataInput 和 DataOutput 接口

DataOutput 接口定义了下面用于以二进制格式写数组、字符、boolean 值和字符串的方法：

writeChars	writeFloat
writeByte	writeDouble
writeInt	writeChar
writeShort	writeBoolean
writeLong	writeUTF

例如，writeInt 总是将一个整数写出为 4 字节的二进制数量值，而不管它有多少位，writeDouble 总是将一个 double 值写出为 8 字节的二进制数量值。这样产生的结果并非人可阅读的，但是对于给定类型的每个值，使用的空间都是相同的，而且将其读回也比解析文本要更快。

> **注释**：根据你所使用的处理器类型，在内存存储整数和浮点数有两种不同的方法。例如，假设你使用的是 4 字节的 int，如果有一个十进制数 1234，也就是十六进制的

> 4D2（1234 = 4×256 + 13×16 + 2），那么它可以按照内存中 4 字节的第一个字节存储最高位字节的方式来存储为 00 00 04 D2，这就是所谓的高位在前顺序（MSB）；我们也可以从最低位字节开始，即 D2 04 00 00，这种方式自然就是所谓的低位在前顺序（LSB）。例如，SPARC 使用的是高位在前顺序，而 Pentium 使用的则是低位在前顺序。这就可能会带来问题，当存储 C 或者 C++ 文件时，数据会精确地按照处理器存储它们的方式来存储，这就使得即使是最简单的数据在从一个平台迁移到另一个平台上时也是一种挑战。在 Java 中，所有的值都按照高位在前的模式写出，不管使用何种处理器，这使得 Java 数据文件可以独立于平台。

writeUTF 方法使用修订版的 8 位 Unicode 转换格式写出字符串。这种方式与直接使用标准的 UTF-8 编码方式不同，其中，Unicode 码元序列首先用 UTF-16 表示，其结果之后使用 UTF-8 规则进行编码。修订后的编码方式对于编码大于 0xFFFF 的字符的处理有所不同，这是为了向后兼容在 Unicode 还没有超过 16 位时构建的虚拟机。

因为没有其他方法会使用 UTF-8 的这种修订，所以你应该只在写出用于 Java 虚拟机的字符串时才使用 writeUTF 方法，例如，当你需要编写一个生成字节码的程序时。对于其他场合，都应该使用 writeChars 方法。

为了读回数据，可以使用在 DataInput 接口中定义的下列方法：

```
readInt          readDouble
readShort        readChar
readLong         readBoolean
readFloat        readUTF
```

DataInputStream 类实现了 DataInput 接口，为了从文件中读入二进制数据，可以将 DataInput-Stream 与某个字节源相组合，例如 FileInputStream：

```
var in = new DataInputStream(new FileInputStream("employee.dat"));
```

与此类似，要想写出二进制数据，你可以使用实现了 DataOutput 接口的 DataOutputStream 类：

```
var out = new DataOutputStream(new FileOutputStream("employee.dat"));
```

API java.io.DataInput 1.0

- boolean readBoolean()
- byte readByte()
- char readChar()
- double readDouble()
- float readFloat()
- int readInt()
- long readLong()
- short readShort()

读入一个给定类型的值。

- void readFully(byte[] b)

  将字节读入到数组 b 中, 其间阻塞直至所有字节都读入。
- void readFully(byte[] b, int off, int len)

  将由 len 指定数量的字节放置到数组 b 从 off 开始的位置, 其间阻塞直至所有字节都读入。
- String readUTF()

  读入由"修订过的 UTF-8"格式的字符构成的字符串。
- int skipBytes(int n)

  跳过 n 个字节, 其间阻塞直至所有字节都被跳过。

API *java.io.DataOutput* 1.0

- void writeBoolean(boolean b)
- void writeByte(int b)
- void writeChar(int c)
- void writeDouble(double d)
- void writeFloat(float f)
- void writeInt(int i)
- void writeLong(long l)
- void writeShort(int s)

  写出一个给定类型的值。
- void writeChars(String s)

  写出字符串中的所有字符。
- void writeUTF(String s)

  写出由"修订过的 UTF-8"格式的字符构成的字符串。

2.2.2 随机访问文件

RandomAccessFile 类可以在文件中的任何位置查找或写入数据。磁盘文件都是随机访问的, 但是与网络套接字通信的输入 / 输出流却不是。你可以打开一个随机访问文件, 只用于读入或者同时用于读写, 你可以通过使用字符串" r"(用于读入访问) 或" rw"(用于读入 / 写出访问) 作为构造器的第二个参数来指定这个选项。

```
var in = new RandomAccessFile("employee.dat", "r");
var inOut = new RandomAccessFile("employee.dat", "rw");
```

当你将已有文件作为 RandomAccessFile 打开时, 这个文件并不会被删除。

随机访问文件有一个表示下一个将被读入或写出的字节所处位置的文件指针, seek 方法可以用来将这个文件指针设置到文件中的任意字节位置, seek 的参数是一个 long 类型的整数, 它的值位于 0 到文件按照字节来度量的长度之间。

getFilePointer 方法将返回文件指针的当前位置。

RandomAccessFile 类同时实现了 DataInput 和 DataOutput 接口。为了读写随机访问文件，可以使用在前面小节中讨论过的诸如 readInt/writeInt 和 readChar/writeChar 之类的方法。

我们现在要剖析一个将雇员记录存储到随机访问文件中的示例程序，其中每条记录都拥有相同的大小，这样我们可以很容易地读入任何记录。假设你希望将文件指针置于第三条记录处，那么你只需将文件指针置于恰当的字节位置，然后就可以开始读入了。

```
long n = 3;
in.seek((n - 1) * RECORD_SIZE);
var e = new Employee();
e.readData(in);
```

如果你希望修改记录，然后将其存回到相同的位置，那么请切记要将文件指针置回到这条记录的开始处：

```
in.seek((n - 1) * RECORD_SIZE);
e.writeData(out);
```

要确定文件中的字节总数，可以使用 length 方法，而记录的总数则等于用字节总数除以每条记录的大小。

```
long nbytes = in.length(); // length in bytes
int nrecords = (int) (nbytes / RECORD_SIZE);
```

整数和浮点值在二进制格式中都具有固定的尺寸，但是在处理字符串时就有些麻烦了，因此我们提供了两个助手方法来读写具有固定尺寸的字符串。

writeFixedString 写出从字符串开头开始的指定数量的码元（如果码元过少，该方法将用 0 值来补齐字符串）。

```
public static void writeFixedString(String s, int size, DataOutput out)
    throws IOException
{
   for (int i = 0; i < size; i++)
   {
      char ch = 0;
      if (i < s.length()) ch = s.charAt(i);
      out.writeChar(ch);
   }
}
```

readFixedString 方法从输入流中读入字符，直至读入 size 个码元，或者直至遇到具有 0 值的字符值，然后跳过输入字段中剩余的 0 值。为了提高效率，这个方法使用了 StringBuilder 类来读入字符串。

```
public static String readFixedString(int size, DataInput in)
    throws IOException
{
   var b = new StringBuilder(size);
   int i = 0;
   var done = false;
   while (!done && i < size)
   {
      char ch = in.readChar();
```

```
            i++;
            if (ch == 0) done = true;
            else b.append(ch);
        }
        in.skipBytes(2 * (size - i));
        return b.toString();
    }
```

我们将 writeFixedString 和 readFixedString 方法放到了 DataIO 助手类的内部。

为了写出一条固定尺寸的记录，我们直接以二进制方式写出所有的字段：

```
DataIO.writeFixedString(e.getName(), Employee.NAME_SIZE, out);
out.writeDouble(e.getSalary());
LocalDate hireDay = e.getHireDay();
out.writeInt(hireDay.getYear());
out.writeInt(hireDay.getMonthValue());
out.writeInt(hireDay.getDayOfMonth());
```

读回数据也很简单：

```
String name = DataIO.readFixedString(Employee.NAME_SIZE, in);
double salary = in.readDouble();
int y = in.readInt();
int m = in.readInt();
int d = in.readInt();
```

让我们来计算每条记录的大小：我们将使用 40 个字符来表示姓名字符串，因此，每条记录包含 100 个字节：

- 40 字符 = 80 字节，用于姓名。
- 1 double = 8 字节，用于薪水。
- 3 int = 12 字节，用于日期。

程序清单 2-2 中所示的程序将三条记录写到了一个数据文件中，然后以逆序将它们从文件中读回。为了高效地执行，这里需要使用随机访问，因为我们需要首先读入第三条记录。

程序清单 2-2 randomAccess/RandomAccessTest.java

```java
 1  package randomAccess;
 2
 3  import java.io.*;
 4  import java.time.*;
 5
 6  /**
 7   * @version 1.14 2018-05-01
 8   * @author Cay Horstmann
 9   */
10  public class RandomAccessTest
11  {
12      public static void main(String[] args) throws IOException
13      {
14          var staff = new Employee[3];
15
16          staff[0] = new Employee("Carl Cracker", 75000, 1987, 12, 15);
17          staff[1] = new Employee("Harry Hacker", 50000, 1989, 10, 1);
```

```java
18         staff[2] = new Employee("Tony Tester", 40000, 1990, 3, 15);
19
20         try (var out = new DataOutputStream(new FileOutputStream("employee.dat")))
21         {
22            // save all employee records to the file employee.dat
23            for (Employee e : staff)
24               writeData(out, e);
25         }
26
27         try (var in = new RandomAccessFile("employee.dat", "r"))
28         {
29            // retrieve all records into a new array
30
31            // compute the array size
32            int n = (int)(in.length() / Employee.RECORD_SIZE);
33            var newStaff = new Employee[n];
34
35            // read employees in reverse order
36            for (int i = n - 1; i >= 0; i--)
37            {
38               newStaff[i] = new Employee();
39               in.seek(i * Employee.RECORD_SIZE);
40               newStaff[i] = readData(in);
41            }
42
43            // print the newly read employee records
44            for (Employee e : newStaff)
45               System.out.println(e);
46         }
47      }
48
49      /**
50       * Writes employee data to a data output
51       * @param out the data output
52       * @param e the employee
53       */
54      public static void writeData(DataOutput out, Employee e) throws IOException
55      {
56         DataIO.writeFixedString(e.getName(), Employee.NAME_SIZE, out);
57         out.writeDouble(e.getSalary());
58
59         LocalDate hireDay = e.getHireDay();
60         out.writeInt(hireDay.getYear());
61         out.writeInt(hireDay.getMonthValue());
62         out.writeInt(hireDay.getDayOfMonth());
63      }
64
65      /**
66       * Reads employee data from a data input
67       * @param in the data input
68       * @return the employee
69       */
70      public static Employee readData(DataInput in) throws IOException
71      {
```

```
72        String name = DataIO.readFixedString(Employee.NAME_SIZE, in);
73        double salary = in.readDouble();
74        int y = in.readInt();
75        int m = in.readInt();
76        int d = in.readInt();
77        return new Employee(name, salary, y, m - 1, d);
78     }
79  }
```

API java.io.RandomAccessFile 1.0

- RandomAccessFile(String file, String mode)
- RandomAccessFile(File file, String mode)

  打开给定的用于随机访问的文件。mode 字符串 "r" 表示只读模式；"rw" 表示读/写模式；"rws" 表示每次更新时，都对数据和元数据的写磁盘操作进行同步的读/写模式；"rwd" 表示每次更新时，只对数据的写磁盘操作进行同步的读/写模式。

- long getFilePointer()

  返回文件指针的当前位置。

- void seek(long pos)

  将文件指针设置到距文件开头 pos 个字节处。

- long length()

  返回文件按照字节来度量的长度。

2.2.3 ZIP 文档

ZIP 文档（通常）以压缩格式存储了一个或多个文件，每个 ZIP 文档都有一个头，包含诸如每个文件名字和所使用的压缩方法等信息。在 Java 中，可以使用 ZipInputStream 来读入 ZIP 文档。你可能需要浏览文档中每个单独的项，getNextEntry 方法就可以返回一个描述这些项的 ZipEntry 类型的对象。该方法会从流中读入数据直至末尾，实际上这里的末尾是指正在读入的项的末尾，然后调用 closeEntry 来读入下一项。在读入最后一项之前，不要关闭 zin。下面是典型的通读 ZIP 文件的代码序列：

```
var zin = new ZipInputStream(new FileInputStream(zipname));
ZipEntry entry;
while ((entry = zin.getNextEntry()) != null)
{
   read the contents of zin
   zin.closeEntry();
}
zin.close();
```

要写出到 ZIP 文件，可以使用 ZipOutputStream，而对于你希望放入到 ZIP 文件中的每一项，都应该创建一个 ZipEntry 对象，并将文件名传递给 ZipEntry 的构造器，它将设置其他诸如文件日期和解压缩方法等参数。如果需要，你可以覆盖这些设置。然后，你需要调用 ZipOutputStream 的 putNextEntry 方法来写出新文件，并将文件数据发送到 ZIP 输出流中。当完成时，需要调用 closeEntry。然后，你需要对所有希望存储的文件都重复这个过程。下面是代码框架：

```
var fout = new FileOutputStream("test.zip");
var zout = new ZipOutputStream(fout);
for all files
{
   var ze = new ZipEntry(filename);
   zout.putNextEntry(ze);
   send data to zout
   zout.closeEntry();
}
zout.close();
```

> **注释**：JAR 文件（在卷 I 第 4 章中讨论过）只是带有一个特殊项的 ZIP 文件，这个项称作清单。你可以使用 JarInputStream 和 JarOutputStream 类来读写清单项。

ZIP 输入流是一个能够展示流的抽象化的强大之处的实例。当你读入以压缩格式存储的数据时，不必担心边请求边解压数据的问题，而且 ZIP 格式的字节源并非必须是文件，也可以是来自网络连接的 ZIP 数据。

> **注释**：2.4.8 节将展示如何使用 Java 7 的 FileSystem 类而无须特殊 API 来访问 ZIP 文档。

API `java.util.zip.ZipInputStream` 1.1

- ZipInputStream(InputStream in)

  创建一个 ZipInputStream，使得我们可以从给定的 InputStream 向其中填充数据。

- ZipEntry getNextEntry()

  为下一项返回 ZipEntry 对象，或者在没有更多的项时返回 null。

- void closeEntry()

  关闭这个 ZIP 文件中当前打开的项。之后可以通过使用 getNextEntry() 读入下一项。

API `java.util.zip.ZipOutputStream` 1.1

- ZipOutputStream(OutputStream out)

  创建一个将压缩数据写出到指定的 OutputStream 的 ZipOutputStream。

- void putNextEntry(ZipEntry ze)

  将给定的 ZipEntry 中的信息写出到输出流中，并定位用于写出数据的流，然后这些数据可以通过 write() 写出到这个输出流中。

- void closeEntry()

  关闭这个 ZIP 文件中当前打开的项。使用 putNextEntry 方法可以开始下一项。

- void setLevel(int level)

  将后续的各个 DEFLATED 项的默认压缩级别设置为从 Deflater.NO_COMPRESSION 到 Deflater.BEST_COMPRESSION 中的某个值，默认值是 Deflater.DEFAULT_COMPRESSION。如果级别无效，则

抛出 IllegalArgumentException。
- void setMethod(int method)
设置用于这个 ZipOutputStream 的默认压缩方法，这个压缩方法会作用于所有没有指定压缩方法的项。method 可以是 DEFLATED 或 STORED。

API java.util.zip.ZipEntry 1.1

- ZipEntry(String name)
用给定的名字构建一个 ZIP 项。
- long getCrc()
返回用于这个 ZipEntry 的 CRC32 校验和的值。
- String getName()
返回这一项的名字。
- long getSize()
返回这一项未压缩的尺寸，或者在未压缩的尺寸不可知的情况下返回 –1。
- boolean isDirectory()
当这一项是目录时返回 true。
- void setMethod(int method)
设置用于这一项的压缩方法，必须是 DEFLATED 或 STORED。
- void setSize(long size)
设置这一项的尺寸，只有在压缩方法是 STORED 时才是必需的。
- void setCrc(long crc)
给这一项设置 CRC32 校验和，这个校验和是使用 CRC32 类计算的。只有在压缩方法是 STORED 时才是必需的。

API java.util.zip.ZipFile 1.1

- ZipFile(String name)
- ZipFile(File file)
创建一个 ZipFile，用于从给定的字符串或 File 对象中读入数据。
- Enumeration entries()
返回一个 Enumeration 对象，它枚举了描述这个 ZipFile 中各个项的 ZipEntry 对象。
- ZipEntry getEntry(String name)
返回给定名字所对应的项，或者在没有对应项的时候返回 null。
- InputStream getInputStream(ZipEntry ze)
返回用于给定项的 InputStream。
- String getName()
返回这个 ZIP 文件的路径。

2.3 对象输入/输出流与序列化

当你需要存储相同类型的数据时，使用固定长度的记录格式是一个不错的选择。但是，在面向对象程序中创建的对象很少全部都具有相同的类型。例如，你可能有一个称为 staff 的数组，它名义上是一个 Employee 记录数组，但是实际上却包含诸如 Manager 这样的子类实例。

我们当然可以自己设计出一种数据格式来存储这种多态集合，但是幸运的是，我们并不需要这么做。Java 语言支持一种称为对象序列化（object serialization）的非常通用的机制，它可以将任何对象写出到输出流中，并在之后将其读回。（你将在本章稍后看到"序列化"这个术语的出处。）

2.3.1 保存和加载序列化对象

为了保存对象数据，首先需要打开一个 ObjectOutputStream 对象：

```
var out = new ObjectOutputStream(new FileOutputStream("employee.dat"));
```

现在，为了保存对象，可以直接使用 ObjectOutputStream 的 writeObject 方法，如下所示：

```
var harry = new Employee("Harry Hacker", 50000, 1989, 10, 1);
var boss = new Manager("Carl Cracker", 80000, 1987, 12, 15);
out.writeObject(harry);
out.writeObject(boss);
```

为了将这些对象读回，首先需要获得一个 ObjectInputStream 对象：

```
var in = new ObjectInputStream(new FileInputStream("employee.dat"));
```

然后，用 readObject 方法以这些对象被写出时的顺序获得它们：

```
var e1 = (Employee) in.readObject();
var e2 = (Employee) in.readObject();
```

但是，对希望在对象输出流中存储或从对象输入流中恢复的所有类都应进行一下修改，这些类必须实现 Serializable 接口：

```
class Employee implements Serializable { ... }
```

Serializable 接口没有任何方法，因此你不需要对这些类做任何改动。在这一点上，它与在卷 I 第 6 章中讨论过的 Cloneable 接口很相似。但是，为了使类可克隆，你仍旧需要覆盖 Object 类中的 clone 方法，而为了使类可序列化，你不需要做任何事。

> 📖 **注释**：你只有在写出对象时才能用 writeObject/readObject 方法，对于基本类型值，你需要使用诸如 writeInt/readInt 或 writeDouble/readDouble 这样的方法。（对象流类都实现了 DataInput/DataOutput 接口。）

在幕后，是 ObjectOutputStream 在浏览对象的所有域，并存储它们的内容。例如，当写出一个 Employee 对象时，其名字、日期和薪水域都会被写出到输出流中。

但是，有一种重要的情况需要考虑：当一个对象被多个对象共享，作为它们各自状态的一部分时，会发生什么呢？

为了说明这个问题，我们对 Manager 类稍微做些修改，假设每个经理都有一个秘书：

```
class Manager extends Employee
{
    private Employee secretary;
    ...
}
```

现在每个 Manager 对象都包含一个表示秘书的 Employee 对象的引用，当然，两个经理可以共用一个秘书，正如图 2-5 和下面的代码所示的那样：

```
var harry = new Employee("Harry Hacker", ...);
var carl = new Manager("Carl Cracker", ...);
carl.setSecretary(harry);
var tony = new Manager("Tony Tester", ...);
tony.setSecretary(harry);
```

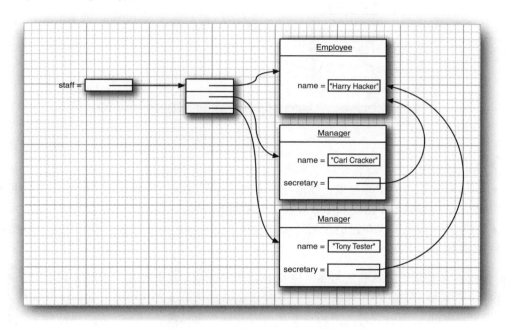

图 2-5　两个经理可以共用一个共有的雇员

保存这样的对象网络是一种挑战，在这里我们当然不能去保存和恢复秘书对象的内存地址，因为当对象被重新加载时，它可能占据的是与原来完全不同的内存地址。

与此不同的是，每个对象都是用一个序列号（serial number）保存的，这就是这种机制之所以称为对象序列化的原因。下面是其算法：

- 对你遇到的每一个对象引用都关联一个序列号（如图 2-6 所示）。
- 对于每个对象，当第一次遇到时，保存其对象数据到输出流中。
- 如果某个对象之前已经被保存过，那么只写出"与之前保存过的序列号为 x 的对象相同"。

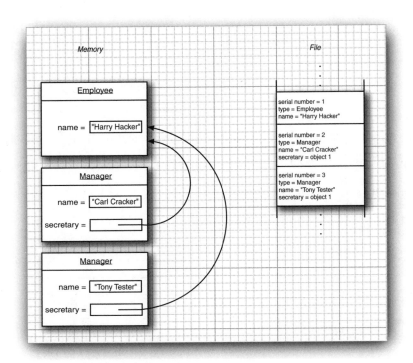

图 2-6 一个对象序列化的实例

在读回对象时,整个过程是反过来的。
- 对于对象输入流中的对象,在第一次遇到其序列号时,构建它,并使用流中数据来初始化它,然后记录这个顺序号和新对象之间的关联。
- 当遇到"与之前保存过的序列号为 x 的对象相同"这一标记时,获取与这个序列号相关联的对象引用。

> 注释:在本章中,我们使用序列化将对象集合保存到磁盘文件中,并按照它们被存储的样子获取它们。序列化的另一种非常重要的应用是通过网络将对象集合传送到另一台计算机上。正如在文件中保存原生的内存地址毫无意义一样,这些地址对于在不同的处理器之间的通信也是毫无意义的。因为序列化用序列号代替了内存地址,所以它允许将对象集合从一台机器传送到另一台机器。

程序清单 2-3 是保存和重新加载 Employee 和 Manager 对象网络的代码(有些对象共享相同的表示秘书的雇员)。注意,秘书对象在重新加载之后是唯一的,当 newStaff[1] 被恢复时,它会反映到经理们的 secretary 域中。

程序清单 2-3 objectStream/ObjectStreamTest.java

```
1 package objectStream;
2
```

```java
3   import java.io.*;
4
5   /**
6    * @version 1.11 2018-05-01
7    * @author Cay Horstmann
8    */
9   class ObjectStreamTest
10  {
11     public static void main(String[] args) throws IOException, ClassNotFoundException
12     {
13        var harry = new Employee("Harry Hacker", 50000, 1989, 10, 1);
14        var carl = new Manager("Carl Cracker", 80000, 1987, 12, 15);
15        carl.setSecretary(harry);
16        var tony = new Manager("Tony Tester", 40000, 1990, 3, 15);
17        tony.setSecretary(harry);
18
19        var staff = new Employee[3];
20
21        staff[0] = carl;
22        staff[1] = harry;
23        staff[2] = tony;
24
25        // save all employee records to the file employee.dat
26        try (var out = new ObjectOutputStream(new FileOutputStream("employee.dat")))
27        {
28           out.writeObject(staff);
29        }
30
31        try (var in = new ObjectInputStream(new FileInputStream("employee.dat")))
32        {
33           // retrieve all records into a new array
34
35           var newStaff = (Employee[]) in.readObject();
36
37           // raise secretary's salary
38           newStaff[1].raiseSalary(10);
39
40           // print the newly read employee records
41           for (Employee e : newStaff)
42              System.out.println(e);
43        }
44     }
45  }
```

API java.io.ObjectOutputStream 1.1

- ObjectOutputStream(OutputStream out)

  创建一个 ObjectOutputStream 使得你可以将对象写出到指定的 OutputStream。

- void writeObject(Object obj)

  写出指定的对象到 ObjectOutputStream，这个方法将存储指定对象的类、类的签名以及这个类及其超类中所有非静态和非瞬时的域的值。

API java.io.ObjectInputStream 1.1

- ObjectInputStream(InputStream in)

  创建一个 ObjectInputStream 用于从指定的 InputStream 中读回对象信息。

- Object readObject()

  从 ObjectInputStream 中读入一个对象。特别是，这个方法会读回对象的类、类的签名以及这个类及其超类中所有非静态和非瞬时的域的值。它执行的反序列化允许恢复多个对象引用。

2.3.2 理解对象序列化的文件格式

对象序列化是以特殊的文件格式存储对象数据的，当然，我们不必了解文件中表示对象的确切字节序列，就可以使用 writeObject/readObject 方法。但是，我们发现研究这种数据格式对于洞察对象流化的处理过程非常有益。因为其细节显得有些专业，所以如果你对其实现不感兴趣，则可以跳过这一节。

每个文件都是以下面这两个字节的"魔幻数字"开始的

```
AC ED
```

后面紧跟着对象序列化格式的版本号，目前是

```
00 05
```

（我们在本节中统一使用十六进制数字来表示字节。）然后是它包含的对象序列，其顺序即它们存储的顺序。

字符串对象被存为

```
74  两字节表示的字符串长度  所有字符
```

例如，字符串"Harry"被存为

```
74 00 05 Harry
```

字符串中的 Unicode 字符被存储为修订过的 UTF-8 格式。

当存储一个对象时，这个对象所属的类也必须存储。这个类的描述包含

- 类名。
- 序列化的版本唯一的 ID，它是数据域类型和方法签名的指纹。
- 描述序列化方法的标志集。
- 对数据域的描述。

指纹是通过对类、超类、接口、域类型和方法签名按照规范方式排序，然后将安全散列算法（SHA）应用于这些数据而获得的。

SHA 是一种可以为较大的信息块提供指纹的快速算法，不论最初的数据块尺寸有多大，这种指纹总是 20 个字节的数据包。它是通过在数据上执行一个灵巧的位操作序列而创建的，这个序列在本质上可以百分之百地保证无论这些数据以何种方式发生变化，其指纹也都会跟着变化。（关于 SHA 的更多细节，可以查看一些参考资料，例如 William Stallings 所著的

Cryptography and Network Security: Principles and Practice（第 7 版，Prentice Hall，2016）。）但是，序列化机制只使用了 SHA 码的前 8 个字节作为类的指纹。即便这样，当类的数据域或方法发生变化时，其指纹跟着变化的可能性还是非常大。

在读入一个对象时，会拿其指纹与它所属的类的当前指纹进行比对，如果它们不匹配，那么就说明这个类的定义在该对象被写出之后发生过变化，因此会产生一个异常。在实际情况下，类当然是会演化的，因此对于程序来说，读入较旧版本的对象可能是必需的。我们将在 2.4.5 节中讨论这个问题。

下面表示了类标识符是如何存储的：

- 72
- 2 字节的类名长度
- 类名
- 8 字节长的指纹
- 1 字节长的标志
- 2 字节长的数据域描述符的数量
- 数据域描述符
- 78（结束标记）
- 超类类型（如果没有就是 70）

标志字节是由在 java.io.ObjectStreamConstants 中定义的 3 位掩码构成的：

```
static final byte SC_WRITE_METHOD = 1;
    // class has a writeObject method that writes additional data
static final byte SC_SERIALIZABLE = 2;
    // class implements the Serializable interface
static final byte SC_EXTERNALIZABLE = 4;
    // class implements the Externalizable interface
```

我们会在本章稍后讨论 Externalizable 接口。可外部化的类提供了定制的接管其实例域输出的读写方法。我们要写出的这些类实现了 Serializable 接口，并且其标志值为 02，而可序列化的 java.util.Date 类定义了它自己的 readObject/writeObject 方法，并且其标志值为 03。

每个数据域描述符的格式如下：

- 1 字节长的类型编码
- 2 字节长的域名长度
- 域名
- 类名（如果域是对象）

其中类型编码是下列取值之一：

```
B       byte
C       char
D       double
F       float
I       int
```

J	long
L	对象
S	short
Z	boolean
[	数组

当类型编码为 L 时，域名后面紧跟域的类型。类名和域名字符串不是以字符串编码 74 开头的，但域类型是。域类型使用的是与域名稍有不同的编码机制，即本地方法使用的格式。

例如，Employee 类的薪水域被编码为：

D 00 06 salary

下面是 Employee 类完整的类描述符：

72 00 08 Employee	
E6 D2 86 7D AE AC 18 1B 02	指纹和标志
00 03	实例域的数量
D 00 06 salary	实例域的类型和名字
L 00 07 hireDay	实例域的类型和名字
74 00 10 Ljava/util/Date;	实例域的类名：Date
L 00 04 name	实例域的类型和名字
74 00 12 Ljava/lang/String;	实例域的类名：String
78	结束标记
70	无超类

这些描述符相当长，如果在文件中再次需要相同的类描述符，可以使用一种缩写版：

71　4 字节长的序列号

这个序列号将引用前面已经描述过的类描述符，我们稍后将讨论编号模式。

对象将被存储为：

73　类描述符　对象数据

例如，下面展示的就是 Employee 对象如何存储：

40 E8 6A 00 00 00 00 00	salary 域的值：double
73	hireDate 域的值：新对象
71 00 7E 00 08	已有的类 java.util.Date
77 08 00 00 00 91 1B 4E B1 80 78	外部存储，稍后讨论细节
74 00 0C Harry Hacker	name 域的值：String

正如你所看见的，数据文件包含了足够的信息来恢复这个 Employee 对象。

数组总是被存储成下面的格式：

75　类描述符　4 字节长的数组项的数量　数组项

在类描述符中的数组类名的格式与本地方法中使用的格式相同（它与在其他的类描述符中的类名稍微有些差异）。在这种格式中，类名以 L 开头，以分号结束。

例如，3 个 Employee 对象构成的数组写出时就像下面一样：

```
75                                          数组
   72 00 0B [LEmployee;                     新类，字符串长度，类名Employee[]

       FC BF 36 11 C5 91 11 C7 02           指纹和标志
       00 00                                实例域的数量
       78                                   结束标记
       70                                   无超类
   00 00 00 03                              数组项的数量
```

注意，Employee 对象数组的指纹与 Employee 类自身的指纹并不相同。

所有对象（包含数组和字符串）和所有的类描述符在存储到输出文件时都被赋予了一个序列号，这个数字以 00 7E 00 00 开头。

我们已经看到过，任何给定类的完整类描述符只保存一次，后续的描述符将引用它。例如，在前面的示例中，对 Date 类的重复引用就被编码为：

`71 00 7E 00 08`

相同的机制还被用于对象。如果要写出一个对之前存储过的对象的引用，那么这个引用也会以完全相同的方式存储，即 71 后面跟随序列号，从上下文中可以很清楚地了解这个特殊的序列引用表示的是类描述符还是对象。

最后，空引用被存储为：

`70`

下面是前面小节中 ObjectRefTest 程序的带注释的输出。如果你喜欢，可以运行这个程序，然后查看其数据文件 employee.dat 的十六进制码，并将其与注释列表比较。在输出中接近结束部分的几行重要编码展示了对之前存储过的对象的引用。

```
AC ED 00 05                                 文件头
75                                          数组staff（序列#1）
   72 00 0B [LEmployee;                     新类、字符串长度、类名Employee[]（序列#0）

       FC BF 36 11 C5 91 11 C7 02           指纹和标志
       00 00                                实例域的数量
       78                                   结束标记
       70                                   无超类
   00 00 00 03                              数组项的数量
73                                          staff[0]：新对象(序列#7)
   72 00 07 Manager                         新类、字符串长度、类名（序列#2）

       36 06 AE 13 63 8F 59 B7 02           指纹和标志
       00 01                                数据的数量
```

L 00 09 secretary	实例域的类型和名字
74 00 0A LEmployee;	实例域的类名：String（序列#3）
78	结束标记
72 00 08 Employee	超类一：新类、字符串长度、类名（序列#4）
E6 D2 86 7D AE AC 18 1B 02	指纹和标志
00 03	实例域的数量
D 00 06 salary	实例域的类型和名字
L 00 07 hireDay	实例域的类型和名字
74 00 10 Ljava/util/Date;	实例域的类名：String（序列#5）
L 00 04 name	实例域的类型和名字
74 00 12 Ljava/lang/String;	实例域的类名：String（序列#6）
78	结束标记
70	无超类
40 F3 88 00 00 00 00 00	salary 域的值：double
73	hireDate 域的值：新对象（序列#9）
72 00 0E java.util.Date	新类、字符串长度、类名（序列#8）
68 6A 81 01 4B 59 74 19 03	指纹和标志
00 00	无实例变量
78	结束标记
70	无超类
77 08	外部存储、字节的数量
00 00 00 83 E9 39 E0 00	日期
78	结束标记
74 00 0C Carl Cracker	name 域的值：String（序列#10）
73	secretary 域的值：新对象（序列#11）
71 00 7E 00 04	已有的类（使用序列#4）
40 E8 6A 00 00 00 00 00	salary 域的值：double
73	hireDate 域的值：新对象（序列#12）
71 00 7E 00 08	已有的类（使用序列#8）
77 08	外部存储、字节的数量
00 00 00 91 1B 4E B1 80	日期
78	结束标记
74 00 0C Harry Hacker	name 域的值：String（序列#13）
71 00 7E 00 0B	staff[1]：已有的对象（使用序列#11）
73	staff[2]：新对象（序列#14）
71 00 7E 00 02	已有的类（使用序列#2）
40 E3 88 00 00 00 00 00	salary 域的值：double

73	hireDay 域的值：新对象（序列#15）
71 00 7E 00 08	已有的类（使用序列#8）
77 08	外部存储、字节的数量
00 00 00 94 6D 3E EC 00 00	日期
78	结束标记
74 00 0B Tony Tester	name 域的值：String（序列#16）
71 00 7E 00 0B	secretary 域的值：已有的对象（使用序列#11）

当然，研究这些编码大概与阅读常用的电话号码簿一样枯燥。了解确切的文件格式确实不那么重要（除非你试图通过修改数据来达到不可告人的目的），但是对象流对其所包含的所有对象都有详细描述，并且这些充足的细节可以用来重构对象和对象数组，因此了解它还是大有益处的。

你应该记住：
- 对象流输出中包含所有对象的类型和数据域。
- 每个对象都被赋予一个序列号。
- 相同对象的重复出现将被存储为对这个对象的序列号的引用。

2.3.3 修改默认的序列化机制

某些数据域是不可以序列化的，例如，只对本地方法有意义的存储文件句柄或窗口句柄的整数值，这种信息在稍后重新加载对象或将其传送到其他机器上时都是没有用处的。事实上，这种域的值如果不恰当，还会引起本地方法崩溃。Java 拥有一种很简单的机制来防止这种域被序列化，那就是将它们标记成 transient 的。如果这些域属于不可序列化的类，你也需要将它们标记成 transient 的。瞬时的域在对象被序列化时总是被跳过的。

序列化机制为单个的类提供了一种方式，去向默认的读写行为添加验证或任何其他想要的行为。可序列化的类可以定义具有下列签名的方法：

```
private void readObject(ObjectInputStream in)
    throws IOException, ClassNotFoundException;
private void writeObject(ObjectOutputStream out)
    throws IOException;
```

之后，数据域就再也不会被自动序列化，取而代之的是调用这些方法。

下面是一个典型的示例。在 java.awt.geom 包中有大量的类都是不可序列化的，例如 Point-2D.Double。现在假设你想要序列化一个 LabeledPoint 类，它存储了一个 String 和一个 Point2D.Double。首先，你需要将 Point2D.Double 标记成 transient，以避免抛出 NotSerializableException。

```
public class LabeledPoint implements Serializable
{
   private String label;
   private transient Point2D.Double point;
   ...
}
```

在 writeObject 方法中，我们首先通过调用 defaultWriteObject 方法写出对象描述符和 String

域 label，这是 ObjectOutputStream 类中的一个特殊的方法，它只能在可序列化类的 writeObject 方法中被调用。然后，我们使用标准的 DataOutput 调用写出点的坐标。

```java
private void writeObject(ObjectOutputStream out)
    throws IOException
{
   out.defaultWriteObject();
   out.writeDouble(point.getX());
   out.writeDouble(point.getY());
}
```

在 readObject 方法中，我们反过来执行上述过程：

```java
private void readObject(ObjectInputStream in)
    throws IOException
{
   in.defaultReadObject();
   double x = in.readDouble();
   double y = in.readDouble();
   point = new Point2D.Double(x, y);
}
```

另一个例子是 java.util.Date 类，它提供了自己的 readObject 和 writeObject 方法，这些方法将日期写出为从纪元（UTC 时间 1970 年 1 月 1 日 0 点）开始的毫秒数。Date 类有一个复杂的内部表示，为了优化查询，它存储了一个 Calendar 对象和一个毫秒计数值。Calendar 的状态是冗余的，因此并不需要保存。

readObject 和 writeObject 方法只需要保存和加载它们的数据域，而不需要关心超类数据和任何其他类的信息。

除了让序列化机制来保存和恢复对象数据，类还可以定义它自己的机制。为了做到这一点，这个类必须实现 Externalizable 接口，这需要它定义两个方法：

```java
public void readExternal(ObjectInputStream in)
    throws IOException, ClassNotFoundException;
public void writeExternal(ObjectOutputStream out)
    throws IOException;
```

与前面一节描述的 readObject 和 writeObject 不同，这些方法对包括超类数据在内的整个对象的存储和恢复负全责。在写出对象时，序列化机制在输出流中仅仅只是记录该对象所属的类。在读入可外部化的类时，对象输入流将用无参构造器创建一个对象，然后调用 readExternal 方法。下面展示了如何为 Employee 类实现这些方法：

```java
public void readExternal(ObjectInput s)
    throws IOException
{
   name = s.readUTF();
   salary = s.readDouble();
   hireDay = LocalDate.ofEpochDay(s.readLong());
}

public void writeExternal(ObjectOutput s)
    throws IOException
```

```
{
    s.writeUTF(name);
    s.writeDouble(salary);
    s.writeLong(hireDay.toEpochDay());
}
```

> ⚠️ **警告**：readObject 和 writeObject 方法是私有的，并且只能被序列化机制调用。与此不同的是，readExternal 和 writeExternal 方法是公共的。特别是，readExternal 还潜在地允许修改现有对象的状态。

2.3.4 序列化单例和类型安全的枚举

在序列化和反序列化时，如果目标对象是唯一的，那么你必须加倍当心，这通常会在实现单例和类型安全的枚举时发生。

如果你使用 Java 语言的 enum 结构，那么你就不必担心序列化，它能够正常工作。但是，假设你在维护遗留代码，其中包含下面这样的枚举类型：

```
public class Orientation
{
    public static final Orientation HORIZONTAL = new Orientation(1);
    public static final Orientation VERTICAL   = new Orientation(2);

    private int value;

    private Orientation(int v) { value = v; }
}
```

这种风格在枚举被添加到 Java 语言中之前是很普遍的。注意，其构造器是私有的。因此，不可能创建出超出 Orientation.HORIZONTAL 和 Orientation.VERTICAL 之外的对象。特别是，你可以使用 == 操作符来测试对象的等同性：

```
if (orientation == Orientation.HORIZONTAL) . . .
```

当类型安全的枚举实现 Serializable 接口时，你必须牢记存在着一种重要的变化，此时，默认的序列化机制是不适用的。假设我们写出一个 Orientation 类型的值，并再次将其读回：

```
Orientation original = Orientation.HORIZONTAL;
ObjectOutputStream out = . . .;
out.write(original);
out.close();
ObjectInputStream in = . . .;
var saved = (Orientation) in.read();
```

现在，下面的测试

```
if (saved == Orientation.HORIZONTAL) . . .
```

将失败。事实上，saved 的值是 Orientation 类型的一个全新的对象，它与任何预定义的常量都不等同。即使构造器是私有的，序列化机制也可以创建新的对象！

为了解决这个问题，你需要定义另外一种称为 readResolve 的特殊序列化方法。如果定义了 readResolve 方法，在对象被序列化之后就会调用它。它必须返回一个对象，而该对象之后会成为 readObject 的返回值。在上面的情况中，readResolve 方法将检查 value 域并返回恰当的枚举常量：

```
protected Object readResolve() throws ObjectStreamException
{
    if (value == 1) return Orientation.HORIZONTAL;
    if (value == 2) return Orientation.VERTICAL;
    throw new ObjectStreamException(); // this shouldn't happen
}
```

请记住向遗留代码中所有类型安全的枚举以及向所有支持单例设计模式的类中添加 readResolve 方法。

2.3.5 版本管理

如果使用序列化来保存对象，就需要考虑在程序演化时会有什么问题。例如，1.1 版本可以读入旧文件吗？仍旧使用 1.0 版本的用户可以读入新版本产生的文件吗？显然，如果对象文件可以处理类的演化问题，那它正是我们想要的。

乍一看，这好像是不可能的。无论类的定义产生了什么样的变化，它的 SHA 指纹也会跟着变化，而我们都知道对象输入流将拒绝读入具有不同指纹的对象。但是，类可以表明它对其早期版本保持兼容，要想这样做，就必须首先获得这个类的早期版本的指纹。我们可以使用 JDK 中的单机程序 serialver 来获得这个数字，例如，运行下面的命令

```
serialver Employee
```

将会打印出

```
Employee: static final long serialVersionUID = -1814239825517340645L;
```

这个类的所有较新的版本都必须把 serialVersionUID 常量定义为与最初版本的指纹相同。

```
class Employee implements Serializable // version 1.1
{
    ...
    public static final long serialVersionUID = -1814239825517340645L;
}
```

如果一个类具有名为 serialVersionUID 的静态数据成员，它就不再需要人工计算指纹，而只需直接使用这个值。

一旦这个静态数据成员被置于某个类的内部，那么序列化系统就可以读入这个类的对象的不同版本。

如果这个类只有方法产生了变化，那么在读入新对象数据时是不会有任何问题的。但是，如果数据域产生了变化，那么就可能会有问题。例如，旧文件对象可能比程序中的对象具有更多或更少的数据域，或者数据域的类型可能有所不同。在这些情况中，对象输入流将尽力将流对象转换成这个类当前的版本。

对象输入流会将这个类当前版本的数据域与被序列化的版本中的数据域进行比较，当

然，对象流只会考虑非瞬时和非静态的数据域。如果这两部分数据域之间名字匹配而类型不匹配，那么对象输入流不会尝试将一种类型转换成另一种类型，因为这两个对象不兼容；如果被序列化的对象具有在当前版本中所没有的数据域，那么对象输入流会忽略这些额外的数据；如果当前版本具有在被序列化的对象中所没有的数据域，那么这些新添加的域将被设置成它们的默认值（如果是对象则是 null，如果是数字则为 0，如果是 boolean 值则是 false）。

下面是一个示例：假设我们已经用雇员类的最初版本（1.0）在磁盘上保存了大量的雇员记录，现在我们在 Employee 类中添加了称为 department 的数据域，从而将其演化到了 2.0 版本。图 2-7 展示了将 1.0 版的对象读入到使用 2.0 版对象的程序中的情形，可以看到 department 域被设置成了 null。图 2-8 展示了相反的情况：一个使用 1.0 版对象的程序读入了 2.0 版的对象，可以看到额外的 department 域被忽略。

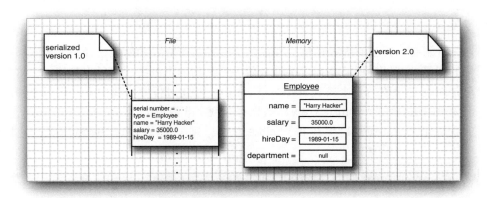

图 2-7 读入具有较少数据域的对象

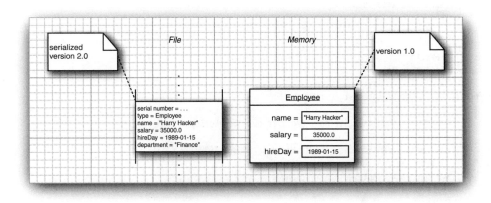

图 2-8 读入具有较多数据域的对象

这种处理是安全的吗？视情况而定。丢掉数据域看起来是无害的，因为接收者仍旧拥有它知道如何处理的所有数据，但是将数据域设置为 null 却有可能并不那么安全。许多类都费尽心思地在其所有的构造器中将所有的数据域都初始化为非 null 的值，以使得其各个方法都不必去处理 null 数据。因此，这个问题取决于类的设计者是否能够在 readObject 方法中实现额外的代码去订正版本不兼容问题，或者是否能够确保所有的方法在处理 null 数据时都足够健壮。

> ✅ **提示**：在将 serialVersionUID 域添加到类中之前，需要问问自己为什么要让这个类是可序列化的。如果序列化只是用于短期持久化，例如在应用服务器中的分布式方法调用，那么就不需要关心版本机制和 serialVersionUID。如果碰巧要扩展一个可序列化的类，但是又从来没想过要持久化该扩展类的任何实例，那么同样不需要关心它们。如果 IDE 总是报有关此问题的烦人的警告消息，那么可以修改 IDE 偏好，将它们关闭，或者添加 @SuppressWarnings("serial") 注解。这样做比添加 serialVersionUID 要更安全，因为也许后续我们会忘记修改 serialVersionUID。

2.3.6 为克隆使用序列化

序列化机制有一种很有趣的用法：即提供了一种克隆对象的简便途径，只要对应的类是可序列化的即可。其做法很简单：直接将对象序列化到输出流中，然后将其读回。这样产生的新对象是对现有对象的一个深拷贝（deep copy）。在此过程中，我们不必将对象写出到文件中，因为可以用 ByteArrayOutputStream 将数据保存到字节数组中。

正如程序清单 2-4 所示，要想得到 clone 方法，只需扩展 SerialCloneable 类，这样就完事了。

程序清单 2-4　serialClone/SerialCloneTest.java

```
 1  package serialClone;
 2
 3  /**
 4   * @version 1.22 2018-05-01
 5   * @author Cay Horstmann
 6   */
 7
 8  import java.io.*;
 9  import java.time.*;
10
11  public class SerialCloneTest
12  {
13     public static void main(String[] args) throws CloneNotSupportedException
14     {
15        var harry = new Employee("Harry Hacker", 35000, 1989, 10, 1);
16        // clone harry
17        var harry2 = (Employee) harry.clone();
18
19        // mutate harry
20        harry.raiseSalary(10);
21
22        // now harry and the clone are different
23        System.out.println(harry);
24        System.out.println(harry2);
```

```java
25        }
26  }
27
28  /**
29   * A class whose clone method uses serialization.
30   */
31  class SerialCloneable implements Cloneable, Serializable
32  {
33      public Object clone() throws CloneNotSupportedException
34      {
35          try {
36              // save the object to a byte array
37              var bout = new ByteArrayOutputStream();
38              try (var out = new ObjectOutputStream(bout))
39              {
40                  out.writeObject(this);
41              }
42
43              // read a clone of the object from the byte array
44              try (var bin = new ByteArrayInputStream(bout.toByteArray()))
45              {
46                  var in = new ObjectInputStream(bin);
47                  return in.readObject();
48              }
49          }
50          catch (IOException | ClassNotFoundException e)
51          {
52              var e2 = new CloneNotSupportedException();
53              e2.initCause(e);
54              throw e2;
55          }
56      }
57  }
58
59  /**
60   * The familiar Employee class, redefined to extend the
61   * SerialCloneable class.
62   */
63  class Employee extends SerialCloneable
64  {
65      private String name;
66      private double salary;
67      private LocalDate hireDay;
68
69      public Employee(String n, double s, int year, int month, int day)
70      {
71          name = n;
72          salary = s;
73          hireDay = LocalDate.of(year, month, day);
74      }
75
76      public String getName()
77      {
78          return name;
```

```
 79       }
 80
 81       public double getSalary()
 82       {
 83          return salary;
 84       }
 85
 86       public LocalDate getHireDay()
 87       {
 88          return hireDay;
 89       }
 90
 91       /**
 92          Raises the salary of this employee.
 93          @byPercent the percentage of the raise
 94       */
 95       public void raiseSalary(double byPercent)
 96       {
 97          double raise = salary * byPercent / 100;
 98          salary += raise;
 99       }
100
101       public String toString()
102       {
103          return getClass().getName()
104             + "[name=" + name
105             + ",salary=" + salary
106             + ",hireDay=" + hireDay
107             + "]";
108       }
109    }
```

我们应该当心这个方法，尽管它很灵巧，但是通常会比显式地构建新对象并复制或克隆数据域的克隆方法慢得多。

2.4 操作文件

你已经学习了如何从文件中读写数据，然而文件管理的内涵远远比读写要广。Path 和 Files 类封装了在用户机器上处理文件系统所需的所有功能。例如，Files 类可以用来移除或重命名文件，或者查询文件最后被修改的时间。换句话说，输入/输出流类关心的是文件的内容，而我们在此处要讨论的类关心的是文件在磁盘上的存储。

Path 接口和 Files 类是在 Java 7 中新添加进来的，它们用起来比自 JDK 1.0 以来就一直使用的 File 类要方便得多。我们认为这两个类会在 Java 程序员中流行起来，因此在这里做深度讨论。

2.4.1 Path

Path（路径）表示的是一个目录名序列，其后还可以跟着一个文件名。路径中的第一个部件可以是根部件，例如 / 或 C:\，而允许访问的根部件取决于文件系统。以根部件开始的路径是绝对路径；否则，就是相对路径。例如，我们要分别创建一个绝对路径和一个相对路径；其中，对于绝对路径，我们假设计算机运行的是类 UNIX 的文件系统：

```
Path absolute = Paths.get("/home", "harry");
Path relative = Paths.get("myprog", "conf", "user.properties");
```

静态的 Paths.get 方法接受一个或多个字符串，并将它们用默认文件系统的路径分隔符（类 UNIX 文件系统是 /，Windows 是 \）连接起来。然后它解析连接起来的结果，如果其表示的不是给定文件系统中的合法路径，那么就抛出 InvalidPathException 异常。这个连接起来的结果就是一个 Path 对象。

get 方法可以获取包含多个部件的单个字符串，例如，可以像下面这样从配置文件中读取路径：

```
String baseDir = props.getProperty("base.dir");
    // May be a string such as /opt/myprog or c:\Program Files\myprog
Path basePath = Paths.get(baseDir); // OK that baseDir has separators
```

> **注释**：路径不必对应着某个实际存在的文件，它仅仅是一个抽象的名字序列。在接下来的小节中将会看到，当你想要创建文件时，首先要创建一个路径，然后才调用方法去创建对应的文件。

组合或解析路径是司空见惯的操作，调用 p.resolve(q) 将按照下列规则返回一个路径：
- 如果 q 是绝对路径，则结果就是 q。
- 否则，根据文件系统的规则，将 "p 后面跟着 q" 作为结果。

例如，假设你的应用系统需要查找相对于给定基目录的工作目录，其中基目录是从配置文件中读取的，就像前一个例子一样。

```
Path workRelative = Paths.get("work");
Path workPath = basePath.resolve(workRelative);
```

resolve 方法有一种快捷方式，它接受一个字符串而不是路径：

```
Path workPath = basePath.resolve("work");
```

还有一个很方便的方法 resolveSibling，它通过解析指定路径的父路径产生其兄弟路径。例如，如果 workPath 是 /opt/myapp/work，那么下面的调用

```
Path tempPath = workPath.resolveSibling("temp");
```

将创建 /opt/myapp/temp。

resolve 的对立面是 relativize，即调用 p.relativize(r) 将产生路径 q，而对 q 进行解析的结果正是 r。例如，以 "/home/harry" 为目标对 "/home/fred/input.txt" 进行相对化操作，会产生 "../fred/input.txt"，其中，我们假设 .. 表示文件系统中的父目录。

normalize 方法将移除所有冗余的 . 和 .. 部件（或者文件系统认为冗余的所有部件）。例如，规范化 /home/harry/../fred/./input.txt 将产生 /home/fred/input.txt。

toAbsolutePath 方法将产生给定路径的绝对路径，该绝对路径从根部件开始，例如 /home/fred/input.txt 或 c:\Users\fred\input.txt。

Path 类有许多有用的方法用来将路径断开。下面的代码示例展示了其中部分最有用的方法：

```
Path p = Paths.get("/home", "fred", "myprog.properties");
Path parent = p.getParent(); // the path /home/fred
Path file = p.getFileName(); // the path myprog.properties
Path root = p.getRoot(); // the path /
```

正如你已经在卷 I 中看到的，还可以从 Path 对象中构建 Scanner 对象：

```
var in = new Scanner(Paths.get("/home/fred/input.txt"));
```

> **注释**：偶尔，你可能需要与遗留系统的 API 交互，它们使用的是 File 类而不是 Path 接口。Path 接口有一个 toFile 方法，而 File 类有一个 toPath 方法。

API *java.nio.file.Paths* 7

- static Path get(String first, String... more)
  通过连接给定的字符串创建一个路径。

API *java.nio.file.Path* 7

- Path resolve(Path other)
- Path resolve(String other)
  如果 other 是绝对路径，那么就返回 other；否则，返回通过连接 this 和 other 获得的路径。
- Path resolveSibling(Path other)
- Path resolveSibling(String other)
  如果 other 是绝对路径，那么就返回 other；否则，返回通过连接 this 的父路径和 other 获得的路径。
- Path relativize(Path other)
  返回用 this 进行解析，相对于 other 的相对路径。
- Path normalize()
  移除诸如 . 和 .. 等冗余的路径元素。
- Path toAbsolutePath()
  返回与该路径等价的绝对路径。
- Path getParent()
  返回父路径，或者在该路径没有父路径时，返回 null。
- Path getFileName()
  返回该路径的最后一个部件，或者在该路径没有任何部件时，返回 null。
- Path getRoot()
  返回该路径的根部件，或者在该路径没有任何根部件时，返回 null。

- toFile()

  从该路径中创建一个 File 对象。

API java.io.File 1.0

- Path toPath() 7

  从该文件中创建一个 Path 对象。

2.4.2 读写文件

Files 类可以使得普通文件操作变得快捷。例如，可以用下面的方式很容易地读取文件的所有内容：

```
byte[] bytes = Files.readAllBytes(path);
```

正如在 2.1.6 节中介绍过的，我们可以如下从文本文件中读取内容：

```
var content = Files.readString(path, charset);
```

但是如果希望将文件当作行序列读入，那么可以调用：

```
List<String> lines = Files.readAllLines(path, charset);
```

相反，如果希望写出一个字符串到文件中，可以调用：

```
Files.writeString(path, content.charset);
```

向指定文件追加内容，可以调用：

```
Files.write(path, content.getBytes(charset), StandardOpenOption.APPEND);
```

还可以用下面的语句将一个行的集合写出到文件中：

```
Files.write(path, lines, charset);
```

这些简便方法适用于处理中等长度的文本文件，如果要处理的文件长度比较大，或者是二进制文件，那么还是应该使用所熟知的输入/输出流或者读入器/写出器：

```
InputStream in = Files.newInputStream(path);
OutputStream out = Files.newOutputStream(path);
Reader in = Files.newBufferedReader(path, charset);
Writer out = Files.newBufferedWriter(path, charset);
```

这些便捷方法可以将你从处理 FileInputStream、FileOutputStream、BufferedReader 和 BufferedWriter 的繁复操作中解脱出来。

API java.nio.file.Files 7

- static byte[] readAllBytes(Path path)
- static String readString(Path path, Charset charset)
- static List<String> readAllLines(Path path, Charset charset)

  读入文件的内容。

- static Path write(Path path, byte[] contents, OpenOption... options)
- static Path write(Path path, String contents, Charset charset, OpenOption... options)

- static Path write(Path path, Iterable<? extends CharSequence> contents, OpenOption options)
  将给定内容写出到文件中，并返回 path。
- static InputStream newInputStream(Path path, OpenOption... options)
- static OutputStream newOutputStream(Path path, OpenOption... options)
- static BufferedReader newBufferedReader(Path path, Charset charset)
- static BufferedWriter newBufferedWriter(Path path, Charset charset, OpenOption... options)
  打开一个文件，用于读入或写出。

2.4.3 创建文件和目录

创建新目录可以调用

```
Files.createDirectory(path);
```

其中，路径中除最后一个部件外，其他部分都必须是已存在的。要创建路径中的中间目录，应该使用

```
Files.createDirectories(path);
```

可以使用下面的语句创建一个空文件：

```
Files.createFile(path);
```

如果文件已经存在了，那么这个调用就会抛出异常。检查文件是否存在和创建文件是原子性的，如果文件不存在，该文件就会被创建，并且其他程序在此过程中是无法执行文件创建操作的。

有些便捷方法可以用来在给定位置或者系统指定位置创建临时文件或临时目录：

```
Path newPath = Files.createTempFile(dir, prefix, suffix);
Path newPath = Files.createTempFile(prefix, suffix);
Path newPath = Files.createTempDirectory(dir, prefix);
Path newPath = Files.createTempDirectory(prefix);
```

其中，dir 是一个 Path 对象，prefix 和 suffix 是可以为 null 的字符串。例如，调用 Files.createTempFile(null, ".txt") 可能会返回一个像 /tmp/1234405522364837194.txt 这样的路径。

在创建文件或目录时，可以指定属性，例如文件的拥有者和权限。但是，指定属性的细节取决于文件系统，本书在此不做讨论。

API java.nio.file.Files 7

- static Path createFile(Path path, FileAttribute<?>... attrs)
- static Path createDirectory(Path path, FileAttribute<?>... attrs)
- static Path createDirectories(Path path, FileAttribute<?>... attrs)
  创建一个文件或目录，createDirectories 方法还会创建路径中所有的中间目录。
- static Path createTempFile(String prefix, String suffix, FileAttribute<?>... attrs)
- static Path createTempFile(Path parentDir, String prefix, String suffix, FileAttribute<?>... attrs)

- static Path createTempDirectory(String prefix, FileAttribute<?>... attrs)
- static Path createTempDirectory(Path parentDir, String prefix, FileAttribute<?>... attrs)

  在适合临时文件的位置，或者在给定的父目录中，创建一个临时文件或目录。返回所创建的文件或目录的路径。

2.4.4 复制、移动和删除文件

将文件从一个位置复制到另一个位置可以直接调用

```
Files.copy(fromPath, toPath);
```

移动文件（即复制并删除原文件）可以调用

```
Files.move(fromPath, toPath);
```

如果目标路径已经存在，那么复制或移动将失败。如果想要覆盖已有的目标路径，可以使用 REPLACE_EXISTING 选项。如果想要复制所有的文件属性，可以使用 COPY_ATTRIBUTES 选项。也可以像下面这样同时选择这两个选项：

```
Files.copy(fromPath, toPath, StandardCopyOption.REPLACE_EXISTING,
    StandardCopyOption.COPY_ATTRIBUTES);
```

你可以将移动操作定义为原子性的，这样就可以保证要么移动操作成功完成，要么源文件继续保持在原来位置。具体可以使用 ATOMIC_MOVE 选项来实现：

```
Files.move(fromPath, toPath, StandardCopyOption.ATOMIC_MOVE);
```

你还可以将一个输入流复制到 Path 中，这表示想要将该输入流存储到硬盘上。类似地，你可以将一个 Path 复制到输出流中。可以使用下面的调用：

```
Files.copy(inputStream, toPath);
Files.copy(fromPath, outputStream);
```

至于其他对 copy 的调用，可以根据需要提供相应的复制选项。

最后，删除文件可以调用：

```
Files.delete(path);
```

如果要删除的文件不存在，这个方法就会抛出异常。因此，可转而使用下面的方法：

```
boolean deleted = Files.deleteIfExists(path);
```

该删除方法还可以用来移除空目录。

请查阅表 2-3 以了解对文件操作而言可用的选项。

表 2-3 用于文件操作的标准选项

选项	描述
StandardOpenOption	与 newBufferedWriter、newInputStream、newOutputStream、write 一起使用
READ	用于读取而打开
WRITE	用于写入而打开
APPEND	如果用于写入而打开，那么在文件末尾追加

（续）

选项	描述
TRUNCATE_EXISTING	如果用于写入而打开，那么移除已有内容
CREATE_NEW	创建新文件并且在文件已存在的情况下会创建失败
CREATE	自动在文件不存在的情况下创建新文件
DELETE_ON_CLOSE	当文件被关闭时，尽"可能"地删除该文件
SPARSE	给文件系统一个提示，表示该文件是稀疏的
DSYNC 或 SYNC	要求对文件数据｜数据和元数据的每次更新都必须同步地写入到存储设备中
StandardCopyOption	与 copy 和 move 一起使用
ATOMIC_MOVE	原子性地移动文件
COPY_ATTRIBUTES	复制文件的属性
REPLACE_EXISTING	如果目标已存在，则替换它
LinkOption	与上面所有方法以及 exists、isDirectory、isRegularFile 等一起使用
NOFOLLOW_LINKS	不要跟踪符号链接
FileVisitOption	与 find、walk、walkFileTree 一起使用
FOLLOW_LINKS	跟踪符号链接

API java.nio.file.Files 7

- static Path copy(Path from, Path to, CopyOption... options)
- static Path move(Path from, Path to, CopyOption... options)

  将 from 复制或移动到给定位置，并返回 to。

- static long copy(InputStream from, Path to, CopyOption... options)
- static long copy(Path from, OutputStream to, CopyOption... options)

  从输入流复制到文件中，或者从文件复制到输出流中，返回复制的字节数。

- static void delete(Path path)
- static boolean deleteIfExists(Path path)

  删除给定文件或空目录。第一个方法在文件或目录不存在情况下抛出异常，而第二个方法在这种情况下会返回 false。

2.4.5 获取文件信息

下面的静态方法都将返回一个 boolean 值，表示检查路径的某个属性的结果：

- exists
- isHidden
- isReadable, isWritable, isExecutable
- isRegularFile, isDirectory, isSymbolicLink

size 方法将返回文件的字节数：

```
long fileSize = Files.size(path);
```

getOwner 方法将文件的拥有者作为 java.nio.file.attribute.UserPrincipal 的一个实例返回。

所有的文件系统都会报告一个基本属性集，它们被封装在 BasicFileAttributes 接口中，这些属性与上述信息有部分重叠。基本文件属性包括：

- 创建文件、最后一次访问以及最后一次修改文件的时间，这些时间都表示成 java.nio.file.attribute.FileTime。
- 文件是常规文件、目录还是符号链接，抑或这三者都不是。
- 文件尺寸。
- 文件主键，这是某种类的对象，具体所属类与文件系统相关，有可能是文件的唯一标识符，也可能不是。

要获取这些属性，可以调用

```
BasicFileAttributes attributes = Files.readAttributes(path, BasicFileAttributes.class);
```

如果你了解到用户的文件系统兼容 POSIX，那么你可以获取一个 PosixFileAttributes 实例：

```
PosixFileAttributes attributes = Files.readAttributes(path, PosixFileAttributes.class);
```

然后从中找到组拥有者，以及文件的拥有者、组和访问权限。我们不会详细讨论其细节，因为这种信息中很多内容在操作系统之间并不具备可移植性。

API java.nio.file.Files 7

- static boolean exists(Path path)
- static boolean isHidden(Path path)
- static boolean isReadable(Path path)
- static boolean isWritable(Path path)
- static boolean isExecutable(Path path)
- static boolean isRegularFile(Path path)
- static boolean isDirectory(Path path)
- static boolean isSymbolicLink(Path path)

  检查由路径指定的文件的给定属性。

- static long size(Path path)

  获取文件按字节数度量的尺寸。

- A readAttributes(Path path, Class<A> type, LinkOption... options)

  读取类型为 A 的文件属性。

API java.nio.file.attribute.BasicFileAttributes 7

- FileTime creationTime()
- FileTime lastAccessTime()
- FileTime lastModifiedTime()
- boolean isRegularFile()

- boolean isDirectory()
- boolean isSymbolicLink()
- long size()
- Object fileKey()

    获取所请求的属性。

2.4.6 访问目录中的项

静态的 Files.list 方法会返回一个可以读取目录中各个项的 Stream<Path> 对象。目录是被惰性读取的，这使得处理具有大量项的目录可以变得更高效。

因为读取目录涉及需要关闭的系统资源，所以应该使用 try 块：

```
try (Stream<Path> entries = Files.list(pathToDirectory))
{
   ...
}
```

list 方法不会进入子目录。为了处理目录中的所有子目录，需要使用 File.walk 方法。

```
try (Stream<Path> entries = Files.walk(pathToRoot))
{
    // Contains all descendants, visited in depth-first order
}
```

下面是解压后的 src.zip 树的遍历样例：

```
java
java/nio
java/nio/DirectCharBufferU.java
java/nio/ByteBufferAsShortBufferRL.java
java/nio/MappedByteBuffer.java
...
java/nio/ByteBufferAsDoubleBufferB.java
java/nio/charset
java/nio/charset/CoderMalfunctionError.java
java/nio/charset/CharsetDecoder.java
java/nio/charset/UnsupportedCharsetException.java
java/nio/charset/spi
java/nio/charset/spi/CharsetProvider.java
java/nio/charset/StandardCharsets.java
java/nio/charset/Charset.java
...
java/nio/charset/CoderResult.java
java/nio/HeapFloatBufferR.java
...
```

正如你所见，无论何时，只要遍历的项是目录，那么在继续访问它的兄弟项之前，会先进入它。

可以通过调用 File.walk(pathToRoot, depth) 来限制想要访问的树的深度。两种 walk 方法都具有 FileVisitOption... 的可变长参数，但是你只能提供一种选项——FOLLOW_LINKS，即跟踪符号链接。

> **注释**：如果要过滤 walk 返回的路径，并且过滤标准涉及与目录存储相关的文件属性，例如尺寸、创建时间和类型（文件、目录、符号链接），那么应该使用 find 方法来替代 walk 方法。可以用某个谓词函数来调用这个方法，该函数接受一个路径和一个 BasicFileAttributes 对象。这样做唯一的优势就是效率高。因为路径总是会被读入，所以这些属性很容易获取。

这段代码使用了 Files.walk 方法来将一个目录复制到另一个目录：

```
Files.walk(source).forEach(p ->
   {
      try
      {
         Path q = target.resolve(source.relativize(p));
         if (Files.isDirectory(p))
            Files.createDirectory(q);
         else
            Files.copy(p, q);
      }
      catch (IOException ex)
      {
         throw new UncheckedIOException(ex);
      }
   });
```

遗憾的是，你无法很容易地使用 Files.walk 方法来删除目录树，因为你必须在删除父目录之前先删除子目录。下一节将展示如何克服此问题。

2.4.7 使用目录流

正如在前一节中所看到的，Files.walk 方法会产生一个可以遍历目录中所有子孙的 Stream<Path> 对象。有时，需要对遍历过程进行更加细粒度的控制。在这种情况下，应该使用 Files.newDirectoryStream 对象，它会产生一个 DirectoryStream。注意，它不是 java.util.stream.Stream 的子接口，而是专门用于目录遍历的接口。它是 Iterable 的子接口，因此可以在增强的 for 循环中使用目录流。下面是其使用模式：

```
try (DirectoryStream<Path> entries = Files.newDirectoryStream(dir))
{
   for (Path entry : entries)
      Process entries
}
```

带资源的 try 语句块用来确保目录流可以被正确关闭。访问目录中的项并没有具体的顺序。

可以用 glob 模式来过滤文件：

```
try (DirectoryStream<Path> entries = Files.newDirectoryStream(dir, "*.java"))
```

表 2-4 展示了所有的 glob 模式。

表 2-4 glob 模式

模式	描述	示例
*	匹配路径组成部分中 0 个或多个字符	*.java 匹配当前目录中的所有 Java 文件
**	匹配跨目录边界的 0 个或多个字符	**.java 匹配在所有子目录中的 Java 文件
?	匹配一个字符	????.java 匹配所有四个字符的 Java 文件 (不包括扩展名)
[...]	匹配一个字符集合，可以使用连线符 [0-9] 和取反符 [!0-9]	Test[0-9A-F].java 匹配 Test*x*.java，其中 *x* 是一个十六进制数字
{...}	匹配由逗号隔开的多个可选项之一	*.{java,class} 匹配所有的 Java 文件和类文件
\	转义上述任意模式中的字符以及 \ 字符	*** 匹配所有文件名中包含 * 的文件

> **警告**：如果使用 Windows 的 glob 语法，则必须对反斜杠转义两次：一次为 glob 语法转义，一次为 Java 字符串转义。例如 Files.newDirectoryStream(dir,"C:\\\\")。

如果想要访问某个目录的所有子孙成员，可以转而调用 walkFileTree 方法，并向其传递一个 FileVisitor 类型的对象，这个对象会得到下列通知：

- 在遇到一个文件或目录时：FileVisitResult visitFile(T path, BasicFileAttributes attrs)
- 在一个目录被处理前：FileVisitResult preVisitDirectory(T dir, IOException ex)
- 在一个目录被处理后：FileVisitResult postVisitDirectory(T dir, IOException ex)
- 在试图访问文件或目录时发生错误，例如没有权限打开目录：FileVisitResult visitFileFailed(path, IOException)

对于上述每种情况，都可以指定是否希望执行下面的操作：

- 继续访问下一个文件：FileVisitResult.CONTINUE
- 继续访问，但是不再访问这个目录下的任何项了：FileVisitResult.SKIP_SUBTREE
- 继续访问，但是不再访问这个文件的兄弟文件（和该文件在同一个目录下的文件）了：FileVisitResult.SKIP_SIBLINGS
- 终止访问：FileVisitResult.TERMINATE

当有任何方法抛出异常时，就会终止访问，而这个异常会从 walkFileTree 方法中抛出。

> **注释**：FileVisitor 接口是泛化类型，但是你不太可能会使用除 FileVisitor<Path> 之外的东西。walkFileTree 方法可以接受 FileVisitor<? Super Path> 类型的参数，但是 Path 并没有多少超类型。

便捷类 SimpleFileVisitor 实现了 FileVisitor 接口，但是其除 visitFileFailed 方法之外的所有方法并不做任何处理而是直接继续访问，而 visitFileFailed 方法会抛出由失败导致的异常，并进而终止访问。

例如，下面的代码展示了如何打印出给定目录下的所有子目录：

```
Files.walkFileTree(Paths.get("/"), new SimpleFileVisitor<Path>()
    {
        public FileVisitResult preVisitDirectory(Path path, BasicFileAttributes attrs)
            throws IOException
```

```
        {
            System.out.println(path);
            return FileVisitResult.CONTINUE;
        }
        public FileVisitResult postVisitDirectory(Path dir, IOException exc)
        {
            return FileVisitResult.CONTINUE;
        }
        public FileVisitResult visitFileFailed(Path path, IOException exc)
            throws IOException
        {
            return FileVisitResult.SKIP_SUBTREE;
        }
    });
```

值得注意的是,我们需要覆盖 postVisitDirectory 方法和 visitFileFailed 方法,否则,访问会在遇到不允许打开的目录或不允许访问的文件时立即失败。

还应该注意的是,路径的众多属性是作为 preVisitDirectory 和 visitFile 方法的参数传递的。访问者不得不通过操作系统调用来获得这些属性,因为它需要区分文件和目录。因此,你就不需要再次执行系统调用了。

如果你需要在进入或离开一个目录时执行某些操作,那么 FileVisitor 接口的其他方法就显得非常有用了。例如,在删除目录树时,需要在移除当前目录的所有文件之后,才能移除该目录。下面是删除目录树的完整代码:

```
// Delete the directory tree starting at root
Files.walkFileTree(root, new SimpleFileVisitor<Path>()
    {
        public FileVisitResult visitFile(Path file, BasicFileAttributes attrs)
            throws IOException
        {
            Files.delete(file);
            return FileVisitResult.CONTINUE;
        }
        public FileVisitResult postVisitDirectory(Path dir, IOException e) throws IOException
        {
            if (e != null) throw e;
            Files.delete(dir);
            return FileVisitResult.CONTINUE;
        }
    });
```

API java.nio.file.Files 7

- static DirectoryStream<Path> newDirectoryStream(Path path)
- static DirectoryStream<Path> newDirectoryStream(Path path, String glob)

  获取给定目录中可以遍历所有文件和目录的迭代器。第二个方法只接受那些与给定的 glob 模式匹配的项。

- static Path walkFileTree(Path start, FileVisitor<? super Path> visitor)

遍历给定路径的所有子孙，并将访问器应用于这些子孙之上。

API java.nio.file.SimpleFileVisitor<T> 7

- static FileVisitResult visitFile(T path, BasicFileAttributes attrs)
  在访问文件或目录时被调用，返回 CONTINUE、SKIP_SUBTREE、SKIP_SIBLINGS 和 TERMINATE 之一，默认实现是不做任何操作而继续访问。
- static FileVisitResult preVisitDirectory(T dir, BasicFileAttributes attrs)
- static FileVisitResult postVisitDirectory(T dir, BasicFileAttributes attrs)
  在访问目录之前和之后被调用，默认实现是不做任何操作而继续访问。
- static FileVisitResult visitFileFailed(T path, IOException exc)
  如果在试图获取给定文件的信息时抛出异常，则该方法被调用。默认实现是重新抛出异常，这会导致访问操作以这个异常而终止。如果你想自己访问，可以覆盖这个方法。

2.4.8 ZIP 文件系统

Paths 类会在默认文件系统中查找路径，即在用户本地磁盘中的文件。你也可以有别的文件系统，其中最有用的之一是 ZIP 文件系统。如果 zipname 是某个 ZIP 文件的名字，那么下面的调用

```
FileSystem fs = FileSystems.newFileSystem(Paths.get(zipname), null);
```

将建立一个文件系统，它包含 ZIP 文档中的所有文件。如果知道文件名，那么从 ZIP 文档中复制出这个文件就会变得很容易：

```
Files.copy(fs.getPath(sourceName), targetPath);
```

其中的 fs.getPath 对于任意文件系统来说都与 Paths.get 类似。

要列出 ZIP 文档中的所有文件，可以遍历文件树：

```
FileSystem fs = FileSystems.newFileSystem(Paths.get(zipname), null);
Files.walkFileTree(fs.getPath("/"), new SimpleFileVisitor<Path>()
    {
        public FileVisitResult visitFile(Path file, BasicFileAttributes attrs)
            throws IOException
        {
            System.out.println(file);
            return FileVisitResult.CONTINUE;
        }
    });
```

这比 2.2.3 节中描述的 API 要好用，后者使用的是多个专门处理 ZIP 文档的新类。

API java.nio.file.FileSystems 7

- static FileSystem newFileSystem(Path path, ClassLoader loader)
  对所安装的文件系统提供者进行迭代，并且如果 loader 不为 null，那么就还会迭代给定的类加载器能够加载的文件系统，返回由第一个可以接受给定路径的文件系统提供

者创建的文件系统。默认情况下，对于 ZIP 文件系统是有一个提供者的，它接受名字以 .zip 或 .jar 结尾的文件。

API java.nio.file.FileSystem 7

- static Path getPath(String first, String... more)
  将给定的字符串连接起来创建一个路径。

2.5 内存映射文件

大多数操作系统都可以利用虚拟内存实现来将一个文件或者文件的一部分"映射"到内存中。然后，这个文件就可以被当作内存数组一样地访问，这比传统的文件操作要快得多。

2.5.1 内存映射文件的性能

在本节的末尾，你可以看到一个计算传统的文件输入和内存映射文件的 CRC32 校验和的程序。在同一台机器上，我们对 JDK 的 jre/lib 目录中 37MB 的 rt.jar 文件用不同的方式来计算校验和，记录下来的时间数据如表 2-5 所示。

表 2-5 文件操作的处理时间数据

方法	时间
普通输入流	110 秒
带缓冲的输入流	9.9 秒
随机访问文件	162 秒
内存映射文件	7.2 秒

正如你所见，在这台特定的机器上，内存映射比使用带缓冲的顺序输入要稍微快一点，但是比使用 RandomAccessFile 快很多。

当然，精确的值因机器不同会产生很大的差异，但是很明显，与随机访问相比，性能提高总是很显著的。另一方面，对于中等尺寸文件的顺序读入则没有必要使用内存映射。

java.nio 包使内存映射变得十分简单，下面就是我们需要做的。

首先，从文件中获得一个通道（channel），通道是用于磁盘文件的一种抽象，它使我们可以访问诸如内存映射、文件加锁机制以及文件间快速数据传递等操作系统特性。

```
FileChannel channel = FileChannel.open(path, options);
```

然后，通过调用 FileChannel 类的 map 方法从这个通道中获得一个 ByteBuffer。你可以指定想要映射的文件区域与映射模式，支持的模式有三种：

- FileChannel.MapMode.READ_ONLY：所产生的缓冲区是只读的，任何对该缓冲区写入的尝试都会导致 ReadOnlyBufferException 异常。

- FileChannel.MapMode.READ_WRITE：所产生的缓冲区是可写的，任何修改都会在某个时刻写回到文件中。注意，其他映射同一个文件的程序可能不能立即看到这些修改，多个程序同时进行文件映射的确切行为是依赖于操作系统的。
- FileChannel.MapMode.PRIVATE：所产生的缓冲区是可写的，但是任何修改对这个缓冲区来说都是私有的，不会传播到文件中。

一旦有了缓冲区，就可以使用 ByteBuffer 类和 Buffer 超类的方法读写数据了。

缓冲区支持顺序和随机数据访问，它有一个可以通过 get 和 put 操作来移动的位置。例如，可以像下面这样顺序遍历缓冲区中的所有字节：

```
while (buffer.hasRemaining())
{
    byte b = buffer.get();
    ...
}
```

或者，像下面这样进行随机访问：

```
for (int i = 0; i < buffer.limit(); i++)
{
    byte b = buffer.get(i);
    ...
}
```

你可以用下面的方法来读写字节数组：

```
get(byte[] bytes)
get(byte[], int offset, int length)
```

最后，还有下面的方法：

```
getInt        getChar
getLong       getFloat
getShort      getDouble
```

用来读入在文件中存储为二进制值的基本类型值。正如我们提到的，Java 对二进制数据使用高位在前的排序机制，但是，如果需要以低位在前的排序方式处理包含二进制数字的文件，那么只需调用

```
buffer.order(ByteOrder.LITTLE_ENDIAN);
```

要查询缓冲区内当前的字节顺序，可以调用：

```
ByteOrder b = buffer.order();
```

> ⚠ **警告**：这一对方法没有使用 set/get 命名惯例。

要向缓冲区写数字，可以使用下列的方法：

```
putInt        putChar
putLong       putFloat
putShort      putDouble
```

在恰当的时机，以及当通道关闭时，会将这些修改写回到文件中。

程序清单 2-5 用于计算文件的 32 位的循环冗余校验和（CRC32），这个数值就是经常用来判断一个文件是否已损坏的校验和，因为文件损坏极有可能导致校验和改变。java.util.zip 包中包含一个 CRC32 类，可以使用下面的循环来计算一个字节序列的校验和：

```
var crc = new CRC32();
while (more bytes)
   crc.update(next byte);
long checksum = crc.getValue();
```

CRC 计算的细节并不重要，我们只是将它作为一个有用的文件操作的实例来使用。（在实践中，每次会以更大的块而不是一个字节为单位来读取和更新数据，而它们的速度差异并不明显。）

应该像下面这样运行程序：

```
java memoryMap.MemoryMapTest filename
```

程序清单 2-5 memoryMap/MemoryMapTest.java

```
 1  package memoryMap;
 2
 3  import java.io.*;
 4  import java.nio.*;
 5  import java.nio.channels.*;
 6  import java.nio.file.*;
 7  import java.util.zip.*;
 8
 9  /**
10   * This program computes the CRC checksum of a file in four ways. <br>
11   * Usage: java memoryMap.MemoryMapTest filename
12   * @version 1.02 2018-05-01
13   * @author Cay Horstmann
14   */
15  public class MemoryMapTest
16  {
17     public static long checksumInputStream(Path filename) throws IOException
18     {
19        try (InputStream in = Files.newInputStream(filename))
20        {
21           var crc = new CRC32();
22
23           int c;
24           while ((c = in.read()) != -1)
25              crc.update(c);
26           return crc.getValue();
27        }
28     }
29
30     public static long checksumBufferedInputStream(Path filename) throws IOException
31     {
32        try (var in = new BufferedInputStream(Files.newInputStream(filename)))
33        {
34           var crc = new CRC32();
35
```

```java
   36        int c;
   37        while ((c = in.read()) != -1)
   38           crc.update(c);
   39        return crc.getValue();
   40     }
   41  }
   42
   43  public static long checksumRandomAccessFile(Path filename) throws IOException
   44  {
   45     try (var file = new RandomAccessFile(filename.toFile(), "r"))
   46     {
   47        long length = file.length();
   48        var crc = new CRC32();
   49
   50        for (long p = 0; p < length; p++)
   51        {
   52           file.seek(p);
   53           int c = file.readByte();
   54           crc.update(c);
   55        }
   56        return crc.getValue();
   57     }
   58  }
   59
   60  public static long checksumMappedFile(Path filename) throws IOException
   61  {
   62     try (FileChannel channel = FileChannel.open(filename))
   63     {
   64        var crc = new CRC32();
   65        int length = (int) channel.size();
   66        MappedByteBuffer buffer = channel.map(FileChannel.MapMode.READ_ONLY, 0, length);
   67
   68        for (int p = 0; p < length; p++)
   69        {
   70           int c = buffer.get(p);
   71           crc.update(c);
   72        }
   73        return crc.getValue();
   74     }
   75  }
   76
   77  public static void main(String[] args) throws IOException
   78  {
   79     System.out.println("Input Stream:");
   80     long start = System.currentTimeMillis();
   81     Path filename = Paths.get(args[0]);
   82     long crcValue = checksumInputStream(filename);
   83     long end = System.currentTimeMillis();
   84     System.out.println(Long.toHexString(crcValue));
   85     System.out.println((end - start) + " milliseconds");
   86
   87     System.out.println("Buffered Input Stream:");
   88     start = System.currentTimeMillis();
   89     crcValue = checksumBufferedInputStream(filename);
```

```
 90        end = System.currentTimeMillis();
 91        System.out.println(Long.toHexString(crcValue));
 92        System.out.println((end - start) + " milliseconds");
 93
 94        System.out.println("Random Access File:");
 95        start = System.currentTimeMillis();
 96        crcValue = checksumRandomAccessFile(filename);
 97        end = System.currentTimeMillis();
 98        System.out.println(Long.toHexString(crcValue));
 99        System.out.println((end - start) + " milliseconds");
100
101        System.out.println("Mapped File:");
102        start = System.currentTimeMillis();
103        crcValue = checksumMappedFile(filename);
104        end = System.currentTimeMillis();
105        System.out.println(Long.toHexString(crcValue));
106        System.out.println((end - start) + " milliseconds");
107     }
108 }
```

API java.io.FileInputStream 1.0

- FileChannel getChannel() 1.4
  返回用于访问这个输入流的通道。

API java.io.FileOutputStream 1.0

- FileChannel getChannel() 1.4
  返回用于访问这个输出流的通道。

API java.io.RandomAccessFile 1.0

- FileChannel getChannel() 1.4
  返回用于访问这个文件的通道。

API java.nio.channels.FileChannel 1.4

- static FileChannel open(Path path, OpenOption... options) 7
  打开指定路径的文件通道，默认情况下，通道打开时用于读入。参数 options 是 Standard-OpenOption 枚举中的 WRITE、APPEND、TRUNCATE_EXISTING、CREATE 值。
- MappedByteBuffer map(FileChannel.MapMode mode, long position, long size)
  将文件的一个区域映射到内存中。参数 mode 是 FileChannel.MapMode 类中的常量 READ_ONLY、READ_WRITE 或 PRIVATE 之一。

API java.nio.Buffer 1.4

- boolean hasRemaining()
  如果当前的缓冲区位置没有到达这个缓冲区的界限位置，则返回 true。
- int limit()

返回这个缓冲区的界限位置，即没有任何值可用的第一个位置。

API **java.nio.ByteBuffer** 1.4

- byte get()
  从当前位置获得一个字节，并将当前位置移动到下一个字节。
- byte get(int index)
  从指定索引处获得一个字节。
- ByteBuffer put(byte b)
  向当前位置推入一个字节，并将当前位置移动到下一个字节。返回对这个缓冲区的引用。
- ByteBuffer put(int index, byte b)
  向指定索引处推入一个字节。返回对这个缓冲区的引用。
- ByteBuffer get(byte[] destination)
- ByteBuffer get(byte[] destination, int offset, int length)
  用缓冲区中的字节来填充字节数组，或者字节数组的某个区域，并将当前位置向前移动读入的字节数个位置。如果缓冲区不够大，那么就不会读入任何字节，并抛出 BufferUnderflowException。返回对这个缓冲区的引用。
- ByteBuffer put(byte[] source)
- ByteBuffer put(byte[] source, int offset, int length)
  将字节数组中的所有字节或者给定区域的字节都推入缓冲区中，并将当前位置向前移动写出的字节数个位置。如果缓冲区不够大，那么就不会读入任何字节，并抛出 BufferUnderflowException。返回对这个缓冲区的引用。
- *Xxx* get*Xxx*()
- *Xxx* get*Xxx*(int index)
- ByteBuffer put*Xxx*(*Xxx* value)
- ByteBuffer put*Xxx*(int index, *Xxx* value)
  获得或放置一个二进制数。*Xxx* 是 Int、Long、Short、Char、Float 或 Double 中的一个。
- ByteBuffer order(ByteOrder order)
- ByteOrder order()
  设置或获得字节顺序，order 的值是 ByteOrder 类的常量 BIG_ENDIAN 或 LITTLE_ENDIAN 中的一个。
- static ByteBuffer allocate(int capacity)
  构建具有给定容量的缓冲区。
- static ByteBuffer wrap(byte[] values)
  构建具有指定容量的缓冲区，该缓冲区是对给定数组的包装。
- CharBuffer asCharBuffer()

构建字符缓冲区，它是对这个缓冲区的包装。对该字符缓冲区的变更将在这个缓冲区中反映出来，但是该字符缓冲区有自己的位置、界限和标记。

API java.nio.CharBuffer 1.4

- char get()
- CharBuffer get(char[] destination)
- CharBuffer get(char[] destination, int offset, int length)

  从这个缓冲区的当前位置开始，获取一个 char 值，或者一个范围内的所有 char 值，然后将位置向前移动以越过所有读入的字符。最后两个方法将返回 this。

- CharBuffer put(char c)
- CharBuffer put(char[] source)
- CharBuffer put(char[] source, int offset, int length)
- CharBuffer put(String source)
- CharBuffer put(CharBuffer source)

  从这个缓冲区的当前位置开始，放置一个 char 值，或者一个范围内的所有 char 值，然后将位置向前移动越过所有被写出的字符。当放置的值是从 CharBuffer 读入时，将读入所有剩余字符。所有方法将返回 this。

2.5.2 缓冲区数据结构

在使用内存映射时，我们创建了单一的缓冲区横跨整个文件或我们感兴趣的文件区域。我们还可以使用更多的缓冲区来读写大小适度的信息块。

本节将简要地介绍 Buffer 对象上的基本操作。缓冲区是由具有相同类型的数值构成的数组，Buffer 类是一个抽象类，它有众多的具体子类，包括 ByteBuffer、CharBuffer、DoubleBuffer、IntBuffer、LongBuffer 和 ShortBuffer。

> 注释：StringBuffer 类与这些缓冲区没有关系。

在实践中，最常用的将是 ByteBuffer 和 CharBuffer。如图 2-9 所示，每个缓冲区都具有：

- 一个容量，它永远不能改变。
- 一个读写位置，下一个值将在此进行读写。
- 一个界限，超过它进行读写是没有意义的。
- 一个可选的标记，用于重复一个读入或写出操作。

这些值满足下面的条件：

$$0 \leq 标记 \leq 读写位置 \leq 界限 \leq 容量$$

使用缓冲区的主要目的是执行"写，然后读入"循环。假设我们有一个缓冲区，在一开始，它的位置为 0，界限等于容量。我们不断地调用 put 将值添加到这个缓冲区中，当我们耗尽所有的数据或者写出的数据量达到容量大小时，就该切换到读入操作了。

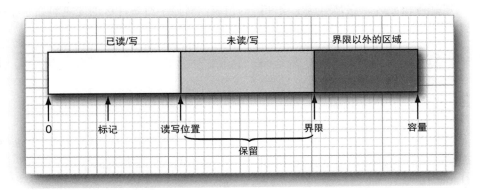

图 2-9 一个缓冲区

这时调用 flip 方法将界限设置到当前位置，并把位置复位到 0。现在在 remaining 方法返回正数时（它返回的值是界限 – 位置），不断地调用 get。在我们将缓冲区中所有的值都读入之后，调用 clear 使缓冲区为下一次写循环做好准备。clear 方法将位置复位到 0，并将界限复位到容量。

如果你想重读缓冲区，可以使用 rewind 或 mark/reset 方法，详细内容请查看 API 注释。

要获取缓冲区，可以调用诸如 ByteBuffer.allocate 或 ByteBuffer.wrap 这样的静态方法。

然后，可以用来自某个通道的数据填充缓冲区，或者将缓冲区的内容写出到通道中。例如：

```
ByteBuffer buffer = ByteBuffer.allocate(RECORD_SIZE);
channel.read(buffer);
channel.position(newpos);
buffer.flip();
channel.write(buffer);
```

这是一种非常有用的方法，可以替代随机访问文件。

API java.nio.Buffer 1.4

- Buffer clear()
  通过将位置复位到 0，并将界限设置到容量，使这个缓冲区为写出做好准备。返回 this。
- Buffer flip()
  通过将界限设置到位置，并将位置复位到 0，使这个缓冲区为读入做好准备。返回 this。
- Buffer rewind()
  通过将读写位置复位到 0，并保持界限不变，使这个缓冲区为重新读入相同的值做好准备。返回 this。
- Buffer mark()

将这个缓冲区的标记设置到读写位置，返回 this。
- `Buffer reset()`
将这个缓冲区的位置设置到标记，从而允许被标记的部分再次被读入或写出，返回 this。
- `int remaining()`
返回剩余可读入或可写出的值的数量，即界限与位置之间的差异。
- `int position()`
- `void position(int newValue)`
返回这个缓冲区的位置。
- `int capacity()`
返回这个缓冲区的容量。

2.6 文件加锁机制

考虑一下多个同时执行的程序需要修改同一个文件的情形，很明显，这些程序需要以某种方式进行通信，不然这个文件很容易被损坏。文件锁可以解决这个问题，它可以控制对文件或文件中某个范围的字节的访问。

假设你的应用程序将用户的偏好存储在一个配置文件中，当用户调用这个应用的两个实例时，这两个实例就有可能会同时希望写配置文件。在这种情况下，第一个实例应该锁定文件，当第二个实例发现文件被锁定时，它必须决策是等待直至文件解锁，还是直接跳过这个写操作过程。

要锁定一个文件，可以调用 FileChannel 类的 lock 或 tryLock 方法：

```
FileChannel = FileChannel.open(path);
FileLock lock = channel.lock();
```

或

```
FileLock lock = channel.tryLock();
```

第一个调用会阻塞直至可获得锁，而第二个调用将立即返回，要么返回锁，要么在锁不可获得的情况下返回 null。这个文件将保持锁定状态，直至通道关闭，或者在锁上调用了 release 方法。

你还可以通过下面的调用锁定文件的一部分：

```
FileLock lock(long start, long size, boolean shared)
```

或

```
FileLock tryLock(long start, long size, boolean shared)
```

如果 shared 标志为 false，则锁定文件的目的是读写；而如果为 true，则这是一个共享锁，允许多个进程从文件中读入，并阻止任何进程获得独占的锁。并非所有的操作系统都支持共享锁，因此你可能会在请求共享锁的时候得到独占的锁。调用 FileLock 类的 isShared 方法可以查询所持有的锁的类型。

> **注释**：如果你锁定了文件的尾部，而这个文件的长度随后增长并超过了锁定的部分，那么增长出来的额外区域是未锁定的，要想锁定所有的字节，可以使用 Long.MAX_VALUE 来表示尺寸。

要确保在操作完成时释放锁，与往常一样，最好在一个带资源的 try 语句中执行释放锁的操作：

```
try (FileLock lock = channel.lock())
{
    access the locked file or segment
}
```

请记住，文件加锁机制是依赖于操作系统的，下面是需要注意的几点：
- 在某些系统中，文件加锁仅仅是建议性的，如果一个应用未能得到锁，它仍旧可以向被另一个应用并发锁定的文件执行写操作。
- 在某些系统中，不能在锁定一个文件的同时将其映射到内存中。
- 文件锁是由整个 Java 虚拟机持有的。如果有两个程序是由同一个虚拟机启动的（例如 Applet 和应用程序启动器），那么它们不可能每一个都获得一个在同一个文件上的锁。当调用 lock 和 tryLock 方法时，如果虚拟机已经在同一个文件上持有了另一个重叠的锁，那么这两个方法将抛出 OverlappingFileLockException。
- 在一些系统中，关闭一个通道会释放由 Java 虚拟机持有的底层文件上的所有锁。因此，在同一个锁定文件上应避免使用多个通道。
- 在网络文件系统上锁定文件是高度依赖于系统的，因此应该尽量避免。

API java.nio.channels.FileChannel 1.4

- FileLock lock()
  在整个文件上获得一个独占的锁，这个方法将阻塞直至获得锁。
- FileLock tryLock()
  在整个文件上获得一个独占的锁，或者在无法获得锁的情况下返回 null。
- FileLock lock(long position, long size, boolean shared)
- FileLock tryLock(long position, long size, boolean shared)
  在文件的一个区域上获得锁。第一个方法将阻塞直至获得锁，而第二个方法将在无法获得锁时返回 null。参数 shared 的值为 true 表示共享锁，为 false 表示独占锁。

API java.nio.channels.FileLock 1.4

- void close() 1.7
  释放这个锁。

2.7 正则表达式

正则表达式（regular expression）用于指定字符串的模式，可以在任何需要定位匹配某种特定模式的字符串的情况下使用正则表达式。例如，我们有一个示例程序就是用来定位 HTML 文件中的所有超链接的，它是通过查找 <a href = "..."> 模式的字符串来实现此目的的。

当然，在指定模式时，... 标记法并不够精确。需要精确地指定什么样的字符序列才是合法的匹配，这就要求无论何时，当你要描述一个模式时，都需要使用某种特定的语法。

在下面各节中，我们将介绍 Java API 用到的正则表达式的语法，并讨论如何使用正则表达式。

2.7.1 正则表达式语法

下面是一个简单的示例，正则表达式

[Jj]ava.+

匹配下列形式的所有字符串：

- 第一个字母是 J 或 j。
- 接下来的三个字母是 ava。
- 字符串的其余部分由一个或多个任意的字符构成。

例如，字符串 "javanese" 就匹配这个特定的正则表达式，但是字符串 "Core Java" 就不匹配。

正如你所见，你需要了解一点这种语法，以理解正则表达式的含义。幸运的是，对于大多数情况，一小部分很直观的语法结构就足够用了。

- 字符类（character class）是一个括在括号中的可选择的字符集，例如，[Jj]、[0-9]、[A-Za-z] 或 [^0-9]。这里 "-" 表示是一个范围（所有 Unicode 值落在两个边界范围之内的字符），而 ^ 表示补集（除了指定字符之外的所有字符）。
- 如果字符类中包含 "-"，那么它必须是第一项或最后一项；如果要包含 "["，那么它必须是第一项；如果要包含 "^"，那么它可以是除开始位置之外的任何位置。其中，你只需要转义 "[" 和 "\"。
- 有许多预定的字符类，例如 \d（数字）和 \p{Sc}（Unicode 货币符号）。请查看表 2-6 和表 2-7。

表 2-6 正则表达式语法

表达式	描述	示例
字符		
c，除 .*+?{|()[\\^$ 之外	字符 c	]
.	任何除行终止符之外的字符，或者在 DOTALL 标志被设置时表示任何字符	
\x{p}	十六进制码为 p 的 Unicode 码点	\x{1D546}

（续）

表达式	描述	示例
\uhhhh, \xhh, \0o, \0oo, \0ooo	具有给定十六进制或八进制值的码元	\uFEFF
\a, \e, \f, \n, \r, \t	响铃符（\x{7}）、转义符（\x{1B}）、换页符（\x{8}）、换行符（\x{A}）、回车符（\x{D}）、指标符（\x{9}）	\n
\c*c*，其中 *c* 在 [A-Z] 的范围内，或者是 @[\]^_? 之一	对应于字符 *c* 的控制字符	\cH 是退格符（\x{8}）
c，其中 *c* 不在 [A-Za-z0-9] 的范围内	字符 *c*	\\
\Q...\E	在左引号和右引号之间的所有字符	\Q(...)\E 匹配字符串（...）
字符类		
[C_1C_2...]，其中 C_i 是多个字符，范围从 c-d，或者是字符类	任何由 C_1, C_2, …表示的字符	[0-9+-]
[^...]	某个字符类的补集	[^\d\s]
[...&&...]	字符集的交集	[\p{L}&&[^A-Za-z]]
\p{...}, \P{...}	某个预定义字符类（参阅表 2-7）；它的补集	\p{L} 匹配一个 Unicode 字母，而 \pL 也匹配这个字母，可以忽略单个字母情况下的括号
\d, \D	数字（[0-9]，或者在 UNICODE_CHARACTER_CLASS 标志被设置时表示 \p{Digit}）；它的补集	\d+ 是一个数字序列
\w, \W	单词字符（[a-zA-Z0-9]，或者在 UNICODE_CHARACTER_CLASS 标志被设置时表示 Unicode 单词字符）；它的补集	
\s, \S	空格（[\n\r\t\f\x{B}]，或者在 UNICODE_CHARACTER_CLASS 标志被设置时表示 \p{IsWhite_Space}）；它的补集	\s*,\s* 是由可选的空格字符包围的逗号
\h, \v, \H, \V	水平空白字符、垂直空白字符，它们的补集	
序列和选择		
XY	任何 X 中的字符串，后面跟随任何 Y 中的字符串	[1-9][0-9]* 表示没有前导零的正整数
X|Y	任何 X 或 Y 中的字符串	http|ftp
群组		
(X)	捕获 X 的匹配	'(['"]*)' 捕获的是被引用的文本
\n	第 n 组	(['"]).*\1 可以匹配 'Fred' 和 "Fred"，但是不能匹配 "Fred'
(?<name>X)	捕获与给定名字匹配的 X	'(?<id>[A-Za-z0-9]+)' 可以捕获名字为 id 的匹配
\k<name>	具有给定名字的组	\k<id> 可以匹配名字为 id 的组
(?:X)	使用括号但是不捕获 X	在 (?:http|ftp)://(.*) 中，在 :// 之后的匹配是 \1

（续）

表达式	描述	示例
$(?f_1f_2\ldots:X)$ $(?f_1\ldots-f_k\ldots:X)$，其中 f_i 在 [dimsuUx] 的范围中	匹配但是不捕获给定标志开或关（在 - 之后）的 X	(?i:jpe?g) 是大小写不敏感的匹配
其他 (?...)	请参阅 Pattern API 文档	
量词		
$X?$	可选 X	\+? 是可选的 + 号
$X*$, $X+$	0 或多个 X，1 或多个 X	[1-9][0-9]+ 是大于等于 10 的整数
$X\{n\}$, $X\{n,\}$, $X\{m,n\}$	n 个 X，至少 n 个 X，m 到 n 个 X	[0-7]{1,3} 是一位到三位的八进制数
$Q?$，其中 Q 是一个量词表达式	勉强量词，在尝试最长匹配之前先尝试最短匹配	.*(<.+?>).* 捕获尖括号括起来的最短序列
$Q+$，其中 Q 是一个量词表达式	占有量词，在不回溯的情况下获取最长匹配	'[^']*+' 匹配单引号引起来的字符串，并且在字符串中没有右单引号的情况下立即匹配失败
边界匹配		
^, $	输入的开头和结尾（或者多行模式中的开头和结尾行）	^Java$ 匹配输入中的 Java 或 Java 构成的行
\A, \Z, \z	输入的开头、输入的结尾、输入的绝对结尾（在多行模式中不会发生变化）	
\b, \B	单词边界，非单词边界	\bJava\b 匹配单词 Java
\R	Unicode 行分隔符	
\G	前一个匹配的结尾	

表 2-7　与 \p 一起使用的预定义字符类名字

字符类名字	解释
posixClass	*posixClass* 是 Lower、Upper、Alpha、Digit、Alnum、Punct、Graph、Print、Cntrl、XDigit、Space、Blank、ASCII 之一，它会依 UNICODE_CHARACTER_CLASS 标志的值而被解释为 POSIX 或 Unicode 类
Is*Script*, sc=*Script*, script=*Script*	Character.UnicodeScript.forName 可以接受的脚本
In*Block*, blk=*Block*, block=*Block*	Character.UnicodeScript.forName 可以接受的块
Category, In*Category*, gc=*Category*, general_category=*Category*	Unicode 通用分类的单字母或双字母名字
Is*Property*	*Property* 是 Alphabetic、Ideographic、Letter、Lowercase、Uppercase、Titlecase、Punctuation、Control、White_Space、Digit、Hex_Digit、Join_Control、Noncharacter_Code_Point、Assigned 之一
java*Method*	调用 Character.is*Method* 方法（必须不是过时的方法）

- 大部分字符都可以与它们自身匹配，例如在前面示例中的 ava 字符。
- . 符号可以匹配任何字符（有可能不包括行终止符，这取决于标志的设置）。

- 使用 \ 作为转义字符，例如，\. 匹配句号而 \\ 匹配反斜线。
- ^ 和 $ 分别匹配一行的开头和结尾。
- 如果 X 和 Y 是正则表达式，那么 XY 表示"任何 X 的匹配后面跟随 Y 的匹配"，X | Y 表示"任何 X 或 Y 的匹配"。
- 可以将量词运用到表达式 X：X+（1 个或多个）、X*（0 个或多个）与 X?（0 个或 1 个）。
- 默认情况下，量词要匹配能够使整个匹配成功的最大可能的重复次数。可以修改这种行为，方法是使用后缀 ?（使用勉强或吝啬匹配，也就是匹配最小的重复次数）或使用后缀 +（使用占有或贪婪匹配，也就是即使让整个匹配失败，也要匹配最大的重复次数）。

    例如，字符串 cab 匹配 [a-z]*ab，但是不匹配 [a-z]*+ab。在第一种情况中，表达式 [a-z]* 只匹配字符 c，使得字符 ab 匹配该模式的剩余部分；但是贪婪版本 [a-z]*+ 将匹配字符 cab，模式的剩余部分将无法匹配。
- 我们使用群组来定义子表达式，其中群组用括号 () 括起来。例如，([+-]?)([0-9]+)。然后可以询问模式匹配器，让其返回每个组的匹配，或者用 \n 来引用某个群组，其中 n 是群组号（从 \1 开始）。

例如，下面是一个有些复杂但是却可能很有用的正则表达式，它描述了十进制和十六进制整数：

    [+-]?[0-9]+|0[Xx][0-9A-Fa-f]+

遗憾的是，在使用正则表达式的各种程序和类库之间，表达式语法并未完全标准化。尽管在基本结构上达成了一致，但是它们在细节上仍旧存在着许多令人抓狂的差异。Java 正则表达式类使用的语法与 Perl 语言使用的语法十分相似，但是并不完全一样。表 2-6 展示的是 Java 语法中的所有结构。关于正则表达式语法的更多信息，可以求教于 Pattern 类的 API 文档和 Jeffrey E. F. Friedl 的 *Mastering Regular Expressions*（O'Reilly and Associates, 2006）。

2.7.2 匹配字符串

正则表达式的最简单用法就是测试某个特定的字符串是否与它匹配。下面展示了如何用 Java 来编写这种测试，首先用表示正则表达式的字符串构建一个 Pattern 对象。然后从这个模式中获得一个 Matcher，并调用它的 matches 方法：

```
Pattern pattern = Pattern.compile(patternString);
Matcher matcher = pattern.matcher(input);
if (matcher.matches()) . . .
```

这个匹配器的输入可以是任何实现了 CharSequence 接口的类的对象，例如 String、StringBuilder 和 CharBuffer。

在编译这个模式时，可以设置一个或多个标志，例如：

```
Pattern pattern = Pattern.compile(expression,
    Pattern.CASE_INSENSITIVE + Pattern.UNICODE_CASE);
```

或者可以在模式中指定它们：

```
String regex = "(?iU:expression)";
```

下面是各个标志。

- Pattern.CASE_INSENSITIVE 或 i：匹配字符时忽略字母的大小写，默认情况下，这个标志只考虑 US ASCII 字符。
- Pattern.UNICODE_CASE 或 u：当与 CASE_INSENSITIVE 组合使用时，用 Unicode 字母的大小写来匹配。
- Pattern.UNICODE_CHARACTER_CLASS 或 U：选择 Unicode 字符类代替 POSIX，其中蕴含了 UNICODE_CASE。
- Pattern.MULTILINE 或 m：^ 和 $ 匹配行的开头和结尾，而不是整个输入的开头和结尾。
- Pattern.UNIX_LINES 或 d：在多行模式中匹配 ^ 和 $ 时，只有 '\n' 被识别成行终止符。
- Pattern.DOTALL 或 s：当使用这个标志时，. 符号匹配所有字符，包括行终止符。
- Pattern.COMMENTS 或 x：空白字符和注释（从 # 到行末尾）将被忽略。
- Pattern.LITERAL：该模式将被逐字地采纳，必须精确匹配，因字母大小写而造成的差异除外。
- Pattern.CANON_EQ：考虑 Unicode 字符规范的等价性，例如，u 后面跟随 ¨（分音符号）匹配 ü。

最后两个标志不能在正则表达式内部指定。

如果想要在集合或流中匹配元素，那么可以将模式转换为谓词：

```
Stream<String> strings = ...;
Stream<String> result = strings.filter(pattern.asPredicate());
```

其结果中包含了匹配正则表达式的所有字符串。

如果正则表达式包含群组，那么 Matcher 对象可以揭示群组的边界。下面的方法

```
int start(int groupIndex)
int end(int groupIndex)
```

将产生指定群组的开始索引和结尾之后的索引。

可以直接通过调用下面的方法抽取匹配的字符串：

```
String group(int groupIndex)
```

群组 0 是整个输入，而用于第一个实际群组的群组索引是 1。调用 groupCount 方法可以获得全部群组的数量。对于具名的组，使用下面的方法

```
int start(String groupName)
int end(String groupName)
String group(String groupName)
```

嵌套群组是按照前括号排序的，例如，假设我们有下面的模式

```
((([1-9]|1[0-2]):([0-5][0-9]))[ap]m
```

和下面的输出

11:59am

那么，匹配器会报告下面的群组：

群组索引	开始	结束	字符串
0	0	7	11:59am
1	0	5	11:59
2	0	2	11
3	3	5	59

程序清单 2-6 的程序提示输入一个模式，然后提示输入用于匹配的字符串，随后将打印出输入是否与模式相匹配。如果输入匹配模式，并且模式包含群组，那么这个程序将用括号打印出群组边界，例如

((11):(59))am

程序清单 2-6　regex/RegexTest.java

```java
package regex;

import java.util.*;
import java.util.regex.*;

/**
 * This program tests regular expression matching. Enter a pattern and strings to match,
 * or hit Cancel to exit. If the pattern contains groups, the group boundaries are displayed
 * in the match.
 * @version 1.03 2018-05-01
 * @author Cay Horstmann
 */
public class RegexTest
{
   public static void main(String[] args) throws PatternSyntaxException
   {
      var in = new Scanner(System.in);
      System.out.println("Enter pattern: ");
      String patternString = in.nextLine();

      Pattern pattern = Pattern.compile(patternString);

      while (true)
      {
         System.out.println("Enter string to match: ");
         String input = in.nextLine();
         if (input == null || input.equals("")) return;
         Matcher matcher = pattern.matcher(input);
         if (matcher.matches())
         {
            System.out.println("Match");
            int g = matcher.groupCount();
            if (g > 0)
            {
```

```
35              for (int i = 0; i < input.length(); i++)
36              {
37                 // Print any empty groups
38                 for (int j = 1; j <= g; j++)
39                    if (i == matcher.start(j) && i == matcher.end(j))
40                       System.out.print("()");
41                 // Print ( for non-empty groups starting here
42                 for (int j = 1; j <= g; j++)
43                    if (i == matcher.start(j) && i != matcher.end(j))
44                       System.out.print('(');
45                 System.out.print(input.charAt(i));
46                 // Print ) for non-empty groups ending here
47                 for (int j = 1; j <= g; j++)
48                    if (i + 1 != matcher.start(j) && i + 1 == matcher.end(j))
49                       System.out.print(')');
50              }
51              System.out.println();
52           }
53        }
54        else
55           System.out.println("No match");
56     }
57  }
58 }
```

2.7.3 找出多个匹配

通常，你不希望用正则表达式来匹配全部输入，而只是想找出输入中一个或多个匹配的子字符串。这时可以使用 Matcher 类的 find 方法来查找匹配内容，如果返回 true，再使用 start 和 end 方法来查找匹配的内容，或使用不带引元的 group 方法来获取匹配的字符串。

```
while (matcher.find())
{
   int start = matcher.start();
   int end = matcher.end();
   String match = input.group();
   ...
}
```

在这种方式中，可以依次处理每个匹配。正如上面的代码片段所示，可以获取匹配的字符串，以及它在输入字符串中的位置。

更优雅的是，可以调用 results 方法来获取一个 Stream<MatchResult>。MatchResult 接口有 group、start 和 end 方法，就像 Matcher 一样。（事实上，Matcher 类实现了这个接口。）下面展示了如何获取所有匹配的列表：

```
List<String> matches = pattern.matcher(input)
   .results()
   .map(Matcher::group)
   .collect(Collectors.toList());
```

如果要处理的是文件中的数据，那么可以使用 Scanner.findAll 方法来获取一个 Stream<Match-

Result>，这样就无须先将内容读取到一个字符串中。可以给 Scanner 传递一个 Pattern 或一个模式字符串：

```
var in = new Scanner(path, StandardCharsets.UTF_8);
Stream<String> words = in.findAll("\\pL+")
   .map(MatchResult::group);
```

程序清单 2-7 对这种机制进行了应用，它定位一个 Web 页面上的所有超文本引用，并打印它们。为了运行这个程序，需要在命令行中提供一个 URL，例如

```
java match.HrefMatch http://horstmann.com
```

程序清单 2-7　match/HrefMatch.java

```java
 1  package match;
 2
 3  import java.io.*;
 4  import java.net.*;
 5  import java.nio.charset.*;
 6  import java.util.regex.*;
 7
 8  /**
 9   * This program displays all URLs in a web page by matching a regular expression that
10   * describes the <a href=...> HTML tag. Start the program as <br>
11   * java match.HrefMatch URL
12   * @version 1.03 2018-03-19
13   * @author Cay Horstmann
14   */
15  public class HrefMatch
16  {
17     public static void main(String[] args)
18     {
19        try
20        {
21           // get URL string from command line or use default
22           String urlString;
23           if (args.length > 0) urlString = args[0];
24           else urlString = "http://openjdk.java.net/";
25
26           // read contents of URL
27           InputStream in = new URL(urlString).openStream();
28           var input = new String(in.readAllBytes(), StandardCharsets.UTF_8);
29
30           // search for all occurrences of pattern
31           var patternString = "<a\\s+href\\s*=\\s*(\"[^\"]*\"|[^\\s>]*)\\s*>";
32           Pattern pattern = Pattern.compile(patternString, Pattern.CASE_INSENSITIVE);
33           pattern.matcher(input)
34              .results()
35              .map(MatchResult::group)
36              .forEach(System.out::println);
37        }
38        catch (IOException | PatternSyntaxException e)
39        {
40           e.printStackTrace();
```

```
41     }
42   }
43 }
```

2.7.4 用分隔符来分割

有时，需要将输入按照匹配的分隔符断开，而其他部分保持不变。Pattern.split 方法可以自动完成这项任务。调用此方法后可以获得一个剔除分隔符之后的字符串数组：

```
String input = . . .;
Pattern commas = Pattern.compile("\\s*,\\s*");
String[] tokens = commas.split(input);
    // "1, 2, 3" turns into ["1", "2", "3"]
```

如果有多个标记，那么可以惰性地获取它们：

```
Stream<String> tokens = commas.splitAsStream(input);
```

如果不关心预编译模式和惰性获取，那么可以使用 String.split 方法：

```
String[] tokens = input.split("\\s*,\\s*");
```

如果输入数据在文件中，那么需要使用扫描器：

```
var in = new Scanner(path, StandardCharsets.UTF_8);
in.useDelimiter("\\s*,\\s*");
Stream<String> tokens = in.tokens();
```

2.7.5 替换匹配

Matcher 类的 replaceAll 方法将正则表达式出现的所有地方都用替换字符串来替换。例如，下面的指令将所有的数字序列都替换成 # 字符。

```
Pattern pattern = Pattern.compile("[0-9]+");
Matcher matcher = pattern.matcher(input);
String output = matcher.replaceAll("#");
```

替换字符串可以包含对模式中群组的引用：$n 表示替换成第 n 个群组，${*name*} 被替换为具有给定名字的组，因此我们需要用 \$ 来表示在替换文本中包含一个 $ 字符。

如果字符串中包含 $ 和 \，但是又不希望它们被解释成群组的替换符，那么就可以调用 matcher.replaceAll(Matcher.quoteReplacement(str))。

如果想要执行比按照群组匹配拼接更复杂的操作，可以提供一个替换函数而不是替换字符串。该函数接受一个 MatchResult 对象，并会产生一个字符串。例如，在下面的代码中，我们将所有单词都替换为至少 4 个字母转换为大写形式的版本：

```
String result = Pattern.compile("\\pL{4,}")
    .matcher("Mary had a little lamb")
    .replaceAll(m -> m.group().toUpperCase());
    // Yields "MARY had a LITTLE LAMB"
```

replaceFirst 方法将只替换模式的第一次出现。

API `java.util.regex.Pattern` 1.4

- static Pattern compile(String expression)
- static Pattern compile(String expression, int flags)

  把正则表达式字符串编译到一个用于快速处理匹配的模式对象中。flags 参数是 CASE_INSENSITIVE、UNICODE_CASE、MULTILINE、UNIX_LINES、DOTALL 和 CANON_EQ 标志中的一个。

- Matcher matcher(CharSequence input)

  返回一个 matcher 对象，你可以用它在输入中定位模式的匹配。

- String[] split(CharSequence input)
- String[] split(CharSequence input, int limit)
- Stream<String> splitAsStream(CharSequence input) 8

  将输入分割成标记，其中模式指定了分隔符的形式。返回标记数组，分隔符并非标记的一部分。第二种形式有一个名为 limit 的参数，表示所产生的字符串的最大数量。如果已经发现了 limit-1 个匹配的分隔符，那么返回的数组中的最后一项就包含所有剩余未分割的输入。如果 limit ≤ 0，那么整个输入都被分割；如果 limit 为 0，那么结尾的空字符串将不会置于返回的数组中。

API `java.util.regex.Matcher` 1.4

- boolean matches()

  如果输入匹配模式，则返回 true。

- boolean lookingAt()

  如果输入的开头匹配模式，则返回 true。

- boolean find()
- boolean find(int start)

  尝试查找下一个匹配，如果找到了另一个匹配，则返回 true。

- int start()
- int end()

  返回当前匹配的开始索引和结尾之后的索引位置。

- String group()

  返回当前的匹配。

- int groupCount()

  返回输入模式中的群组数量。

- int start(int groupIndex)
- int start(String name) 8
- int end(int groupIndex)
- int end(String name) 8

  返回当前匹配中给定群组的开始和结尾之后的位置。群组是由从 1 开始的索引指定的，或者用 0 表示整个匹配，或者用表示具名群组的字符串来指定。
- String group(int groupIndex)
- String group(String name) 7

  返回匹配给定群组的字符串。该群组是由从 1 开始的索引指定的，或者用 0 表示整个匹配，或者用表示具名群组的字符串来指定。

- String replaceAll(String replacement)
- String replaceFirst(String replacement)

  返回从匹配器输入获得的通过将所有匹配或第一个匹配用替换字符串替换之后的字符串。

  替换字符串可以包含用 $n 表示的对群组的引用，这时需要用 \$ 来表示字符串中包含一个 $ 符号。

- static String quoteReplacement(String str) 5.0

  引用 str 中的所有 \ 和 $。

- String replaceAll(Function<MatchResult,String> replacer) 9

  将每个匹配都替换为 replacer 函数应用于 MatchResult 上所产生的结果。

- Stream<MatchResult> results() 9

  产生一个包含所有匹配结果的流。

API java.util.regex.MatchResult 5

- String group()
- String group(int group)

  产生匹配的字符串，或者匹配给定群组的字符串。

- int start()
- int end()
- int start(int group)
- int end(int group)

  产生匹配字符串或匹配给定群组的字符串的开始与结尾的偏移量。

API java.util.Scanner 5.0

- Stream<MatchResult> findAll(Pattern pattern) 9

  产生一个流，其中包含了这个扫描器所产生的输入中针对给定模式的所有匹配。

你现在已经看到了在 Java 中输入输出操作是如何实现的，也对作为"新 I/O"规范一部分的正则表达式有了概略的了解。在下一章中，我们将转而研究对 XML 数据的处理。

第 3 章 XML

- ▲ XML 概述
- ▲ XML 文档的结构
- ▲ 解析 XML 文档
- ▲ 验证 XML 文档
- ▲ 使用 XPath 来定位信息
- ▲ 使用命名空间
- ▲ 流机制解析器
- ▲ 生成 XML 文档
- ▲ XSL 转换

Don Box 等人在其合著的 *Essential XML*（Addison-Wesley 出版社 2000 年出版）的前言中半开玩笑地说道："可扩展标记语言（Extensible Markup Language，XML）已经取代了 Java、设计模式、对象技术，成为软件行业解决世界饥荒的方案。"这种炒作早就不新鲜了，但是正如你将在本章中看到的，XML 是一种非常有用的描述结构化信息的技术。XML 工具使处理和转化信息变得十分容易。但是，XML 并不是万能药，我们需要领域相关的标准和代码库才能有效地使用 XML。此外，XML 非但没有使 Java 技术过时，还与 Java 配合得很好。从 20 世纪 90 年代末以来，IBM、Apache 和其他许多公司一直在帮助开发用于 XML 处理的高质量 Java 库，其中大部分重要的代码库都整合到了 Java 平台中。

本章将介绍 XML，并涵盖了 Java 库的 XML 特性。一如既往，我们将指出何时大量地使用 XML 是正确的；而何时必须有保留地使用 XML，通过利用良好的设计和代码，来采用老办法解决问题。

3.1 XML 概述

在卷 I 第 13 章中，你已经看见过用属性文件（property file）来描述程序配置。属性文件包含了一组名/值对，例如：

```
fontname=Times Roman
fontsize=12
windowsize=400 200
color=0 50 100
```

可以用 Properties 类在单个方法调用中读入这样的属性文件。这是一个很好的特性，但这还不够。在许多情况下，想要描述的信息的结构比较复杂，属性文件不能很方便地处理它。例如，对于下面例子中的 fontname/fontsize 项，使用以下的单一项将更符合面向对象的要求：

```
font=Times Roman 12
```

但是，这时对字体描述的解析就变得很讨厌了，必须确定字体名在何处结束，字体大小在何处开始。

属性文件采用的是一种单一的平面层次结构。你常常会看到程序员用如下的键名来努力解决这种局限性：

```
title.fontname=Helvetica
title.fontsize=36
body.fontname=Times Roman
body.fontsize=12
```

属性文件格式的另一个缺点是要求键是唯一的。如果要存放一个值序列，则需要另一个变通方法，例如：

```
menu.item.1=Times Roman
menu.item.2=Helvetica
menu.item.3=Goudy Old Style
```

XML 格式解决了这些问题，因为它能够表示层次结构，这比属性文件的平面表结构更灵活。

描述程序配置的 XML 文件可能会像这样：

```
<config>
   <entry id="title">
      <font>
         <name>Helvetica</name>
         <size>36</size>
      </font>
   </entry>
   <entry id="body">
      <font>
         <name>Times Roman</name>
         <size>12</size>
      </font>
   </entry>
   <entry id="background">
      <color>
         <red>0</red>
         <green>50</green>
         <blue>100</blue>
      </color>
   </entry>
</config>
```

XML 格式能够表达层次结构，并且重复的元素不会被曲解。

正如上面看到的，XML 文件的格式非常直观，它与 HTML 文件非常相似。这是有原因的，因为 XML 和 HTML 格式是古老的标准通用标记语言（Standard Generalized Markup Language，SGML）的衍生语言。

SGML 从 20 世纪 70 年代开始就用于描述复杂文件的结构。它的使用在一些要求对海量文献进行持续维护的产业中取得了成功，特别是在飞机制造业中。但是，SGML 相当复杂，所以它从未风行。造成 SGML 如此复杂的主要原因是 SGML 有两个相互矛盾的目标。它既想要确保文档能够根据其文档类型的规则来形成，又想要通过可以减少数据键入的快捷方式使数据项变得容易表示。XML 设计成了一个用于因特网的 SGML 的简化版本。和通常情况一样，越简单的东西越好，XML 立即得到了长期以来一直在躲避 SGML 的用户的热情追捧。

> **注释**：在 http://www.xml.com/axml/axml.html 处可以找到一个由 Tim Bray 注释的 XML 标准的极佳版本。

尽管 HTML 和 XML 同宗同源，但是两者之间存在着重要的区别：
- 与 HTML 不同，XML 是大小写敏感的。例如，`<H1>` 和 `<h1>` 是不同的 XML 标签。
- 在 HTML 中，如果从上下文中可以分清哪里是段落或列表项的结尾，那么结束标签（如 `</p>` 或 `</li>`）就可以省略，而在 XML 中结束标签绝对不能省略。
- 在 XML 中，只有单个标签而没有相对应的结束标签的元素必须以 / 结尾，比如 `<img src="coffeecup.png"/>`。这样，解析器就知道不需要查找 `</img>` 标签了。
- 在 XML 中，属性值必须用引号括起来。在 HTML 中，引号是可有可无的。例如，`<applet code="MyApplet.class" width=300 height=300>` 对 HTML 来说是合法的，但是对 XML 来说则是不合法的。在 XML 中，必须使用引号，比如，width = "300"。
- 在 HTML 中，属性名可以没有值。例如，`<input type="radio" name="language" value="Java" checked>`。在 XML 中，所有属性必须都有属性值。比如，checked= "true" 或 checked="checked"。

3.2 XML 文档的结构

XML 文档应当以一个文档头开始，例如：

```
<?xml version="1.0"?>
```

或者

```
<?xml version="1.0" encoding="UTF-8"?>
```

严格来说，文档头是可选的，但是强烈推荐使用文档头。

> **注释**：因为建立 SGML 是为了处理真正的文档，因此 XML 文件被称为文档，尽管许多 XML 文件是用来描述通常不被称作文档的数据集的。

文档头之后通常是文档类型定义（Document Type Definition，DTD），例如：

```
<!DOCTYPE web-app PUBLIC
    "-//Sun Microsystems, Inc.//DTD Web Application 2.2//EN"
    "http://java.sun.com/j2ee/dtds/web-app_2_2.dtd">
```

文档类型定义是确保文档正确的一个重要机制，但是它不是必需的。我们将在本章的后面讨论这个问题。

最后，XML 文档的正文包含根元素，根元素包含其他元素。例如：

```
<?xml version="1.0"?>
<!DOCTYPE config . . .>
<config>
```

```
<entry id="title">
   <font>
      <name>Helvetica</name>
      <size>36</size>
   </font>
</entry>
...
</config>
```

元素可以有子元素（child element）、文本或两者皆有。在上述例子中，font 元素有两个子元素，它们是 name 和 size。name 元素包含文本"Helvetica"。

> **提示**：在设计 XML 文档结构时，最好让元素要么包含子元素，要么包含文本。换句话说，你应该避免下面的情况：
> ```
> <font>
>    Helvetica
>    <size>36</size>
> </font>
> ```
> 在 XML 规范中，这叫作混合式内容（mixed content）。在本章中，稍后你将会看到，如果避免了混合式内容，就可以简化解析过程。

XML 元素可以包含属性，例如：

```
<size unit="pt">36</size>
```

何时用元素，何时用属性，在 XML 设计人员中存在一些分歧。例如，将 font 做如下描述：

```
<font name="Helvetica" size="36"/>
```

似乎比下面的描述更简单一些：

```
<font>
   <name>Helvetica</name>
   <size>36</size>
</font>
```

但是，属性的灵活性要差很多。假设你想把单位添加到 size 的值中去，如果使用属性，那么就必须把单位添加到属性值中去：

```
<font name="Helvetica" size="36 pt"/>
```

嗨！现在必须对字符串 "36 pt" 进行解析，而这正是 XML 被设计用来避免的那种麻烦。而向 size 元素中添加一个属性看起来会清晰得多：

```
<font>
   <name>Helvetica</name>
   <size unit="pt">36</size>
</font>
```

一条常用的经验法则是，属性只应该用来修改值的解释，而不是用来指定值。如果你发现自己陷入了争论，在纠结于某个设置是否是对某个值的解释所做的修改，那么你就应该对属性说"不",转而使用元素,许多有用的文档根本就不使用属性。

> 📘 **注释**:在 HTML 中,属性的使用规则很简单:凡是不显示在网页上的都是属性。例如在下面的超链接中:
>
> `<a href="http://java.sun.com">Java Technology</a>`
>
> 字符串 Java Technology 要在网页上显示,但是这个链接的 URL 并不是显示页面的一部分。然而,这个规则对于大多数 XML 并不那么管用,因为 XML 文件中的数据并非像通常意义那样是让人浏览的。

元素和文本是 XML 文档"主要的支撑要素",你可能还会遇到的其他一些标记,说明如下:

- 字符引用 (character reference) 的形式是 &# 十进制值;或 &#x 十六进制值;。例如,字符 é 可以用下面两种形式表示:

    `&#233; &#xE9;`

- 实体引用 (entity reference) 的形式是 &*name*;。下面这些实体引用:

    `&lt; &gt; &amp; &quot; &apos;`

    都有预定义的含义:小于、大于、&、引号、省略号等字符。还可以在 DTD 中定义其他的实体引用。

- CDATA 部分 (CDATA Section) 用 `<![CDATA[` 和 `]]>` 来限定其界限。它们是字符数据的一种特殊形式。可以使用它们来囊括那些含有 <、>、& 之类字符的字符串,而不必将它们解释为标记,例如:

    `<![CDATA[< & > are my favorite delimiters]]>`

    CDATA 部分不能包含字符串]]>。使用这一特性时要特别小心,因为它常用来当作将遗留数据偷偷纳入 XML 文档的一个后门。

- 处理指令 (processing instruction) 是那些专门在处理 XML 文档的应用程序中使用的指令,它们由 `<?` 和 `?>` 来限定其界限,例如:

    `<?xml-stylesheet href="mystyle.css" type="text/css"?>`

    每个 XML 都以一个处理指令开头:

    `<?xml version="1.0"?>`

- 注释 (comment) 用 `<!-` 和 `-->` 限定其界限,例如:

    `<!-- This is a comment. -->`

    注释不应该含有字符串 --。注释只能是给文档的读者提供的信息,其中绝不应该含有隐藏的命令,命令应该是用处理指令来实现。

3.3 解析 XML 文档

要处理 XML 文档,就要先解析(parse)它。解析器是这样一个程序:它读入一个文件,确认这个文件具有正确的格式，然后将其分解成各种元素，使得程序员能够访问这些元素。Java 库提供了两种 XML 解析器：
- 像文档对象模型（Document Object Model，DOM）解析器这样的树型解析器（tree parser），它们将读入的 XML 文档转换成树结构。
- 像 XML 简单 API（Simple API for XML，SAX）解析器这样的流机制解析器（streaming parser），它们在读入 XML 文档时生成相应的事件。

DOM 解析器对于实现我们的大多数目的来说都更容易一些，所以我们首先介绍它。如果要处理很长的文档，用它生成树结构将会消耗大量内存，或者如果只是对于某些元素感兴趣，而不关心它们的上下文，那么在这些情况下应该考虑使用流机制解析器。更多的信息可以查看 3.7 节。

DOM 解析器的接口已经被 W3C 标准化了。org.w3c.dom 包中包含了这些接口类型的定义，比如：Document 和 Element 等。不同的提供者，比如 Apache 组织和 IBM，都编写了实现这些接口的 DOM 解析器。Java XML 处理 API（Java API for XML Processing，JAXP）库使得我们实际上可以以插件形式使用这些解析器中的任意一个。但是 JDK 中也包含了从 Apache 解析器导出的 DOM 解析器。

要读入一个 XML 文档，首先需要一个 DocumentBuilder 对象，可以从 DocumentBuilder Factory 中得到这个对象，例如：

```
DocumentBuilderFactory factory = DocumentBuilderFactory.newInstance();
DocumentBuilder builder = factory.newDocumentBuilder();
```

现在，可以从文件中读入某个文档：

```
File f = . . .;
Document doc = builder.parse(f);
```

或者，可以用一个 URL：

```
URL u = . . .;
Document doc = builder.parse(u);
```

甚至可以指定一个任意的输入流：

```
InputStream in = . . .;
Document doc = builder.parse(in);
```

> **注释**：如果使用输入流作为输入源，那么对于那些以该文档的位置为相对路径而被引用的文档，解析器将无法定位，比如在同一个目录中的 DTD。但是，可以通过安装一个"实体解析器"（entity resolver）来解决这个问题。请查看 www.xml.com/pub/a/2004/03/03/catalogs.html 或 www.ibm.com/developerworks/xml/library/x-mxd3.html，以了解更多信息。

Document 对象是 XML 文档的树型结构在内存中的表示方式，它由实现了 Node 接口及其各种子接口的类的对象构成。图 3-1 显示了各个子接口的层次结构。

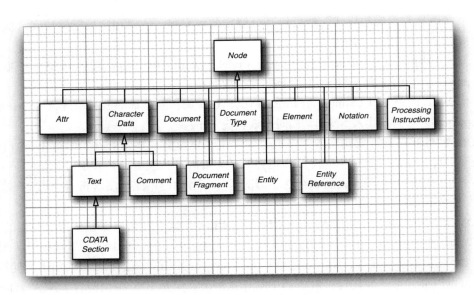

图 3-1 Node 接口及其子接口

可以通过调用 getDocumentElement 方法来启动对文档内容的分析，它将返回根元素。

```
Element root = doc.getDocumentElement();
```

例如，如果要处理下面的文档：

```
<?xml version="1.0"?>
<font>
   . . .
</font>
```

那么，调用 getDocumentElement 方法可以返回 font 元素。getTagName 方法可以返回元素的标签名。在前面这个例子中，root.getTagName() 返回字符串 "font"。

如果要得到该元素的子元素（可能是子元素、文本、注释或其他节点），请使用 getChildNodes 方法，这个方法会返回一个类型为 NodeList 的集合。这个类型在标准的 Java 集合类创建之前就已经被标准化了，因此它具有一种不同的访问协议；item 方法将得到指定索引值的项；getLength 方法则提供了项的总数。因此，我们可以像下面这样枚举所有子元素：

```
NodeList children = root.getChildNodes();
for (int i = 0; i < children.getLength(); i++)
{
   Node child = children.item(i);
   . . .
}
```

分析子元素时要很仔细。例如，假设你正在处理以下文档：

```
<font>
   <name>Helvetica</name>
   <size>36</size>
</font>
```

你预期 font 有两个子元素，但是解析器却报告说有 5 个：
- <font> 和 <name> 之间的空白字符
- name 元素
- </name> 和 <size> 之间的空白字符
- size 元素
- </size> 和 </font> 之间的空白字符

图 3-2 显示了其 DOM 树。

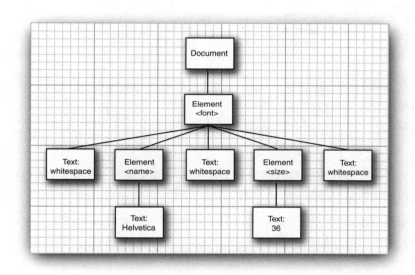

图 3-2　一棵简单的 DOM 树

如果只希望得到子元素，那么可以忽略空白字符：

```
for (int i = 0; i < children.getLength(); i++)
{
   Node child = children.item(i);
   if (child instanceof Element)
   {
      var childElement = (Element) child;
      . . .
   }
}
```

现在，只会看到两个元素，它们的标签名是 name 和 size。

正如将在下一节中所看到的那样，如果你的文档有 DTD，那么你就可以做得更好。这时，解析器知道哪些元素没有文本节点的子元素，而且它会帮你剔除空白字符。

在分析 name 和 size 元素时，你肯定想获取它们包含的文本字符串。这些文本字符串本身都包含在 Text 类型的子节点中。既然知道了这些 Text 节点是唯一的子元素，就可以用 getFirstChild 方法而不用再遍历另一个 NodeList。然后可以用 getData 方法获取存储在 Text 节点中的字符串。

```
for (int i = 0; i < children.getLength(); i++)
{
   Node child = children.item(i);
   if (child instanceof Element)
   {
      var childElement = (Element) child;
      var textNode = (Text) childElement.getFirstChild();
      String text = textNode.getData().trim();
      if (childElement.getTagName().equals("name"))
         name = text;
      else if (childElement.getTagName().equals("size"))
         size = Integer.parseInt(text);
   }
}
```

> **提示**：对 getData 的返回值调用 trim 方法是个好主意。如果 XML 文件的作者将起始和结束的标签放在不同的行上，例如：
>
> ```
> <size>
>    36
> </size>
> ```
>
> 那么，解析器将会把所有的换行符和空格都包含到文本节点中去。调用 trim 方法可以把位于实际数据前后的空白字符删掉。

也可以用 getLastChild 方法得到最后一项子元素，用 getNextSibling 得到下一个兄弟节点。这样，另一种遍历子节点集的方法就是：

```
for (Node childNode = element.getFirstChild();
     childNode != null;
     childNode = childNode.getNextSibling())
{
   ...
}
```

如果要枚举节点的属性，可以调用 getAttributes 方法。它返回一个 NamedNodeMap 对象，其中包含了描述属性的 Node 对象。可以用和遍历 NodeList 一样的方式在 NamedNodeMap 中遍历各子节点。然后，调用 getNodeName 和 getNodeValue 方法可以得到属性名和属性值。

```
NamedNodeMap attributes = element.getAttributes();
for (int i = 0; i < attributes.getLength(); i++)
{
   Node attribute = attributes.item(i);
   String name = attribute.getNodeName();
   String value = attribute.getNodeValue();
   ...
}
```

或者，如果知道属性名，则可以直接获取相应的属性值：

```
String unit = element.getAttribute("unit");
```

现在你已经知道怎么分析 DOM 树了。程序清单 3-1 中的程序将这些技术都运用了一遍，将一个 XML 文档转换成了 JSON 格式。

程序清单 3-1 dom/JSONConverter.java

```java
package dom;

import java.io.*;
import java.util.*;

import javax.xml.parsers.*;

import org.w3c.dom.*;
import org.w3c.dom.CharacterData;
import org.xml.sax.*;

/**
 * This program displays an XML document as a tree in JSON format.
 * @version 1.2 2018-04-02
 * @author Cay Horstmann
 */
public class JSONConverter
{
   public static void main(String[] args)
         throws SAXException, IOException, ParserConfigurationException
   {
      String filename;
      if (args.length == 0)
      {
         try (var in = new Scanner(System.in))
         {
            System.out.print("Input file: ");
            filename = in.nextLine();
         }
      }
      else
         filename = args[0];
      DocumentBuilderFactory factory = DocumentBuilderFactory.newInstance();
      DocumentBuilder builder = factory.newDocumentBuilder();

      Document doc = builder.parse(filename);
      Element root = doc.getDocumentElement();
      System.out.println(convert(root, 0));
   }

   public static StringBuilder convert(Node node, int level)
   {
      if (node instanceof Element)
      {
         return elementObject((Element) node, level);
      }
      else if (node instanceof CharacterData)
      {
         return characterString((CharacterData) node, level);
      }
      else
```

```java
53         return pad(new StringBuilder(), level).append(
54            jsonEscape(node.getClass().getName()));
55      }
56   }
57
58   private static Map<Character, String> replacements = Map.of('\b', "\\b", '\f', "\\f",
59      '\n', "\\n", '\r', "\\r", '\t', "\\t", '"', "\\\"", '\\', "\\\\");
60
61   private static StringBuilder jsonEscape(String str)
62   {
63      var result = new StringBuilder("\"");
64      for (int i = 0; i < str.length(); i++)
65      {
66         char ch = str.charAt(i);
67         String replacement = replacements.get(ch);
68         if (replacement == null) result.append(ch);
69         else result.append(replacement);
70      }
71      result.append("\"");
72      return result;
73   }
74
75   private static StringBuilder characterString(CharacterData node, int level)
76   {
77      var result = new StringBuilder();
78      StringBuilder data = jsonEscape(node.getData());
79      if (node instanceof Comment) data.insert(1, "Comment: ");
80      pad(result, level).append(data);
81      return result;
82   }
83
84   private static StringBuilder elementObject(Element elem, int level)
85   {
86      var result = new StringBuilder();
87      pad(result, level).append("{\n");
88      pad(result, level + 1).append("\"name\": ");
89      result.append(jsonEscape(elem.getTagName()));
90      NamedNodeMap attrs = elem.getAttributes();
91      if (attrs.getLength() > 0)
92      {
93         pad(result.append(",\n"), level + 1).append("\"attributes\": ");
94         result.append(attributeObject(attrs));
95      }
96      NodeList children = elem.getChildNodes();
97      if (children.getLength() > 0)
98      {
99         pad(result.append(",\n"), level + 1).append("\"children\": [\n");
100        for (int i = 0; i < children.getLength(); i++)
101        {
102           if (i > 0) result.append(",\n");
103           result.append(convert(children.item(i), level + 2));
104        }
105        result.append("\n");
106        pad(result, level + 1).append("]\n");
```

```
107         }
108         pad(result, level).append("}");
109         return result;
110     }
111
112     private static StringBuilder pad(StringBuilder builder, int level)
113     {
114         for (int i = 0; i < level; i++) builder.append("  ");
115         return builder;
116     }
117
118     private static StringBuilder attributeObject(NamedNodeMap attrs)
119     {
120         var result = new StringBuilder("{");
121         for (int i = 0; i < attrs.getLength(); i++)
122         {
123             if (i > 0) result.append(", ");
124             result.append(jsonEscape(attrs.item(i).getNodeName()));
125             result.append(": ");
126             result.append(jsonEscape(attrs.item(i).getNodeValue()));
127         }
128         result.append("}");
129         return result;
130     }
131 }
```

该树形结构清楚地显示了子元素是怎样被包含空白字符和注释的文本包围起来的。你可以清楚地看到换行符和回车符显示成 \n。

无须熟悉 JSON 就可以理解这个程序是如何操作 DOM 树的，你只需观察以下几点：

- 我们使用了一个 DocumentBuilder 来从文件中读取一个 Document。
- 对于每一个元素，我们打印了标签名、属性和元素。
- 对于字符数据，我们用这些数据产生了一个字符串。如果数据来自于注释，那么我们就会添加 "Comment:" 前缀。

API javax.xml.parsers.DocumentBuilderFactory 1.4

- static DocumentBuilderFactory newInstance()
  返回 DocumentBuilderFactory 类的一个实例。
- DocumentBuilder newDocumentBuilder()
  返回 DocumentBuilder 类的一个实例。

API javax.xml.parsers.DocumentBuilder 1.4

- Document parse(File f)
- Document parse(String url)
- Document parse(InputStream in)
  解析来自给定文件、URL 或输入流的 XML 文档，返回解析后的文档。

API *org.w3c.dom.Document* 1.4

- Element getDocumentElement()
  返回文档的根元素。

API *org.w3c.dom.Element* 1.4

- String getTagName()
  返回元素的名字。
- String getAttribute(String name)
  返回给定名字的属性值，没有该属性时返回空字符串。

API *org.w3c.dom.Node* 1.4

- NodeList getChildNodes()
  返回包含该节点所有子元素的节点列表。
- Node getFirstChild()
- Node getLastChild()
  获取该节点的第一个或最后一个子节点，在该节点没有子节点时返回 null。
- Node getNextSibling()
- Node getPreviousSibling()
  获取该节点的下一个或上一个兄弟节点，在该节点没有兄弟节点时返回 null。
- Node getParentNode()
  获取该节点的父节点，在该节点是文档节点时返回 null。
- NamedNodeMap getAttributes()
  返回含有描述该节点所有属性的 Attr 节点的映射表。
- String getNodeName()
  返回该节点的名字。当该节点是 Attr 节点时，该名字就是属性名。
- String getNodeValue()
  返回该节点的值。当该节点是 Attr 节点时，该值就是属性值。

API *org.w3c.dom.CharacterData* 1.4

- String getData()
  返回存储在节点中的文本。

API *org.w3c.dom.NodeList* 1.4

- int getLength()
  返回列表中的节点数。
- Node item(int index)
  返回给定索引值处的节点。索引值范围在 0 到 getLength()-1 之间。

API **org.w3c.dom.NamedNodeMap** 1.4

- int getLength()
  返回该节点映射表中的节点数。
- Node item(int index)
  返回给定索引值处的节点。索引值范围在 0 到 getLength()-1 之间。

3.4 验证 XML 文档

在前一节中，我们了解了如何遍历 DOM 文档的树形结构。然而，如果仅仅按照这种方法来操作，会发现需要大量冗长的编程和错误检查工作。不但需要处理元素间的空白字符，还要检查该文档包含的节点是否和期望的一样。例如，在读入下面这个元素时：

```
<font>
   <name>Helvetica</name>
   <size>36</size>
</font>
```

将首先得到第一个子节点，这是一个含有空白字符 "\n" 的文本节点。跳过文本节点找到第一个元素节点。然后，要检查它的标签名是不是 "name"，还要检查它是否有一个 Text 类型的子节点。接下来，转到下一个非空白字符的子节点，并进行同样的检查。那么，当文档作者改变了子元素的顺序或是加入另一个子元素时又会怎样呢？要是对所有的错误检查都进行编码，就会显得太琐碎麻烦了，而跳过这些检查又显得不慎重。

幸好，XML 解析器的一个很大的好处就是它能自动校验某个文档是否具有正确的结构。这样，解析就变得简单多了。例如，如果知道 font 片段已经通过了验证，那么不用进一步检查就能得到其两个孙节点，并把它们转换成 Text 节点，得到它们的文本数据。

如果要指定文档结构，可以提供一个文档类型定义（DTD）或一个 XML Schema 定义。DTD 或 schema 包含了用于解释文档应如何构成的规则，这些规则指定了每个元素的合法子元素和属性。例如，某个 DTD 可能含有一项规则：

```
<!ELEMENT font (name,size)>
```

这项规则表示，一个 font 元素必须总是有两个子元素，分别是 name 和 size。将同样的约束用 XML Schema 表示如下：

```
<xsd:element name="font">
   <xsd:sequence>
      <xsd:element name="name" type="xsd:string"/>
      <xsd:element name="size" type="xsd:int"/>
   </xsd:sequence>
</xsd:element>
```

与 DTD 相比，XML Schema 可以表达更加复杂的验证条件（比如 size 元素必须包含一个整数）。与 DTD 语法不同，XML Schema 自身使用的就是 XML，这为处理 Schema 文件带来了方便。

在下一节中，我们将详细讨论 DTD。接着简要介绍 XML Schema 的一些基础知识。最后，我们会展示一个完整的应用程序来演示验证是如何简化 XML 编程的。

3.4.1 文档类型定义

提供 DTD 的方式有多种。可以像下面这样将其纳入到 XML 文档中：

```
<?xml version="1.0"?>
<!DOCTYPE config [
   <!ELEMENT config . . .>
   more rules
   . . .
]>
<config>
   . . .
</config>
```

正如你看到的，这些规则被纳入到 DOCTYPE 声明中，位于由 [...] 限定界限的块中。文档类型必须匹配根元素的名字，比如我们例子中的 configuration。

在 XML 文档内部提供 DTD 不是很普遍，因为 DTD 会使文件长度变得很长。把 DTD 存储在外部会更具意义，SYSTEM 声明可以用来实现这个目标。可以指定一个包含 DTD 的 URL，例如：

```
<!DOCTYPE config SYSTEM "config.dtd">
```

或者

```
<!DOCTYPE config SYSTEM "http://myserver.com/config.dtd">
```

> ⚠️ **警告**：如果使用的是 DTD 的相对 URL（比如 "config.dtd"），那么要给解析器一个 File 或 URL 对象，而不是 InputStream。如果必须从一个输入流来解析，那么请提供一个实体解析器（请看下面的说明）。

最后，有一个来源于 SGML 的用于识别"众所周知的" DTD 的机制，下面是一个例子：

```
<!DOCTYPE web-app
   PUBLIC "-//Sun Microsystems, Inc.//DTD Web Application 2.2//EN"
   "http://java.sun.com/j2ee/dtds/web-app_2_2.dtd">
```

如果 XML 处理器知道如何定位带有公共标识符的 DTD，那么就不需要 URL 了。

> 📝 **注释**：DTD 的系统标识符 URL 可能实际无法工作，或者会显著地降低性能。后者有一个例子，即 XHTML1.0 Strict DTD 的系统标识符，可以在 http://www.w3.org/TR/xhtml1/DTD/xhtml1-strict.dtd 处找到它的信息。如果要解析一个 XHTML 文件，可能会花费一两分钟来处理 DTD。
>
> 有一种解决方案是使用**实体解析器**，它会将公共标识符映射为本地文件。在 Java 9 之前，我们不得不提供一个实现了 EntityResolver 接口并实现了 resolveEntity 方法的某个类的对象。

但是，现在我们可以使用 XML 目录来管理这种映射。我们需要提供一个或多个具有下面这种形式的目录文件：

```xml
<?xml version="1.0"?>
<!DOCTYPE catalog PUBLIC "-//OASIS//DTD XML Catalogs V1.0//EN"
    "http://www.oasis-open.org/committees/entity/release/1.0/catalog.dtd">
<catalog xmlns="urn:oasis:names:tc:entity:xmlns:xml:catalog" prefer="public">
  <public publicId="..." uri="..."/>
  ...
</catalog>
```

然后像下面这样构造和安装一个解析器：

```
builder.setEntityResolver(CatalogManager.catalogResolver(
    CatalogFeatures.defaults(),
    Paths.get("catalog.xml").toAbsolutePath().toUri()));
```

请参阅程序清单 3.6 中完整的示例。

除了在程序中设置目录文件的位置，还可以在命令行中用 javax.xml.catalog.files 系统属性来设置它，我们需要提供由分号分隔的 file 的绝对 URL。

既然你已经知道解析器怎样定位 DTD 了，那么下面就让我们来看看不同种类的规则。

ELEMENT 规则用于指定某个元素可以拥有什么样的子元素。可以指定一个正则表达式，它由表 3-1 中所示的组成部分构成。

表 3-1　用于元素内容的规则

规则	含义
$E*$	0 或多个 E
$E+$	1 或多个 E
$E?$	0 或 1 个 E
$E_1\|E_2\|...\|E_n$	$E_1, E_2, ..., E_n$ 中的一个
$E_1, E_2, ..., E_n$	E_1 后面跟着 $E_2, ..., E_n$
#PCDATA	文本
(#PCDATA$\|E_1\|E_2\|...\|E_n$)*	0 或多个文本且 $E_1, E_2, ..., E_n$ 以任意顺序排列（混合式内容）
ANY	允许有任意子元素
EMPTY	不允许有子元素

下面是一些简单而典型的例子。下面的规则声明了 menu 元素包含 0 或多个 item 元素：

```
<!ELEMENT menu (item)*>
```

下面这组规则声明 font 是用一个 name 后跟一个 size 来描述的，它们都包含文本：

```
<!ELEMENT font (name,size)>
<!ELEMENT name (#PCDATA)>
<!ELEMENT size (#PCDATA)>
```

缩写 PCDATA 表示被解析的字符数据。这些数据之所以被称为"被解析的"是因为解析器通过寻找表示一个新标签起始的 < 字符或表示一个实体起始的 & 字符，来解释这些文本字符串。

元素的规格说明可以包含嵌套的和复杂的正则表达式，例如，下面是一个描述了本书中每一章的结构的规则：

```
<!ELEMENT chapter (intro,(heading,(para|image|table|note)+)+)
```

每章都以简介开头，其后是 1 或多个小节，每个小节由一个标题和 1 个或多个段落、图片、表格或说明构成。

然而，有一种常见的情况是无法把规则定义得像你希望的那样灵活的。当一个元素可以包含文本时，那么就只有两种合法的情况。要么该元素只包含文本，比如：

```
<!ELEMENT name (#PCDATA)>
```

要么该元素包含任意顺序的文本和标签的组合，比如：

```
<!ELEMENT para (#PCDATA|em|strong|code)*>
```

指定其他任何类型的包含 #PCDATA 的规则都是不合法的。例如，以下规则是非法的：

```
<!ELEMENT captionedImage (image,#PCDATA)>
```

必须重写这项规则，以引入另一个 caption 元素或者允许使用 image 元素和文本的任意组合。

这种限制简化了 XML 解析器在解析混合式内容（标签和文本的混合）时的工作。因为在允许使用混合式内容时难免会失控，所以最好在设计 DTD 时，让其中所有的元素要么包含其他元素，要么只有文本。

> **注释**：实际上，在 DTD 规则中并不能为元素指定任意的正则表达式，XML 解析器会拒绝某些导致非确定性的复杂规则。例如，正则表达式 ((x,y)|(x,z)) 就是非确定性的。当解析器看到 x 时，它不知道在两个选择中应该选取哪一个。这个表达式可以改写成确定性的形式，如 (x,(y|z))。然而，有一些表达式不能被改写，如 ((x,y)*|x?)。Java XML 库中的解析器在遇到有歧义的 DTD 时，不会给出警告。在解析时，它仅仅在两者中选取第一个匹配项，这将导致它会拒绝一些正确的输入。当然，解析器有权这么做，因为 XML 标准允许解析器假设 DTD 都是非二义性的。

还可以指定描述合法的元素属性的规则，其通用语法为：

```
<!ATTLIST element attribute type default>
```

表 3-2 显示了合法的属性类型（type），表 3-3 显示了属性默认值（default）的语法。

表 3-2　属性类型

类型	含义
CDATA	任意字符串
$(A_1\|A_2\|\ldots\|A_n)$	字符串属性 $A_1, A_2, \ldots, A_n$ 之一
NMTOKEN, NMTOKENS	1 或多个名字标记
ID	1 个唯一的 ID
IDREF, IDREFS	1 或多个对唯一 ID 的引用
ENTITY, ENTITIES	1 或多个未解析的实体

表 3-3 属性的默认值

默认值	含义
#REQUIRED	属性是必需的
#IMPLIED	属性是可选的
A	属性是可选的；若未指定，解析器报告的属性是 A
#FIXED A	属性必须是未指定的或者是 A；在这两种情况下，解析器报告的属性都是 A

以下是两个典型的属性规格说明：

```
<!ATTLIST font style (plain|bold|italic|bold-italic) "plain">
<!ATTLIST size unit CDATA #IMPLIED>
```

第一个规格说明描述了 font 元素的 style 属性。它有 4 个合法的属性值，默认值是 plain。第二个规格说明表示 size 元素的 unit 属性可以包含任意的字符数据序列。

> **注释**：一般情况下，我们推荐用元素而非属性来描述数据。按照这个推荐，font style 应该是一个独立的元素，例如 `<font><style>plain</style>...</font>`。然而，对于枚举类型，属性有一个不可否认的优点，那就是解析器能够校验其取值是否合法。例如，如果 font style 是一个属性，那么解析器就会检查它是不是 4 个允许的值之一，并且如果没有为其提供属性值，那么解析器还会为其提供一个默认值。

CDATA 属性值的处理与前面看到的对 #PCDATA 的处理有着微妙的差别，并且与 `<![CDATA[...]]>` 部分没有多大关系。属性值首先被规范化，也就是说，解析器要先处理对字符和实体的引用（比如 é 或 <），并且要用空格来替换空白字符。

NMTOKEN（即名字标记）与 CDATA 相似，但是大多数非字母数字字符和内部的空白字符是不允许使用的，而且解析器会删除起始和结尾的空白字符。NMTOKENS 是一个以空白字符分隔的名字标记列表。

ID 结构是很有用的，ID 是在文档中必须唯一的名字标记，解析器会检查其唯一性。在下一个示例程序中，你会看到它的应用。IDREF 是对同一文档中已存在的 ID 的引用，解析器也会对它进行检查。IDREFS 是以空白字符分隔的 ID 引用的列表。

ENTITY 属性值将引用一个"未解析的外部实体"。这是从 SGML 那里沿用下来的，在实际应用中很少见到。在 http://www.xml.com/axml/axml.html 处的被注解的 XML 规范中有该属性的一个例子。

DTD 也可以定义实体，或者定义解析过程中被替换的缩写。你可以在 Firefox 浏览器的用户界面描述中找到一个很好的使用实体的例子。这些描述被格式化为 XML 格式，包含了如下的实体定义：

```
<!ENTITY back.label "Back">
```

其他地方的文本可以包含对这个实体的引用，例如：

```
<menuitem label="&back.label;"/>
```

解析器会用替代字符串来替换该实体引用。如果要对应用程序进行国际化处理，只需修改实体定义中的字符串即可。其他的实体使用方法更加复杂,且不太常用,详细说明参见 XML 规范。

这样我们就结束了对 DTD 的介绍。既然你已经知道如何使用 DTD 了,那么你就可以配置你的解析器以充分利用它们了。首先,通知文档生成工厂打开验证特性。

```
factory.setValidating(true);
```

这样,该工厂生成的所有文档生成器都将根据 DTD 来验证它们的输入。验证的最大好处是可以忽略元素内容中的空白字符。例如,考虑下面的 XML 代码片段:

```
<font>
    <name>Helvetica</name>
    <size>36</size>
</font>
```

一个不进行验证的解析器会报告 font、name 和 size 元素之间的空白字符,因为它无法知道 font 的子元素是:

```
(name,size)
(#PCDATA,name,size)*
```

还是:

```
ANY
```

一旦 DTD 指定了子元素是(name,size),解析器就知道它们之间的空白字符不是文本。调用下面的代码:

```
factory.setIgnoringElementContentWhitespace(true);
```

这样,生成器将不会报告文本节点中的空白字符。这意味着,你可以依赖 font 节点拥有 2 个子元素这一事实。你再也不用编写下面这样的单调冗长的循环代码了:

```
for (int i = 0; i < children.getLength(); i++)
{
    Node child = children.item(i);
    if (child instanceof Element)
    {
        var childElement = (Element) child;
        if (childElement.getTagName().equals("name")) . . .;
        else if (childElement.getTagName().equals("size")) . . .;
    }
}
```

而只需仅仅通过如下代码访问第一个和第二个子元素:

```
var nameElement = (Element) children.item(0);
var sizeElement = (Element) children.item(1);
```

这就是 DTD 如此有用的原因。你不会为了检查规则而使程序负担过重。在得到文档之前,解析器已经做完了这些工作。

当解析器报告错误时,应用程序希望对该错误执行某些操作。例如,记录到日志中,把它显示给用户,或是抛出一个异常以放弃解析。因此,只要使用验证,就应该安装一个错误处理器，这需要提供一个实现了 ErrorHandler 接口的对象。这个接口有三个方法：

```
void warning(SAXParseException exception)
void error(SAXParseException exception)
void fatalError(SAXParseException exception)
```

可以通过 DocumentBuilder 类的 setErrorHandler 方法来安装错误处理器：

```
builder.setErrorHandler(handler);
```

API javax.xml.parsers.DocumentBuilder 1.4

- void setEntityResolver(EntityResolver resolver)
  设置解析器，来定位要解析的 XML 文档中引用的实体。
- void setErrorHandler(ErrorHandler handler)
  设置用来报告在解析过程中出现的错误和警告的处理器。

API org.xml.sax.EntityResolver 1.4

- public InputSource resolveEntity(String publicID, String systemID)
  返回一个输入源，它包含了被给定 ID 所引用的数据，或者，当解析器不知道如何解析这个特定名字时，返回 null。如果没有提供公共 ID，那么参数 publicID 可以为 null。

API org.xml.sax.InputSource 1.4

- InputSource(InputStream in)
- InputSource(Reader in)
- InputSource(String systemID)
  从流、读入器或系统 ID（通常是相对或绝对 URL）中构建输入源。

API org.xml.sax.ErrorHandler 1.4

- void fatalError(SAXParseException exception)
- void error(SAXParseException exception)
- void warning(SAXParseException exception)
  覆盖这些方法以提供对致命错误、非致命错误和警告进行处理的处理器。

API org.xml.sax.SAXParseException 1.4

- int getLineNumber()
- int getColumnNumber()
  返回引起异常的已处理的输入信息末尾的行号和列号。

API javax.xml.catalog.CatalogManager 9

- static CatalogResolver catalogResolver(CatalogFeatures features, URI... uris)
  产生一个解析器，它将使用由所提供的 URI 指定的位置上的目录文件。这个类实现了 EntityResolver 接口，StAX、Schema 校验和 XSL 转换用到的类也实现了该接口。

API javax.xml.catalog.CatalogFeatures 9

- static CatalogFeatures defaults()
  用默认设置产生一个实例。

API javax.xml.parsers.DocumentBuilderFactory 1.4

- boolean isValidating()
- void setValidating(boolean value)
  获取和设置工厂的 validating 属性。当它设为 true 时，该工厂生成的解析器会验证它们的输入信息。
- boolean isIgnoringElementContentWhitespace()
- void setIgnoringElementContentWhitespace(boolean value)
  获取和设置工厂的 ignoringElementContentWhitespace 属性。当它设为 true 时，该工厂生成的解析器会忽略不含混合内容（即，元素与 #PCDATA 混合）的元素节点之间的空白字符。

3.4.2 XML Schema

因为 XML Schema 比起 DTD 语法要复杂许多，所以我们只涉及其基本知识。更多信息请参考 http://www.w3.org/TR/xmlschema-0 上的指南。

如果要在文档中引用 Schema 文件，需要在根元素中添加属性，例如：

```
<?xml version="1.0"?>
<config xmlns:xsi="http://www.w3.org/2001/XMLSchema-instance"
    xsi:noNamespaceSchemaLocation="config.xsd">
    ...
</config>
```

这个声明说明 Schema 文件 config.xsd 会被用来验证该文档。如果使用命名空间，语法就更加复杂了。详情请参见 XML Schema 指南（前缀 xsi 是一个命名空间别名（namespace alias），请查看 3.6 节以了解更多信息）。

Schema 为每个元素和属性都定义了类型。类型中的简单类型是对内容有限制的字符串，其他都是复杂类型。具有简单类型的元素可以没有任何属性和子元素。否则，它就必然是复杂类型。与此相反，属性总是简单类型。

一些简单类型已经被内建到了 XML Schema 内，包括：

```
xsd:string
xsd:int
xsd:boolean
```

> 注释：我们用前缀 xsd: 来表示 XSL Schema 定义的命名空间。一些作者代之以 xs:。

可以定义自己的简单类型。例如，下面是一个枚举类型：

```
<xsd:simpleType name="StyleType">
    <xsd:restriction base="xsd:string">
        <xsd:enumeration value="PLAIN" />
```

```
        <xsd:enumeration value="BOLD" />
        <xsd:enumeration value="ITALIC" />
        <xsd:enumeration value="BOLD_ITALIC" />
    </xsd:restriction>
</xsd:simpleType>
```

当定义元素时,要指定它的类型:

```
<xsd:element name="name" type="xsd:string"/>
<xsd:element name="size" type="xsd:int"/>
<xsd:element name="style" type="StyleType"/>
```

类型约束了元素的内容。例如,下面的元素将被验证为具有正确格式:

```
<size>10</size>
<style>PLAIN</style>
```

但是,下面的元素会被解析器拒绝:

```
<size>default</size>
<style>SLANTED</style>
```

可以把类型组合成复杂类型,例如:

```
<xsd:complexType name="FontType">
    <xsd:sequence>
        <xsd:element ref="name"/>
        <xsd:element ref="size"/>
        <xsd:element ref="style"/>
    </xsd:sequence>
</xsd:complexType>
```

FontType 是 name、size 和 style 元素的序列。在这个类型定义中,我们使用了 ref 属性来引用在 Schema 中位于别处的定义。也可以嵌套定义,像这样:

```
<xsd:complexType name="FontType">
    <xsd:sequence>
        <xsd:element name="name" type="xsd:string"/>
        <xsd:element name="size" type="xsd:int"/>
        <xsd:element name="style">
            <xsd:simpleType>
                <xsd:restriction base="xsd:string">
                    <xsd:enumeration value="PLAIN" />
                    <xsd:enumeration value="BOLD" />
                    <xsd:enumeration value="ITALIC" />
                    <xsd:enumeration value="BOLD_ITALIC" />
                </xsd:restriction>
            </xsd:simpleType>
        </xsd:element>
    </xsd:sequence>
</xsd:complexType>
```

请注意 style 元素的匿名类型定义。

xsd:sequence 结构和 DTD 中的连接符号等价,而 xsd:choice 结构和 | 操作符等价,例如:

```
<xsd:complexType name="contactinfo">
    <xsd:choice>
        <xsd:element ref="email"/>
```

```
        <xsd:element ref="phone"/>
    </xsd:choice>
</xsd:complexType>
```

这和 DTD 中的类型 email|phone 类型是等价的。

如果要允许重复元素，可以使用 minoccurs 和 maxoccurs 属性，例如，与 DTD 类型 item* 等价的形式如下：

```
<xsd:element name="item" type="..." minoccurs="0" maxoccurs="unbounded">
```

如果要指定属性，可以把 xsd:attribute 元素添加到 complexType 定义中去：

```
<xsd:element name="size">
    <xsd:complexType>
    ...
        <xsd:attribute name="unit" type="xsd:string" use="optional" default="cm"/>
    </xsd:complexType>
</xsd:element>
```

这与下面的 DTD 语句等价：

```
<!ATTLIST size unit CDATA #IMPLIED "cm">
```

可以把 Schema 的元素和类型定义封装在 xsd:schema 元素中：

```
<xsd:schema xmlns:xsd="http://www.w3.org/2001/XMLSchema">
    ...
</xsd:schema>
```

解析带有 Schema 的 XML 文件和解析带有 DTD 的文件相似，但有 2 点差别：

1. 必须打开对命名空间的支持，即使在 XML 文件里可能不会用到它。

```
factory.setNamespaceAware(true);
```

2. 必须通过如下的"魔咒"来准备好处理 Schema 的工厂。

```
final String JAXP_SCHEMA_LANGUAGE =
    "http://java.sun.com/xml/jaxp/properties/schemaLanguage";
final String W3C_XML_SCHEMA = "http://www.w3.org/2001/XMLSchema";
factory.setAttribute(JAXP_SCHEMA_LANGUAGE, W3C_XML_SCHEMA);
```

3.4.3 一个实践示例

在本节中，我们将要介绍一个实用的示例程序，用来说明在实际环境中 XML 的用法。

假设有一个应用程序需要配置数据，这些数据可以指定任意对象，而不只是文本字符串。我们提供了两种机制来实例化对象：使用构造器和使用工厂方法。下面展示了如何使用构造器来创建 Color 对象：

```
<construct class="java.awt.Color">
    <int>55</int>
    <int>200</int>
    <int>100</int>
</construct>
```

下面是使用工厂方法的例子：

```
<factory class="java.util.logging.Logger" method="getLogger">
    <string>com.horstmann.corejava</string>
</factory>
```

如果忽略工厂方法名，那么其默认值就是 getInstance。

正如你所见，有多个元素用来描述字符串和整数。我们还支持 boolean 类型，其他基本类型也都可以按照相同的方式添加进来。

只是为了显摆一下，我们给出了第二种针对基本类型的机制：

```
<value type="int">30</value>
```

配置是由多个项构成的序列。每一项都有一个 ID 和一个对象：

```
<config>
    <entry id="background">
        <construct class="java.awt.Color">
            <value type="int">55</value>
            <value type="int">200</value>
            <value type="int">100</value>
        </construct>
    </entry>
    ...
</config>
```

解析器会检查这些 ID 是否唯一。

DTD 显示在程序清单 3-4 中，很简单。

程序清单 3-5 包含了一个等价的 Schema。在这个 Schema 中，我们可以提供额外的检查：一个 int 或 boolean 元素只能包含整数或布尔值。注意，这里使用了 xsd:group 结构来定义会反复使用的复杂类型的各个部件。

程序清单 3-2 中的程序展示了如何解析配置文件。程序清单 3-3 中定义配置样例。

如果选择了包含字符串 -Schema 的文件，那么该程序使用 Schema 而不是 DTD。

如果选择了包含字符串 -Schema 的文件，那么该程序除了 DTD，还可以处理 Schema。

这个例子是 XML 的典型用法。XML 格式十分健壮，足以表达复杂的关系。在此基础上，通过接管有效性检查和提供默认值等例行工作，XML 解析器添加了新的价值。

程序清单 3-2　read/XML ReadTest.java

```java
1  package read;
2
3  import java.io.*;
4  import java.lang.reflect.*;
5  import java.util.*;
6
7  import javax.xml.parsers.*;
8
9  import org.w3c.dom.*;
10 import org.xml.sax.*;
11
12 /**
13   * This program shows how to use an XML file to describe Java objects
```

```java
14   * @version 1.0 2018-04-03
15   * @author Cay Horstmann
16   */
17  public class XMLReadTest
18  {
19     public static void main(String[] args) throws ParserConfigurationException,
20           SAXException, IOException, ReflectiveOperationException
21     {
22        String filename;
23        if (args.length == 0)
24        {
25           try (var in = new Scanner(System.in))
26           {
27              System.out.print("Input file: ");
28              filename = in.nextLine();
29           }
30        }
31        else
32           filename = args[0];
33
34        DocumentBuilderFactory factory = DocumentBuilderFactory.newInstance();
35        factory.setValidating(true);
36
37        if (filename.contains("-schema"))
38        {
39           factory.setNamespaceAware(true);
40           final String JAXP_SCHEMA_LANGUAGE =
41                 "http://java.sun.com/xml/jaxp/properties/schemaLanguage";
42           final String W3C_XML_SCHEMA = "http://www.w3.org/2001/XMLSchema";
43           factory.setAttribute(JAXP_SCHEMA_LANGUAGE, W3C_XML_SCHEMA);
44        }
45
46        factory.setIgnoringElementContentWhitespace(true);
47
48        DocumentBuilder builder = factory.newDocumentBuilder();
49
50        builder.setErrorHandler(new ErrorHandler()
51           {
52              public void warning(SAXParseException e) throws SAXException
53              {
54                 System.err.println("Warning: " + e.getMessage());
55              }
56
57              public void error(SAXParseException e) throws SAXException
58              {
59                 System.err.println("Error: " + e.getMessage());
60                 System.exit(0);
61              }
62
63              public void fatalError(SAXParseException e) throws SAXException
64              {
65                 System.err.println("Fatal error: " + e.getMessage());
66                 System.exit(0);
67              }
```

```java
 68            });
 69
 70        Document doc = builder.parse(filename);
 71        Map<String, Object> config = parseConfig(doc.getDocumentElement());
 72        System.out.println(config);
 73    }
 74
 75    private static Map<String, Object> parseConfig(Element e)
 76            throws ReflectiveOperationException
 77    {
 78        var result = new HashMap<String, Object>();
 79        NodeList children = e.getChildNodes();
 80        for (int i = 0; i < children.getLength(); i++)
 81        {
 82            var child = (Element) children.item(i);
 83            String name = child.getAttribute("id");
 84            Object value = parseObject((Element) child.getFirstChild());
 85            result.put(name, value);
 86        }
 87        return result;
 88    }
 89
 90    private static Object parseObject(Element e)
 91            throws ReflectiveOperationException
 92    {
 93        String tagName = e.getTagName();
 94        if (tagName.equals("factory")) return parseFactory(e);
 95        else if (tagName.equals("construct")) return parseConstruct(e);
 96        else
 97        {
 98            String childData = ((CharacterData) e.getFirstChild()).getData();
 99            if (tagName.equals("int"))
100                return Integer.valueOf(childData);
101            else if (tagName.equals("boolean"))
102                return Boolean.valueOf(childData);
103            else
104                return childData;
105        }
106    }
107
108    private static Object parseFactory(Element e)
109            throws ReflectiveOperationException
110    {
111        String className = e.getAttribute("class");
112        String methodName = e.getAttribute("method");
113        Object[] args = parseArgs(e.getChildNodes());
114        Class<?>[] parameterTypes = getParameterTypes(args);
115        Method method = Class.forName(className).getMethod(methodName, parameterTypes);
116        return method.invoke(null, args);
117    }
118
119    private static Object parseConstruct(Element e)
120            throws ReflectiveOperationException
121    {
```

```java
122         String className = e.getAttribute("class");
123         Object[] args = parseArgs(e.getChildNodes());
124         Class<?>[] parameterTypes = getParameterTypes(args);
125         Constructor<?> constructor = Class.forName(className).getConstructor(parameterTypes);
126         return constructor.newInstance(args);
127     }
128
129     private static Object[] parseArgs(NodeList elements)
130             throws ReflectiveOperationException
131     {
132         var result = new Object[elements.getLength()];
133         for (int i = 0; i < result.length; i++)
134             result[i] = parseObject((Element) elements.item(i));
135         return result;
136     }
137
138     private static Map<Class<?>, Class<?>> toPrimitive = Map.of(
139         Integer.class, int.class,
140         Boolean.class, boolean.class);
141
142     private static Class<?>[] getParameterTypes(Object[] args)
143     {
144         var result = new Class<?>[args.length];
145         for (int i = 0; i < result.length; i++)
146         {
147             Class<?> cl = args[i].getClass();
148             result[i] = toPrimitive.get(cl);
149             if (result[i] == null) result[i] = cl;
150         }
151         return result;
152     }
153 }
```

程序清单 3-3　read/config.xml

```xml
 1 <?xml version="1.0"?>
 2 <!DOCTYPE config SYSTEM "config.dtd">
 3 <config>
 4   <entry id="background">
 5     <construct class="java.awt.Color">
 6       <int>55</int>
 7       <int>200</int>
 8       <int>100</int>
 9     </construct>
10   </entry>
11   <entry id="currency">
12     <factory class="java.util.Currency">
13       <string>USD</string>
14     </factory>
15   </entry>
16 </config>
```

程序清单 3-4　read/config.dtd

```
1  <!ELEMENT config (entry)*>
2
3  <!ELEMENT entry (string|int|boolean|construct|factory)>
4  <!ATTLIST entry id ID #IMPLIED>
5
6  <!ELEMENT construct (string|int|boolean|construct|factory)*>
7  <!ATTLIST construct class CDATA #IMPLIED>
8
9  <!ELEMENT factory (string|int|boolean|construct|factory)*>
10 <!ATTLIST factory class CDATA #IMPLIED>
11 <!ATTLIST factory method CDATA "getInstance">
12
13 <!ELEMENT string (#PCDATA)>
14 <!ELEMENT int (#PCDATA)>
15 <!ELEMENT boolean (#PCDATA)>
```

程序清单 3-5　read/config.xsd

```
1  <xsd:schema xmlns:xsd="http://www.w3.org/2001/XMLSchema">
2    <xsd:element name="config">
3      <xsd:complexType>
4        <xsd:sequence>
5          <xsd:element name="entry" minOccurs="0" maxOccurs="unbounded">
6            <xsd:complexType>
7              <xsd:group ref="Object"/>
8              <xsd:attribute name="id" type="xsd:ID"/>
9            </xsd:complexType>
10         </xsd:element>
11       </xsd:sequence>
12     </xsd:complexType>
13   </xsd:element>
14
15   <xsd:element name="construct">
16     <xsd:complexType>
17       <xsd:group ref="Arguments"/>
18       <xsd:attribute name="class" type="xsd:string"/>
19     </xsd:complexType>
20   </xsd:element>
21
22   <xsd:element name="factory">
23     <xsd:complexType>
24       <xsd:group ref="Arguments"/>
25       <xsd:attribute name="class" type="xsd:string"/>
26       <xsd:attribute name="method" type="xsd:string" default="getInstance"/>
27     </xsd:complexType>
28   </xsd:element>
29
30   <xsd:group name="Object">
31     <xsd:choice>
32       <xsd:element ref="construct"/>
33       <xsd:element ref="factory"/>
```

```
34          <xsd:element name="string" type="xsd:string"/>
35          <xsd:element name="int" type="xsd:int"/>
36          <xsd:element name="boolean" type="xsd:boolean"/>
37      </xsd:choice>
38  </xsd:group>
39
40  <xsd:group name="Arguments">
41      <xsd:sequence>
42          <xsd:group ref="Object" minOccurs="0" maxOccurs="unbounded"/>
43      </xsd:sequence>
44  </xsd:group>
45  </xsd:schema>
```

3.5 使用 XPath 来定位信息

如果要定位某个 XML 文档中的一段特定信息，那么，通过遍历 DOM 树的众多节点来进行查找会显得有些麻烦。XPath 语言使得访问树节点变得很容易。例如，假设有如下 HTML 文档：

```
<html>
    <head>
        . . .
        <title>. . .</title>
        . . .
    </database>
    . . .
</html>
```

可以通过对 XPath 表达式 /html/head/title/text() 求值来得到标题的文本。

使用 Xpath 执行下列操作比普通的 DOM 方式要简单得多：

1. 获得文档根节点。
2. 获取第一个子节点，并将其转型为一个 Element 对象。
3. 在其所有子节点中定位 title 元素。
4. 获取其第一个子元素，并将其转型为一个 CharacterData 节点。
5. 获取其数据。

XPath 可以描述 XML 文档中的一个节点集，例如，下面的 XPath：

/html/body/form

描述了 XHTML 文件中 body 元素的子元素中所有的 form 元素。可以用 [] 操作符来选择特定元素：

/html/body/form[1]

这表示的是第一个 form（索引号从 1 开始）。

使用 @ 操作符可以得到属性值。XPath 表达式

/html/body/form[1]/@action

描述了第一个表中的 action 属性。XPath 表达式

```
/html/body/form/@action
```

描述了 body 元素的子元素中所有 form 元素的所有 action 属性节点。

XPath 有很多有用的函数，例如：

```
count(/html/body/form)
```

返回 body 根元素的 form 子元素的数量。精细的 XPath 表达式还有很多，请参见 http://www.w3c.org/TR/xpath 的规范，或者在 http://www.zvon.org/xxl/XPathTutorial/General/examples.html 上的一个非常好的在线指南。

要计算 XPath 表达式，首先需要从 XPathFactory 创建一个 XPath 对象：

```
XPathFactory xpfactory = XPathFactory.newInstance();
path = xpfactory.newXPath();
```

然后，调用 evaluate 方法来计算 XPath 表达式：

```
String username = path.evaluate("/html/head/title/text()", doc);
```

可以用同一个 XPath 对象来计算多个表达式。

这种形式的 evaluate 方法将返回一个字符串。这很适合用来获取文本，比如前面的例子中的 title 元素的文本子节点。如果 XPath 表达式产生了一组节点，请做如下调用：

```
XPathNodes result = path.evaluateExpression("/html/body/form", doc, XPathNodes.class);
```

XPathNodes 类与 NodeList 类相似，但是它扩展了 Iterable 接口，使得我们可以使用增强型 for 循环。

这个方法是在 Java 9 中添加进来的，在老版本中，需要使用下面这条语句：

```
var nodes = (NodeList) path.evaluate("/html/body/form", doc, XPathConstants.NODESET);
```

如果结果只有一个节点，则使用下面的调用：

```
Node node = path.evaluateExpression("/html/body/form[1]", doc, Node.class);
node = (Node) path.evaluate("/html/body/form[1]", doc, XPathConstants.NODE);
```

如果结果是一个数字，则使用：

```
int count = path.evaluateExpression("count(/html/body/form)", doc, Integer.class);
count = ((Number) path.evaluate("count(/html/body/form)",
    doc, XPathConstants.NUMBER)).intValue();
```

不必从文档的根节点开始搜索，可以从任意一个节点或节点列表开始。例如，如果有前一次计算得到的节点，那么就可以调用：

```
String result = path.evaluate(expression, node);
```

如果不知道 XPath 表达式的计算结果是什么（可能该表达式来自于用户），那么就调用

```
XPathEvaluationResult<?> result = path.evaluateExpression(expression, doc);
```

表达式 result.type() 是下列 XPathEvaluationResult.XPathResultType 枚举常量之一：

```
STRING
```

```
NODESET
NODE
NUMBER
BOOLEAN
```

调用 result.value() 可以获取结果值。

程序清单 3-6 展示了对任意的 XPath 表达式的计算过程。加载一个 XML 文件，输入一个表达式，该表达式的结果就会显示出来。

程序清单 3-6   xpath/XPathTest.java

```java
 1  package xpath;
 2
 3  import java.io.*;
 4  import java.nio.file.*;
 5  import java.util.*;
 6
 7  import javax.xml.catalog.*;
 8  import javax.xml.parsers.*;
 9  import javax.xml.xpath.*;
10
11  import org.w3c.dom.*;
12  import org.xml.sax.*;
13
14  /**
15   * This program evaluates XPath expressions.
16   * @version 1.1 2018-04-06
17   * @author Cay Horstmann
18   */
19  public class XPathTest
20  {
21     public static void main(String[] args) throws Exception
22     {
23        DocumentBuilderFactory factory = DocumentBuilderFactory.newInstance();
24        DocumentBuilder builder = factory.newDocumentBuilder();
25
26        // Avoid a delay in parsing an XHTML file--see the first note in
27        // Section 3.3.1
28        builder.setEntityResolver(CatalogManager.catalogResolver(
29           CatalogFeatures.defaults(),
30           Paths.get("xpath/catalog.xml").toAbsolutePath().toUri()));
31
32        XPathFactory xpfactory = XPathFactory.newInstance();
33        XPath path = xpfactory.newXPath();
34        try (var in = new Scanner(System.in))
35        {
36           String filename;
37           if (args.length == 0)
38           {
39              System.out.print("Input file: ");
40              filename = in.nextLine();
41           }
42           else
43              filename = args[0];
```

```
44
45            Document doc = builder.parse(filename);
46            var done = false;
47            while (!done)
48            {
49               System.out.print("XPath expression (empty line to exit): " );
50               String expression = in.nextLine();
51               if (expression.trim().isEmpty()) done = true;
52               else
53               {
54                  try
55                  {
56                     XPathEvaluationResult<?> result
57                           = path.evaluateExpression(expression, doc);
58                     if (result.type() == XPathEvaluationResult.XPathResultType.NODESET)
59                     {
60                        for (Node n : (XPathNodes) result.value())
61                           System.out.println(description(n));
62                     }
63                     else if (result.type() == XPathEvaluationResult.XPathResultType.NODESET)
64                        System.out.println((Node) result.value());
65                     else
66                        System.out.println(result.value());
67                  }
68                  catch (XPathExpressionException e)
69                  {
70                     System.out.println(e.getMessage());
71                  }
72               }
73            }
74         }
75   }
76
77   public static String description(Node n)
78   {
79      if (n instanceof Element) return "Element " + n.getNodeName();
80      else if (n instanceof Attr) return "Attribute " + n;
81      else return n.toString();
82   }
83 }
```

API **javax.xml.xpath.XPathFactory** 5.0

- static XPathFactory newInstance()

  返回用于创建 XPath 对象的 XPathFactory 实例。

- XPath newXpath()

  构建用于计算 XPath 表达式的 XPath 对象。

API **javax.xml.xpath.XPath** 5.0

- String evaluate(String expression, Object startingPoint)

  从给定的起点计算表达式。起点可以是一个节点或节点列表。如果结果是一个节点或节点集，则返回的字符串由所有文本节点子元素的数据构成。
- Object evaluate(String expression, Object startingPoint, QName resultType)
  从给定的起点计算表达式。起点可以是一个节点或节点列表。resultType 是 XPathConstants 类的常量 STRING、NODE、NODESET、NUMBER 或 BOOLEAN 之一。返回值是 String、Node、NodeList、Number 或 Boolean。
- <T> T evaluateExpression(String expression, Object item, Class<T> type) 9
  计算给定表达式，并产生给定类型值的结果。
- XPathEvaluationResult<?> evaluateExpression(String expression, InputSource source) 9
  计算给定表达式。

API *javax.xml.xpath.XPathEvaluationResult<T>* 9

- XPathEvaluationResult.XPathResultType type()
  返回枚举常量 STRING、NODESET、NODE、NUMBER 和 BOOLEAN 之一。
- T value()
  返回结果值。

3.6 使用命名空间

Java 语言使用包来避免名字冲突。程序员可以为不同的类使用相同的名字，只要它们不在同一个包中即可。XML 也有类似的命名空间（namespace）机制，可以用于元素名和属性名。名字空间是由统一资源标识符（Uniform Resource Identifier，URI）来标识的，比如：

```
http://www.w3.org/2001/XMLSchema
uuid:1c759aed-b748-475c-ab68-10679700c4f2
urn:com:books-r-us
```

HTTP 的 URL 格式是最常见的标识符。注意，URL 只用作标识符字符串，而不是一个文件的定位符。例如，名字空间标识符：

```
http://www.horstmann.com/corejava
http://www.horstmann.com/corejava/index.html
```

表示了不同的命名空间，尽管 Web 服务器将为这两个 URL 提供同一个文档。

在命名空间的 URL 所表示的位置上不需要有任何文档，XML 解析器不会尝试去该处查找任何东西。然而，为了给可能会遇到不熟悉的命名空间的程序员提供一些帮助，人们习惯于将解释该命名空间的文档放在 URL 位置上。例如，如果把浏览器指向 XML Schema 的命名空间 URL（http://www.w3.org/2001/XMLSchema），就会发现一个描述 XML Schema 标准的文档。

为什么要用 HTTP URL 作为命名空间的标识符？这是因为这样容易确保它们是独一无二的。如果使用实际的 URL，那么主机部分的唯一性就将由域名系统来保证。然后，你的组织可以安排 URL 余下部分的唯一性，这和 Java 包名中的反向域名是一个原理。

尽管长名字空间的唯一性很好，但是你肯定不想处理超出必需范围的长标识符。在 Java

编程语言中，可以用 import 机制来指定很长的包名，然后就可以只使用较短的类名了。在 XML 中有类似的机制，比如：

```
<element xmlns="namespaceURI">
    children
</element>
```

现在，该元素和它的子元素都是给定命名空间的一部分了。

子元素可以提供自己的命名空间，例如：

```
<element xmlns="namespaceURI1">
    <child xmlns="namespaceURI2">
        grandchildren
    </child>
    more children
</element>
```

这时，第一个子元素和孙元素都是第二个命名空间的一部分。

无论是只需要一个命名空间，还是命名空间本质上是嵌套的，这个简单机制都工作得很好。如若不然，就需要使用第二种机制，而 Java 中并没有类似的机制。你可以用一个前缀来表示命名空间，即为特定文档选取的一个短的标识符。下面是一个典型的例子：

```
<xsd:schema xmlns:xsd="http://www.w3.org/2001/XMLSchema">
    <xsd:element name="config"/>
    . . .
</xsd:schema>
```

下面的属性：

xmlns:prefix="namespaceURI"

用于定义命名空间和前缀。在我们的例子中，前缀是字符串 xsd。这样，xsd:schema 实际上指的是命名空间 http://www.w3.org/2001/XMLSchema 中的 schema。

> **注释**：只有子元素继承了它们父元素的命名空间，而不带显式前缀的属性并不是命名空间的一部分。请看下面这个特意构造出来的例子：
>
> ```
> <configuration xmlns="http://www.horstmann.com/corejava"
>     xmlns:si="http://www.bipm.fr/enus/3_SI/si.html">
>     <size value="210" si:unit="mm"/>
>     . . .
> </configuration>
> ```
>
> 在这个示例中，元素 configuration 和 size 是 URI 为 http://www.horstmann.com/corejava 的命名空间的一部分。属性 si:unit 是 URI 为 http://www.bipm.fr/enus/3_SI/si.html 的命名空间的一部分。然而，属性 value 不是任何命名空间的一部分。

可以控制解析器对命名空间的处理。默认情况下，Java XML 库的 DOM 解析器并非"命名空间感知的"。

要打开命名空间处理特性，请调用 DocumentBuilderFactory 类的 setNamespace Aware 方法：

```
factory.setNamespaceAware(true);
```

这样，该工厂产生的所有生成器便都支持命名空间了。每个节点有三个属性：
- 带有前缀的限定名（qualified），由 getNodeName 和 getTagName 等方法返回。
- 命名空间 URI，由 getNamespaceURI 方法返回。
- 不带前缀和命名空间的本地名（local name），由 getLocalName 方法返回。

下面是一个例子。假设解析器看到了以下元素：

```
<xsd:schema xmlns:xsd="http://www.w3.org/2001/XMLSchema">
```

它会报告如下信息：
- 限定名 = xsd:schema
- 命名空间 URI = http://www.w3.org/2001/XMLSchema
- 本地名 = schema

> 注释：如果对命名空间的感知特性被关闭，getLocalName 和 getNamespaceURI 方法将返回 null。

API org.w3c.dom.Node 1.4

- String getLocalName()
  返回本地名（不带前缀），或者在解析器不感知命名空间时，返回 null。
- String getNamespaceURI()
  返回命名空间 URI，或者在解析器不感知命名空间时，返回 null。

API javax.xml.parsers.DocumentBuilderFactory 1.4

- boolean isNamespaceAware()
- void setNamespaceAware(boolean value)
  获取或设置工厂的 namespaceAware 属性。当设为 true 时，工厂产生的解析器是命名空间感知的。

3.7 流机制解析器

DOM 解析器会完整地读入 XML 文档，然后将其转换成一个树形的数据结构。对于大多数应用，DOM 都运行得很好。但是，如果文档很大，并且处理算法又非常简单，可以在运行时解析节点，而不必看到完整的树形结构，那么 DOM 可能就会显得效率低下了。在这种情况下，我们应该使用流机制解析器（streaming parser）。

在下面的小节中，我们将讨论 Java 类库提供的流机制解析器：老而弥坚的 SAX 解析器和添加到 Java 6 中的更现代化的 StAX 解析器。SAX 解析器使用的是事件回调（event callback），而 StAX 解析器提供了遍历解析事件的迭代器，后者用起来通常更方便一些。

3.7.1 使用 SAX 解析器

SAX 解析器在解析 XML 输入数据的各个组成部分时会报告事件，但不会以任何方式存储文档，而是由事件处理器建立相应的数据结构。实际上，DOM 解析器是在 SAX 解析器的基础上构建的，它在接收到解析器事件时构建 DOM 树。

在使用 SAX 解析器时，需要一个处理器来为各种解析器事件定义事件动作。ContentHandler 接口定义了若干个在解析文档时解析器会调用的回调方法。下面是最重要的几个：

- startElement 和 endElement 在每当遇到起始或终止标签时调用。
- characters 在每当遇到字符数据时调用。
- startDocument 和 endDocument 分别在文档开始和结束时各调用一次。

例如，在解析以下片段时：

```
<font>
    <name>Helvetica</name>
    <size units="pt">36</size>
</font>
```

解析器会产生以下回调：

1. startElement，元素名：font
2. startElement，元素名：name
3. characters，内容：Helvetica
4. endElement，元素名：name
5. startElement，元素名：size，属性：units="pt"
6. characters，内容：36
7. endElement，元素名：size
8. endElement，元素名：font

处理器必须覆盖这些方法，让它们执行在解析文件时我们想要让它们执行的动作。本节最后的程序会打印出一个 HTML 文件中的所有链接 <a href="...">。它直接覆盖了处理器的 startElement 方法，以检查名字为 a，且属性名为 href 的链接，其潜在用途包括用于实现"网络爬虫"，即一个沿着链接到达越来越多网页的程序。

> **注释**：遗憾的是，HTML 不必是合法的 XML，大多数 HTML 页面都与良构的 XML 差别很大，以至于示例程序无法解析它们。但是，W3C 编写的大部分页面都是用 XHTML 编写的，XHTML 是一种 HTML 方言，且是良构的 XML，你可以用这些页面来测试示例程序。例如，运行：
>
>   java SAXTest http://www.w3.org/MarkUp
>
> 将看到那个页面上所有链接的 URL 列表。

示例程序是一个很好的使用 SAX 的例子。我们根本不在乎 a 元素出现的上下文环境，而且不必存储树形结构。

下面是如何得到 SAX 解析器的代码：

```
SAXParserFactory factory = SAXParserFactory.newInstance();
SAXParser parser = factory.newSAXParser();
```

现在可以处理文档了：

```
parser.parse(source, handler);
```

这里的 source 可以是一个文件、一个 URL 字符串或者是一个输入流。handler 属于 Default-Handler 的一个子类，DefaultHandler 类为以下四个接口定义了空的方法：

```
ContentHandler
DTDHandler
EntityResolver
ErrorHandler
```

示例程序定义了一个处理器，它覆盖了 ContentHandler 接口的 startElement 方法，以观察带有 href 属性的 a 元素。

```
var handler = new DefaultHandler()
   {
      public void startElement(String namespaceURI, String lname, String qname,
         Attributes attrs) throws SAXException
      {
         if (lname.equalsIgnoreCase("a") && attrs != null)
         {
         for (int i = 0; i < attrs.getLength(); i++)
         {
            String aname = attrs.getLocalName(i);
            if (aname.equalsIgnoreCase("href"))
               System.out.println(attrs.getValue(i));
         }
         }
      }
   };
```

startElement 方法有 3 个描述元素名的参数，其中 qname 参数以 prefix:localname 的形式报告限定名。如果命名空间处理特性已经打开，那么 namespaceURI 和 lname 参数提供的就是命名空间和本地（非限定）名。

与 DOM 解析器一样，命名空间处理特性默认是关闭的，可以调用工厂类的 setNamespaceAware 方法来激活命名空间处理特性：

```
SAXParserFactory factory = SAXParserFactory.newInstance();
factory.setNamespaceAware(true);
SAXParser saxParser = factory.newSAXParser();
```

在这个程序中，我们还处理了另一个常见的问题。XHTML 文件总是以一个包含对 DTD 引用的标签开头，解析器会加载这个 DTD。可以理解的是，W3C 肯定不乐意对诸如 www.w3.org/TR/xhtml/DTD/xhtml-strict.dtd 这样的文件提供千万亿次的下载。总有一天他们会完全拒绝提供这些文件，但到写本章时为止，他们还在并不情愿地提供 DTD 下载。如果你不需要验证文件，只需调用：

```
factory.setFeature("http://apache.org/xml/features/nonvalidating/load-external-dtd", false);
```

程序清单 3-7 包含了网络爬虫程序的代码。在本章的后续部分，将会看到 SAX 的另一个有趣用法，即将非 XML 数据源转换成 XML 的一种简单方式是报告 XML 解析器将要报告的

SAX 事件。详情请参见 3.9 节。

程序清单 3-7　　sax/SAXTest.java

```java
package sax;

import java.io.*;
import java.net.*;
import javax.xml.parsers.*;
import org.xml.sax.*;
import org.xml.sax.helpers.*;

/**
 * This program demonstrates how to use a SAX parser. The program prints all
 * hyperlinks of an XHTML web page. <br>
 * Usage: java sax.SAXTest URL
 * @version 1.01 2018-05-01
 * @author Cay Horstmann
 */
public class SAXTest
{
   public static void main(String[] args) throws Exception
   {
      String url;
      if (args.length == 0)
      {
         url = "http://www.w3c.org";
         System.out.println("Using " + url);
      }
      else url = args[0];

      var handler = new DefaultHandler()
         {
            public void startElement(String namespaceURI, String lname,
                  String qname, Attributes attrs)
            {
               if (lname.equals("a") && attrs != null)
               {
                  for (int i = 0; i < attrs.getLength(); i++)
                  {
                     String aname = attrs.getLocalName(i);
                     if (aname.equals("href"))
                        System.out.println(attrs.getValue(i));
                  }
               }
            }
         };

      SAXParserFactory factory = SAXParserFactory.newInstance();
      factory.setNamespaceAware(true);
      factory.setFeature(
         "http://apache.org/xml/features/nonvalidating/load-external-dtd",
         false);
      SAXParser saxParser = factory.newSAXParser();
      InputStream in = new URL(url).openStream();
```

```
52        saxParser.parse(in, handler);
53     }
54 }
```

API javax.xml.parsers.SAXParserFactory 1.4

- static SAXParserFactory newInstance()
  返回 SAXParserFactory 类的一个实例。
- SAXParser newSAXParser()
  返回 SAXParser 类的一个实例。
- boolean isNamespaceAware()
- void setNamespaceAware(boolean value)
  获取和设置工厂的 namespaceAware 属性。当设为 true 时，该工厂生成的解析器是命名空间感知的。
- boolean isValidating()
- void setValidating(boolean value)
  获取和设置工厂的 validating 属性。当设为 true 时，该工厂生成的解析器将要验证其输入。

API javax.xml.parsers.SAXParser 1.4

- void parse(File f, DefaultHandler handler)
- void parse(String url, DefaultHandler handler)
- void parse(InputStream in, DefaultHandler handler)
  解析来自给定文件、URL 或输入流的 XML 文档，并把解析事件报告给指定的处理器。

API org.xml.sax.ContentHandler 1.4

- void startDocument()
- void endDocument()
  在文档的开头和结尾处被调用。
- void startElement(String uri, String lname, String qname, Attributes attr)
- void endElement(String uri, String lname, String qname)
  在元素的开头和结尾处被调用。如果解析器是名字空间感知的，那么它会报告名字空间的 URI、无前缀的本地名字，以及带前缀的限定名。
- void characters(char[] data, int start, int length)
  解析器报告字符数据时被调用。

API org.xml.sax.Attributes 1.4

- int getLength()

返回存储在该属性集合中的属性数量。

- String getLocalName(int index)

  返回给定索引的属性的本地名（无前缀），或在解析器不是命名空间感知的情况下返回空字符串。

- String getURI(int index)

  返回给定索引的属性的命名空间 URI，或者，当该节点不是命名空间的一部分，或解析器并非命名空间感知时返回空字符串。

- String getQName(int index)

  返回给定索引的属性的限定名（带前缀），或当解析器不报告限定名时返回空字符串。

- String getValue(int index)
- String getValue(String qname)
- String getValue(String uri, String lname)

  根据给定索引、限定名或命名空间 URI+ 本地名来返回属性值；当该值不存在时，返回 null。

3.7.2 使用 StAX 解析器

StAX 解析器是一种"拉解析器"（pull parser），与安装事件处理器不同，你只需使用下面这样的基本循环来迭代所有的事件：

```
InputStream in = url.openStream();
XMLInputFactory factory = XMLInputFactory.newInstance();
XMLStreamReader parser = factory.createXMLStreamReader(in);
while (parser.hasNext())
{
    int event = parser.next();
    Call parser methods to obtain event details
}
```

例如，在解析下面的片段时

```
<font>
    <name>Helvetica</name>
    <size units="pt">36</size>
</font>
```

解析器将产生下面的事件：

1. START_ELEMENT，元素名：font
2. CHARACTERS，内容：空白字符
3. START_ELEMENT，元素名：name
4. CHARACTERS，内容：Helvetica
5. END_ELEMENT，元素名：name
6. CHARACTERS，内容：空白字符
7. START_ELEMENT，元素名：size

8. CHARACTERS，内容：36
9. END_ELEMENT，元素名：size
10. CHARACTERS，内容：空白字符
11. END_ELEMENT，元素名：font

要分析这些属性值，需要调用 XMLStreamReader 类中恰当的方法，例如：

```
String units = parser.getAttributeValue(null, "units");
```

它可以获取当前元素的 units 属性。

默认情况下，命名空间处理是启用的，可以通过像下面这样修改工厂来使其无效：

```
XMLInputFactory factory = XMLInputFactory.newInstance();
factory.setProperty(XMLInputFactory.IS_NAMESPACE_AWARE, false);
```

程序清单 3-8 包含了用 StAX 解析器实现的网络爬虫程序。正如你所见，这段代码比等效的 SAX 代码要简短了许多，因为此时我们不必操心事件处理问题。

程序清单 3-8 stax/StAXTest.java

```java
 1 package stax;
 2
 3 import java.io.*;
 4 import java.net.*;
 5 import javax.xml.stream.*;
 6
 7 /**
 8  * This program demonstrates how to use a StAX parser. The program prints all
 9  * hyperlinks links of an XHTML web page. <br>
10  * Usage: java stax.StAXTest URL
11  * @author Cay Horstmann
12  * @version 1.1 2018-05-01
13  */
14 public class StAXTest
15 {
16    public static void main(String[] args) throws Exception
17    {
18       String urlString;
19       if (args.length == 0)
20       {
21          urlString = "http://www.w3c.org";
22          System.out.println("Using " + urlString);
23       }
24       else urlString = args[0];
25       var url = new URL(urlString);
26       InputStream in = url.openStream();
27       XMLInputFactory factory = XMLInputFactory.newInstance();
28       XMLStreamReader parser = factory.createXMLStreamReader(in);
29       while (parser.hasNext())
30       {
31          int event = parser.next();
32          if (event == XMLStreamConstants.START_ELEMENT)
33          {
34             if (parser.getLocalName().equals("a"))
```

```
35                {
36                    String href = parser.getAttributeValue(null, "href");
37                    if (href != null)
38                        System.out.println(href);
39                }
40            }
41        }
42    }
43 }
```

API **javax.xml.stream.XMLInputFactory** 6

- static XMLInputFactory newInstance()
  返回 XMLInputFactory 类的一个实例。

- void setProperty(String name, Object value)
  设置这个工厂的属性，或者在要设置的属性不支持设置成给定值时，抛出 Illegal-ArgumentException。JDK 的实现支持下列 Boolean 类型的属性：

"javax.xml.stream.isValidating"	为 false（默认值）时，不验证文档（规范不要求必须支持）。
"javax.xml.stream.isNamespaceAware"	为 true（默认值）时，将处理命名空间（规范不要求必须支持）。
"javax.xml.stream.isCoalescing"	为 false（默认值）时，邻近的字符数据不进行连接。
"javax.xml.stream.isReplacingEntityReferences"	为 true（默认值）时，实体引用将作为字符数据被替换和报告。
"javax.xml.stream.isSupportingExternalEntities"	为 true（默认值）时，外部实体将被解析。规范对于这个属性没有给出默认值。
"javax.xml.stream.supportDTD"	为 true（默认值）时，DTD 将作为事件被报告。

- XMLStreamReader createXMLStreamReader(InputStream in)
- XMLStreamReader createXMLStreamReader(InputStream in, String characterEncoding)
- XMLStreamReader createXMLStreamReader(Reader in)
- XMLStreamReader createXMLStreamReader(Source in)
  创建一个从给定的流、阅读器或 JAXP 源读入的解析器。

API **javax.xml.stream.XMLStreamReader** 6

- boolean hasNext()
  如果有另一个解析事件，则返回 true。

- int next()
  将解析器的状态设置为下一个解析事件，并返回下列常量之一：START_ELEMENT、END_ELEMENT、CHARACTERS、START_DOCUMENT、END_DOCUMENT、CDATA、COMMENT、SPACE（可忽略的空白字符）、

PROCESSING_INSTRUCTION、ENTITY_REFERENCE、DTD。
- boolean isStartElement()
- boolean isEndElement()
- boolean isCharacters()
- boolean isWhiteSpace()

  如果当前事件是一个开始元素、结束元素、字符数据或空白字符，则返回 true。
- QName getName()
- String getLocalName()

  获取在 START_ELEMENT 或 END_ELEMENT 事件中的元素的名字。
- String getText()

  返回一个 CHARACTERS、COMMENT 或 CDATA 事件中的字符，或一个 ENTITY_REFERENCE 的替换值，或者一个 DTD 的内部子集。
- int getAttributeCount()
- QName getAttributeName(int index)
- String getAttributeLocalName(int index)
- String getAttributeValue(int index)

  如果当前事件是 START_ELEMENT，则获取属性数量和属性的名字与值。
- String getAttributeValue(String namespaceURI, String name)

  如果当前事件是 START_ELEMENT，则获取具有给定名称的属性的值。如果 namespaceURI 为 null，则不检查名字空间。

3.8 生成 XML 文档

现在你已经知道怎样编写读取 XML 的 Java 程序了。下面让我们开始介绍它的反向过程，即产生 XML 输出。当然，你可以直接通过一系列 print 调用，打印出各元素、属性和文本内容，以此来编写 XML 文件，但这并不是一个好主意。这样的代码会非常冗长复杂，对于属性值和文本内容中的那些特殊符号（如：" 和 <），一不注意就会出错。

一种更好的方式是用文档的内容构建一棵 DOM 树，然后再写出该树的所有内容。下面的小节将讨论其细节。

3.8.1 不带命名空间的文档

要建立一棵 DOM 树，可以从一个空的文档开始。通过调用 DocumentBuilder 类的 newDocument 方法可以得到一个空文档。

```
Document doc = builder.newDocument();
```

使用 Document 类的 createElement 方法可以构建文档里的元素：

```
Element rootElement = doc.createElement(rootName);
Element childElement = doc.createElement(childName);
```

使用 createTextNode 方法可以构建文本节点：

```
Text textNode = doc.createTextNode(textContents);
```

使用以下方法可以给文档添加根元素，给父结点添加子节点：

```
doc.appendChild(rootElement);
rootElement.appendChild(childElement);
childElement.appendChild(textNode);
```

在建立 DOM 树时，可能还需要设置元素属性，这只需调用 Element 类的 setAttribute 方法即可：

```
rootElement.setAttribute(name, value);
```

3.8.2 带命名空间的文档

如果要使用命名空间，那么创建文档的过程就会稍微有些差异。

首先需要将生成器工厂设置为是命名空间感知的，然后创建生成器：

```
DocumentBuilderFactory factory = DocumentBuilderFactory.newInstance();
factory.setNamespaceAware(true);
builder = factory.newDocumentBuilder();
```

再使用 createElementNS 而不是 createElement 来创建所有节点：

```
String namespace = "http://www.w3.org/2000/svg";
Element rootElement = doc.createElementNS(namespace, "svg");
```

如果节点具有带命名空间前缀的限定名，那么所有必需的带有 xmlns 前缀的属性都会被自动创建。例如，如果需要在 HTML 中包含 SVG，那么就可以像下面这样构建元素：

```
Element svgElement = doc.createElement(namespace, "svg:svg")
```

当该元素被写入 XML 文件时，它会转变为：

```
<svg:svg xmlns:svg="http://www.w3.org/2000/svg">
```

如果需要设置的元素属性的名字位于命名空间中，那么可以使用 Element 类的 setAttributeNS 方法：

```
rootElement.setAttributeNS(namespace, qualifiedName, value);
```

3.8.3 写出文档

有些奇怪的是，把 DOM 树写出到输出流中并非一件易事。最容易的方式是使用可扩展的样式表语言转换（Extensible Stylesheet Language Transformations, XSLT）API。关于 XSLT 的更多信息请参见 3.9 节。当下，我们先考虑根据生成 XML 输出的"魔咒"而编写的代码。

我们把"不做任何操作"的转换应用于文档，并且捕获它的输出。为了将 DOCTYPE 节点纳入输出，我们还需要将 SYSTEM 和 PUBLIC 标识符设置为输出属性。

```
// construct the do-nothing transformation
```

```
Transformer t = TransformerFactory.newInstance().newTransformer();
// set output properties to get a DOCTYPE node
t.setOutputProperty(OutputKeys.DOCTYPE_SYSTEM, systemIdentifier);
t.setOutputProperty(OutputKeys.DOCTYPE_PUBLIC, publicIdentifier);
// set indentation
t.setOutputProperty(OutputKeys.INDENT, "yes");
t.setOutputProperty(OutputKeys.METHOD, "xml");
t.setOutputProperty("{http://xml.apache.org/xslt}indent-amount", "2");
// apply the do-nothing transformation and send the output to a file
t.transform(new DOMSource(doc), new StreamResult(new FileOutputStream(file)));
```

另一种方式是使用 LSSerializer 接口。为了获取实例，可以使用下面的魔咒：

```
DOMImplementation impl = doc.getImplementation();
var implLS = (DOMImplementationLS) impl.getFeature("LS", "3.0");
LSSerializer ser = implLS.createLSSerializer();
```

如果需要空格和换行，可以设置下面的标志：

```
ser.getDomConfig().setParameter("format-pretty-print", true);
```

然后可以易如反掌地将文档转换为字符串：

```
String str = ser.writeToString(doc);
```

如果想要将输出直接写入到文件中，则需要一个 LSOutput：

```
LSOutput out = implLS.createLSOutput();
out.setEncoding("UTF-8");
out.setByteStream(Files.newOutputStream(path));
ser.write(doc, out);
```

API javax.xml.parsers.DocumentBuilder 1.4

- Document newDocument()
  返回一个空文档。

API org.w3c.dom.Document 1.4

- Element createElement(String name)
- Element createElementNS(String uri, String qname)
  返回具有给定名字的元素。
- Text createTextNode(String data)
  返回具有给定数据的文本节点。

API org.w3c.dom.Node 1.4

- Node appendChild(Node child)
  在该节点的子节点列表中追加一个节点。返回被追加的节点。

API org.w3c.dom.Element 1.4

- void setAttribute(String name, String value)
- void setAttributeNS(String uri, String qname, String value)

将有给定名字的属性设置为指定的值。

如果限定名有别名前缀,则 uri 不能为 null。

API javax.xml.transform.TransformerFactory 1.4

- static TransformerFactory newInstance()
  返回 TransformerFactory 类的一个实例。
- Transformer newTransformer()
  返回 Transformer 类的一个实例,它实现了标识符转换(不做任何事情的转换)。

API javax.xml.transform.Transformer 1.4

- void setOutputProperty(String name, String value)
  设置输出属性。标准输出属性参见 http://www.w3.org/TR/xslt#output,其中最有用的几个如下所示:
  doctype-public   DOCTYPE 声明中使用的公共 ID
  doctype-system   DOCTYPE 声明中使用的系统 ID
  Indent   "yes" 或者 "no"
  method   "xml""html"、"text" 或定制的字符串
- void transform(Source from, Result to)
  转换一个 XML 文档。

API javax.xml.transform.dom.DOMSource 1.4

- DOMSource(Node n)
  从给定的节点中构建一个源。通常,n 是文档节点。

API javax.xml.transform.stream.StreamResult 1.4

- StreamResult(File f)
- StreamResult(OutputStream out)
- StreamResult(Writer out)
- StreamResult(String systemID)
  从文件、流、写出器或系统 ID(通常是相对或绝对 URL)中构建流结果。

3.8.4 使用 StAX 写出 XML 文档

在前一节中,你看到了如何通过写出 DOM 树的方法来产生 XML 文件。如果这个 DOM 树没有其他任何用途,那么这种方式就不是很高效。

StAX API 使我们可以直接将 XML 树写出,这需要从某个 OutputStream 中构建一个 XML-StreamWriter,就像下面这样:

```
XMLOutputFactory factory = XMLOutputFactory.newInstance();
XMLStreamWriter writer = factory.createXMLStreamWriter(out);
```

要产生 XML 文件头，需要调用

```
writer.writeStartDocument()
```

然后调用

```
writer.writeStartElement(name);
```

添加属性需要调用

```
writer.writeAttribute(name, value);
```

现在，可以通过再次调用 writeStartElement 添加新的子节点，或者用下面的调用写出字符：

```
writer.writeCharacters(text);
```

在写完所有子节点之后，调用

```
writer.writeEndElement();
```

这会导致当前元素被关闭。

要写出没有子节点的元素（例如 <img .../>），可以使用下面的调用

```
writer.writeEmptyElement(name);
```

最后，在文档的结尾，调用

```
writer.writeEndDocument();
```

将关闭所有打开的元素。

你仍旧需要关闭 XMLStreamWriter，并且需要人为关闭它，因为 XMLStreamWriter 接口没有扩展 AutoCloseable 接口。

与使用 DOM/XSLT 的方式一样，我们不必担心属性值和字符数据中的转义字符。但是，我们仍旧有可能会产生非良构的 XML，例如具有多个根节点的文档。并且，StAX 当前的版本还没有任何对产生缩进输出的支持。

程序清单 3-9 中的程序展示了写出 XML 的两种方式。

程序清单 3-9　write/XMLWriteTest.java

```
1  package write;
2
3  import java.io.*;
4  import java.nio.file.*;
5  import java.util.*;
6
7  import javax.xml.parsers.*;
8  import javax.xml.stream.*;
9  import javax.xml.transform.*;
10 import javax.xml.transform.dom.*;
11 import javax.xml.transform.stream.*;
12
13 import org.w3c.dom.*;
14
15 /**
16  * This program shows how to write an XML file. It produces modern art in SVG
```

```
17      * format.
18      * @version 1.12 2016-04-27
19      * @author Cay Horstmann
20      */
21     public class XMLWriteTest
22     {
23        public static void main(String[] args) throws Exception
24        {
25           Document doc = newDrawing(600, 400);
26           writeDocument(doc, "drawing1.svg");
27           writeNewDrawing(600, 400, "drawing2.svg");
28        }
29
30        private static Random generator = new Random();
31
32        /**
33         * Creates a new random drawing.
34         * @return the DOM tree of the SVG document
35         */
36        public static Document newDrawing(int drawingWidth, int drawingHeight)
37              throws ParserConfigurationException
38        {
39           DocumentBuilderFactory factory = DocumentBuilderFactory.newInstance();
40           factory.setNamespaceAware(true);
41           DocumentBuilder builder = factory.newDocumentBuilder();;
42           var namespace = "http://www.w3.org/2000/svg";
43           Document doc = builder.newDocument();
44           Element svgElement = doc.createElementNS(namespace, "svg");
45           doc.appendChild(svgElement);
46           svgElement.setAttribute("width", "" + drawingWidth);
47           svgElement.setAttribute("height", "" + drawingHeight);
48           int n = 10 + generator.nextInt(20);
49           for (int i = 1; i <= n; i++)
50           {
51              int x = generator.nextInt(drawingWidth);
52              int y = generator.nextInt(drawingHeight);
53              int width = generator.nextInt(drawingWidth - x);
54              int height = generator.nextInt(drawingHeight - y);
55              int r = generator.nextInt(256);
56              int g = generator.nextInt(256);
57              int b = generator.nextInt(256);
58
59              Element rectElement = doc.createElementNS(namespace, "rect");
60              rectElement.setAttribute("x", "" + x);
61              rectElement.setAttribute("y", "" + y);
62              rectElement.setAttribute("width", "" + width);
63              rectElement.setAttribute("height", "" + height);
64              rectElement.setAttribute("fill",
65                    String.format("#%02x%02x%02x", r, g, b));
66              svgElement.appendChild(rectElement);
67           }
68           return doc;
69        }
70
```

```java
71      /**
72       * Saves a document using DOM/XSLT
73       */
74      public static void writeDocument(Document doc, String filename)
75              throws TransformerException, IOException
76      {
77         Transformer t = TransformerFactory.newInstance().newTransformer();
78         t.setOutputProperty(OutputKeys.DOCTYPE_SYSTEM,
79              "http://www.w3.org/TR/2000/CR-SVG-20000802/DTD/svg-20000802.dtd");
80         t.setOutputProperty(OutputKeys.DOCTYPE_PUBLIC,
81              "-//W3C//DTD SVG 20000802//EN");
82         t.setOutputProperty(OutputKeys.INDENT, "yes");
83         t.setOutputProperty(OutputKeys.METHOD, "xml");
84         t.setOutputProperty("{http://xml.apache.org/xslt}indent-amount", "2");
85         t.transform(new DOMSource(doc), new StreamResult(
86              Files.newOutputStream(Paths.get(filename))));
87      }
88
89      /**
90       * Writes an SVG document of the current drawing.
91       * @param writer the document destination
92       * @throws IOException
93       */
94      public static void writeNewDrawing(int drawingWidth, int drawingHeight,
95              String filename) throws XMLStreamException, IOException
96      {
97         XMLOutputFactory factory = XMLOutputFactory.newInstance();
98         XMLStreamWriter writer = factory.createXMLStreamWriter(
99              Files.newOutputStream(Paths.get(filename)));
100        writer.writeStartDocument();
101        writer.writeDTD("<!DOCTYPE svg PUBLIC \"-//W3C//DTD SVG 20000802//EN\" "
102             + "\"http://www.w3.org/TR/2000/CR-SVG-20000802/DTD/svg-20000802.dtd\">");
103        writer.writeStartElement("svg");
104        writer.writeDefaultNamespace("http://www.w3.org/2000/svg");
105        writer.writeAttribute("width", "" + drawingWidth);
106        writer.writeAttribute("height", "" + drawingHeight);
107        int n = 10 + generator.nextInt(20);
108        for (int i = 1; i <= n; i++)
109        {
110           int x = generator.nextInt(drawingWidth);
111           int y = generator.nextInt(drawingHeight);
112           int width = generator.nextInt(drawingWidth - x);
113           int height = generator.nextInt(drawingHeight - y);
114           int r = generator.nextInt(256);
115           int g = generator.nextInt(256);
116           int b = generator.nextInt(256);
117           writer.writeEmptyElement("rect");
118           writer.writeAttribute("x", "" + x);
119           writer.writeAttribute("y", "" + y);
120           writer.writeAttribute("width", "" + width);
121           writer.writeAttribute("height", "" + height);
122           writer.writeAttribute("fill", String.format("#%02x%02x%02x", r, g, b));
123        }
124        writer.writeEndDocument(); // closes svg element
```

```
125     }
126 }
```

API *javax.xml.stream.XMLOutputFactory* 6

- static XMLOutputFactory newInstance()

  返回 XMLOutputFactory 类的一个实例。

- XMLStreamWriter createXMLStreamWriter(OutputStream in)
- XMLStreamWriter createXMLStreamWriter(OutputStream in, String characterEncoding)
- XMLStreamWriter createXMLStreamWriter(Writer in)
- XMLStreamWriter createXMLStreamWriter(Result in)

  创建写出到给定流、写出器或 JAXP 结果的写出器。

API *javax.xml.stream.XMLStreamWriter* 6

- void writeStartDocument()
- void writeStartDocument(String xmlVersion)
- void writeStartDocument(String encoding, String xmlVersion)

  在文档的顶部写入 XML 处理指令。注意，encoding 参数只是用于写入这个属性，它不会设置输出的字符编码机制。

- void setDefaultNamespace(String namespaceURI)
- void setPrefix(String prefix, String namespaceURI)

  设置默认的命名空间，或者具有前缀的命名空间。这种声明的作用域只是当前元素，如果没有写明具体元素，其作用域为文档的根。

- void writeStartElement(String localName)
- void writeStartElement(String namespaceURI, String localName)

  写出一个开始标签，其中 namespaceURI 将用相关联的前缀来代替。

- void writeEndElement()

  关闭当前元素。

- void writeEndDocument()

  关闭所有打开的元素。

- void writeEmptyElement(String localName)
- void writeEmptyElement(String namespaceURI, String localName)

  写出一个自闭合的标签，其中 namespaceURI 将用相关联的前缀来代替。

- void writeAttribute(String localName, String value)
- void writeAttribute(String namespaceURI, String localName, String value)

  写出一个用于当前元素的属性，其中 namespaceURI 将用相关联的前缀来代替。

- void writeCharacters(String text)

  写出字符数据。

- void writeCData(String text)

  写出 CDATA 块。

- void writeDTD(String dtd)

  写出 dtd 字符串，该字符串需要包含一个 DOCTYPE 声明。

- void writeComment(String comment)

  写出一个注释。

- void close()

  关闭这个写出器。

3.8.5 示例：生成 SVG 文件

程序清单 3-9 是一个生成 XML 输出的典型程序。该程序绘制了一幅现代派绘画，即一组随机的彩色矩形（参见图 3-3）。我们使用可伸缩向量图形（Scalable Vector Graphics，SVG）来保存作品。SVG 是 XML 格式的，它使用设备无关的方式描述复杂图形。你可以在 http://www.w3c.org/ Graphics/SVG 找到更多关于 SVG 的信息。要查看 SVG 文件，只需使用任意的现在主流的浏览器。

该程序演示了两种产生 XML 的方式：通过构建并保存 DOM 树，以及通过直接用 StAX API 写出 XML。

我们并没有涉及 SVG 的细节。就我们的目的而言，我们只需要知道怎样表示一组彩色的矩形。下面是一个例子：

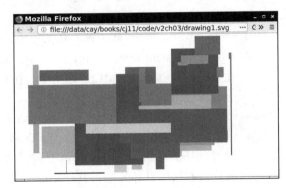

图 3-3　生成的现代艺术品

```
<?xml version="1.0" encoding="UTF-8"?>
<!DOCTYPE svg PUBLIC "-//W3C//DTD SVG 20000802//EN"
    "http://www.w3.org/TR/2000/CR-SVG-20000802/DTD/svg-20000802.dtd">
<svg xmlns="http://www.w3.org/2000/svg" width="300" height="150">
    <rect x="231" y="61" width="9" height="12" fill="#6e4a13"/>
    <rect x="107" y="106" width="56" height="5" fill="#c406be"/>
    . . .
</svg>
```

正如你看到的，每个矩形都被描述成了一个 rect 节点。它有位置、宽度、高度和填充色等属性，其中填充色以十六进制 RGB 值表示。

> 📝 **注释：** SVG 大量使用了属性。实际上，某些属性相当复杂。例如，下面的 path 元素：
>
> ```
> <path d="M 100 100 L 300 100 L 200 300 z">
> ```
>
> M 是指"moveto"命令、L 是指"lineto"、z 是指"closepath"(!)。显然，该数据格式的设计者不太信任 XML 表示结构化数据的能力。在你自己的 XML 格式中，你可能想使用元素来替代复杂的属性。

3.9 XSL 转换

XSL 转换（XSLT）机制可以指定将 XML 文档转换为其他格式的规则，例如，转换为纯文本、XHTML 或任何其他的 XML 格式。XSLT 通常用来将某种机器可读的 XML 格式转译为另一种机器可读的 XML 格式，或者将 XML 转译为适于人类阅读的表示格式。

你需要提供 XSLT 样式表，它描述了 XML 文档向某种其他格式转换的规则。XSLT 处理器将读入 XML 文档和这个样式表，并产生所要的输出（参见图 3-4）。

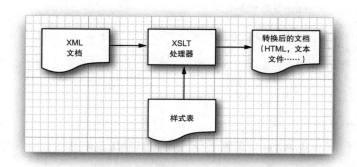

图 3-4　应用 XSL 转换

XSLT 规范很复杂，已经有很多书描述了该主题。我们不可能讨论 XSLT 的全部特性，所以我们只能介绍一个有代表性的例子。你可以在 Don Box 等人合著的 *Essential XML* 一书中找到更多的信息。XSLT 规范可以在 http://www.w3.org/TR/xslt 获得。

假设我们想要把有雇员记录的 XML 文件转换成 HTML 文件。请看这个输入文件：

```
<staff>
   <employee>
      <name>Carl Cracker</name>
      <salary>75000</salary>
      <hiredate year="1987" month="12" day="15"/>
   </employee>
   <employee>
      <name>Harry Hacker</name>
      <salary>50000</salary>
      <hiredate year="1989" month="10" day="1"/>
   </employee>
   <employee>
      <name>Tony Tester</name>
      <salary>40000</salary>
      <hiredate year="1990" month="3" day="15"/>
   </employee>
</staff>
```

我们希望的输出是一张 HTML 表格：

```
<table border="1">
<tr>
```

```
<td>Carl Cracker</td><td>$75000.0</td><td>1987-12-15</td>
</tr>
<tr>
<td>Harry Hacker</td><td>$50000.0</td><td>1989-10-1</td>
</tr>
<tr>
<td>Tony Tester</td><td>$40000.0</td><td>1990-3-15</td>
</tr>
</table>
```

具有转换模板的样式表形式如下：

```
<?xml version="1.0" encoding="ISO-8859-1"?>
<xsl:stylesheet
    xmlns:xsl="http://www.w3.org/1999/XSL/Transform"
    version="1.0">
  <xsl:output method="html"/>
```
$template_1$

$template_2$
```
 . . .
</xsl:stylesheet>
```

在我们的例子中，xsl:output 元素将方法设定为 HTML，而其他有效的方法设置是 xml 和 text。

下面是一个典型的模板：

```
<xsl:template match="/staff/employee">
    <tr><xsl:apply-templates/></tr>
</xsl:template>
```

match 属性的值是一个 XPath 表达式。该模板声明，每当看到 XPath 集 /staff/ employee 中的一个节点时，将做以下操作：

1. 产生字符串 `<tr>`。
2. 在处理其子节点时，持续应用该模板。
3. 当处理完所有子节点后，产生字符串 `</tr>`。

换句话说，该模板会生成围绕每条雇员记录的 HTML 表格的行标记。

XSLT 处理器以检查根元素开始其处理过程。每当一个节点匹配某个模板时，就会应用该模板（如果匹配多个模板，就会使用最佳匹配的那个，详情请参见 http://www.w3.org/TR/xslt）。如果没有匹配的模板，处理器会执行默认操作。对于文本节点，默认操作是把它的内容囊括到输出中去。对于元素，默认操作是不产生任何输出，但会继续处理其子节点。

下面是一个用来转换雇员记录文件中的 name 节点的模板：

```
<xsl:template match="/staff/employee/name">
    <td><xsl:apply-templates/></td>
</xsl:template>
```

正如你所见，模板产生定界符 `<td>...</td>`，并且让处理器递归访问 name 元素的子节点。它只有一个子节点，即文本节点。当处理器访问该节点时，它会提取出其中的文本内容（当然，前提是没有其他匹配的模板）。

如果想要把属性值复制到输出中去，就不得不再做一些稍微复杂的操作了。下面是一个例子：

```
<xsl:template match="/staff/employee/hiredate">
    <td><xsl:value-of select="@year"/>-<xsl:value-of
    select="@month"/>-<xsl:value-of select="@day"/></td>
</xsl:template>
```

当处理 hiredate 节点时，该模板会产生：

1. 字符串 \<td\>

2. year 属性的值

3. 一个连字符

4. month 属性的值

5. 一个连字符

6. day 属性的值

7. 字符串 \</td\>

xsl:value-of 语句用于计算节点集的字符串值，其中，节点集由 select 属性的 XPath 值指定。在这个例子中，路径是相对于当前正在处理的节点的相对路径。节点集通过将各个节点的字符串值连接起来而被转换成一个字符串。属性节点的字符串值就是它的值，文本节点的字符串值是它的内容，元素节点的字符串值是它的所有子节点（而不是其属性）的字符串值的连接。

程序清单 3-10 包含了将带有雇员记录的 XML 文件转换成 HTML 表格的样式表。

程序清单 3-10　transform/makehtml.xsl

```
 1  <?xml version="1.0" encoding="ISO-8859-1"?>
 2
 3  <xsl:stylesheet
 4      xmlns:xsl="http://www.w3.org/1999/XSL/Transform"
 5      version="1.0">
 6
 7      <xsl:output method="html"/>
 8
 9      <xsl:template match="/staff">
10          <table border="1"><xsl:apply-templates/></table>
11      </xsl:template>
12
13      <xsl:template match="/staff/employee">
14          <tr><xsl:apply-templates/></tr>
15      </xsl:template>
16
17      <xsl:template match="/staff/employee/name">
18          <td><xsl:apply-templates/></td>
19      </xsl:template>
20
21      <xsl:template match="/staff/employee/salary">
22          <td>$<xsl:apply-templates/></td>
23      </xsl:template>
```

```
24
25      <xsl:template match="/staff/employee/hiredate">
26         <td><xsl:value-of select="@year"/>-<xsl:value-of
27      select="@month"/>-<xsl:value-of select="@day"/></td>
28      </xsl:template>
29
30   </xsl:stylesheet>
```

程序清单 3-11 显示了一组不同的转换，其输入是相同的 XML 文件，输出是我们熟悉的属性文件格式的纯文本。

```
employee.1.name=Carl Cracker
employee.1.salary=75000.0
employee.1.hiredate=1987-12-15
employee.2.name=Harry Hacker
employee.2.salary=50000.0
employee.2.hiredate=1989-10-1
employee.3.name=Tony Tester
employee.3.salary=40000.0
employee.3.hiredate=1990-3-15
```

程序清单 3-11　transform/makeprop.xsl

```
1    <?xml version="1.0"?>
2
3    <xsl:stylesheet
4       xmlns:xsl="http://www.w3.org/1999/XSL/Transform"
5       version="1.0">
6
7       <xsl:output method="text" omit-xml-declaration="yes"/>
8
9       <xsl:template match="/staff/employee">
10   employee.<xsl:value-of select="position()"
11   />.name=<xsl:value-of select="name/text()"/>
12   employee.<xsl:value-of select="position()"
13   />.salary=<xsl:value-of select="salary/text()"/>
14   employee.<xsl:value-of select="position()"
15   />.hiredate=<xsl:value-of select="hiredate/@year"
16   />-<xsl:value-of select="hiredate/@month"
17   />-<xsl:value-of select="hiredate/@day"/>
18      </xsl:template>
19
20   </xsl:stylesheet>
```

该示例使用 position() 函数来产生以其父节点的角度来看的当前节点的位置。我们只要切换样式表就可以得到一个完全不同的输出。这样，就可以安全地使用 XML 来描述数据了，即便一些应用程序需要的是其他格式的数据，我们只要用 XSLT 来产生对应的可替代格式即可。

在 Java 平台中产生 XML 的转换极其简单，只需为每个样式表设置一个转换器工厂，然后得到一个转换器对象，并告诉它把一个源转换成结果。

```
var styleSheet = new File(filename);
var styleSource = new StreamSource(styleSheet);
Transformer t = TransformerFactory.newInstance().newTransformer(styleSource);
t.transform(source, result);
```

transform 方法的参数是 Source 和 Result 接口的实现类的对象。Source 接口有 4 个实现类:

```
DOMSource
SAXSource
StAXSource
StreamSource
```

你可以从一个文件、流、阅读器或 URL 中构建 StreamSource 对象,或者从 DOM 树节点中构建 DOMSource 对象。例如,在上一节中,我们调用了如下的标识转换:

```
t.transform(new DOMSource(doc), result);
```

在示例程序中,我们做了一些更有趣的事情。我们并不是从一个现有的 XML 文件开始工作,而是产生一个 SAX XML 阅读器,通过产生适合的 SAX 事件,给人以解析 XML 文件的错觉。实际上,XML 阅读器读入的是一个如第 2 章所描述的扁平文件,输入文件看上去是这样的:

```
Carl Cracker|75000.0|1987|12|15
Harry Hacker|50000.0|1989|10|1
Tony Tester|40000.0|1990|3|15
```

处理输入时,XML 阅读器将产生 SAX 事件。下面是实现了 XMLReader 接口的 EmployeeReader 类的 parse 方法的一部分代码:

```
var attributes = new AttributesImpl();
handler.startDocument();
handler.startElement("", "staff", "staff", attributes);
while ((line = in.readLine()) != null)
{
   handler.startElement("", "employee", "employee", attributes);
   var tokenizer = new StringTokenizer(line, "|");
   handler.startElement("", "name", "name", attributes);
   String s = tokenizer.nextToken();
   handler.characters(s.toCharArray(), 0, s.length());
   handler.endElement("", "name", "name");
   . . .
   handler.endElement("", "employee", "employee");
}
handler.endElement("", rootElement, rootElement);
handler.endDocument();
```

用于转换器的 SAXSource 是从 XML 阅读器中构建的:

```
t.transform(new SAXSource(new EmployeeReader(),
   new InputSource(new FileInputStream(filename))), result);
```

这是将非 XML 的遗留数据转换成 XML 的一个小技巧。当然,大多数 XSLT 应用程序都已经有了 XML 格式的输入数据,只需要在一个 StreamSource 对象上调用 transform 方法即可,例如:

```
t.transform(new StreamSource(file), result);
```

其转换结果是 Result 接口的实现类的一个对象。Java 库提供了 3 个类:

```
DOMResult
SAXResult
StreamResult
```

要把结果存储到 DOM 树中,请使用 DocumentBuilder 产生一个新的文档节点,并将其包装到 DOMResult 中:

```
Document doc = builder.newDocument();
t.transform(source, new DOMResult(doc));
```

要将输出保存到文件中,请使用 StreamResult:

```
t.transform(source, new StreamResult(file));
```

程序清单 3-12 包含了完整的源代码。

程序清单 3-12　transform/TransformTest.java

```
 1  package transform;
 2
 3  import java.io.*;
 4  import java.nio.file.*;
 5  import java.util.*;
 6  import javax.xml.transform.*;
 7  import javax.xml.transform.sax.*;
 8  import javax.xml.transform.stream.*;
 9  import org.xml.sax.*;
10  import org.xml.sax.helpers.*;
11
12  /**
13   * This program demonstrates XSL transformations. It applies a transformation to a set of
14   * employee records. The records are stored in the file employee.dat and turned into XML
15   * format. Specify the stylesheet on the command line, e.g.<br>
16   *    java transform.TransformTest transform/makeprop.xsl
17   * @version 1.04 2018-04-10
18   * @author Cay Horstmann
19   */
20  public class TransformTest
21  {
22     public static void main(String[] args) throws Exception
23     {
24        Path path;
25        if (args.length > 0) path = Paths.get(args[0]);
26        else path = Paths.get("transform", "makehtml.xsl");
27        try (InputStream styleIn = Files.newInputStream(path))
28        {
29           var styleSource = new StreamSource(styleIn);
30
31           Transformer t = TransformerFactory.newInstance().newTransformer(styleSource);
32           t.setOutputProperty(OutputKeys.INDENT, "yes");
33           t.setOutputProperty(OutputKeys.METHOD, "xml");
34           t.setOutputProperty("{http://xml.apache.org/xslt}indent-amount", "2");
35
36           try (InputStream docIn = Files.newInputStream(Paths.get("transform", "employee.dat")))
```

```java
37          {
38             t.transform(new SAXSource(new EmployeeReader(), new InputSource(docIn)),
39                new StreamResult(System.out));
40          }
41       }
42    }
43 }
44
45 /**
46  * This class reads the flat file employee.dat and reports SAX parser events to act as if it
47  * was parsing an XML file.
48  */
49 class EmployeeReader implements XMLReader
50 {
51    private ContentHandler handler;
52
53    public void parse(InputSource source) throws IOException, SAXException
54    {
55       InputStream stream = source.getByteStream();
56       var in = new BufferedReader(new InputStreamReader(stream));
57       String rootElement = "staff";
58       var atts = new AttributesImpl();
59
60       if (handler == null) throw new SAXException("No content handler");
61
62       handler.startDocument();
63       handler.startElement("", rootElement, rootElement, atts);
64       String line;
65       while ((line = in.readLine()) != null)
66       {
67          handler.startElement("", "employee", "employee", atts);
68          var t = new StringTokenizer(line, "|");
69
70          handler.startElement("", "name", "name", atts);
71          String s = t.nextToken();
72          handler.characters(s.toCharArray(), 0, s.length());
73          handler.endElement("", "name", "name");
74
75          handler.startElement("", "salary", "salary", atts);
76          s = t.nextToken();
77          handler.characters(s.toCharArray(), 0, s.length());
78          handler.endElement("", "salary", "salary");
79
80          atts.addAttribute("", "year", "year", "CDATA", t.nextToken());
81          atts.addAttribute("", "month", "month", "CDATA", t.nextToken());
82          atts.addAttribute("", "day", "day", "CDATA", t.nextToken());
83          handler.startElement("", "hiredate", "hiredate", atts);
84          handler.endElement("", "hiredate", "hiredate");
85          atts.clear();
86
87          handler.endElement("", "employee", "employee");
88       }
89
90       handler.endElement("", rootElement, rootElement);
```

```
 91          handler.endDocument();
 92       }
 93
 94       public void setContentHandler(ContentHandler newValue)
 95       {
 96          handler = newValue;
 97       }
 98
 99       public ContentHandler getContentHandler()
100       {
101          return handler;
102       }
103
104       // the following methods are just do-nothing implementations
105       public void parse(String systemId) throws IOException, SAXException {}
106       public void setErrorHandler(ErrorHandler handler) {}
107       public ErrorHandler getErrorHandler() { return null; }
108       public void setDTDHandler(DTDHandler handler) {}
109       public DTDHandler getDTDHandler() { return null; }
110       public void setEntityResolver(EntityResolver resolver) {}
111       public EntityResolver getEntityResolver() { return null; }
112       public void setProperty(String name, Object value) {}
113       public Object getProperty(String name) { return null; }
114       public void setFeature(String name, boolean value) {}
115       public boolean getFeature(String name) { return false; }
116    }
```

API javax.xml.transform.TransformerFactory 1.4

- Transformer newTransformer(Source styleSheet)

  返回一个 transformer 类的实例,用来从指定的源中读取样式表。

API javax.xml.transform.stream.StreamSource 1.4

- StreamSource(File f)
- StreamSource(InputStream in)
- StreamSource(Reader in)
- StreamSource(String systemID)

  自一个文件、流、阅读器或系统 ID(通常是相对或绝对 URL)构建一个数据流源。

API javax.xml.transform.sax.SAXSource 1.4

- SAXSource(XMLReader reader, InputSource source)

  构建一个 SAX 数据源,以便从给定输入源中获取数据,并使用给定的阅读器来解析输入数据。

API org.xml.sax.XMLReader 1.4

- void setContentHandler(ContentHandler handler)

  设置在输入被解析时会被告知解析事件的处理器。

- void parse(InputSource source)

  解析来自给定输入源的输入数据，并将解析事件发送到内容处理器。

API javax.xml.transform.dom.DOMResult 1.4

- DOMResult(Node n)

  自给定节点构建一个数据源。通常，n 是一个新文档节点。

API org.xml.sax.helpers.AttributesImpl 1.4

- void addAttribute(String uri, String lname, String qname, String type, String value)

  将一个属性添加到该属性集合。

  lname 参数是无前缀的本地名，而 qname 参数是带前缀的限定名，type 参数是 "CDATA" "ID" "IDREF" "IDREFS" "NMTOKEN" "NMTOKENS" "ENTITY" "ENTITIES" 或 "NOTATION" 之一

- void clear()

  删除当前属性集合中的所有属性。

我们以该示例结束对 Java 库中的 XML 支持特性的讨论。现在，你应该对 XML 的强大功能有了很好的了解，尤其是它的自动解析、验证和强大的转换机制。当然，所有这些技术只有在你很好地设计了 XML 格式之后才能发挥作用。你必须确保那些格式足够丰富，能够表达全部业务需求，随着时间的推移也依旧稳定，你的业务伙伴也愿意接受你的 XML 文档。这些问题要远比处理解析器、DTD 或转换更具挑战。

在下一章，我们将讨论在 Java 平台上的网络编程，从最基础的网络套接字开始，逐渐过渡到用于 E-mail 和万维网的更高层协议。

第 4 章 网 络

- ▲ 连接到服务器
- ▲ 实现服务器
- ▲ 获取 Web 数据
- ▲ HTTP 客户端
- ▲ 发送 E-mail

本章的开头部分将首先回顾一下网络方面的基本概念，然后进一步介绍如何编写连接网络服务的 Java 程序，并演示网络客户端和服务器是如何实现的，最后将介绍如何通过 Java 程序发送 E-mail，以及如何从 Web 服务器获得信息。

4.1 连接到服务器

在下面各节中，你将会学习如何连接到服务器，先是手工用 telnet 连接，然后是用 Java 程序连接。

4.1.1 使用 telnet

telnet 是一种用于网络编程的非常强大的调试工具，可以在命令 shell 中输入 telnet 来启动它。

> **注释**：在 Windows 中，需要激活 telnet。要激活它，需要到"控制面板"，选择"程序"，点击"打开/关闭 Windows 特性"，然后选择"Telnet 客户端"复选框。Windows 防火墙将会阻止我们在本章中使用的很多网络端口，你可能需要管理员账户才能解除对它们的禁用。

你可能曾经使用过 telnet 来连接远程计算机，但其实你也可以用它与因特网主机所提供的其他服务进行通信。下面是一个可以操作的例子。请输入：

    telnet time-a.nist.gov 13

如图 4-1 所示，你可以得到与下面这一行相似的信息：

    57488 16-04-10 04:23:00 50 0 0 610.5 UTC(NIST) *

上面例子说明了什么？它说明你已经连接到了大多数 UNIX 计算机都支持的"当日时

图 4-1 "当日时间"服务的输出

间"服务。而你刚才所连接的那台服务器就是由国家标准与技术研究所运维的，这家研究所负责提供铯原子钟的计量时间。（当然，由于网络延迟的缘故，原子钟反馈过来的时间并不完全准确。）

按照惯例，"当日时间"服务总是连接到端口 13。

> **注释**：在网络术语中，端口并不是指物理设备，而是为了便于实现服务器与客户端之间的通信所使用的抽象概念（见图 4-2）。

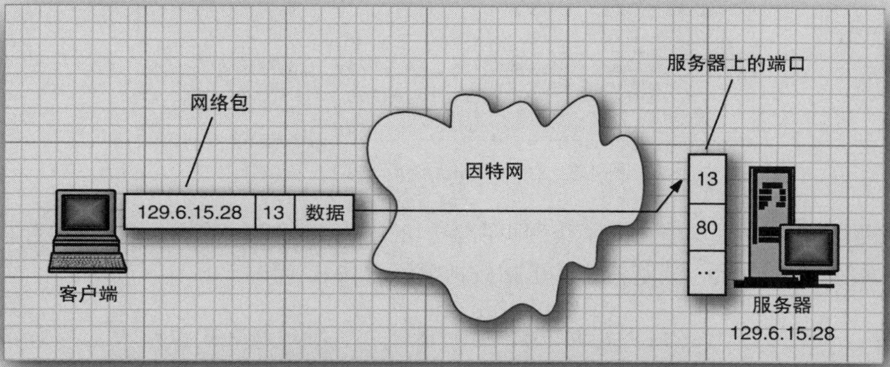

图 4-2　连接到服务器端口的客户端

运行在远程计算机上的服务器软件不停地等待那些希望与端口 13 连接的网络请求。当远程计算机上的操作系统接收到一个请求与端口 13 连接的网络数据包时，它便唤醒正在监听网络连接请求的服务器进程，并为两者建立连接。这种连接将一直保持下去，直到被其中任何一方中止。

当你开始用 time-a.nist.gov 在端口 13 上建立 telnet 会话时，网络软件中有一段代码非常清楚地知道应该将字符串 "time-a.nist.gov" 转换为正确的 IP 地址 129.6.15.28。随后，telnet 软件发送一个连接请求给该地址，请求一个到端口 13 的连接。一旦建立连接，远程程序便发送回一行数据，然后关闭该连接。当然，一般而言，客户端和服务器在其中一方关闭连接之前，会进行更多的对话。

下面是另一个同类型的试验，但它更加有趣。请执行以下操作：

```
telnet horstmann.com 80
```

然后非常仔细地键入以下内容：

```
GET / HTTP/1.1
Host: horstmann.com
blank line
```

也就是在末尾按两次 Enter 键。

图 4-3 显示了以上操作的响应结果。它看上去应该是你非常熟悉的——你得到的是一个

HTML 格式的文本页，即 Cay Horstmann 的主页。

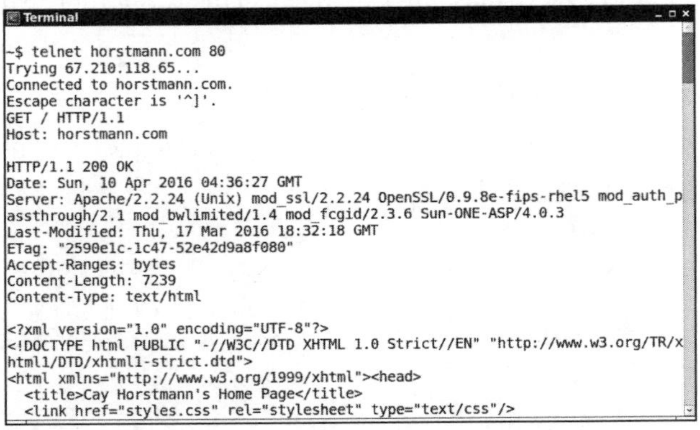

图 4-3 使用 telnet 访问 HTTP 端口

上述操作与 Web 浏览器访问某个网页所经历的过程是完全一致的，它使用 HTTP 向服务器请求 Web 页面。当然，浏览器能够更精致地显示 HTML 代码。

> **注释**：如果一台 Web 服务器用相同的 IP 地址为多个域提供宿主环境，那么在连接这台 Web Server 时，就必须提供 Host 键 / 值对。如果服务器只为单个域提供宿主环境，则可以忽略该键 / 值对。

4.1.2 用 Java 连接到服务器

程序清单 4-1 是我们的第一个网络程序。它的作用与我们使用 telnet 工具是相同的，即连接到某个端口并打印出它所找到的信息。

程序清单 4-1 socket/SocketTest.java

```java
package socket;

import java.io.*;
import java.net.*;
import java.nio.charset.*;
import java.util.*;

/**
 * This program makes a socket connection to the atomic clock in Boulder, Colorado, and prints
 * the time that the server sends.
 * @version 1.22 2018-03-17
 * @author Cay Horstmann
 */
public class SocketTest
{
    public static void main(String[] args) throws IOException
    {
        try (var s = new Socket("time-a.nist.gov", 13);
```

```
19                var in = new Scanner(s.getInputStream(), StandardCharsets.UTF_8))
20          {
21              while (in.hasNextLine())
22              {
23                  String line = in.nextLine();
24                  System.out.println(line);
25              }
26          }
27      }
28  }
```

下面是这个简单程序的几行关键代码：

```
var s = new Socket("time-a.nist.gov", 13);
InputStream inStream = s.getInputStream();
```

第一行代码用于打开一个套接字，它也是网络软件中的一个抽象概念，负责启动该程序内部和外部之间的通信。我们将远程地址和端口号传递给套接字的构造器，如果连接失败，它将抛出一个 UnknownHostException 异常；如果存在其他问题，它将抛出一个 IOException 异常。因为 UnknownHostException 是 IOException 的一个子类，况且这只是一个示例程序，所以我们在这里仅仅捕获超类的异常。

一旦套接字被打开，java.net.Socket 类中的 getInputStream 方法就会返回一个 InputStream 对象，该对象可以像其他任何流对象一样使用。而一旦获取了这个流，该程序将直接把每一行打印到标准输出。这个过程将一直持续到流发送完毕且服务器断开连接为止。

该程序只适用于非常简单的服务器，比如"当日时间"之类的服务。在比较复杂的网络程序中，客户端发送请求数据给服务器，而服务器可能在响应结束时并不立刻断开连接。在本章的若干个示例程序中，都会看到我们是如何实现这种行为的。

Socket 类非常简单易用，因为 Java 库隐藏了建立网络连接和通过连接发送数据的复杂过程。实际上，java.net 包提供的编程接口与操作文件时所使用的接口基本相同。

> **注释**：本书所介绍的内容仅覆盖了 TCP（传输控制协议）网络协议。Java 平台另外还支持 UDP（用户数据报协议）协议，该协议可以用于发送数据包（也称为数据报），它所需付出的开销要比 TCP 少得多。UDP 有一个重要的缺点：数据包无须按照顺序传递到接收应用程序，它们甚至可能在传输过程中全部丢失。UDP 让数据包的接收者自己负责对它们进行排序，并请求发送者重新发送那些丢失的数据包。UDP 比较适合于那些可以忍受数据包丢失的应用，例如用于音频流和视频流的传输，或者用于连续测量的应用领域。

API java.net.Socket 1.0

- Socket(String host, int port)
  构建一个套接字，用来连接给定的主机和端口。
- InputStream getInputStream()

- OutputStream getOutputStream()

  获取可以从套接字中读取数据的流，以及可以向套接字写出数据的流。

4.1.3 套接字超时

从套接字读取信息时，在有数据可供访问之前，读操作将会被阻塞。如果此时主机不可达，那么应用将要等待很长的时间，并且因为受底层操作系统的限制而最终会导致超时。

对于不同的应用，应该确定合理的超时值。然后调用 setSoTimeout 方法设置这个超时值（单位：毫秒）。

```
var s = new Socket(. . .);
s.setSoTimeout(10000); // time out after 10 seconds
```

如果已经为套接字设置了超时值，并且之后的读操作和写操作在没有完成之前就超过了时间限制，那么这些操作就会抛出 SocketTimeoutException 异常。你可以捕获这个异常，并对超时做出反应。

```
try
{
   InputStream in = s.getInputStream(); // read from in
   . . .
}
catch (SocketTimeoutException e)
{
   react to timeout
}
```

另外还有一个超时问题是必须解决的。下面这个构造器：

```
Socket(String host, int port)
```

会一直无限期地阻塞下去，直到建立了到达主机的初始连接为止。

可以通过先构建一个无连接的套接字，然后再使用一个超时来进行连接的方式解决这个问题。

```
var s = new Socket();
s.connect(new InetSocketAddress(host, port), timeout);
```

如果你希望允许用户在任何时刻都可以中断套接字连接，请查看 4.2.4 节。

API **java.net.Socket** 1.0

- Socket() 1.1

  创建一个还未被连接的套接字。

- void connect(SocketAddress address) 1.4

  将该套接字连接到给定的地址。

- void connect(SocketAddress address, int timeoutInMilliseconds) 1.4

  将套接字连接到给定的地址。如果在给定的时间内没有响应，则返回。

- void setSoTimeout(int timeoutInMilliseconds) 1.1

设置该套接字上读请求的阻塞时间。如果超出给定时间，则抛出一个 SocketTimeoutException 异常。

- boolean isConnected() 1.4

  如果该套接字已被连接，则返回 true。

- boolean isClosed() 1.4

  如果套接字已经被关闭，则返回 true。

4.1.4 因特网地址

通常，不用过多考虑因特网地址的问题，它们是用一串数字表示的主机地址，一个因特网地址由 4 个字节组成（在 IPv6 中是 16 个字节），比如 129.6.15.28。但是，如果需要在主机名和因特网地址之间进行转换，那么就可以使用 InetAddress 类。

只要主机操作系统支持 IPv6 格式的因特网地址，java.net 包也将支持它。

静态的 getByName 方法可以返回代表某个主机的 InetAddress 对象。例如，

```
InetAddress address = InetAddress.getByName("time-a.nist.gov");
```

将返回一个 InetAddress 对象，该对象封装了一个 4 字节的序列：129.6.15.28。然后，可以使用 getAddress 方法来访问这些字节：

```
byte[] addressBytes = address.getAddress();
```

一些访问量较大的主机名通常会对应于多个因特网地址，以实现负载均衡。例如，在撰写本书时，主机名 google.com 就对应着 12 个不同的因特网地址。当访问主机时，会随机选取其中的一个。可以通过调用 getAllByName 方法来获得所有主机：

```
InetAddress[] addresses = InetAddress.getAllByName(host);
```

最后需要说明的是，有时我们可能需要本地主机的地址。如果只是要求得到 localhost 的地址，那总会得到本地回环地址 127.0.0.1，但是其他程序无法用这个地址来连接到这台机器上。此时，可以使用静态的 getLocalHost 方法来得到本地主机的地址：

```
InetAddress address = InetAddress.getLocalHost();
```

程序清单 4-2 是一段比较简单的程序代码。如果不在命令行中设置任何参数，那么它将打印出本地主机的因特网地址。反之，如果在命令行中指定了主机名，那么它将打印出该主机的所有因特网地址，例如：

```
java inetAddress/InetAddressTest www.horstmann.com
```

程序清单 4-2   inetAddress/InetAddressTest.java

```
 1  package inetAddress;
 2
 3  import java.io.*;
 4  import java.net.*;
 5
 6  /**
 7   * This program demonstrates the InetAddress class. Supply a host name as command-line
```

```
  8    * argument, or run without command-line arguments to see the address of the local host.
  9    * @version 1.02 2012-06-05
 10    * @author Cay Horstmann
 11    */
 12   public class InetAddressTest
 13   {
 14      public static void main(String[] args) throws IOException
 15      {
 16         if (args.length > 0)
 17         {
 18            String host = args[0];
 19            InetAddress[] addresses = InetAddress.getAllByName(host);
 20            for (InetAddress a : addresses)
 21               System.out.println(a);
 22         }
 23         else
 24         {
 25            InetAddress localHostAddress = InetAddress.getLocalHost();
 26            System.out.println(localHostAddress);
 27         }
 28      }
 29   }
```

API java.net.InetAddress 1.0

- static InetAddress getByName(String host)
- static InetAddress[] getAllByName(String host)

  为给定的主机名创建一个 InetAddress 对象，或者一个包含了该主机名所对应的所有因特网地址的数组。

- static InetAddress getLocalHost()

  为本地主机创建一个 InetAddress 对象。

- byte[] getAddress()

  返回一个包含数字型地址的字节数组。

- String getHostAddress()

  返回一个由十进制数组成的字符串，各数字间用圆点符号隔开，例如，"129.6.15.28"。

- String getHostName()

  返回主机名。

4.2 实现服务器

在上一节中，我们已经实现了一个基本的网络客户端，并且用它从因特网上获取了数据。在这一节中，我们将实现一个简单的服务器，它可以向客户端发送信息。

4.2.1 服务器套接字

一旦启动了服务器程序，它便会等待某个客户端连接到它的端口。在我们的示例程序中，我们选择端口号 8189，因为所有标准服务都不使用这个端口。ServerSocket 类用于建立套接字。在我们的示例中，下面这行命令：

```
var s = new ServerSocket(8189);
```

用于建立一个负责监控端口 8189 的服务器。以下命令：

```
Socket incoming = s.accept();
```

用于告诉程序不停地等待，直到有客户端连接到这个端口。一旦有人通过网络发送了正确的连接请求，并以此连接到了端口上，该方法就会返回一个表示连接已经建立的 Socket 对象。你可以使用这个对象来得到输入流和输出流，代码如下：

```
InputStream inStream = incoming.getInputStream();
OutputStream outStream = incoming.getOutputStream();
```

服务器发送给服务器输出流的所有信息都会成为客户端程序的输入，同时来自客户端程序的所有输出都会被包含在服务器输入流中。

因为在本章的所有示例程序中，我们都要通过套接字来发送文本，所以我们将流转换成扫描器和写入器。

```
var in = new Scanner(inStream, StandardCharsets.UTF_8);
var out = new PrintWriter(new OutputStreamWriter(outStream, StandardCharsets.UTF_8),
    true /* autoFlush */);
```

以下代码将给客户端发送一条问候信息：

```
out.println("Hello! Enter BYE to exit.");
```

当使用 telnet 通过端口 8189 连接到这个服务器程序时，将会在终端屏幕上看到上述问候信息。

在这个简单的服务器程序中，它只是读取客户端输入，每次读取一行，并回送这一行。这表明程序接收到了客户端的输入。当然，实际应用中的服务器都会对输入进行计算并返回处理结果。

```
String line = in.nextLine();
out.println("Echo: " + line);
if (line.trim().equals("BYE")) done = true;
```

在代码的最后，我们关闭了连接进来的套接字。

```
incoming.close();
```

这就是整个示例代码的大致情况。每一个服务器程序，比如一个 HTTP Web 服务器，都会不间断地执行下面这个循环：

1. 通过输入数据流从客户端接收一个命令（"get me this information"）。
2. 解码这个客户端命令。
3. 收集客户端所请求的信息。
4. 通过输出数据流发送信息给客户端。

程序清单 4-3 给出了这个程序的完整代码。

程序清单 4-3 server/EchoServer.java

```java
1  package server;
2
3  import java.io.*;
4  import java.net.*;
5  import java.nio.charset.*;
6  import java.util.*;
7
8  /**
9   * This program implements a simple server that listens to port 8189 and echoes back all
10  * client input.
11  * @version 1.22 2018-03-17
12  * @author Cay Horstmann
13  */
14  public class EchoServer
15  {
16     public static void main(String[] args) throws IOException
17     {
18        // establish server socket
19        try (var s = new ServerSocket(8189))
20        {
21           // wait for client connection
22           try (Socket incoming = s.accept())
23           {
24              InputStream inStream = incoming.getInputStream();
25              OutputStream outStream = incoming.getOutputStream();
26
27              try (var in = new Scanner(inStream, StandardCharsets.UTF_8))
28              {
29                 var out = new PrintWriter(
30                    new OutputStreamWriter(outStream, StandardCharsets.UTF_8),
31                    true /* autoFlush */);
32
33                 out.println("Hello! Enter BYE to exit.");
34
35                 // echo client input
36                 var done = false;
37                 while (!done && in.hasNextLine())
38                 {
39                    String line = in.nextLine();
40                    out.println("Echo: " + line);
41                    if (line.trim().equals("BYE")) done = true;
42                 }
43              }
44           }
45        }
46     }
47  }
```

想要试一下这个例子，就请编译并运行这个程序。然后使用 telnet 连接到服务器 localhost（或 IP 地址 127.0.0.1）和端口 8189。

如果你直接连接到因特网上，那么世界上任何人都可以访问到你的回送服务器，只要他们知道你的 IP 地址和端口号。

当你连接到该端口时，将看到如图 4-4 所示的信息：

Hello! Enter BYE to exit.

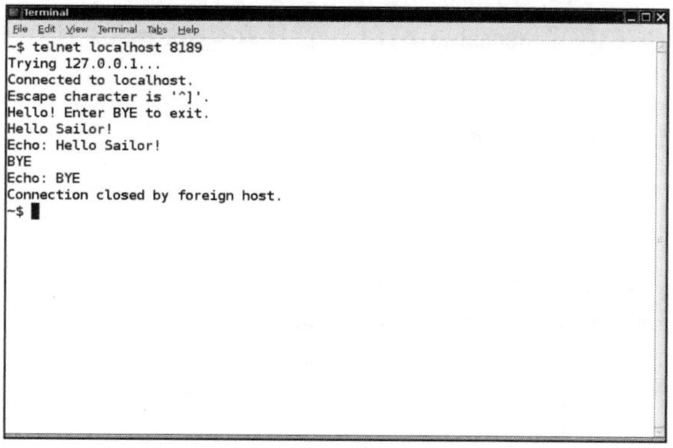

图 4-4　访问一个回送服务器

可以随意键入一条信息，然后观察屏幕上的回送信息。输入 BYE（全为大写字母）可以断开连接，同时，服务器程序也会终止运行。

API java.net.ServerSocket 1.0

- ServerSocket(int port)
  创建一个监听端口的服务器套接字。
- Socket accept()
  等待连接。该方法阻塞（即，使之空闲）当前线程直到建立连接为止。该方法返回一个 Socket 对象，程序可以通过这个对象与连接中的客户端进行通信。
- void close()
  关闭服务器套接字。

4.2.2　为多个客户端服务

前面例子中的简单服务器存在一个问题。假设我们希望有多个客户端同时连接到我们的服务器上。通常，服务器总是不间断地运行在服务器计算机上，来自整个因特网的用户希望同时使用服务器。前面的简单服务器会提供对客户端连接的支持，使得任何一个客户端都可以因长时间地连接服务而独占服务，其实我们可以运用线程的魔力把这个问题解决得更好。

每当程序建立一个新的套接字连接，也就是说当调用 accept() 时，将会启动一个新的线程来处理服务器和该客户端之间的连接，而主程序将立即返回并等待下一个连接。为了实现这种机制，服务器应该具有类似以下代码的循环操作：

```
while (true)
{
    Socket incoming = s.accept();
```

```
      var r = new ThreadedEchoHandler(incoming);
      var t = new Thread(r);
      t.start();
   }
```

ThreadedEchoHandler 类实现了 Runnable 接口,而且在它的 run 方法中包含了与客户端循环通信的代码。

```
class ThreadedEchoHandler implements Runnable
{
   . . .
   public void run()
   {
      try (InputStream inStream = incoming.getInputStream();
           OutputStream outStream = incoming.getOutputStream())
      {
         Process input and send response
      }
      catch(IOException e)
      {
         Handle exception
      }
   }
}
```

由于每一个连接都会启动一个新的线程,因而多个客户端就可以同时连接到服务器了。对此可以做个简单的测试:

1. 编译和运行服务器程序(程序清单 4-4)。
2. 如图 4-5 所示打开数个 telnet 窗口。
3. 在这些窗口之间切换,并键入命令。注意你可以同时通过这些窗口进行通信。
4. 当完成之后,切换到启动服务器程序的窗口,并使用 CTRL+C 强行关闭它。

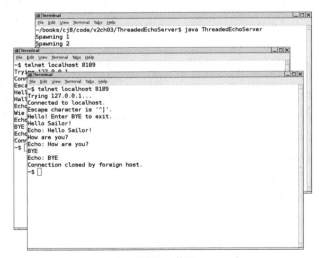

图 4-5 多个同时通信的 telnet 窗口

> **注释**：在这个程序中，我们为每个连接生成一个单独的线程。这种方法并不能满足高性能服务器的要求。为使服务器实现更高的吞吐量，可以使用 java.nio 包中一些特性。详情请参见以下链接：http://www.ibm.com/developerworks/java/library/j-javaio。

程序清单 4-4　threaded/ThreadedEchoServer.java

```java
 1  package threaded;
 2
 3  import java.io.*;
 4  import java.net.*;
 5  import java.nio.charset.*;
 6  import java.util.*;
 7
 8  /**
 9   * This program implements a multithreaded server that listens to port 8189 and echoes back
10   * all client input.
11   * @author Cay Horstmann
12   * @version 1.23 2018-03-17
13   */
14  public class ThreadedEchoServer
15  {
16     public static void main(String[] args )
17     {
18        try (var s = new ServerSocket(8189))
19        {
20           int i = 1;
21
22           while (true)
23           {
24              Socket incoming = s.accept();
25              System.out.println("Spawning " + i);
26              Runnable r = new ThreadedEchoHandler(incoming);
27              var t = new Thread(r);
28              t.start();
29              i++;
30           }
31        }
32        catch (IOException e)
33        {
34           e.printStackTrace();
35        }
36     }
37  }
38
39  /**
40   * This class handles the client input for one server socket connection.
41   */
42  class ThreadedEchoHandler implements Runnable
43  {
44     private Socket incoming;
45
46     /**
47      * Constructs a handler.
```

```
48         @param incomingSocket the incoming socket
49      */
50     public ThreadedEchoHandler(Socket incomingSocket)
51     {
52         incoming = incomingSocket;
53     }
54
55     public void run()
56     {
57         try (InputStream inStream = incoming.getInputStream();
58              OutputStream outStream = incoming.getOutputStream();
59              var in = new Scanner(inStream, StandardCharsets.UTF_8);
60              var out = new PrintWriter(
61                  new OutputStreamWriter(outStream, StandardCharsets.UTF_8),
62                  true /* autoFlush */))
63         {
64             out.println( "Hello! Enter BYE to exit." );
65
66             // echo client input
67             var done = false;
68             while (!done && in.hasNextLine())
69             {
70                 String line = in.nextLine();
71                 out.println("Echo: " + line);
72                 if (line.trim().equals("BYE"))
73                     done = true;
74             }
75         }
76         catch (IOException e)
77         {
78             e.printStackTrace();
79         }
80     }
81 }
```

4.2.3 半关闭

半关闭（half-close）提供了这样一种能力：套接字连接的一端可以终止其输出，同时仍旧可以接收来自另一端的数据。

这是一种很典型的情况，例如我们在向服务器传输数据，但是一开始并不知道要传输多少数据。在向文件写数据时，我们只需在数据写入后关闭文件即可。但是，如果关闭一个套接字，那么与服务器的连接将立刻断开，因而也就无法读取服务器的响应了。

使用半关闭的方法就可以解决上述问题。可以通过关闭一个套接字的输出流来表示发送给服务器的请求数据已经结束，但是必须保持输入流处于打开状态。

如下代码演示了如何在客户端使用半关闭方法：

```
try (var socket = new Socket(host, port))
{
    var in = new Scanner(socket.getInputStream(), StandardCharsets.UTF_8);
    var writer = new PrintWriter(socket.getOutputStream());
```

```
// send request data
writer.print(. . .);
writer.flush();
socket.shutdownOutput();
// now socket is half-closed
// read response data
while (in.hasNextLine() != null)
{
    String line = in.nextLine();
    . . .
}
```

服务器端将读取输入信息，直至到达输入流的结尾，然后它再发送响应。

当然，该协议只适用于一站式（one-shot）的服务，例如 HTTP 服务，在这种服务中，客户端连接服务器，发送一个请求，捕获响应信息，然后断开连接。

API java.net.Socket 1.0

- void shutdownOutput() 1.3

  将输出流设为"流结束"。

- void shutdownInput() 1.3

  将输入流设为"流结束"。

- boolean isOutputShutdown() 1.4

  如果输出已被关闭，则返回 true。

- boolean isInputShutdown() 1.4

  如果输入已被关闭，则返回 true。

4.2.4 可中断套接字

当连接到一个套接字时，当前线程将会被阻塞直到建立连接或产生超时为止。同样地，当通过套接字读数据时，当前线程也会被阻塞直到操作成功或产生超时为止。

在交互式的应用中，也许会考虑为用户提供一个选项，用以取消那些看似不会产生结果的连接。但是，当线程因套接字无法响应而发生阻塞时，则无法通过调用 interrupt 来解除阻塞。

为了中断套接字操作，可以使用 java.nio 包提供的一个特性——SocketChannel 类。可以使用如下方法打开 SocketChannel：

```
SocketChannel channel = SocketChannel.open(new InetSocketAddress(host, port));
```

通道（channel）并没有与之相关联的流。实际上，它所拥有的 read 和 write 方法都是通过使用 Buffer 对象来实现的（关于 NIO 缓冲区的相关信息请参见第 2 章）。ReadableByteChannel 接口和 WritableByteChannel 接口都声明了这两个方法。

如果不想处理缓冲区，可以使用 Scanner 类从 SocketChannel 中读取信息，因为 Scanner 有一个带 ReadableByteChannel 参数的构造器：

```
var in = new Scanner(channel, StandardCharsets.UTF_8);
```

通过调用静态方法 Channels.newOutputStream，可以将通道转换成输出流。

```
OutputStream outStream = Channels.newOutputStream(channel);
```

上述操作就是所有要做的事情。当线程正在执行打开、读取或写入操作时，如果线程发生中断，那么这些操作将不会陷入阻塞，而是以抛出异常的方式结束。

程序清单 4-5 的程序对比了可中断套接字和阻塞套接字：服务器将连续发送数字，并在每发送十个数字之后停滞一下。点击两个按钮中的任何一个，都会启动一个线程来连接服务器并打印输出。第一个线程使用可中断套接字，而第二个线程使用阻塞套接字。如果在第一批的十个数字的读取过程中点击"Cancel"按钮，这两个线程都会中断。

程序清单 4-5　interruptible/InterruptibleSocketTest.java

```java
 1  package interruptible;
 2
 3  import java.awt.*;
 4  import java.awt.event.*;
 5  import java.util.*;
 6  import java.net.*;
 7  import java.io.*;
 8  import java.nio.charset.*;
 9  import java.nio.channels.*;
10  import javax.swing.*;
11
12  /**
13   * This program shows how to interrupt a socket channel.
14   * @author Cay Horstmann
15   * @version 1.05 2018-03-17
16   */
17  public class InterruptibleSocketTest
18  {
19     public static void main(String[] args)
20     {
21        EventQueue.invokeLater(() ->
22           {
23              var frame = new InterruptibleSocketFrame();
24              frame.setTitle("InterruptibleSocketTest");
25              frame.setDefaultCloseOperation(JFrame.EXIT_ON_CLOSE);
26              frame.setVisible(true);
27           });
28     }
29  }
30
31  class InterruptibleSocketFrame extends JFrame
32  {
33     private Scanner in;
34     private JButton interruptibleButton;
35     private JButton blockingButton;
36     private JButton cancelButton;
37     private JTextArea messages;
38     private TestServer server;
```

```java
39      private Thread connectThread;
40
41      public InterruptibleSocketFrame()
42      {
43         var northPanel = new JPanel();
44         add(northPanel, BorderLayout.NORTH);
45
46         final int TEXT_ROWS = 20;
47         final int TEXT_COLUMNS = 60;
48         messages = new JTextArea(TEXT_ROWS, TEXT_COLUMNS);
49         add(new JScrollPane(messages));
50
51         interruptibleButton = new JButton("Interruptible");
52         blockingButton = new JButton("Blocking");
53
54         northPanel.add(interruptibleButton);
55         northPanel.add(blockingButton);
56
57         interruptibleButton.addActionListener(event ->
58            {
59               interruptibleButton.setEnabled(false);
60               blockingButton.setEnabled(false);
61               cancelButton.setEnabled(true);
62               connectThread = new Thread(() ->
63                  {
64                     try
65                     {
66                        connectInterruptibly();
67                     }
68                     catch (IOException e)
69                     {
70                        messages.append("\nInterruptibleSocketTest.connectInterruptibly: " + e);
71                     }
72                  });
73               connectThread.start();
74            });
75
76         blockingButton.addActionListener(event ->
77            {
78               interruptibleButton.setEnabled(false);
79               blockingButton.setEnabled(false);
80               cancelButton.setEnabled(true);
81               connectThread = new Thread(() ->
82                  {
83                     try
84                     {
85                        connectBlocking();
86                     }
87                     catch (IOException e)
88                     {
89                        messages.append("\nInterruptibleSocketTest.connectBlocking: " + e);
90                     }
91                  });
92               connectThread.start();
```

```java
 93          });
 94
 95       cancelButton = new JButton("Cancel");
 96       cancelButton.setEnabled(false);
 97       northPanel.add(cancelButton);
 98       cancelButton.addActionListener(event ->
 99          {
100             connectThread.interrupt();
101             cancelButton.setEnabled(false);
102          });
103       server = new TestServer();
104       new Thread(server).start();
105       pack();
106    }
107
108    /**
109     * Connects to the test server, using interruptible I/O
110     */
111    public void connectInterruptibly() throws IOException
112    {
113       messages.append("Interruptible:\n");
114       try (SocketChannel channel
115             = SocketChannel.open(new InetSocketAddress("localhost", 8189)))
116       {
117          in = new Scanner(channel, StandardCharsets.UTF_8);
118          while (!Thread.currentThread().isInterrupted())
119          {
120             messages.append("Reading ");
121             if (in.hasNextLine())
122             {
123                String line = in.nextLine();
124                messages.append(line);
125                messages.append("\n");
126             }
127          }
128       }
129       finally
130       {
131          EventQueue.invokeLater(() ->
132             {
133                messages.append("Channel closed\n");
134                interruptibleButton.setEnabled(true);
135                blockingButton.setEnabled(true);
136             });
137       }
138    }
139
140    /**
141     * Connects to the test server, using blocking I/O
142     */
143    public void connectBlocking() throws IOException
144    {
145       messages.append("Blocking:\n");
146       try (var sock = new Socket("localhost", 8189))
```

```java
147         {
148            in = new Scanner(sock.getInputStream(), StandardCharsets.UTF_8);
149            while (!Thread.currentThread().isInterrupted())
150            {
151               messages.append("Reading ");
152               if (in.hasNextLine())
153               {
154                  String line = in.nextLine();
155                  messages.append(line);
156                  messages.append("\n");
157               }
158            }
159         }
160         finally
161         {
162            EventQueue.invokeLater(() ->
163               {
164                  messages.append("Socket closed\n");
165                  interruptibleButton.setEnabled(true);
166                  blockingButton.setEnabled(true);
167               });
168         }
169      }
170
171   /**
172    * A multithreaded server that listens to port 8189 and sends numbers to the client,
173    * simulating a hanging server after 10 numbers.
174    */
175   class TestServer implements Runnable
176   {
177      public void run()
178      {
179         try (var s = new ServerSocket(8189))
180         {
181            while (true)
182            {
183               Socket incoming = s.accept();
184               Runnable r = new TestServerHandler(incoming);
185               new Thread(r).start();
186            }
187         }
188         catch (IOException e)
189         {
190            messages.append("\nTestServer.run: " + e);
191         }
192      }
193   }
194
195   /**
196    * This class handles the client input for one server socket connection.
197    */
198   class TestServerHandler implements Runnable
199   {
200      private Socket incoming;
```

```
201        private int counter;
202
203        /**
204         * Constructs a handler.
205         * @param i the incoming socket
206         */
207        public TestServerHandler(Socket i)
208        {
209           incoming = i;
210        }
211
212        public void run()
213        {
214           try
215           {
216              try
217              {
218                 OutputStream outStream = incoming.getOutputStream();
219                 var out = new PrintWriter(
220                    new OutputStreamWriter(outStream, StandardCharsets.UTF_8),
221                    true /* autoFlush */);
222                 while (counter < 100)
223                 {
224                    counter++;
225                    if (counter <= 10) out.println(counter);
226                    Thread.sleep(100);
227                 }
228              }
229              finally
230              {
231                 incoming.close();
232                 messages.append("Closing server\n");
233              }
234           }
235           catch (Exception e)
236           {
237              messages.append("\nTestServerHandler.run: " + e);
238           }
239        }
240     }
241 }
```

但是，在第一批十个数字之后，就只能中断第一个线程了，第二个线程将保持阻塞直到服务器最终关闭连接（参见图 4-6）。

API **java.net.InetSocketAddress** 1.4

- InetSocketAddress(String hostname, int port)
  用给定的主机和端口参数创建一个地址对象，并在创建过程中解析主机名。如果主机名不能被解析，那么该地址对象的 unresolved 属性将被设为 true。

- boolean isUnresolved()
  如果不能解析该地址对象，则返回 true。

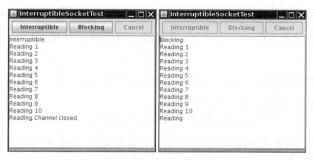

图 4-6 中断一个套接字

API java.nio.channels.SocketChannel 1.4

- static SocketChannel open(SocketAddress address)

  打开一个套接字通道,并将其连接到远程地址。

API java.nio.channels.Channels 1.4

- static InputStream newInputStream(ReadableByteChannel channel)

  创建一个输入流,用以从指定的通道读取数据。

- static OutputStream newOutputStream(WritableByteChannel channel)

  创建一个输出流,用以向指定的通道写入数据。

4.3 获取 Web 数据

为了在 Java 程序中访问 Web 服务器,你可能希望在更高的级别上进行处理,而不只是创建套接字连接和发送 HTTP 请求。在下面的各个小节中,我们将讨论专用于此目的的 Java 类库中的各个类。

4.3.1 URL 和 URI

URL 和 URLConnection 类封装了大量复杂的实现细节,这些细节涉及如何从远程站点获取信息。例如,可以自一个字符串构建一个 URL 对象:

```
var url = new URL(urlString);
```

如果只是想获得该资源的内容,可以使用 URL 类中的 openStream 方法。该方法将产生一个 InputStream 对象,然后就可以按照一般的用法来使用这个对象了,比如用它构建一个 Scanner 对象:

```
InputStream inStream = url.openStream();
var in = new Scanner(inStream, StandardCharsets.UTF_8);
```

java.net 包对统一资源定位符 (Uniform Resource Locator,URL) 和统一资源标识符 (Uniform Resource Identifier,URI) 进行了非常有用的区分。

URI 是个纯粹的语法结构，包含用来指定 Web 资源的字符串的各种组成部分。URL 是 URI 的一个特例，它包含了用于定位 Web 资源的足够信息。其他 URI，比如

```
mailto:cay@horstmann.com
```

则不属于定位符，因为根据该标识符我们无法定位任何数据。像这样的 URI 我们称之为 URN（uniform resource name，统一资源名称）。

在 Java 类库中，URI 类并不包含任何用于访问资源的方法，它的唯一作用就是解析。但是，URL 类可以打开一个连接到资源的流。因此，URL 类只能作用于那些 Java 类库知道该如何处理的模式，例如 http:、https:、ftp:、本地文件系统（file:）和 JAR 文件（jar:）。

要想了解为什么对 URI 进行解析并非小事一桩，那么考虑一下 URL 会变得多么复杂。例如，

```
http://google.com?q=Beach+Chalet
ftp://username:password@ftp.yourserver.com/pub/file.txt
```

URI 规范给出了标记这些标识符的规则。一个 URI 具有以下句法：

[*scheme:*]*schemeSpecificPart*[*#fragment*]

上式中，[...] 表示可选部分，并且 : 和 # 可以被包含在标识符内。

包含 *scheme:* 部分的 URI 称为绝对 URI。否则，称为相对 URI。

如果绝对 URI 的 *schemeSpecificPart* 不是以 / 开头的，我们就称它是不透明的。例如：

```
mailto:cay@horstmann.com
```

所有绝对的透明 URI 和所有相对 URI 都是分层的（hierarchical）。例如：

```
http://horstmann.com/index.html
../../java/net/Socket.html#Socket()
```

一个分层 URI 的 *schemeSpecificPart* 具有以下结构：

[//*authority*][*path*][?*query*]

在这里，[...] 同样表示可选的部分。

对于那些基于服务器的 URI，authority 部分具有以下形式：

[*user-info@*]*host*[*:port*]

port 必须是一个整数。

RFC 2396（标准化 URI 的文献）还支持一种基于注册表的机制，此时 authority 采用了一种不同的格式。不过，这种情况并不常见。

URI 类的作用之一是解析标识符并将它分解成各种不同的组成部分。你可以用以下方法读取它们：

```
getScheme
getSchemeSpecificPart
getAuthority
getUserInfo
getHost
getPort
getPath
```

```
getQuery
getFragment
```

URI 类的另一个作用是处理绝对标识符和相对标识符。如果存在一个如下的绝对 URI：

```
http://docs.mycompany.com/api/java/net/ServerSocket.html
```

和一个如下的相对 URI：

```
../../java/net/Socket.html#Socket()
```

那么可以用它们组合出一个绝对 URI：

```
http://docs.mycompany.com/api/java/net/Socket.html#Socket()
```

这个过程称为解析相对 URL。

与此相反的过程称为相对化（relativization）。例如，假设有一个基本 URI：

```
http://docs.mycompany.com/api
```

和另一个 URI：

```
http://docs.mycompany.com/api/java/lang/String.html
```

那么相对化之后的 URI 就是：

```
java/lang/String.html
```

URI 类同时支持以下两个操作：

```
relative = base.relativize(combined);
combined = base.resolve(relative);
```

4.3.2 使用 URLConnection 获取信息

如果想从某个 Web 资源获取更多信息，那么应该使用 URLConnection 类，通过它能够得到比基本的 URL 类更多的控制功能。

当操作一个 URLConnection 对象时，必须像下面这样非常小心地安排操作步骤：

1. 调用 URL 类中的 openConnection 方法获得 URLConnection 对象：

```
URLConnection connection = url.openConnection();
```

2. 使用以下方法来设置任意的请求属性：

```
setDoInput
setDoOutput
setIfModifiedSince
setUseCaches
setAllowUserInteraction
setRequestProperty
setConnectTimeout
setReadTimeout
```

我们将在本节的稍后部分以及 API 说明中讨论这些方法。

3. 调用 connect 方法连接远程资源：

```
connection.connect();
```

除了与服务器建立套接字连接外，该方法还可用于向服务器查询头信息（header information）。

4. 与服务器建立连接后，你可以查询头信息。getHeaderFieldKey 和 getHeaderField 这两个方法枚举了消息头的所有字段。getHeaderFields 方法返回一个包含了消息头中所有字段的标准 Map 对象。为了方便使用，以下方法可以查询各标准字段：

```
getContentType
getContentLength
getContentEncoding
getDate
getExpiration
getLastModified
```

5. 最后，访问资源数据。使用 getInputStream 方法获取一个输入流用以读取信息（这个输入流与 URL 类中的 openStream 方法所返回的流相同）。另一个方法 getContent 在实际操作中并不是很有用。由标准内容类型（比如 text/plain 和 image/gif）所返回的对象需要使用 com.sun 层次结构中的类来进行处理。也可以注册自己的内容处理器，但是在本书中我们不讨论这项技术。

> ⚠️ **警告**：一些程序员在使用 URLConnection 类的过程中形成了错误的观念，他们认为 URLConnection 类中的 getInputStream 和 getOutputStream 方法与 Socket 类中的这些方法相似，但是这种想法并不十分正确。URLConnection 类具有很多表象之下的神奇功能，尤其在处理请求和响应消息头时。正因为如此，严格遵循建立连接的每个步骤显得非常重要。

下面将详细介绍一下 URLConnection 类中的一些方法。有几个方法可以在与服务器建立连接之前设置连接属性，其中最重要的是 setDoInput 和 setDoOutput。在默认情况下，建立的连接只产生从服务器读取信息的输入流，并不产生任何执行写操作的输出流。如果想获得输出流（例如，用于向一个 Web 服务器提交数据），那么需要调用：

```
connection.setDoOutput(true);
```

接下来，也许想设置某些请求头（request header）。请求头是与请求命令一起被发送到服务器的。例如：

```
GET www.server.com/index.html HTTP/1.0
Referer: http://www.somewhere.com/links.html
Proxy-Connection: Keep-Alive
User-Agent: Mozilla/5.0 (X11; U; Linux i686; en-US; rv:1.8.1.4)
Host: www.server.com
Accept: text/html, image/gif, image/jpeg, image/png, */*
Accept-Language: en
Accept-Charset: iso-8859-1,*,utf-8
Cookie: orangemilano=192218887821987
```

setIfModifiedSince 方法用于告诉连接你只对自某个特定日期以来被修改过的数据感兴趣。

最后我们再介绍一个总览全局的方法：setRequestProperty，它可以用来设置对特定协议起作用的任何"名 – 值（name/value）对"。关于 HTTP 请求头的格式，请参见 RFC 2616，其中的某些参数没有很好地建档，它们通常在程序员之间口头传授。例如，如果你想访问一个有密码保护的 Web 页，那么就必须按如下步骤操作：

1. 将用户名、冒号和密码以字符串形式连接在一起。

```
String input = username + ":" + password;
```

2. 计算上一步骤所得字符串的 Base64 编码。(Base64 编码用于将字节序列编码成可打印的 ASCII 字符序列。)

```
Base64.Encoder encoder = Base64.getEncoder();
String encoding = encoder.encodeToString(input.getBytes(StandardCharsets.UTF_8));
```

3. 用 "Authorization" 这个名字和 "Basic"+encoding 的值调用 setRequestProperty 方法。

```
connection.setRequestProperty("Authorization", "Basic " + encoding);
```

> **提示**：我们上面介绍的是如何访问一个有密码保护的 Web 页。如果想要通过 FTP 访问一个有密码保护的文件时，则需要采用一种完全不同的方法，即构建如下格式的 URL：
>
> ftp://username:password@ftp.yourserver.com/pub/file.txt

一旦调用了 connect 方法，就可以查询响应头信息了。首先，我们将介绍如何枚举所有响应头的字段。似乎是为了展示自己的个性，该类的实现者引入了另一种迭代协议。调用如下方法：

```
String key = connection.getHeaderFieldKey(n);
```

可以获得响应头的第 n 个键，其中 n 从 1 开始！如果 n 为 0 或大于消息头的字段总数，该方法将返回 null 值。没有哪种方法可以返回字段的数量，必须反复调用 getHeaderFieldKey 方法直到返回 null 为止。同样地，调用以下方法：

```
String value = connection.getHeaderField(n);
```

可以得到第 n 个值。

getHeaderFields 方法可以返回一个封装了响应头字段的 Map 对象。

```
Map<String,List<String>> headerFields = connection.getHeaderFields();
```

下面是一组来自典型的 HTTP 请求的响应头字段。

```
Date: Wed, 27 Aug 2008 00:15:48 GMT
Server: Apache/2.2.2 (Unix)
Last-Modified: Sun, 22 Jun 2008 20:53:38 GMT
Accept-Ranges: bytes
Content-Length: 4813
Connection: close
Content-Type: text/html
```

> **注释**：可以用 connection.getHeaderField(0) 或 headerFields.get(null) 获取响应状态行（例如 "HTTP/1.1 200 OK"）。

为了简便起见，Java 提供了 6 个方法用以访问最常用的消息头类型的值，并在需要的时候将它们转换成数字类型，这些方法的详细信息请参见表 4-1。返回类型为 long 的方法返回的是从格林尼治时间 1970 年 1 月 1 日开始计算的秒数。

表 4-1 用于访问响应头值的简便方法

键名	方法名	返回类型
Date	getDate	long
Expires	getExpiration	long
Last-Modified	getLastModified	long
Content-Length	getContentLength	int
Content-Type	getContentType	String
Content-Encoding	getContentEncoding	String

通过程序清单 4-6 的程序，可以对 URL 连接做一些试验。程序运行起来后，请在命令行中输入一个 URL 以及用户名和密码（可选），例如：

```
java urlConnection.URLConnectionTest http://www.yourserver.com user password
```

该程序将输出以下内容：

- 消息头中的所有键和值。
- 表 4-1 中 6 个简便方法的返回值。
- 被请求资源的前 10 行信息。

程序清单 4-6 urlConnection/URLConnectionTest.java

```java
 1  package urlConnection;
 2
 3  import java.io.*;
 4  import java.net.*;
 5  import java.nio.charset.*;
 6  import java.util.*;
 7
 8  /**
 9   * This program connects to an URL and displays the response header data and the first
10   * 10 lines of the requested data.
11   *
12   * Supply the URL and an optional username and password (for HTTP basic authentication) on the
13   * command line.
14   * @version 1.12 2018-03-17
15   * @author Cay Horstmann
16   */
17  public class URLConnectionTest
18  {
19     public static void main(String[] args)
20     {
21        try
22        {
23           String urlName;
24           if (args.length > 0) urlName = args[0];
```

```java
25          else urlName = "http://horstmann.com";
26
27          var url = new URL(urlName);
28          URLConnection connection = url.openConnection();
29
30          // set username, password if specified on command line
31
32          if (args.length > 2)
33          {
34             String username = args[1];
35             String password = args[2];
36             String input = username + ":" + password;
37             Base64.Encoder encoder = Base64.getEncoder();
38             String encoding = encoder.encodeToString(input.getBytes(StandardCharsets.UTF_8));
39             connection.setRequestProperty("Authorization", "Basic " + encoding);
40          }
41
42          connection.connect();
43
44          // print header fields
45
46          Map<String, List<String>> headers = connection.getHeaderFields();
47          for (Map.Entry<String, List<String>> entry : headers.entrySet())
48          {
49             String key = entry.getKey();
50             for (String value : entry.getValue())
51                System.out.println(key + ": " + value);
52          }
53
54          // print convenience functions
55
56          System.out.println("----------");
57          System.out.println("getContentType: " + connection.getContentType());
58          System.out.println("getContentLength: " + connection.getContentLength());
59          System.out.println("getContentEncoding: " + connection.getContentEncoding());
60          System.out.println("getDate: " + connection.getDate());
61          System.out.println("getExpiration: " + connection.getExpiration());
62          System.out.println("getLastModifed: " + connection.getLastModified());
63          System.out.println("----------");
64
65          String encoding = connection.getContentEncoding();
66          if (encoding == null) encoding = "UTF-8";
67          try (var in = new Scanner(connection.getInputStream(), encoding))
68          {
69             // print first ten lines of contents
70
71             for (int n = 1; in.hasNextLine() && n <= 10; n++)
72                System.out.println(in.nextLine());
73             if (in.hasNextLine()) System.out.println(". . .");
74          }
75       }
76       catch (IOException e)
77       {
78          e.printStackTrace();
```

```
79        }
80    }
81 }
```

API java.net.URL 1.0

- InputStream openStream()
  打开一个用于读取资源数据的输入流。
- URLConnection openConnection()
  返回一个 URLConnection 对象，该对象负责管理与资源之间的连接。

API java.net.URLConnection 1.0

- void setDoInput(boolean doInput)
- boolean getDoInput()
  如果 doInput 为 true，那么用户可以接收来自该 URLConnection 的输入。
- void setDoOutput(boolean doOutput)
- boolean getDoOutput()
  如果 doOutput 为 true，那么用户可以将输出发送到该 URLConnection。
- void setIfModifiedSince(long time)
- long getIfModifiedSince()
  属性 ifModifiedSince 用于配置该 URLConnection 对象，使它只获取那些自从某个给定时间以来被修改过的数据。调用方法时需要传入的 time 参数指的是从格林尼治时间 1970 年 1 月 1 日午夜开始计算的秒数。
- void setConnectTimeout(int timeout) 5.0
- int getConnectTimeout() 5.0
  设置或得到连接超时时限（单位：毫秒）。如果在连接建立之前就已经达到了超时的时限，那么相关联的输入流的 connect 方法就会抛出一个 SocketTimeoutException 异常。
- void setReadTimeout(int timeout) 5.0
- int getReadTimeout() 5.0
  设置读取数据的超时时限（单位：毫秒）。如果在一个读操作成功之前就已经达到了超时的时限，那么 read 方法就会抛出一个 SocketTimeoutException 异常。
- void setRequestProperty(String key, String value)
  设置请求头的一个字段。
- Map<String,List<String>> getRequestProperties() 1.4
  返回请求头属性的一个映射表。相同的键对应的所有值被放置在同一个列表中。
- void connect()
  连接远程资源并获取响应头信息。
- Map<String,List<String>> getHeaderFields() 1.4

返回响应头的一个映射表。相同的键对应的所有值被放置在同一个列表中。
- String getHeaderFieldKey(int n)
  得到响应头第 n 个字段的键。如果 n 小于等于 0 或大于响应头字段的总数，则该方法返回 null 值。
- String getHeaderField(int n)
  得到响应头第 n 个字段的值。如果 n 小于等于 0 或大于响应头字段的总数，则该方法返回 null 值。
- int getContentLength()
  如果内容长度可获得，则返回该长度值，否则返回 –1。
- String getContentType()
  获取内容的类型，比如 text/plain 或 image/gif。
- String getContentEncoding()
  获取内容的编码机制，比如 gzip。这个值不太常用，因为默认的 identity 编码机制并不是用 Content-Encoding 头来设定的。
- long getDate()
- long getExpiration()
- long getLastModifed()
  获取创建日期、过期日以及最后一次被修改的日期。这些日期指的是从格林尼治时间 1970 年 1 月 1 日午夜开始计算的秒数。
- InputStream getInputStream()
- OutputStream getOutputStream()
  返回从资源读取信息或向资源写入信息的流。
- Object getContent()
  选择适当的内容处理器，以便读取资源数据并将它转换成对象。该方法对于读取诸如 text/plain 或 image/gif 之类的标准内容类型并没有什么用处，除非你安装了自己的内容处理器。

4.3.3 提交表单数据

在上一节中，我们介绍了如何从 Web 服务器读取数据。现在，我们将介绍如何让程序再将数据反馈回 Web 服务器和那些被 Web 服务器调用的程序。

为了将信息从 Web 浏览器发送到 Web 服务器，用户需要填写一个类似图 4-7 中所示的表单。当用户点击提交按钮时，文本框中的文本以及复选框、单选按钮和其他输入元素的设定值都被发送到了 Web 服务器。此时，Web 服务器调用程序对用户的输入进行处理。

有许多技术可以让 Web 服务器实现对程序的调用。其中最广人所知的是 Java Servlet、JavaServer Face、微软的 ASP（Active Server Pages，动态服务器主页）以及 CGI（Common Gateway Interface，通用网关接口）脚本。

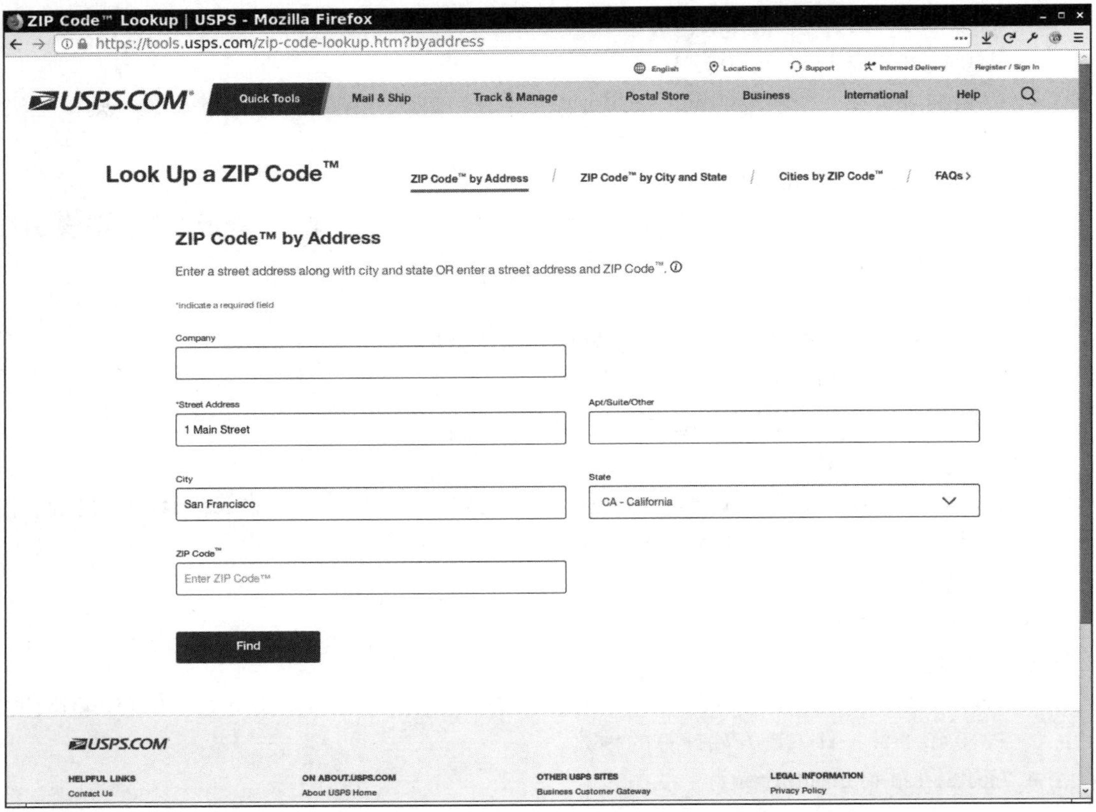

图 4-7 HTML 表单

服务器端程序用于处理表单数据并生成另一个 HTML 页，该页会被 Web 服务器发回给浏览器，这个操作过程我们在图 4-8 中做了说明。返回给浏览器的响应页可以包含新的信息（例如，信息检索程序中的响应页）或者只是一个确认。之后，Web 浏览器将显示响应页。

我们不会在本书中介绍应该如何实现服务器端程序，而是将侧重点放在如何编写客户端程序使之与已有的服务器端程序进行交互。

当表单数据被发送到 Web 服务器时，数据到底由谁来解释并不重要，可能是 Servlet 或 CGI 脚本，也可能是其他服务器端技术。客户端以标准格式将数据发送给 Web 服务器，而 Web 服务器则负责将数据传递给具体的程序以产生响应。

在向 Web 服务器发送信息时，通常有两个命令会被用到：GET 和 POST。

在使用 GET 命令时，只需将参数附在 URL 的结尾处即可。这种 URL 的格式如下：

http://host/path?query

其中，每个参数都具有"名字 = 值"的形式，而这些参数之间用 & 字符分隔开。参数的值将遵循下面的规则，使用 URL 编码模式进行编码：

- 保留字符 A 到 Z、a 到 z、0 到 9，以及 . - ~ _。

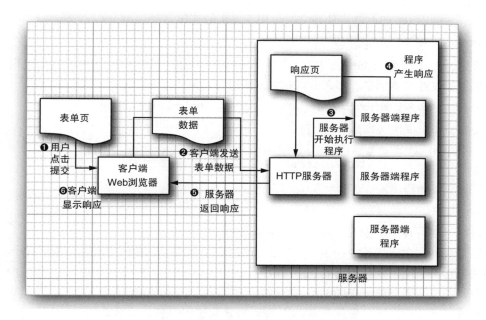

图 4-8　执行服务器端脚本过程中的数据流

- 用 + 字符替换所有的空格。
- 将其他所有字符编码为 UTF-8，并将每个字节都编码为 % 后面紧跟一个两位的十六进制数字。

例如，若要发送街道名 San Francisco，CA，可以使用 San+Francisco%2c+CA，因为十六进制数 2c（即十进制数 44）是"，"的 UTF-8 码值。

这种编码方式使得在任何中间程序中都不会混入空格和其他特殊字符。

例如，就在写作本书的时候，Google Map 网站（www.google.com/maps）可以接受带有两个名为 q 和 hl 参数的查询请求，这两个参数分别表示查询的位置和响应中所使用的人类语言。为了得到 1 Market Street, San Franciso, CA 的地图，并且让响应使用德语，只需访问下面的 URL 即可：

http://www.google.com/maps?q=1+Market+Street+San+Francisco&hl=de

在浏览器中出现很长的查询字符串很让人郁闷，而且老式的浏览器和代理对在 GET 请求中能够包含的字符数量做出了限制。正因为此，POST 请求经常用来处理具有大量数据的表单。在 POST 请求中，我们不会在 URL 上附着参数，而是从 URLConnection 中获得输出流，并将名/值对写入到该输出流中。我们仍旧需要对这些值进行 URL 编码，并用 & 字符将它们隔开。

下面，我们将详细介绍这个过程。在提交数据给服务器端程序之前，首先需要创建一个 URLConnection 对象。

```
var url = new URL("http://host/path");
URLConnection connection = url.openConnection();
```

然后，调用 setDoOutput 方法建立一个用于输出的连接。

```
connection.setDoOutput(true);
```

接着，调用 getOutputStream 方法获得一个流，可以通过这个流向服务器发送数据。如果要向服务器发送文本信息，那么可以非常方便地将流包装在 PrintWriter 对象中。

```
var out = new PrintWriter(connection.getOutputStream(), StandardCharsets.UTF_8);
```

现在，可以向服务器发送数据了。

```
out.print(name1 + "=" + URLEncoder.encode(value1, StandardCharsets.UTF_8) + "&");
out.print(name2 + "=" + URLEncoder.encode(value2, StandardCharsets.UTF_8));
```

之后，关闭输出流：

```
out.close();
```

最后，调用 getInputStream 方法读取服务器的响应。

下面我们来实际操作一个例子。地址为 https://tools.usps.com/zip-code-lookup.htm?byaddress 的网站包含一个用于查找街道地址的邮政编码的表单（见图 4-7）。要想在 Java 程序中使用这个表单，需要知道 POST 请求的 URL 和参数。

你可以通过查看这个表单的 HTML 源码来获取这些信息，但是通常用网络监视器来"窥视"发出的请求会更容易一些。作为其开发工具包的组成部分，大多数浏览器都具有网络监视器。例如，图 4-9 展示了 Firefox 网络监视器向我们的示例网站提交数据时的截屏。你可以发现其中的提交 URL 以及参数名和参数值。

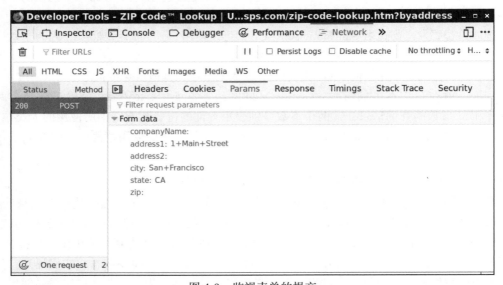

图 4-9　监视表单的提交

在提交表单数据时，HTTP 头包含了内容类型：

```
Content-Type: application/x-www-form-urlencoded
```

你还可以以其他格式提交表单。例如，发送用 JavaScript 对象表示法（JSON）表示的数据，将内容类型设置为 application/json。

POST 的头还必须包括内容长度，例如：

Content-Length: 124

程序清单 4-7 用于将 POST 数据发送给任何脚本，它将数据放在如下的 .properties 文件：

```
url=https://tools.usps.com/tools/app/ziplookup/zipByAddress
User-Agent=HTTPie/0.9.2
address1=1 Market Street
address2=
city=San Francisco
state=CA
companyName=
. . .
```

程序清单 4-7　post/PostTest.java

```java
 1  package post;
 2
 3  import java.io.*;
 4  import java.net.*;
 5  import java.nio.charset.*;
 6  import java.nio.file.*;
 7  import java.util.*;
 8
 9  /**
10   * This program demonstrates how to use the URLConnection class for a POST request.
11   * @version 1.42 2018-03-17
12   * @author Cay Horstmann
13   */
14  public class PostTest
15  {
16     public static void main(String[] args) throws IOException
17     {
18        String propsFilename = args.length > 0 ? args[0] : "post/post.properties";
19        var props = new Properties();
20        try (InputStream in = Files.newInputStream(Paths.get(propsFilename)))
21        {
22           props.load(in);
23        }
24        String urlString = props.remove("url").toString();
25        Object userAgent = props.remove("User-Agent");
26        Object redirects = props.remove("redirects");
27        CookieHandler.setDefault(new CookieManager(null, CookiePolicy.ACCEPT_ALL));
28        String result = doPost(new URL(urlString), props,
29           userAgent == null ? null : userAgent.toString(),
30           redirects == null ? -1 : Integer.parseInt(redirects.toString()));
31        System.out.println(result);
32     }
33
34     /**
35      * Do an HTTP POST.
```

```java
  36         * @param url the URL to post to
  37         * @param nameValuePairs the query parameters
  38         * @param userAgent the user agent to use, or null for the default user agent
  39         * @param redirects the number of redirects to follow manually, or -1 for automatic
  40         * redirects
  41         * @return the data returned from the server
  42         */
  43        public static String doPost(URL url, Map<Object, Object> nameValuePairs, String userAgent,
  44              int redirects) throws IOException
  45        {
  46           var connection = (HttpURLConnection) url.openConnection();
  47           if (userAgent != null)
  48              connection.setRequestProperty("User-Agent", userAgent);
  49
  50           if (redirects >= 0)
  51              connection.setInstanceFollowRedirects(false);
  52
  53           connection.setDoOutput(true);
  54
  55           try (var out = new PrintWriter(connection.getOutputStream()))
  56           {
  57              var first = true;
  58              for (Map.Entry<Object, Object> pair : nameValuePairs.entrySet())
  59              {
  60                 if (first) first = false;
  61                 else out.print('&');
  62                 String name = pair.getKey().toString();
  63                 String value = pair.getValue().toString();
  64                 out.print(name);
  65                 out.print('=');
  66                 out.print(URLEncoder.encode(value, StandardCharsets.UTF_8));
  67              }
  68           }
  69           String encoding = connection.getContentEncoding();
  70           if (encoding == null) encoding = "UTF-8";
  71
  72           if (redirects > 0)
  73           {
  74              int responseCode = connection.getResponseCode();
  75              if (responseCode == HttpURLConnection.HTTP_MOVED_PERM
  76                    || responseCode == HttpURLConnection.HTTP_MOVED_TEMP
  77                    || responseCode == HttpURLConnection.HTTP_SEE_OTHER)
  78              {
  79                 String location = connection.getHeaderField("Location");
  80                 if (location != null)
  81                 {
  82                    URL base = connection.getURL();
  83                    connection.disconnect();
  84                    return doPost(new URL(base, location), nameValuePairs, userAgent,
  85                       redirects - 1);
  86                 }
  87              }
  88           }
  89           else if (redirects == 0)
```

```
 90        {
 91           throw new IOException("Too many redirects");
 92        }
 93
 94        var response = new StringBuilder();
 95        try (var in = new Scanner(connection.getInputStream(), encoding))
 96        {
 97           while (in.hasNextLine())
 98           {
 99              response.append(in.nextLine());
100              response.append("\n");
101           }
102        }
103        catch (IOException e)
104        {
105           InputStream err = connection.getErrorStream();
106           if (err == null) throw e;
107           try (var in = new Scanner(err))
108           {
109              response.append(in.nextLine());
110              response.append("\n");
111           }
112        }
113
114        return response.toString();
115     }
116  }
```

这个程序移除了 url 和 User-Agent 项，并将其他内容都发送到了 doPost 方法。

在 doPost 方法中，我们首先打开连接并设置用户代理。（邮政编码服务在默认的 User-Agent 请求参数包含字符串 Java 时无法工作，这可能是因为邮政局不想为程序自动产生的请求服务。）

然后调用 setDoOutput(true) 并打开输出流。然后，枚举 Map 对象中的所有键和值。对每一个键 – 值对，我们发送 key、= 字符、value 和 & 分隔符：

```
out.print(key);
out.print('=');
out.print(URLEncoder.encode(value, StandardCharsets.UTF_8));
if (more pairs) out.print('&');
```

在从写出请求切换到读取响应的任何部分时，就会发生与服务器的实际交互。Content-Length 头被设置为输出的尺寸，而 Content-Type 头被设置为 application/x-www-form-urlencoded，除非指定了不同的内容类型。这些头信息和数据都被发送给服务器，然后，响应头和服务器响应会被读取，并可以被查询。在我们的示例程序中，这种切换发生在对 connection.getContentEncoding() 的调用中。

在读取响应过程中会碰到一个问题。如果服务器端出现错误，那么调用 connection.getInputStream() 时就会抛出一个 FileNotFoundException 异常。但是，此时服务器仍然会向浏览器返回一个错误页面（例如，常见的"错误 404– 找不到该页"）。为了捕捉这个错误页，可以调用 getErrorStream 方法：

```
InputStream err = connection.getErrorStream();
```

> **注释**：getErrorStream 方法与这个程序中的许多其他方法一样，属于 URLConnection 类的子类 HttpURLConnection。如果要创建以 http:// 或 https:// 开头的 URL，那么可以将所产生的连接对象强制转型为 HttpURLConnection。

在将 POST 数据发送给服务器时，服务器端程序产生的响应可能是 redirect:，后面跟着一个完全不同的 URL，该 URL 应该被调用以获取实际的信息。服务器可以这么做，因为这些信息位于他处，或者提供了一个可以作为书签标记的 URL。HttpURLConnection 类在大多数情况下可以处理这种重定向。

> **注释**：如果 cookie 需要在重定向中从一个站点发送给另一个站点，那么可以像下面这样配置一个全局的 cookie 处理器：
> ```
> CookieHandler.setDefault(new CookieManager(null, CookiePolicy.ACCEPT_ALL));
> ```
> 然后，cookie 就可以被正确地包含在重定向请求中了。

尽管重定向通常是自动处理的，但是有些情况下，你需要自己完成重定向。例如，在 HTTP 和 HTTPS 之间的自动重定向因为安全原因而不被支持。重定向还会因更细微的原因而失败。例如，早期版本的邮政编码服务就使用了重定向。回忆一下，我们设置了 User-Agent 请求参数，以便让邮局认为我们不是在通过 Java API 发送请求。尽管可以在最初的请求中将用户代理设置为其他的字符串，但是这项设置在自动重定向中并没有用到。自动重定向总是会发送包含单词 Java 的通用用户代理字符串。

在这些情况下，可以人工实现重定向。在连接到服务器之前，将关闭自动重定向：

```
connection.setInstanceFollowRedirects(false);
```

在发送请求之后，获取响应码：

```
int responseCode = connection.getResponseCode();
```

检查它是否是下列值之一：

```
HttpURLConnection.HTTP_MOVED_PERM
HttpURLConnection.HTTP_MOVED_TEMP
HttpURLConnection.HTTP_SEE_OTHER
```

如果是这些值之一，那么获取 Location 响应头，以获得重定向的 URL。然后，断开连接，并创建到新的 URL 的连接：

```
String location = connection.getHeaderField("Location");
if (location != null)
{
   URL base = connection.getURL();
   connection.disconnect();
   connection = (HttpURLConnection) new URL(base, location).openConnection();
   . . .
}
```

每当需要从某个现有的 Web 站点查询信息时，该程序所展示的处理技术就会显得很有用。只需找出需要发送的参数，然后从回复信息中剔除 HTML 和其他不必要的信息。

API java.net.HttpURLConnection 1.0

- InputStream getErrorStream()
  返回一个流，通过这个流可以读取 Web 服务器的错误信息。

API java.net.URLEncoder 1.0

- static String encode(String s, String encoding) 1.4
  采用指定的字符编码模式（推荐使用 "UTF-8"）对字符串 s 进行编码，并返回它的 URL 编码形式。在 URL 编码中，'A'-'Z'，'a'-'z'，'0'-'9'，'-'，'_'，'.' 和 '*' 等字符保持不变，空格被编码成 '+'，所有其他字符被编码成 "%XY" 形式的字节序列，其中 0xXY 为该字节十六进制数。

API java.net.URLDecoder 1.2

- static string decode(String s, String encoding) 1.4
  采用指定编码模式对已编码字符串 s 进行解码，并返回结果。

4.4 HTTP 客户端

URLConnection 类是在 HTTP 成为 Web 普适协议之前设计的，它提供了对大量协议的支持，但是它对 HTTP 的支持有些笨重。当做出决定要支持 HTTP/2 时，情况就很清楚了，它最好是提供一个新的客户端接口，而不是对现有 API 做重构。HttpClient 提供了更便捷的 API 和对 HTTP/2 的支持。在 Java 9 和 10 中，其 API 类位于 jdk.incubator.http 包中，使该 API 有机会成为根据用户反馈不断演化的产物。到了 Java 11，HttpClient 位于 java.net.http 包中。

> 注释：在使用 Java 9 和 10 时，需要用下面的命令行选项来运行程序：
>   --add-modules jdk.incubator.httpclient

与 URLConnection 类相比，HTTP 客户端 API 从设计初始就提供了一种更简单的连接到 Web 服务器的机制。

HttpCLient 对象可以发出请求并接收响应。可以通过下面的调用获取客户端：

HttpClient client = HttpClient.newHttpClient()

或者，如果需要配置客户端，可以使用像下面这样的构建器 API：

HttpClient client = HttpClient.newBuilder()
    .followRedirects(HttpClient.Redirect.ALWAYS)
    .build();

即，获取一个构建器，调用其方法定制需要待构建的项，然后调用 build 方法来终结构建过程。这是一种构建不可修改对象的常见模式。

还可以遵循构建器模式来定制请求，下面是一个 Get 请求：

```
HttpRequest request = HttpRequest.newBuilder()
    .uri(new URI("http://horstmann.com"))
    .GET()
    .build();
```

URI 是指"统一资源标识符"，在使用 HTTP 时，它与 URL 相同。但是，在 Java 中，URL 类确实有一些用来打开到某个 URL 的连接的方法，而 URI 类只关心语法（模式、主机、端口、路径、查询、片段等）。

对于 POST 请求，需要一个"体发布器"（body publisher），它会将请求数据转换为要推送的数据。有针对字符串、字节数组和文件的体发布器。例如，如果请求是 JSON 格式的，那么只需将 JSON 字符串提供给某个字符串体发布器：

```
HttpRequest request = HttpRequest.newBuilder()
    .uri(new URI(url))
    .header("Content-Type", "application/json")
    .POST(HttpRequest.BodyPublishers.ofString(jsonString))
    .build();
```

遗憾的是，该 API 不支持对常见内容类型做上面所要求的格式化处理。程序清单 4-8 中的样例程序提供了用于表单数据和文件上传的体发布器。

在发送请求时，必须告诉客户端如何处理响应。如果只是想将体当作字符串处理，那么就可以像下面这样用 HttpResponse.BodyHandlers.ofString() 来发送请求：

```
HttpResponse<String> response = client.send(request, HttpResponse.BodyHandlers.ofString());
```

HttpResponse 类是一个泛化类，它的类型参数表示体的类型。可以直接获取响应体字符串：

```
String bodyString = response.body();
```

还有其他的响应体处理器，可以将响应作为字节数组或输入流来获取。BodyHandlers.ofFile(filePath) 会产生一个处理器，将响应存储到给定的文件中，BodyHandlers.ofFileDownload(directoryPath) 会用 Content-Disposition 头中的信息将响应存入给定的目录中。最后，从 BodyHandlers.dicarding() 中获得的处理器会直接丢弃响应。

处理响应的内容并不在该 API 的考虑范围内。例如，如果收到了 JSON 数据，那么就需要某个 JSON 库来解析其中的内容。

HttpResponse 对象还会产生状态码与响应头。

```
int status = response.statusCode();
HttpHeaders responseHeaders = response.headers();
```

可以将 HttpHeader 对象转换为一个映射表：

```
Map<String, List<String>> headerMap = responseHeaders.map();
```

这个映射表的值是列表，因为在 HTTP 中，每个键都可以有多个值。

如果只想要某个特定键的值，并且知道它没有多个值，那么可以调用 firstValue 方法：

```
Optional<String> lastModified = headerMap.firstValue("Last-Modified");
```

这样可以得到该响应的值，或者在没有提供该值时，返回空的 Optional 对象。

可以异步地处理响应。在构建客户端时，可以提供一个执行器：

```
ExecutorService executor = Executors.newCachedThreadPool();
HttpClient client = HttpClient.newBuilder().executor(executor).build();
```

构建一个请求，然后在该客户端上调用 sendAsync 方法，就会收到一个 CompletableFuture<HttpResponse<T>> 对象，其中 T 是体处理器的类型。需要使用卷 I 第 12 章描述的 CompletableFuture API：

```
HttpRequest request = HttpRequest.newBuilder().uri(uri).GET().build();
client.sendAsync(request, HttpResponse.BodyHandlers.ofString())
    .thenAccept(response -> . . .);
```

> ✓ 提示：为了启用针对 HttpClient 记录日志的功能，需要在 JDK 的 net.properties 文件中添加下面的行：
>
> jdk.httpclient.HttpClient.log=all
>
> 除了 all，还可以指定为一个由逗号分隔的列表，其中包含 headers、requests、content、errors、ssl、trace 和 frames，后面还可以选择跟着 :control、:data、:window 或 :all。中间不要使用任何空格。
>
> 然后，将名为 jdk.httpclient.HttpClient 的日志记录器的日志级别设置为 INFO，例如，在 JDK 的 logging.properties 文件中添加下面的行：
>
> jdk.httpclient.HttpClient.level=INFO

程序清单 4-8　client/HttpClientTest.java

```
 1  package client;
 2
 3  import java.io.*;
 4  import java.math.*;
 5  import java.net.*;
 6  import java.nio.charset.*;
 7  import java.nio.file.*;
 8  import java.util.*;
 9
10  import java.net.http.*;
11  import java.net.http.HttpRequest.*;
12
13  class MoreBodyPublishers
14  {
15     public static BodyPublisher ofFormData(Map<Object, Object> data)
16     {
17        var first = true;
18        var builder = new StringBuilder();
19        for (Map.Entry<Object, Object> entry : data.entrySet())
20        {
21           if (first) first = false;
```

```java
22            else builder.append("&");
23          builder.append(URLEncoder.encode(entry.getKey().toString(),
24             StandardCharsets.UTF_8));
25          builder.append("=");
26          builder.append(URLEncoder.encode(entry.getValue().toString(),
27             StandardCharsets.UTF_8));
28       }
29       return BodyPublishers.ofString(builder.toString());
30    }
31
32    private static byte[] bytes(String s) { return s.getBytes(StandardCharsets.UTF_8); }
33
34    public static BodyPublisher ofMimeMultipartData(Map<Object, Object> data, String boundary)
35          throws IOException
36    {
37       var byteArrays = new ArrayList<byte[]>();
38       byte[] separator = bytes("--" + boundary + "\nContent-Disposition: form-data; name=");
39       for (Map.Entry<Object, Object> entry : data.entrySet())
40       {
41          byteArrays.add(separator);
42
43          if (entry.getValue() instanceof Path)
44          {
45             var path = (Path) entry.getValue();
46             String mimeType = Files.probeContentType(path);
47             byteArrays.add(bytes("\"" + entry.getKey() + "\"; filename=\"" + path.getFileName()
48                + "\"\nContent-Type: " + mimeType + "\n\n"));
49             byteArrays.add(Files.readAllBytes(path));
50          }
51          else
52             byteArrays.add(bytes("\"" + entry.getKey() + "\"\n\n" + entry.getValue() + "\n"));
53       }
54       byteArrays.add(bytes("--" + boundary + "--"));
55       return BodyPublishers.ofByteArrays(byteArrays);
56    }
57
58    public static BodyPublisher ofSimpleJSON(Map<Object, Object> data)
59    {
60       var builder = new StringBuilder();
61       builder.append("{");
62       var first = true;
63       for (Map.Entry<Object, Object> entry : data.entrySet())
64       {
65          if (first) first = false;
66          else
67             builder.append(",");
68          builder.append(jsonEscape(entry.getKey().toString())).append(": ")
69             .append(jsonEscape(entry.getValue().toString()));
70       }
71       builder.append("}");
72       return BodyPublishers.ofString(builder.toString());
73    }
74
75    private static Map<Character, String> replacements = Map.of('\b', "\\b", '\f', "\\f",
```

```java
                '\n', "\\n", '\r', "\\r", '\t', "\\t", '"', "\\\"", '\\', "\\\\");

        private static StringBuilder jsonEscape(String str)
        {
            var result = new StringBuilder("\"");
            for (int i = 0; i < str.length(); i++)
            {
                char ch = str.charAt(i);
                String replacement = replacements.get(ch);
                if (replacement == null) result.append(ch);
                else result.append(replacement);
            }
            result.append("\"");
            return result;
        }
    }

    public class HttpClientTest
    {
        public static void main(String[] args)
                throws IOException, URISyntaxException, InterruptedException
        {
            System.setProperty("jdk.httpclient.HttpClient.log", "headers,errors");
            String propsFilename = args.length > 0 ? args[0] : "client/post.properties";
            Path propsPath = Paths.get(propsFilename);
            var props = new Properties();
            try (InputStream in = Files.newInputStream(propsPath))
            {
                props.load(in);
            }
            String urlString = "" + props.remove("url");
            String contentType = "" + props.remove("Content-Type");
            if (contentType.equals("multipart/form-data"))
            {
                var generator = new Random();
                String boundary = new BigInteger(256, generator).toString();
                contentType += ";boundary=" + boundary;
                props.replaceAll((k, v) ->
                    v.toString().startsWith("file://")
                        ? propsPath.getParent().resolve(Paths.get(v.toString().substring(7)))
                        : v);
            }
            String result = doPost(urlString, contentType, props);
            System.out.println(result);
        }

        public static String doPost(String url, String contentType, Map<Object, Object> data)
                throws IOException, URISyntaxException, InterruptedException
        {
            HttpClient client = HttpClient.newBuilder()
                .followRedirects(HttpClient.Redirect.ALWAYS).build();

            BodyPublisher publisher = null;
            if (contentType.startsWith("multipart/form-data"))
```

```
130     {
131        String boundary = contentType.substring(contentType.lastIndexOf("=") + 1);
132        publisher = MoreBodyPublishers.ofMimeMultipartData(data, boundary);
133     }
134     else if (contentType.equals("application/x-www-form-urlencoded"))
135        publisher = MoreBodyPublishers.ofFormData(data);
136     else
137     {
138        contentType = "application/json";
139        publisher = MoreBodyPublishers.ofSimpleJSON(data);
140     }
141
142     HttpRequest request  = HttpRequest.newBuilder()
143        .uri(new URI(url))
144        .header("Content-Type", contentType)
145        .POST(publisher)
146        .build();
147     HttpResponse<String> response
148        = client.send(request, HttpResponse.BodyHandlers.ofString());
149     return response.body();
150  }
151 }
```

API java.net.http.HttpClient 11

- static HttpClient newHttpClient()
  用默认配置产生一个 HttpClient 对象。
- static HttpClient.Builder newBuilder()
  产生一个用于构建 HttpClient 对象的构建器。
- <T> HttpResponse<T> send(HttpRequest request, HttpResponse.BodyHandler<T> responseBodyHandler)
- <T> CompletableFuture<HttpResponse<T>> sendAsync(HttpRequest request, HttpResponse.BodyHandler<T> responseBodyHandler)
  产生一个同步或异步的请求，并使用给定的处理器来处理响应体。

API java.net.http.HttpClient.Builder 11

- HttpClient build()
  用由当前构建器配置的属性产生一个 HttpClient 对象。
- HttpClient.Builder followRedirects(HttpClient.Redirect policy)
  将重定向策略设置为 HttpClient.Redirect 枚举中的 ALWAYS、NEVER 或 NORMAL 之一（仅拒绝从 HTTPS 重定向到 HTTP）
- HttpClient.Builder executor(Executor executor)
  设置用于异步请求的执行器。

API java.net.http.HttpRequest 11

- HttpRequest.Builder newBuilder()

产生一个用于构建 HttpRequest 对象的构建器。

API java.net.http.HttpRequest.Builder 11

- HttpRequest build()
  用由当前构建器配置的属性产生一个 HttpRequest 对象。
- HttpRequest.Builder uri(URI uri)
  为当前请求设置 URI。
- HttpRequest.Builder header(String name, String value)
  为当前请求设置请求头。
- HttpRequest.Builder GET()
- HttpRequest.Builder DELETE()
- HttpRequest.Builder POST(HttpRequest.BodyPublisher bodyPublisher)
- HttpRequest.Builder PUT(HttpRequest.BodyPublisher bodyPublisher)
  为当前请求设置请求方法和请求体。

API java.net.http.HttpResponse<T> 11

- T body()
  产生当前响应的体。
- int statusCode()
  产生当前响应的状态码。
- HttpHeaders headers()
  产生响应头。

API java.net.http.HttpHeaders 11

- Map<String,List<String>> map()
  产生这些头的映射。
- Optional<String> firstValue(String name)
  在头中具有给定名的第一个值，如果存在的话。

4.5 发送 E-mail

过去，编写程序通过创建到邮件服务器上 SMTP 专用的端口 25 来发送邮件是一件很简单的事。简单邮件传输协议用于描述 E-mail 消息的格式。一旦连接到服务器，就可以发送一个邮件报头（采用 SMTP 格式，该格式很容易生成）。紧随其后的是邮件消息。

以下是操作的详细过程。

1. 打开一个到达主机的套接字：

```
var s = new Socket("mail.yourserver.com", 25); // 25 is SMTP
var out = new PrintWriter(s.getOutputStream(), StandardCharsets.UTF_8);
```

2. 发送以下信息到打印流:

```
HELO sending host
MAIL FROM: sender e-mail address
RCPT TO: recipient e-mail address
DATA
Subject: subject
(blank line)
mail message (any number of lines)
.
QUIT
```

SMTP 规范(RFC 821)规定,每一行都要以 \r 再紧跟一个 \n 来结尾。

SMTP 曾经总是例行公事般地路由任何人的 E-mail,但是,在蠕虫泛滥的今天,许多服务器都内置了检查功能,并且只接受来自授信用户或授信 IP 地址范围的请求。其中,认证通常是通过安全套接字连接来实现的。

实现人工认证模式的代码非常冗长乏味,因此,我们将展示如何利用 JavaMail API 在 Java 程序中发送 E-mail。

可以从 www.oracle.com/technetwork/java/javamail 处下载 JavaMail,然后将它解压到硬盘上的某处。

如果要使用 JavaMail,则需要设置一些和邮件服务器相关的属性。例如,在使用 GMail 时,需要设置:

```
mail.transport.protocol=smtps
mail.smtps.auth=true
mail.smtps.host=smtp.gmail.com
mail.smtps.user=accountname@gmail.com
```

我们的示例程序是从一个属性文件中读取这些属性值的。

出于安全的原因,我们没有将密码放在属性文件中,而是要求提示用户需要输入。

首先要读入属性文件,然后像下面这样获取一个邮件会话:

```
Session mailSession = Session.getDefaultInstance(props);
```

接着,用恰当的发送者、接受者、主题和消息文本来创建消息:

```
var message = new MimeMessage(mailSession);
message.setFrom(new InternetAddress(from));
message.addRecipient(RecipientType.TO, new InternetAddress(to));
message.setSubject(subject);
message.setText(builder.toString());
```

然后将消息发送走:

```
Transport tr = mailSession.getTransport();
tr.connect(null, password);
tr.sendMessage(message, message.getAllRecipients());
tr.close();
```

程序清单 4-9 中的程序是从具有下面这种格式的文本文件中读取消息的:

Sender

Recipient
Subject
Message text (any number of lines)

要运行该程序，需要从 https://javaee.github.io/javamail 下载 JavaMail 的实现，还需要 Java 激活框架（Java Activation Framework）的 JAR 文件，可以从 http://www.oracle.com/technetwork/java/javase/jaf-135115.html 处获得，或者可以在 Maven Central 中搜索。然后运行：

```
java -classpath .:javax.mail.jar:activation-1.1.1.jar path/to/message.txt
```

到撰写本章时为止，GMail 还不会检查信息的真实性，即你可以输入任何你喜欢的发送者。（当你下一次收到来自 president@whitehouse.gov 的 E-mail 消息邀请你盛装出席白宫南草坪的活动时，请牢记这一点，谨防上当。）

> ✓ 提示：如果你搞不清楚为什么你的邮件连接无法正常工作，那么可以调用：
> ```
> mailSession.setDebug(true);
> ```
> 并检查消息。而且，JavaMail API FAQ 也有些挺有用的调试提示。

程序清单 4-9　mail/MailTest.java

```java
 1  package mail;
 2
 3  import java.io.*;
 4  import java.nio.charset.*;
 5  import java.nio.file.*;
 6  import java.util.*;
 7  import javax.mail.*;
 8  import javax.mail.internet.*;
 9  import javax.mail.internet.MimeMessage.RecipientType;
10
11  /**
12   * This program shows how to use JavaMail to send mail messages.
13   * @author Cay Horstmann
14   * @version 1.01 2018-03-17
15   */
16  public class MailTest
17  {
18      public static void main(String[] args) throws MessagingException, IOException
19      {
20          var props = new Properties();
21          try (InputStream in = Files.newInputStream(Paths.get("mail", "mail.properties")))
22          {
23              props.load(in);
24          }
25          List<String> lines = Files.readAllLines(Paths.get(args[0]), StandardCharsets.UTF_8);
26
27          String from = lines.get(0);
28          String to = lines.get(1);
29          String subject = lines.get(2);
30
31          var builder = new StringBuilder();
```

```
32      for (int i = 3; i < lines.size(); i++)
33      {
34         builder.append(lines.get(i));
35         builder.append("\n");
36      }
37
38      Console console = System.console();
39      var password = new String(console.readPassword("Password: "));
40
41      Session mailSession = Session.getDefaultInstance(props);
42      // mailSession.setDebug(true);
43      var message = new MimeMessage(mailSession);
44      message.setFrom(new InternetAddress(from));
45      message.addRecipient(RecipientType.TO, new InternetAddress(to));
46      message.setSubject(subject);
47      message.setText(builder.toString());
48      Transport tr = mailSession.getTransport();
49      try
50      {
51         tr.connect(null, password);
52         tr.sendMessage(message, message.getAllRecipients());
53      }
54      finally
55      {
56         tr.close();
57      }
58   }
59 }
```

在本章中，你已经看到了如何用 Java 编写网络客户端和服务器，以及如何从 Web 服务器上获取数据。下一章将讨论数据库连接，你将会学习如何通过使用 JDBC API 来实现用 Java 操作关系型数据库。

第 5 章 数据库编程

- ▲ JDBC 的设计
- ▲ 结构化查询语言
- ▲ JDBC 配置
- ▲ 使用 JDBC 语句
- ▲ 执行查询操作
- ▲ 可滚动和可更新的结果集
- ▲ 行集
- ▲ 元数据
- ▲ 事务
- ▲ Web 和企业应用中的连接管理

1996 年，Sun 公司发布了第 1 版的 Java 数据库连接（JDBC）API，使编程人员可以通过这个 API 接口连接到数据库，并使用结构化查询语言（即 SQL）完成对数据库的查找与更新。（SQL 通常发音为"sequel"，它是数据库访问的业界标准。）JDBC 自此成为 Java 类库中最常使用的 API 之一。

JDBC 的版本已更新过数次。在本书出版之际，最新版的 JDBC 4.3 也被囊括到了 Java 9 中。

在本章中，我们将阐述 JDBC 幕后的关键思想，并将介绍（或者是复习）一下 SQL（Structured Query Language，结构化查询语言），它是关系数据库的业界标准。我们还将提供足够的细节，使你可以将 JDBC 融入常见的编程场景中。

> **注释**：根据 Oracle 的声明，JDBC 是一个注册了商标的术语，而并非 Java Database Connectivity 的首字母缩写。对它的命名体现了对 ODBC 的致敬，后者是微软开创的标准数据库 API，并因此而并入了 SQL 标准中。

5.1 JDBC 的设计

从一开始，Java 技术开发人员就意识到了 Java 在数据库应用方面的巨大潜力。从 1995 年开始，他们就致力于扩展 Java 标准类库，使之可以运用 SQL 访问数据库。他们最初希望通过扩展 Java，就可以让人们"纯"用 Java 语言与任何数据库进行通信。但是，他们很快发现这是一项无法完成的任务：因为业界存在许多不同的数据库，且它们所使用的协议也各不相同。尽管很多数据库供应商都表示支持 Java 提供一套数据库访问的标准网络协议，但是每一家企业都希望 Java 能采用自己的网络协议。

所有的数据库供应商和工具开发商都认为，如果 Java 能够为 SQL 访问提供一套"纯"Java API，同时提供一个驱动管理器，以允许第三方驱动程序可以连接到特定的数据库，那它就会显得非常有用。这样，数据库供应商就可以提供自己的驱动程序，将其插入到驱动管理器中。这将成为一种向驱动管理器注册第三方驱动程序的简单机制。

这种接口组织方式遵循了微软公司非常成功的 ODBC 模式，ODBC 为 C 语言访问数据库提供了一套编程接口。JDBC 和 ODBC 都基于同一个思想：根据 API 编写的程序都可以与驱动管理器进行通信，而驱动管理器则通过驱动程序与实际的数据库进行通信。

所有这些都意味着 JDBC API 是大部分程序员不得不使用的接口。

5.1.1 JDBC 驱动程序类型

JDBC 规范将驱动程序归结为以下几类：

- 第 1 类驱动程序将 JDBC 翻译成 ODBC，然后使用 ODBC 驱动程序与数据库进行通信。较早版本的 Java 包含了一个这样的驱动程序：JDBC/ODBC 桥，不过在使用这个桥接器之前需要对 ODBC 进行相应的部署和正确的设置。在 JDBC 面世之初，桥接器可以方便地用于测试，却不太适用于产品的开发。现在，有很多更好的驱动程序可用，所以 JDK 已经不再提供 JDBC/ODBC 桥了。
- 第 2 类驱动程序是由部分 Java 程序和部分本地代码组成的，用于与数据库的客户端 API 进行通信。在使用这种驱动程序之前，客户端不仅需要安装 Java 类库，还需要安装一些与平台相关的代码。
- 第 3 类驱动程序是纯 Java 客户端类库，它使用一种与具体数据库无关的协议将数据库请求发送给服务器构件，然后该构件再将数据库请求翻译成数据库相关的协议。这简化了部署，因为平台相关的代码只位于服务器端。
- 第 4 类驱动程序是纯 Java 类库，它将 JDBC 请求直接翻译成数据库相关的协议。

> 📘 **注释**：JDBC 规范可以在 https://jcp.org/aboutJava/communityprocess/mrel/jsr221/index3.html 处获得。

大部分数据库供应商都为他们的产品提供第 3 类或第 4 类驱动程序。与数据库供应商提供的驱动程序相比，许多第三方公司专门开发了很多更符合标准的产品，它们支持更多的平台、运行性能也更佳，某些情况下甚至具有更高的可靠性。

总之，JDBC 最终是为了实现以下目标：

- 通过使用标准的 SQL 语句，甚至是专门的 SQL 扩展，程序员就可以利用 Java 语言开发访问数据库的应用，同时还依旧遵守 Java 语言的相关约定。
- 数据库供应商和数据库工具开发商可以提供底层的驱动程序。因此，他们可以优化各自数据库产品的驱动程序。

> 📘 **注释**：也许你会问为什么 Java 没有采用 ODBC 模型，下面就是在 1996 年举行的 JavaOne 研讨会上给出的说法：
> - ODBC 很难学会。
> - ODBC 中有几个命令需要配置很多复杂的选项，而在 Java 编程语言中所采用的风格是要让方法简单而直观，但数量巨大。

- ODBC 依赖于 void* 指针和其他 C 语言特性，而这些特性 Java 编程语言并不具备。
- 与纯 Java 的解决方案相比，基于 ODBC 的解决方案天生就缺乏安全性，且难于部署。

5.1.2 JDBC 的典型用法

在传统的客户端/服务器模型中，通常是在服务器端部署数据库，而在客户端安装富 GUI 程序（参见图 5-1）。在此模型中，JDBC 驱动程序应该部署在客户端。

但是，如今三层模型更加常见。在三层应用模型中，客户端不直接调用数据库，而是调用服务器上的中间件层，由中间件层完成数据库查询操作。这种三层模型有以下优点：它将可视化表示（位于客户端）从业务逻辑（位于中间层）和原始数据（位于数据库）中分离出来。因此，我们可以从不同的客户端，如 Java 桌面应用、浏览器或者移动 App，来访问相同的数据和相同的业务规则。

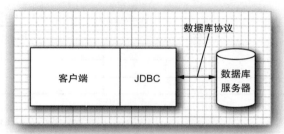

图 5-1 传统的客户端/服务器应用

客户端和中间层之间的通信在典型情况下是通过 HTTP 来实现的。JDBC 管理着中间层和后台数据库之间的通信，图 5-2 展示了这种通信模型的基本架构。

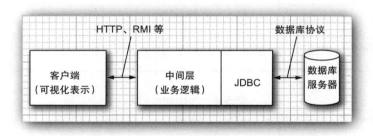

图 5-2 三层结构的应用

5.2 结构化查询语言

SQL 是对所有现代关系型数据库都至关重要的命令行语言，JDBC 则使得我们可以通过 SQL 与数据库进行通信。桌面数据库通常都有一个图形用户界面；通过这种界面，用户可以直接操作数据。但是，基于服务器的数据库只能使用 SQL 进行访问。

我们可以将 JDBC 包看作是一个用于将 SQL 语句传递给数据库的应用编程接口（API）。在本节中，我们将简单介绍一下 SQL。如果之前没有接触过 SQL，你会发现这些介绍是远远不够的，你可以参阅关于 SQL 的其他著作。我们推荐 Alan Beaulieu 所著的 *Learning SQL*（2009 年由 OReilly 出版社出版），或者还可以参考在线图书 *Learn SQL The Hard Way*，该书可在 http://sql.learncodethehardway.org 处获得。

可以将数据库想象成一组由行和列构成的具名表，其中每一列都有列名（column name），而每一行则包含了一个相关的数据集。

作为本书的数据库实例，我们将使用一组数据库表来描述一组经典的计算机著作（请参见表 5-1 ～表 5-4）。

表 5-1 Authors 表

Author_ID	Name	Fname
ALEX	Alexander	Christopher
BROO	Brooks	Frederick P.
...	...	...

表 5-2 Books 表

Title	ISBN	Publisher_ID	Price
A Guide to the SQL Standard	0-201-96426-0	0201	47.95
A Pattern Language: Towns, Buildings, Construction	0-19-501919-9	019	65.00
...	...	...	...

表 5-3 BooksAuthors 表

ISBN	Author_ID	Seq_No
0-201-96426-0	DATE	1
0-201-96426-0	DARW	2
0-19-501919-9	ALEX	1
...	...	...

表 5-4 Publishers 表

Publisher_ID	Name	URL
0201	Addison-Wesley	www.aw-bc.com
0407	John Wiley & Sons	www.wiley.com
...	...	...

图 5-3 显示的是一个 Books 表的视图，而图 5-4 显示了对 Books 表和 Publishers 表执行连接操作后的结果。Books 表和 Publishers 表都包含了一个表示出版社的 ID 字段。当我们利用出版社编号对这两个表进行连接操作时，我们就得到了由连接后的表格的值所组成的查询结果。结果中的每一行都包含了图书的信息、出版社名称及其 Web 页的 URL 地址。注意，有的出版社名称和 URL 地址会重复出现在数行中，因为这些行都对应于同一个出版社。

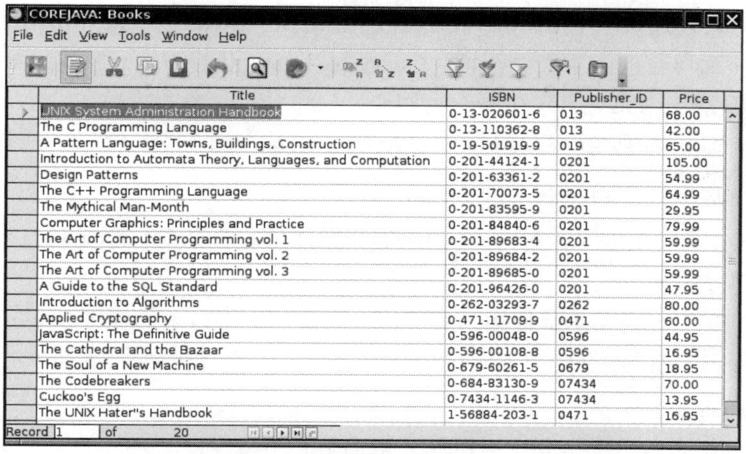

CORE JAVA: Books

Title	ISBN	Publisher_ID	Price
UNIX System Administration Handbook	0-13-020601-6	013	68.00
The C Programming Language	0-13-110362-8	013	42.00
A Pattern Language: Towns, Buildings, Construction	0-19-501919-9	019	65.00
Introduction to Automata Theory, Languages, and Computation	0-201-44124-1	0201	105.00
Design Patterns	0-201-63361-2	0201	54.99
The C++ Programming Language	0-201-70073-5	0201	64.99
The Mythical Man-Month	0-201-83595-9	0201	29.95
Computer Graphics: Principles and Practice	0-201-84840-6	0201	79.99
The Art of Computer Programming vol. 1	0-201-89683-4	0201	59.99
The Art of Computer Programming vol. 2	0-201-89684-2	0201	59.99
The Art of Computer Programming vol. 3	0-201-89685-0	0201	59.99
A Guide to the SQL Standard	0-201-96426-0	0201	47.95
Introduction to Algorithms	0-262-03293-7	0262	80.00
Applied Cryptography	0-471-11709-9	0471	60.00
JavaScript: The Definitive Guide	0-596-00048-0	0596	44.95
The Cathedral and the Bazaar	0-596-00108-8	0596	16.95
The Soul of a New Machine	0-679-60261-5	0679	18.95
The Codebreakers	0-684-83130-9	07434	70.00
Cuckoo's Egg	0-7434-1146-3	07434	13.95
The UNIX Hater"s Handbook	1-56884-203-1	0471	16.95

图 5-3　包含图书信息的示例表格

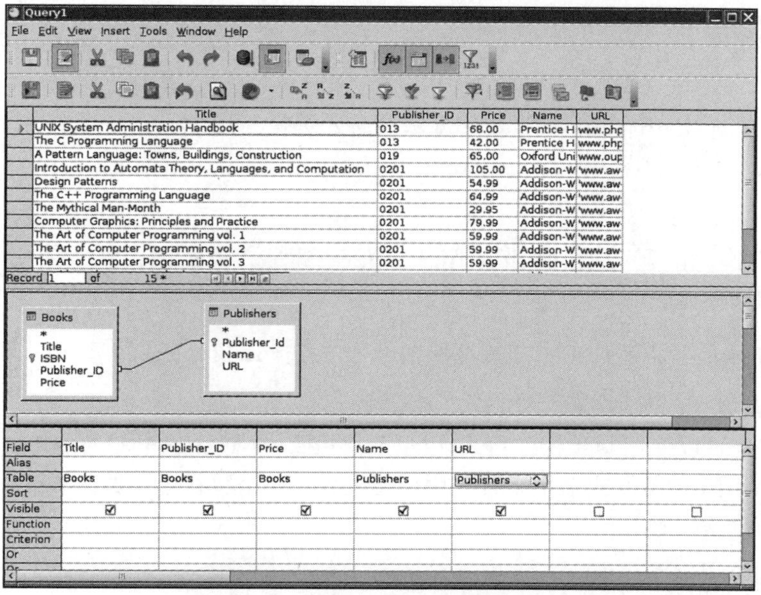

图 5-4　对两个表进行连接操作

对表格进行连接操作的好处是能够避免在数据库表中出现不必要的重复数据。例如，有一种比较简陋的数据库设计是在 Books 表中设置出版社名称和 URL 地址字段。但是这样一来，数据库本身，而非查询结果，将出现许多重复数据。如果出版社的 Web 地址发生了改变，就需要更新所有的重复数据。显然，这在一定程度上很容易导致错误。在关系模型中，我们将数据分布到多个表中，使得所有信息都不会出现不必要的重复。例如，每个出版社的 URL 地址只在出版社表中出现一次。如果需要将此信息与其他信息组合，我们只需对表进行连接操作。

在上述两幅图中，可以看到一个用于查看和链接表的图形工具。许多数据库提供商都具有相应的工具，通过连接列名和在表单中填入信息，让用户能够以某种简单的形式来表示其各种查询。这种工具通常称为实例查询（Query by Example，QBE）工具。而使用 SQL 的查询则是利用 SQL 语法以文本方式编写的。例如，

```
SELECT Books, Books.Publisher_Id, Books.Price, Publishers.Name, Publishers.URL
FROM Books, Publishers
WHERE Books.Publisher_Id = Publishers.Publisher_Id
```

在本节的余下部分中，我们将介绍如何编写这样的查询语句。如果你已经熟悉 SQL 了，就可以跳过这部分内容。

按照惯例，SQL 关键字全部使用大写字母。当然，也可以不这样做。

SELECT 语句相当灵活。仅使用下面这个查询语句，就可以查出 Books 表中的所有记录：

```
SELECT * FROM Books
```

在每一个 SQL 的 SELECT 语句中，FROM 子句都是必不可少的。FROM 子句用于告知数据库应该在哪个表上查询数据。

我们还可以选择所需要的列：

```
SELECT ISBN, Price, Title
FROM Books
```

并且还可以在查询语句中使用 WHERE 子句来限定所要选择的行：

```
SELECT ISBN, Price, Title
FROM Books
WHERE Price <= 29.95
```

请小心使用"相等"这个比较操作。与 Java 编程语言不同，SQL 使用 = 和 <> 而非 == 和 != 来进行相等性比较。

> **注释**：有些数据库供应商的产品支持在进行不等于比较时使用 !=。这不符合标准 SQL 的语法，所以我们建议不要使用这种方法。

WHERE 子句也可以使用 LIKE 操作符来实现模式匹配。不过，这里的通配符并不是通常使用的 * 和 ?，而是用 % 表示 0 或多个字符，用下划线表示单个字符。例如，

```
SELECT ISBN, Price, Title
FROM Books
WHERE Title NOT LIKE '%n_x%'
```

这条语句排除了所有书名中包含 UNIX 或者 Linux 的图书。

请注意，字符串都是用单引号括起来的，而非双引号。字符串中的单引号则需要用一对单引号代替。例如，

```
SELECT Title
FROM Books
WHERE Title LIKE '%''%'
```

上述语句会返回所有包含单引号的书名。

你也可以从多个表中选取数据：

```
SELECT * FROM Books, Publishers
```

如果没有 WHERE 子句，上述查询语句就意义不大了，它只是罗列了两个表中所有记录的组合。在我们这个例子中，Books 表有 20 行记录，Publishers 表有 8 行记录，合并的结果将产生 20×8 条记录，其中不乏大量重复数据。实际上我们需要对查询结果进行限制，只对那些图书与出版社相匹配的数据感兴趣。

```
SELECT * FROM Books, Publishers
WHERE Books.Publisher_Id = Publishers.Publisher_Id
```

这条语句的查询结果共有 20 行记录，每一条记录对应于一本书，因为每本书都在 Publishers 表中只对应一个出版社。

每当查询语句涉及多个表时，相同的列名可能会出现在两个不同的地方。在我们的例子中也存在这种情况，Books 表和 Publishers 表都拥有一个列名为 PublisherId 的列。当出现歧义时，可以在每个列名前添加它所在表的表名作为前缀，比如 Books/Publisher_Id。

也可以使用 SQL 来改变数据库中的数据。例如，假设现在要将所有书名中包含"C++"的图书降价 5 美元，可以执行以下语句：

```
UPDATE Books
SET Price = Price - 5.00
WHERE Title LIKE '%C++%'
```

类似地，要删除所有的 C++ 图书，可以使用下面的 DELETE 查询：

```
DELETE FROM Books
WHERE Title LIKE '%C++%'
```

此外，SQL 中还有许多内置函数，用于对某一列计算平均值、查找最大值和最小值以及其他许多功能。在此我们就不讨论了。

典型情况下，可以使用 INSERT 语句向表中插入值：

```
INSERT INTO Books
VALUES ('A Guide to the SQL Standard', '0-201-96426-0', '0201', 47.95)
```

我们必须为每一条插入到表中的记录使用一次 INSERT 语句。

当然，在查询、修改和插入数据之前，必须要有存储数据的位置。可以使用 CREATE TABLE 语句创建一个新表，还可以为每一列指定列名和数据类型。

```
CREATE TABLE Books
(
    Title CHAR(60),
    ISBN CHAR(13),
    Publisher_Id CHAR(6),
    Price DECIMAL(10,2)
)
```

表 5-5 给出了最常见的 SQL 数据类型。

表 5-5 SQL 数据类型

数据类型	说明
INTEGER 或 INT	通常为 32 位的整数
SMALLINT	通常为 16 位的整数

（续）

数据类型	说明
NUMERIC(m,n), DECIMAL(m,n) 或 DEC(m,n)	m 位长的定点十进制数，其中小数点后为 n 位
FLOAT(n)	运算精度为 n 位二进制数的浮点数
REAL	通常为 32 位浮点数
DOUBLE	通常为 64 位浮点数
CHARACTER(n) 或 CHAR(n)	固定长度为 n 的字符串
VARCHAR(n)	最大长度为 n 的可变长字符串
BOOLEAN	布尔值
DATE	日历日期（与具体的实现相关）
TIME	当前时间（与具体的实现相关）
TIMESTAMP	当前日期和时间（与具体的实现相关）
BLOB	二进制大对象
CLOB	字符大对象

在本书中，我们不再介绍更多的子句，比如可以应用于 CREATE TABLE 语句的主键子句和约束子句。

5.3 JDBC 配置

当然，你需要有一个可获得其 JDBC 驱动程序的数据库程序。目前这方面有许多出色的程序可供选择，比如 IBM DB2、Microsoft SQL Server、MySQL、Oracle 和 PostgreSQL。

为了练习本部分内容，你还需要创建一个数据库，我们假定你将这个数据库命名为 COREJAVA。你要自己创建，或者让数据库管理员创建这个数据库，并让你拥有适当权限，因为你需要拥有对这个数据库进行创建、更新和删除表的权限。

如果你以前从未安装过采用客户端/服务器模式的数据库，那么就会发现配置这样一个数据库会稍显复杂并且难于诊断故障的原因。如果安装的数据库无法正常运行，那么最好请专家来帮忙。

如果第一次接触数据库，我们建议使用 Apache Derby，它可以从 http://db.apache.org/derby 处下载到，在某些 JDK 版本中也包含了它。

在编写第一个数据库程序之前，你需要收集大量的信息和文件，下面将讨论这些内容。

5.3.1 数据库 URL

在连接数据库时，我们必须使用各种与数据库类型相关的参数，例如主机名、端口号和数据库名。

JDBC 使用了一种与普通 URL 相类似的语法来描述数据源。下面是这种语法的两个实例：

```
jdbc:derby://localhost:1527/COREJAVA;create=true
jdbc:postgresql:COREJAVA
```

上述 JDBC URL 指定了名为 COREJAVA 的一个 Derby 数据库和一个 PostgreSQL 数据库。JDBC URL 的一般语法为：

jdbc:*subprotocol*:*other stuff*

其中，*subprotocol* 用于选择连接到数据库的具体驱动程序。

other stuff 参数的格式随所使用的 *subprotocol* 不同而不同。如果要了解具体格式，你需要查阅数据库供应商提供的相关文档。

5.3.2 驱动程序 JAR 文件

你需要获得包含了你所使用的数据库的驱动程序的 JAR 文件。如果你使用的是 Derby，那么就需要 derbyclient.jar；如果你使用的是其他的数据库，那么就需要去寻找恰当的驱动程序。例如，PostgreSQL 的驱动程序可以在 http://jdbc.postgresql.org 处找到。

在运行访问数据库的程序时，需要将驱动程序的 JAR 文件包括到类路径中（编译时并不需要这个 JAR 文件）。

在从命令行启动程序时，只需要使用下面的命令：

java -classpath *driverPath*:. *ProgramName*

在 Windows 上，可以使用分号将当前路径（即由.字符表示的路径）与驱动程序 JAR 文件分隔开。

5.3.3 启动数据库

数据库服务器在连接之前需要先启动，启动的细节取决于所使用的数据库。

在使用 Derby 数据库时，需要遵循下面的步骤：

1. 打开命令 shell，并转到将来存放数据库文件的目录中。

2. 定位 derbyrun.jar。对于某些 JDK 版本，它包含在 jdk/db/lib 目录中，如果没有包含，那就安装 Apache Derby，并定位安装目录的 JAR 文件。我们用 derby 来表示包含 lib/derbyrun.jar 的目录。

3. 运行下面的命令：

java -jar *derby*/lib/derbyrun.jar server start

4. 仔细检查数据库是否正确工作了。然后创建一个名为 ij.properties 并包含下面各行的文件：

```
ij.driver=org.apache.derby.jdbc.ClientDriver
ij.protocol=jdbc:derby://localhost:1527/
ij.database=COREJAVA;create=true
```

在另一个命令 shell 中，通过执行下面的命令来运行 Derby 的交互式脚本执行工具（称为 ij）：

java -jar *derby*/lib/derbyrun.jar ij -p ij.properties

现在，可以发布像下面这样的 SQL 命令了：

```
CREATE TABLE Greetings (Message CHAR(20));
INSERT INTO Greetings VALUES ('Hello, World!');
SELECT * FROM Greetings;
DROP TABLE Greetings;
```

注意，每条命令都需要以分号结尾。要退出编辑器，可以键入

```
EXIT;
```

5. 在使用完数据库之后，可以用下面的命令关闭服务器：

```
java -jar derby/lib/derbyrun.jar server shutdown
```

如果使用其他的数据库，则需要查看文档，以了解如何启动和关闭数据库服务器，以及如何连接到数据库和发布 SQL 命令。

5.3.4 注册驱动器类

许多 JDBC 的 JAR 文件（例如 Derby 驱动程序）会自动注册驱动器类，在这种情况下，可以跳过本节所描述的手动注册步骤。包含 META-INF/services/java.sql.Driver 文件的 JAR 文件可以自动注册驱动器类，解压缩驱动程序 JAR 文件就可以检查其是否包含该文件。

如果驱动程序 JAR 文件不支持自动注册，那就需要找出数据库提供商使用的 JDBC 驱动器类的名字。典型的驱动器名字如下：

```
org.apache.derby.jdbc.ClientDriver
org.postgresql.Driver
```

通过使用 DriverManager，可以用两种方式来注册驱动器。一种方式是在 Java 程序中加载驱动器类，例如：

```
Class.forName("org.postgresql.Driver"); // force loading of driver class
```

这条语句将使得驱动器类被加载，由此将执行可以注册驱动器的静态初始化器。

另一种方式是设置 jdbc.drivers 属性。可以用命令行参数来指定这个属性，例如：

```
java -Djdbc.drivers=org.postgresql.Driver ProgramName
```

或者在应用中用下面这样的调用来设置系统属性

```
System.setProperty("jdbc.drivers", "org.postgresql.Driver");
```

在这种方式中可以提供多个驱动器，用冒号将它们分隔开，例如

```
org.postgresql.Driver:org.apache.derby.jdbc.ClientDriver
```

5.3.5 连接到数据库

在 Java 程序中，我们可以用下面这样的代码打开一个数据库连接：

```
String url = "jdbc:postgresql:COREJAVA";
String username = "dbuser";
String password = "secret";
Connection conn = DriverManager.getConnection(url, username, password);
```

驱动管理器会遍历所有注册过的驱动程序，以便找到一个能够使用数据库 URL 中指定的子协议的驱动程序。

getConnection 方法返回一个 Connection 对象。在下一节中，我们将详细介绍如何使用 Connection 对象来执行 SQL 语句。

要连接到数据库，我们还需要知道数据库的名字和密码。

> **注释**：在默认情况下，Derby 允许我们使用任何用户名进行连接，并且不检查密码。它会为每个用户生成一个单独的表集合，而默认的用户名是 app。

程序清单 5-1 中的测试程序将所有这些步骤放到了一起：它从名为 database.properties 的文件中加载连接参数，并连接到数据库。示例代码中提供的 database.properties 文件包含的是关于 Derby 数据库的连接信息，如果使用其他的数据库，则需要将与数据库相关的连接信息放到这个文件中。下面是一个用于连接到 PostgreSQL 数据库的示例：

```
jdbc.drivers=org.postgresql.Driver
jdbc.url=jdbc:postgresql:COREJAVA
jdbc.username=dbuser
jdbc.password=secret
```

在连接到数据库之后，这个测试程序执行了下面的 SQL 语句：

```
CREATE TABLE Greetings (Message CHAR(20))
INSERT INTO Greetings VALUES ('Hello, World!')
SELECT * FROM Greetings
```

SELECT 语句的结果将被打印出来，你应该可以看到如下的输出：

```
Hello, World!
```

然后，通过执行下面的语句移除这张表：

```
DROP TABLE Greetings
```

要运行这个测试程序，需要按照前面所描述的方式启动数据库，并像下面这样启动这个程序：

```
java -classpath .:driverJAR test.TestDB
```

（Windows 用户需要注意，用；代替：来分隔路径元素。）

> **提示**：调试与 JDBC 相关的问题时，有种方法是启用 JDBC 的跟踪机制。调用 DriverManager.setLogWriter 方法可以将跟踪信息发送给 PrintWriter，而 PrintWriter 将输出 JDBC 活动的详细列表。大多数 JDBC 驱动程序的实现都提供了用于跟踪的附加机制，例如，在使用 Derby 时，可以在 JDBC 的 URL 中添加 traceFile 选项，如 jdbc:derby://localhost:1527/ COREJAVA;create=true;traceFile=trace.out。

程序清单 5-1　test/TestDB.java

```
1 package test;
2
```

```java
3   import java.nio.file.*;
4   import java.sql.*;
5   import java.io.*;
6   import java.util.*;
7
8   /**
9    * This program tests that the database and the JDBC driver are correctly configured.
10   * @version 1.03 2018-05-01
11   * @author Cay Horstmann
12   */
13  public class TestDB
14  {
15     public static void main(String args[]) throws IOException
16     {
17        try
18        {
19           runTest();
20        }
21        catch (SQLException ex)
22        {
23           for (Throwable t : ex)
24              t.printStackTrace();
25        }
26     }
27
28     /**
29      * Runs a test by creating a table, adding a value, showing the table contents, and
30      * removing the table.
31      */
32     public static void runTest() throws SQLException, IOException
33     {
34        try (Connection conn = getConnection();
35              Statement stat = conn.createStatement())
36        {
37           stat.executeUpdate("CREATE TABLE Greetings (Message CHAR(20))");
38           stat.executeUpdate("INSERT INTO Greetings VALUES ('Hello, World!')");
39
40           try (ResultSet result = stat.executeQuery("SELECT * FROM Greetings"))
41           {
42              if (result.next())
43                 System.out.println(result.getString(1));
44           }
45           stat.executeUpdate("DROP TABLE Greetings");
46        }
47     }
48
49     /**
50      * Gets a connection from the properties specified in the file database.properties.
51      * @return the database connection
52      */
53     public static Connection getConnection() throws SQLException, IOException
54     {
55        var props = new Properties();
56        try (InputStream in = Files.newInputStream(Paths.get("database.properties")))
```

```
57        {
58           props.load(in);
59        }
60        String drivers = props.getProperty("jdbc.drivers");
61        if (drivers != null) System.setProperty("jdbc.drivers", drivers);
62        String url = props.getProperty("jdbc.url");
63        String username = props.getProperty("jdbc.username");
64        String password = props.getProperty("jdbc.password");
65
66        return DriverManager.getConnection(url, username, password);
67     }
68  }
```

API `java.sql.DriverManager` 1.1

- static Connection getConnection(String url, String user, String password)
  建立一个到指定数据库的连接，并返回一个 Connection 对象。

5.4 使用 JDBC 语句

在下面各节中，你将会看到如何使用 JDBC Statement 来执行 SQL 语句，获得执行结果，以及处理错误。然后，我们将向你展示一个操作数据库的简单示例。

5.4.1 执行 SQL 语句

在执行 SQL 语句之前，首先需要创建一个 Statement 对象。要创建 Statement 对象，需要使用调用 DriverManager.getConnection 方法所获得的 Connection 对象。

```
Statement stat = conn.createStatement();
```

接着，把要执行的 SQL 语句放入字符串中，例如：

```
String command = "UPDATE Books"
   + " SET Price = Price - 5.00"
   + " WHERE Title NOT LIKE '%Introduction%'";
```

然后，调用 Statement 接口中的 executeUpdate 方法：

```
stat.executeUpdate(command);
```

executeUpdate 方法将返回受 SQL 语句影响的行数，或者对不返回行数的语句返回 0。例如，在先前的例子中调用 executeUpdate 方法将返回那些降价 5 美元的行数。

executeUpdate 方法既可以执行诸如 INSERT、UPDATE 和 DELETE 之类的操作，也可以执行诸如 CREATE TABLE 和 DROP TABLE 之类的数据定义语句。但是，执行 SELECT 查询时必须使用 executeQuery 方法。另外还有一个 execute 语句可以执行任意的 SQL 语句，此方法通常只用于由用户提供的交互式查询。

当我们执行查询操作时，通常感兴趣的是查询结果。executeQuery 方法会返回一个 ResultSet 类型的对象，可以通过它来每次一行地迭代遍历所有查询结果。

```
ResultSet rs = stat.executeQuery("SELECT * FROM Books");
```

分析结果集时通常可以使用类似如下的循环语句代码：

```
while (rs.next())
{
   look at a row of the result set
}
```

> **警告**：ResultSet 接口的迭代协议与 java.util.Iterator 接口稍有不同。对于 ResultSet 接口，迭代器初始化时被设定在第一行之前的位置，必须调用 next 方法将它移动到第一行。另外，它没有 hasNext 方法，我们需要不断地调用 next，直至该方法返回 false。

结果集中行的顺序是任意排列的。除非使用 ORDER BY 子句指定行的顺序，否则不能为行序强加任何意义。

查看每一行时，可能希望知道其中每一列的内容，有许多访问器（accessor）方法可以用于获取这些信息。

```
String isbn = rs.getString(1);
double price = rs.getDouble("Price");
```

不同的数据类型有不同的访问器，比如 getString 和 getDouble。每个访问器都有两种形式，一种接受数字型参数，另一种接受字符串参数。当使用数字型参数时，我们指的是该数字所对应的列。例如，rs.getString(1) 返回的是当前行中第一列的值。

> **警告**：与数组的索引不同，数据库的列序号是从 1 开始计算的。

当使用字符串参数时，指的是结果集中以该字符串为列名的列。例如，rs.getDouble("Price") 返回列名为 Price 的列所对应的值。使用数字型参数效率更高一些，但是使用字符串参数可以使代码易于阅读和维护。

当 get 方法的类型和列的数据类型不一致时，每个 get 方法都会进行合理的类型转换。例如，调用 rs.getString("Price") 时，该方法会将 Price 列的浮点值转换成字符串。

API *java.sql.Connection* 1.1

- Statement createStatement()
  创建一个 Statement 对象，用以执行不带参数的 SQL 查询和更新。
- void close()
  立即关闭当前的连接，并释放由它所创建的 JDBC 资源。

API *java.sql.Statement* 1.1

- ResultSet executeQuery(String sqlQuery)
  执行给定字符串中的 SQL 语句，并返回一个用于查看查询结果的 ResultSet 对象。
- int executeUpdate(String sqlStatement)
- long executeLargeUpdate(String sqlStatement) 8

执行字符串中指定的 INSERT、UPDATE 或 DELETE 等 SQL 语句。还可以执行数据定义语言（Data Definition Language，DDL）的语句，如 CREATE TABLE。返回受影响的行数，如果是没有更新计数的语句，则返回 0。

- boolean execute(String sqlStatement)

  执行字符串中指定的 SQL 语句。可能会产生多个结果集和更新计数。如果第一个执行结果是结果集，则返回 true；反之，返回 false。调用 getResultSet 或 getUpdateCount 方法可以得到第一个执行结果。请参见 5.5.4 节中关于处理多结果集的详细信息。

- ResultSet getResultSet()

  返回前一条查询语句的结果集。如果前一条语句未产生结果集，则返回 null 值。对于每一条执行过的语句，该方法只能被调用一次。

- int getUpdateCount()

- long getLargeUpdateCount() 8

  返回受前一条更新语句影响的行数。如果前一条语句未更新数据库，则返回 -1。对于每一条执行过的语句，该方法只能被调用一次。

- void close()

  关闭该语句对象以及它所对应的结果集。

- boolean isClosed() 6

  如果该语句被关闭，则返回 true。

- void closeOnCompletion() 7

  一旦该语句的所有结果集都被关闭，则关闭该语句。

API java.sql.ResultSet 1.1

- boolean next()

  将结果集中的当前行向前移动一行。如果已经到达最后一行的后面，则返回 false。注意，初始情况下必须调用该方法才能转到第一行。

- *Xxx* get*Xxx*(int columnNumber)

- *Xxx* get*Xxx*(String columnLabel)

  （X*xx* 指数据类型，例如 int、double、String 和 Date 等。）

- \<T\> T getObject(int columnIndex, Class\<T\> type) 7

- \<T\> T getObject(String columnLabel, Class\<T\> type) 7

- void updateObject(int columnIndex, Object x, SQLType targetSqlType) 8

- void updateObject(String columnLabel, Object x, SQLType targetSqlType) 8

  用给定的列序号或列标签返回或更新该列的值，并将值转换成指定的类型。列标签是 SQL 的 AS 子句中指定的标签，在没有使用 AS 时，它就是列名。

- int findColumn(String columnName)

  根据给定的列名，返回该列的序号。

- void close()

  立即关闭当前的结果集。

- boolean isClosed() 6

  如果该语句被关闭，则返回 true。

5.4.2 管理连接、语句和结果集

每个 Connection 对象都可以创建一个或多个 Statement 对象。同一个 Statement 对象可以用于多个不相关的命令和查询。但是，一个 Statement 对象最多只能有一个打开的结果集。如果需要执行多个查询操作，且需要同时分析查询结果，那么必须创建多个 Statement 对象。

需要说明的是，每个链接上的语句数是有限制的。使用 DatabaseMetaData 接口中的 getMaxStatements 方法可以获取 JDBC 驱动程序支持的同时打开的语句对象的总数。

实际上，我们通常并不需要同时处理多个结果集。如果结果集相互关联，我们可以使用组合查询，这样就只需要分析一个结果。对数据库进行组合查询比使用 Java 程序遍历多个结果集要高效得多。

我们应该确保在一个 Statement 对象上触发新的查询或更新语句之前结束对所有结果集的处理，因为前序查询的所有结果集都会被自动关闭。

使用完 ResultSet、Statement 或 Connection 对象后，应立即调用 close 方法。这些对象都使用了规模较大的数据结构，它们会占用数据库存服务器上的有限资源。

Statement 对象的 close 方法将自动关闭所有与其相关联的结果集。同样地，调用 Connection 类的 close 方法将关闭该连接上的所有语句。

反过来的情况是，可以在 Statement 上调用 closeOnCompletion 方法，在其所有结果集都被关闭后，该语句会立即被自动关闭。

如果所用连接都是短时的，那么无须操心语句和结果集的关闭。只需将 close 语句放在带资源的 try 语句中，以便确保连接对象不可能继续保持打开状态。

```
try (Connection conn = . . .)
{
    Statement stat = conn.createStatement();
    ResultSet result = stat.executeQuery(queryString);
    process query result
}
```

5.4.3 分析 SQL 异常

每个 SQLException 都有一个由多个 SQLException 对象构成的链，这些对象可以通过 getNextException 方法获取。这个异常链是每个异常都具有的由 Throwable 对象构成的"成因"链之外的异常链（请参见卷 I 第 7 章以了解 Java 异常的详细信息），因此，我们需要用两个嵌套的循环来完整枚举所有的异常。幸运的是，SQLException 类得到了增强，实现了 Iterable<Throwable> 接口，其 iterator() 方法可以产生一个 Iterator<Throwable>，这个迭代器可以迭代这两个链，首先迭代第一个 SQLException 的成因链，然后迭代下一个 SQLException，以此类推。我们可以

直接使用下面这个改进的 for 循环：

```
for (Throwable t : sqlException)
{
    do something with t
}
```

可以在 SQLException 上调用 getSQLState 和 getErrorCode 方法来进一步分析它，其中第一个方法将产生符合 X/Open 或 SQL:2003 标准的字符串（调用 DatabaseMetaData 接口的 getSQLStateType 方法可以查出驱动程序所使用的标准）。而错误代码是与具体的提供商相关的。

SQL 异常按照层次结构树的方式组织到了一起（如图 5-5 所示），这使得我们可以按照与提供商无关的方式来捕获具体的错误类型。

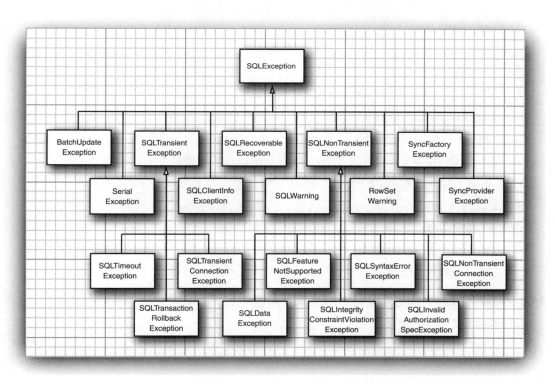

图 5-5　SQL 异常类型

另外，数据库驱动程序可以将非致命问题作为警告报告，我们可以从连接、语句和结果集中获取这些警告。SQLWarning 类是 SQLException 的子类（尽管 SQLWarning 不会被当作异常抛出），我们可以调用 getSQLState 和 getErrorCode 来获取有关警告的更多信息。与 SQL 异常类似，警告也是串成链的。要获得所有的警告，可以使用下面的循环：

```
SQLWarning w = stat.getWarning();
while (w != null)
{
```

```
    do something with w
    w = w.nextWarning();
}
```

当数据从数据库中读出并意外被截断时，SQLWarning 的 DataTruncation 子类就派上用场了。如果数据截断发生在更新语句中，那么 DataTruncation 对象将会被当作异常抛出。

API *java.sql.SQLException* 1.1

- SQLException getNextException()
  返回链接到该 SQL 异常的下一个 SQL 异常，或者在到达链尾时返回 null。
- Iterator<Throwable> iterator() 6
  获取迭代器，可以迭代链接的 SQL 异常和它们的成因。
- String getSQLState()
  获取 "SQL 状态"，即标准化的错误代码。
- int getErrorCode()
  获取提供商相关的错误代码。

API *java.sql.SQLWarning* 1.1

- SQLWarning getNextWarning()
  返回链接到该警告的下一个警告，或者在到达链尾时返回 null。

API *java.sql.Connection* 1.1
　　java.sql.Statement 1.1
　　java.sql.ResultSet 1.1

- SQLWarning getWarnings()
  返回未处理警告中的第一个，或者在没有未处理警告时返回 null。

API *java.sql.DataTruncation* 1.1

- boolean getParameter()
  如果在参数上进行了数据截断，则返回 true；如果在列上进行了数据截断，则返回 false。
- int getIndex()
  返回被截断的参数或列的索引。
- int getDataSize()
  返回应该被传输的字节数量，或者在该值未知的情况下返回 –1。
- int getTransferSize()
  返回实际被传输的字节数量，或者在该值未知的情况下返回 –1。

5.4.4 组装数据库

至此，大家也许都迫不及待地想编写一个真正实用的 JDBC 程序了。如果我们可以编写一段程序来执行之前所介绍的那些巧妙的查询，那当然很好。不过，在此之前我们还有一个问题没有解决：目前数据库中还没有数据。我们需要组装数据库，并且也确实存在一种简单方法可以实现此目的：用一系列的 SQL 指令来创建数据表并向其中插入数据。大多数数据库程序都可以处理来自文本文件中的一系列 SQL 指令，但是在语句终止符和其他一些文法问题上，这些数据库程序之间存在着令人讨厌的差异。

正是由于这个原因，我们使用 JDBC 创建了一个简单的程序，它从文件中读取 SQL 指令，其中一条指令占据一行，然后执行它们。

该程序专门用于从下列格式的文本文件中读取数据：

```
CREATE TABLE Publishers (Publisher_Id CHAR(6), Name CHAR(30), URL CHAR(80));
INSERT INTO Publishers VALUES ('0201', 'Addison-Wesley', 'www.aw-bc.com');
INSERT INTO Publishers VALUES ('0471', 'John Wiley & Sons', 'www.wiley.com');
. . .
```

程序清单 5-2 是用来读取 SQL 语句文件以及执行这些语句的程序代码。通读这些代码并不重要，我们在这里只是提供了这样的程序，使你能够组装数据库并运行本章剩余部分的代码。

请确认你的数据库服务器是在运行的，然后可以使用如下方法运行该程序：

```
java -classpath driverPath:. exec.ExecSQL Books.sql
java -classpath driverPath:. exec.ExecSQL Authors.sql
java -classpath driverPath:. exec.ExecSQL Publishers.sql
java -classpath driverPath:. exec.ExecSQL BooksAuthors.sql
```

在运行程序之前，请检查一下 database.properties 文件是否已经针对你的运行环境进行了正确设置。请查看 5.3.5 节。

> **注释**：你的数据库可能也包含直接读取 SQL 文件的工具，例如，在使用 Derby 时，可以运行下面的命令：
>
> ```
> java -jar derby/lib/derbyrun.jar ij -p ij.properties Books.sql
> ```
>
> （ij.properties 文件在 5.3.3 节中描述过。）
>
> 在用于 ExecSQL 命令的数据格式中，我们允许每行的结尾都可以有一个可选的分号，因为大多数数据库工具都希望使用这种格式。

下面将简要介绍一下 ExecSQL 程序的操作步骤。

1. 连接数据库。getConnection 方法读取 database.properties 文件中的属性信息，并将属性 jdbc.drivers 添加到系统属性中。驱动程序管理器使用属性 jdbc.drivers 加载相应的驱动程序。getConnection 方法使用 jdbc.url、jdbc.username 和 jdbc.password 等属性打开数据库连接。

2. 打开包含 SQL 语句的文件。如果未提供任何文件名，则在控制台中提示用户输入语句。

3. 使用泛化的 execute 方法执行每条语句。如果它返回 true，则说明该语句产生了一个结果集。我们为图书数据库提供的 4 个 SQL 文件都以一条 SELECT * 语句结束，这样就可以看到数据是否已成功插入到了数据库中。

4. 如果产生了结果集，则打印出结果。因为这是一个泛化的结果集，所以我们必须使用元数据来确定该结果的列数。更多的信息请查看 5.8 节。

5. 如果运行过程中出现 SQL 异常，则打印出这个异常以及所有可能包含在其中的与其链接在一起的相关异常。

6. 关闭数据库连接。

程序清单 5-2 给出了该程序的代码。

程序清单 5-2　exec/ExecSQL.java

```java
 1  package exec;
 2
 3  import java.io.*;
 4  import java.nio.charset.*;
 5  import java.nio.file.*;
 6  import java.util.*;
 7  import java.sql.*;
 8
 9  /**
10   * Executes all SQL statements in a file. Call this program as <br>
11   * java -classpath driverPath:. ExecSQL commandFile
12   *
13   * @version 1.33 2018-05-01
14   * @author Cay Horstmann
15   */
16  class ExecSQL
17  {
18     public static void main(String args[]) throws IOException
19     {
20        try (Scanner in = args.length == 0 ? new Scanner(System.in)
21              : new Scanner(Paths.get(args[0]), StandardCharsets.UTF_8))
22        {
23           try (Connection conn = getConnection();
24                 Statement stat = conn.createStatement())
25           {
26              while (true)
27              {
28                 if (args.length == 0) System.out.println("Enter command or EXIT to exit:");
29
30                 if (!in.hasNextLine()) return;
31
32                 String line = in.nextLine().trim();
33                 if (line.equalsIgnoreCase("EXIT")) return;
34                 if (line.endsWith(";")) // remove trailing semicolon
35                    line = line.substring(0, line.length() - 1);
36                 try
37                 {
38                    boolean isResult = stat.execute(line);
39                    if (isResult)
40                    {
41                       try (ResultSet rs = stat.getResultSet())
42                       {
43                          showResultSet(rs);
44                       }
```

```java
45                  }
46                  else
47                  {
48                      int updateCount = stat.getUpdateCount();
49                      System.out.println(updateCount + " rows updated");
50                  }
51              }
52              catch (SQLException e)
53              {
54                  for (Throwable t : e)
55                      t.printStackTrace();
56              }
57          }
58      }
59   }
60   catch (SQLException e)
61   {
62      for (Throwable t : e)
63          t.printStackTrace();
64   }
65  }
66
67  /**
68   * Gets a connection from the properties specified in the file database.properties
69   * @return the database connection
70   */
71  public static Connection getConnection() throws SQLException, IOException
72  {
73      var props = new Properties();
74      try (InputStream in = Files.newInputStream(Paths.get("database.properties")))
75      {
76          props.load(in);
77      }
78      String drivers = props.getProperty("jdbc.drivers");
79      if (drivers != null) System.setProperty("jdbc.drivers", drivers);
80
81      String url = props.getProperty("jdbc.url");
82      String username = props.getProperty("jdbc.username");
83      String password = props.getProperty("jdbc.password");
84
85      return DriverManager.getConnection(url, username, password);
86  }
87
88  /**
89   * Prints a result set.
90   * @param result the result set to be printed
91   */
92  public static void showResultSet(ResultSet result) throws SQLException
93  {
94      ResultSetMetaData metaData = result.getMetaData();
95      int columnCount = metaData.getColumnCount();
96
97      for (int i = 1; i <= columnCount; i++)
98      {
```

```
 99            if (i > 1) System.out.print(", ");
100            System.out.print(metaData.getColumnLabel(i));
101         }
102         System.out.println();
103
104         while (result.next())
105         {
106            for (int i = 1; i <= columnCount; i++)
107            {
108               if (i > 1) System.out.print(", ");
109               System.out.print(result.getString(i));
110            }
111            System.out.println();
112         }
113      }
114   }
```

5.5 执行查询操作

在这一节中，我们将编写一段用于对 COREJAVA 数据库执行查询操作的程序。为了使程序可以正常运行，必须按照上一节中的说明用表组装 COREJAVA 数据库。

在查询数据库时，可以选择作者和出版社，或者将这两项中的一项设置为"Any"。

还可以修改数据库中的数据。选择一家出版社，然后输入金额。该出版社对应的所有价格都将按照填入的金额进行调整，同时程序将显示被修改的行数。修改完价格以后，可以运行一个查询操作，以核实新的价格。

5.5.1 预备语句

在这个程序中，我们使用了一个新的特性，即预备语句（prepared statement）。如果我们要查询某个出版社的所有图书而不考虑具体的作者，那么该查询的 SQL 语句如下：

```
SELECT Books.Price, Books
FROM Books, Publishers
WHERE Books.Publisher_Id = Publishers.Publisher_Id
AND Publishers.Name = the name from the list box
```

我们没有必要在每次触发一个这样的查询时都建立新的查询语句，而是可以准备一个带有宿主变量的查询语句，每次查询时只需为该变量填入不同的字符串就可以反复多次地使用该语句。这一技术改进了查询性能，每当数据库执行一个查询时，它总是首先通过计算来确定查询策略，以便高效地执行查询操作。通过事先准备好查询并多次重用它，我们就可以确保查询所需的准备步骤只被执行一次。

在预备查询语句中，每个宿主变量都用"?"来表示。如果存在一个以上的变量，那么在设置变量值时必须注意"?"的位置。例如，如果我们的预备查询为如下形式：

```
String publisherQuery
   = "SELECT Books.Price, Books"
```

```
    + " FROM Books, Publishers"
    + " WHERE Books.Publisher_Id = Publishers.Publisher_Id AND Publishers.Name = ?";
PreparedStatement stat = conn.prepareStatement(publisherQuery);
```

在执行预备语句之前,必须使用 set 方法将变量绑定到实际的值上。和 ResultSet 接口中的 get 方法类似,针对不同的数据类型也有不同的 set 方法。在本例中,我们为出版社名称设置了一个字符串值。

```
stat.setString(1, publisher);
```

第一个参数指的是需要设置的宿主变量的位置,位置 1 表示第一个 "?"。第二个参数指的是赋予宿主变量的值。

如果想要重用已经执行过的预备查询语句,那么除非使用 set 方法或调用 clearParameters 方法,否则所有宿主变量的绑定都不会改变。这就意味着,在从一个查询到另一个查询的过程中,只需使用 setXxx 方法重新绑定那些需要改变的变量即可。

一旦为所有变量都绑定了具体的值,就可以执行预备语句了:

```
ResultSet rs = stat.executeQuery();
```

✓ **提示**:通过连接字符串来手动构建查询显得非常枯燥乏味,而且存在潜在的危险。你必须注意像引号这样的特殊字符,而且如果查询中涉及用户的输入,那就还需要警惕注入攻击。因此,只要查询涉及变量,就应该使用预备语句。

价格更新操作可以由 UPDATE 语句实现。请注意,我们调用的是 executeUpdate 方法,而非 executeQuery 方法,因为 UPDATE 语句不返回结果集。executeUpdate 的返回值为被修改过的行数。

```
int r = stat.executeUpdate();
System.out.println(r + " rows updated");
```

📄 **注释**:在相关的 Connection 对象关闭之后,PreparedStatement 对象也就变得无效了。不过,许多数据库通常都会自动缓存预备语句。如果相同的查询被预备两次,数据库通常会直接重用查询策略。因此,无须过多考虑调用 prepareStatement 的开销。

下面简要说明示例程序的结构:

- 通过执行两个查询得到数据库中所有的作者和出版社名称,作者和出版社数组列表由此组装而成。
- 涉及作者的查询比较复杂。因为一本书可能有多个作者,BooksAuthors 表给出了作者和图书之间的对应关系。例如,ISBN 号为 0-201-96426-0 的图书有两个作者,其代号为:DATE 和 DARW。以下为 BooksAuthors 表中的两行记录:

```
0-201-96426-0, DATE, 1
0-201-96426-0, DARW, 2
```

BooksAuthors 表中第三列指的是作者的顺序(我们不能只使用表中行的位置,在关系表中没有固定的行顺序)。因此,查询时需要连接 Books 表、BooksAuthors 表和 Authors 表,以便和用户所选的作者名进行比较。

```
SELECT Books.Price, Books FROM Books, BooksAuthors, Authors, Publishers
WHERE Authors.Author_Id = BooksAuthors.Author_Id AND BooksAuthors.ISBN = Books.ISBN
AND Books.Publisher_Id = Publishers.Publisher_Id AND Authors.Name = ?
AND Publishers.Name = ?
```

> ✅ **提示**：许多程序员都不喜欢使用如此复杂的 SQL 语句。比较常见的方法是使用大量的 Java 代码来迭代多个结果集，但是这种方法效率非常低。通常，使用数据库的查询代码要比使用 Java 程序好得多——这是数据库的核心竞争力之一。一般而言，可以使用 SQL 解决的问题，就不要使用 Java 程序。

- change Prices 方法执行了一条 UPDATE 语句。注意，UPDATE 语句中的 WHERE 子句需要使用出版社代码，而我们只知道出版社名称。这个问题可以使用嵌套子查询来解决。

```
UPDATE Books
SET Price = Price + ?
WHERE Books.Publisher_Id = (SELECT Publisher_Id FROM Publishers WHERE Name = ?)
```

程序清单 5-3 给出了程序的完整代码。

程序清单 5-3　query/QueryTest.java

```java
 1  package query;
 2
 3  import java.io.*;
 4  import java.nio.file.*;
 5  import java.sql.*;
 6  import java.util.*;
 7
 8  /**
 9   * This program demonstrates several complex database queries.
10   * @version 1.31 2018-05-01
11   * @author Cay Horstmann
12   */
13  public class QueryTest
14  {
15     private static final String allQuery = "SELECT Books.Price, Books.Title FROM Books";
16
17     private static final String authorPublisherQuery = "SELECT Books.Price, Books.Title"
18        + " FROM Books, BooksAuthors, Authors, Publishers"
19        + " WHERE Authors.Author_Id = BooksAuthors.Author_Id AND BooksAuthors.ISBN = Books.ISBN"
20        + " AND Books.Publisher_Id = Publishers.Publisher_Id AND Authors.Name = ?"
21        + " AND Publishers.Name = ?";
22
23     private static final String authorQuery
24        = "SELECT Books.Price, Books.Title FROM Books, BooksAuthors, Authors"
25        + " WHERE Authors.Author_Id = BooksAuthors.Author_Id"
26        + " AND BooksAuthors.ISBN = Books.ISBN"
27        + " AND Authors.Name = ?";
28
29     private static final String publisherQuery
30        = "SELECT Books.Price, Books.Title FROM Books, Publishers"
31        + " WHERE Books.Publisher_Id = Publishers.Publisher_Id AND Publishers.Name = ?";
32
```

```java
   private static final String priceUpdate = "UPDATE Books SET Price = Price + ? "
      + " WHERE Books.Publisher_Id = (SELECT Publisher_Id FROM Publishers WHERE Name = ?)";

   private static Scanner in;
   private static ArrayList<String> authors = new ArrayList<>();
   private static ArrayList<String> publishers = new ArrayList<>();

   public static void main(String[] args) throws IOException
   {
      try (Connection conn = getConnection())
      {
         in = new Scanner(System.in);
         authors.add("Any");
         publishers.add("Any");
         try (Statement stat = conn.createStatement())
         {
            // Fill the authors array list
            var query = "SELECT Name FROM Authors";
            try (ResultSet rs = stat.executeQuery(query))
            {
               while (rs.next())
                  authors.add(rs.getString(1));
            }

            // Fill the publishers array list
            query = "SELECT Name FROM Publishers";
            try (ResultSet rs = stat.executeQuery(query))
            {
               while (rs.next())
                  publishers.add(rs.getString(1));
            }
         }
         var done = false;
         while (!done)
         {
            System.out.print("Q)uery C)hange prices E)xit: ");
            String input = in.next().toUpperCase();
            if (input.equals("Q"))
               executeQuery(conn);
            else if (input.equals("C"))
               changePrices(conn);
            else
               done = true;
         }
      }
      catch (SQLException e)
      {
         for (Throwable t : e)
            System.out.println(t.getMessage());
      }
   }

   /**
    * Executes the selected query.
```

```
 87        * @param conn the database connection
 88        */
 89       private static void executeQuery(Connection conn) throws SQLException
 90       {
 91          String author = select("Authors:", authors);
 92          String publisher = select("Publishers:", publishers);
 93          PreparedStatement stat;
 94          if (!author.equals("Any") && !publisher.equals("Any"))
 95          {
 96             stat = conn.prepareStatement(authorPublisherQuery);
 97             stat.setString(1, author);
 98             stat.setString(2, publisher);
 99          }
100          else if (!author.equals("Any") && publisher.equals("Any"))
101          {
102              stat = conn.prepareStatement(authorQuery);
103             stat.setString(1, author);
104          }
105          else if (author.equals("Any") && !publisher.equals("Any"))
106          {
107             stat = conn.prepareStatement(publisherQuery);
108             stat.setString(1, publisher);
109          }
110          else
111             stat = conn.prepareStatement(allQuery);
112
113          try (ResultSet rs = stat.executeQuery())
114          {
115             while (rs.next())
116                System.out.println(rs.getString(1) + ", " + rs.getString(2));
117          }
118       }
119
120       /**
121        * Executes an update statement to change prices.
122        * @param conn the database connection
123        */
124       public static void changePrices(Connection conn) throws SQLException
125       {
126          String publisher = select("Publishers:", publishers.subList(1, publishers.size()));
127          System.out.print("Change prices by: ");
128          double priceChange = in.nextDouble();
129          PreparedStatement stat = conn.prepareStatement(priceUpdate);
130          stat.setDouble(1, priceChange);
131          stat.setString(2, publisher);
132          int r = stat.executeUpdate();
133          System.out.println(r + " records updated.");
134       }
135
136       /**
137        * Asks the user to select a string.
138        * @param prompt the prompt to display
139        * @param options the options from which the user can choose
140        * @return the option that the user chose
```

```java
141     */
142    public static String select(String prompt, List<String> options)
143    {
144       while (true)
145       {
146          System.out.println(prompt);
147          for (int i = 0; i < options.size(); i++)
148             System.out.printf("%2d) %s%n", i + 1, options.get(i));
149          int sel = in.nextInt();
150          if (sel > 0 && sel <= options.size())
151             return options.get(sel - 1);
152       }
153    }
154
155    /**
156     * Gets a connection from the properties specified in the file database.properties.
157     * @return the database connection
158     */
159    public static Connection getConnection() throws SQLException, IOException
160    {
161       var props = new Properties();
162       try (InputStream in = Files.newInputStream(Paths.get("database.properties")))
163       {
164          props.load(in);
165       }
166
167       String drivers = props.getProperty("jdbc.drivers");
168       if (drivers != null) System.setProperty("jdbc.drivers", drivers);
169
170       String url = props.getProperty("jdbc.url");
171       String username = props.getProperty("jdbc.username");
172       String password = props.getProperty("jdbc.password");
173
174       return DriverManager.getConnection(url, username, password);
175    }
176 }
```

API *java.sql.Connection* 1.1

- PreparedStatement prepareStatement(String sql)

  返回一个含预编译语句的 PreparedStatement 对象。字符串 sql 代表一个 SQL 语句，该语句可以包含一个或多个由?字符指明的参数占位符。

API *java.sql.PreparedStatement* 1.1

- void set*Xxx*(int n, *Xxx* x)

  (*Xxx* 指 int、double、String、Date 之类的数据类型) 设置第 n 个参数值为 x。

- void clearParameters()

  清除预备语句中的所有当前参数。

- ResultSet executeQuery()

执行预备 SQL 查询，并返回一个 ResultSet 对象。
- int executeUpdate()

执行预备 SQL 语句 INSERT、UPDATE 或 DELETE，这些语句由 PreparedStatement 对象表示。该方法返回在执行上述语句过程中所有受影响的记录总数。如果执行的是数据定义语言（DDL）中的语句，如 CREATE TABLE，则该方法返回 0。

5.5.2 读写 LOB

除了数字、字符串和日期之外，许多数据库还可以存储大对象，例如图片或其他数据。在 SQL 中，二进制大对象称为 BLOB，字符型大对象称为 CLOB。

要读取 LOB，需要执行 SELECT 语句，然后在 ResultSet 上调用 getBlob 或 getClob 方法，这样就可以获得 Blob 或 Clob 类型的对象。要从 Blob 中获取二进制数据，可以调用 getBytes 或 getBinaryStream。例如，如果你有一张保存图书封面图像的表，那么就可以像下面这样获取一张图像：

```
PreparedStatement stat = conn.prepareStatement("SELECT Cover FROM BookCovers WHERE ISBN=?");
...
stat.set(1, isbn);
try (ResultSet result = stat.executeQuery())
{
   if (result.next())
   {
      Blob coverBlob = result.getBlob(1);
      Image coverImage = ImageIO.read(coverBlob.getBinaryStream());
   }
}
```

类似地，如果获取了 Clob 对象，那么就可以通过调用 getSubString 或 getCharacterStream 方法来获取其中的字符数据。

要将 LOB 置于数据库中，需要在 Connection 对象上调用 createBlob 或 createClob，然后获取一个用于该 LOB 的输出流或写出器，写出数据，并将该对象存储到数据库中。例如，下面展示了如何存储一张图像：

```
Blob coverBlob = connection.createBlob();
int offset = 0;
OutputStream out = coverBlob.setBinaryStream(offset);
ImageIO.write(coverImage, "PNG", out);
PreparedStatement stat = conn.prepareStatement("INSERT INTO Cover VALUES (?, ?)");
stat.set(1, isbn);
stat.set(2, coverBlob);
stat.executeUpdate();
```

API **java.sql.ResultSet** 1.1

- Blob getBlob(int columnIndex) 1.2
- Blob getBlob(String columnLabel) 1.2
- Clob getClob(int columnIndex) 1.2

- Clob getClob(String columnLabel) 1.2

  获取给定列的 BLOB 或 CLOB。

API *java.sql.Blob* 1.2

- long length()

  获取该 BLOB 的长度。

- byte[] getBytes(long startPosition, long length)

  获取该 BLOB 中给定范围的数据。

- InputStream getBinaryStream()
- InputStream getBinaryStream(long startPosition, long length)

  返回一个输入流，用于读取该 BLOB 中全部或给定范围的数据。

- OutputStream setBinaryStream(long startPosition) 1.4

  返回一个输出流，用于从给定位置开始写入该 BLOB。

API *java.sql.Clob* 1.4

- long length()

  获取该 CLOB 中的字符总数。

- String getSubString(long startPosition, long length)

  获取该 CLOB 中给定范围的字符。

- Reader getCharacterStream()
- Reader getCharacterStream(long startPosition, long length)

  返回一个读入器（而不是流），用于读取 CLOB 中全部或给定范围的数据。

- Writer setCharacterStream(long startPosition) 1.4

  返回一个写出器（而不是流），用于从给定位置开始写入该 CLOB。

API *java.sql.Connection* 1.1

- Blob createBlob() 6
- Clob createClob() 6

  创建一个空的 BLOB 或 CLOB。

5.5.3 SQL 转义

"转义"语法是各种数据库普遍支持的特性，但是数据库使用的是与数据库相关的语法变体，因此，将转义语法转译为特定数据库的语法是 JDBC 驱动程序的任务之一。

转义主要用于下列场景：

- 日期和时间字面常量
- 调用标量函数
- 调用存储过程

- 外连接
- 在 LIKE 子句中的转义字符

日期和时间字面常量随数据库的不同而变化很大。要嵌入日期或时间字面常量，需要按照 ISO 8601 格式（http://www.cl.cam.ac.uk/~mgk25/iso-time.html）指定它的值，之后驱动程序会将其转译为本地格式。应该使用 d、t、ts 来表示 DATE、TIME 和 TIMESTAMP 值：

```
{d '2008-01-24'}
{t '23:59:59'}
{ts '2008-01-24 23:59:59.999'}
```

标量函数（scalar function）是指仅返回单个值的函数。在数据库中包含大量的函数，但是不同的数据库中这些函数名存在着差异。JDBC 规范提供了标准的名字，并将其转译为数据库相关的名字。要调用函数，需要像下面这样嵌入标准的函数名和参数：

```
{fn left(?, 20)}
{fn user()}
```

在 JDBC 规范中可以找到它支持的函数名的完整列表。

存储过程（stored procedure）是在数据库中执行的用数据库相关的语言编写的过程。要调用存储过程，需要使用 call 转义命令，在存储过程没有任何参数时，可以不用加上括号。另外，应该用 = 来捕获存储过程的返回值：

```
{call PROC1(?, ?)}
{call PROC2}
{call ? = PROC3(?)}
```

两个表的外连接（outer join）并不要求每个表的所有行都要根据连接条件进行匹配，例如，假设有如下的查询：

```
SELECT * FROM {oj Books LEFT OUTER JOIN Publishers
ON Books.Publisher_Id = Publisher.Publisher_Id}
```

这个查询的执行结果中将包含有 Publisher_Id 在 Publishers 表中没有任何匹配的书，其中，Publisher_ID 为 NULL 值的行，就表示不存在任何匹配。如果应该使用 RIGHT OUTER JOIN，就可以囊括没有任何匹配图书的出版商，而使用 FULL OUTER JOIN 可以同时返回这两类没有任何匹配的信息。由于并非所有的数据库对于这些连接都使用标准的写法，因此需要使用转义语法。

最后一种情况，_ 和 % 字符在 LIKE 子句中具有特殊含义，用来匹配一个字符或一个字符序列。目前并不存在任何在字面上使用它们的标准方式，所以如果想要匹配所有包含 _ 字符的字符串，就必须使用下面的结构：

```
... WHERE ? LIKE %!_% {escape '!'}
```

这里我们将 ! 定义为转义字符，而 !_ 组合表示字面常量下划线。

5.5.4 多结果集

在执行存储过程，或者在使用允许在单个查询中提交多个 SELECT 语句的数据库时，一个查询有可能会返回多个结果集。下面是获取所有结果集的步骤：

1. 使用 execute 方法来执行 SQL 语句。
2. 获取第一个结果集或更新计数。
3. 重复调用 getMoreResults 方法以移动到下一个结果集。
4. 当不存在更多的结果集或更新计数时，完成操作。

如果由多结果集构成的链中的下一项是结果集，execute 和 getMoreResults 方法将返回 true，而如果在链中的下一项不是更新计数，getUpdateCount 方法将返回 -1。

下面的循环可以遍历所有的结果：

```
boolean isResult = stat.execute(command);
boolean done = false;
while (!done)
{
   if (isResult)
   {
      ResultSet result = stat.getResultSet();
      do something with result
   }
   else
   {
      int updateCount = stat.getUpdateCount();
      if (updateCount >= 0)
         do something with updateCount
      else
         done = true;
   }
   if (!done) isResult = stat.getMoreResults();
}
```

API *java.sql.Statement* 1.1

- boolean getMoreResults()
- boolean getMoreResults(int current) 6

    获取该语句的下一个结果集，Current 参数是 CLOSE_CURRENT_RESULT（默认值）、KEEP_CURRENT_RESULT 或 CLOSE_ALL_RESULTS 之一。如果存在下一个结果集，并且它确实是一个结果集，则返回 true。

5.5.5 获取自动生成的键

大多数数据库都支持某种在数据库中对行自动编号的机制。但是，不同的提供商所提供的机制之间存在着很大的差异，而这些自动编号的值经常用作主键。尽管 JDBC 没有提供独立于提供商的自动生成键的解决方案，但是它提供了获取自动生成键的有效途径。当我们向数据表中插入一个新行，且其键自动生成时，可以用下面的代码来获取这个键：

```
stat.executeUpdate(insertStatement, Statement.RETURN_GENERATED_KEYS);
ResultSet rs = stat.getGeneratedKeys();
if (rs.next())
```

```
{
    int key = rs.getInt(1);
    ...
}
```

API *java.sql.Statement* 1.1

- boolean execute(String statement, int autogenerated) 1.4
- int executeUpdate(String statement, int autogenerated) 1.4

  像前面描述的那样执行给定的 SQL 语句，如果 autogenerated 被设置为 Statement.RETURN_GENERATED_KEYS，并且该语句是一条 INSERT 语句，那么第一列中就是自动生成的键。

5.6 可滚动和可更新的结果集

我们前面已经介绍过，使用 ResultSet 接口中的 next 方法可以迭代遍历结果集中的所有行。对于一个只需要分析数据的程序来说，这显然已经足够了。但是，如果是用于展示一张表或查询结果的可视化数据显示（参见图 5-4），我们通常会希望用户可以在结果集上前后移动。对于可滚动结果集而言，我们可以在其中向前或向后移动，甚至可以跳到任意位置。

另外，一旦向用户显示了结果集中的内容，他们就可能希望编辑这些内容。在可更新的结果集中，可以以编程方式来更新其中的项，使得数据库可以自动更新数据。我们将在下面的小节中讨论这些功能。

5.6.1 可滚动的结果集

默认情况下，结果集是不可滚动和不可更新的。为了从查询中获取可滚动的结果集，必须使用下面的方法得到一个不同的 Statement 对象：

```
Statement stat = conn.createStatement(type, concurrency);
```

如果要获得预备语句，请调用下面的方法：

```
PreparedStatement stat = conn.prepareStatement(command, type, concurrency);
```

表 5-6 和表 5-7 列出了 type 和 concurrency 的所有可能值，可以有以下几种选择：
- 是否希望结果集是可滚动的？如果不需要，则使用 ResultSet.TYPE_FORWARD_ONLY。
- 如果结果集是可滚动的，且数据库在查询生成结果集之后发生了变化，那么是否希望结果集反映出这些变化？（在我们的讨论中，我们假设将可滚动的结果集设置为 ResultSet.TYPE_SCROLL_INSENSITIVE。这个设置将使结果集"感应"不到查询结束后出现的数据库变化。）
- 是否希望通过编辑结果集就可以更新数据库？（详细说明请参见下一节内容。）

表 5-6 ResultSet 类的 type 值

值	解释
TYPE_FORWARD_ONLY	结果集不能滚动（默认值）

（续）

值	解释
TYPE_SCROLL_INSENSITIVE	结果集可以滚动，但对数据库变化不敏感
TYPE_SCROLL_SENSITIVE	结果集可以滚动，且对数据库变化敏感

表 5-7　ResultSet 类的 Concurrency 值

值	解释
CONCUR_READ_ONLY	结果集不能用于更新数据库（默认值）
CONCUR_UPDATABLE	结果集可以用于更新数据库

例如，如果只想滚动遍历结果集，而不想编辑它的数据，那么可以使用以下语句：

```
Statement stat = conn.createStatement(
    ResultSet.TYPE_SCROLL_INSENSITIVE, ResultSet.CONCUR_READ_ONLY);
```

现在，通过调用以下方法获得的所有结果集都将是可滚动的。

```
ResultSet rs = stat.executeQuery(query);
```

可滚动的结果集有一个游标，用以指示当前位置。

> **注释**：并非所有的数据库驱动程序都支持可滚动和可更新的结果集。（使用 Database-MetaData 接口中的 supportsResultSetType 和 supportsResultSetConcurrency 方法，我们可以获知在使用特定的驱动程序时，某个数据库究竟支持哪些结果集类型以及哪些并发模式。）即便是数据库支持所有的结果集模式，某个特定的查询也可能无法产生带有所要求的所有属性的结果集。（例如，一个复杂查询的结果集就有可能是不可更新的结果集。）在这种情况下，executeQuery 方法将返回一个功能较少的 ResultSet 对象，并添加一个 SQLWarning 到连接对象中。（参见 5.4.3 节有关如何获取警告信息的内容）或者，也可以使用 ResultSet 接口中的 getType 和 getConcurrency 方法查看结果集实际支持的模式。如果不检查结果集的功能就发起一个不支持的操作，比如对不可滚动的结果集调用 previous 方法，那么该操作将抛出一个 SQLException 异常。

在结果集上滚动是非常简单的，可以使用

```
if (rs.previous()) . . .;
```

向后滚动。如果游标位于一个实际的行上，那么该方法将返回 true；如果游标位于第一行之前，那么返回 false。

可以使用以下调用将游标向后或向前移动多行：

```
rs.relative(n);
```

如果 n 为正数，游标将向前移动。如果 n 为负数，游标将向后移动。如果 n 为 0，那么调用该方法将不起任何作用。如果试图将游标移动到当前行集的范围之外，即根据 n 值的正负号，游标需要被设置在最后一行之后或第一行之前，那么，该方法将返回 false，且不移动游标。如果游标位于一个实际的行上,那么该方法将返回 true。

或者,还可以将游标设置到指定的行号上:

```
rs.absolute(n);
```

调用以下方法将返回当前行的行号:

```
int currentRow = rs.getRow();
```

结果集中第一行的行号为 1。如果返回值为 0,那么游标当前不在任何行上,它要么位于第一行之前,要么位于最后一行之后。

first、last、beforeFirst 和 afterLast 这些简便方法用于将游标移动到第一行、最后一行、第一行之前或最后一行之后。

最后,isFirst、isLast、isBeforeFirst 和 isAfterLast 用于测试游标是否位于这些特殊位置上。

使用可滚动的结果集是非常简单的,将查询数据放入缓存中的复杂工作是由数据库驱动程序在后台完成的。

5.6.2 可更新的结果集

如果希望编辑结果集中的数据,并且将结果集上的数据变更自动反映到数据库中,那么就必须使用可更新的结果集。可更新的结果集并非必须是可滚动的,但如果将数据提供给用户去编辑,那么通常也会希望结果集是可滚动的。

如果要获得可更新的结果集,应该使用以下方法创建一条语句:

```
Statement stat = conn.createStatement(
    ResultSet.TYPE_SCROLL_INSENSITIVE, ResultSet.CONCUR_UPDATABLE);
```

这样,调用 executeQuery 方法返回的结果集就将是可更新的结果集。

> **注释**:并非所有的查询都会返回可更新的结果集。如果查询涉及多个表的连接操作,那么它所产生的结果集将是不可更新的。如果查询只涉及一个表,或者在查询时是使用主键连接多个表的,那么它所产生的结果集将是可更新的结果集。可以调用 ResultSet 接口中的 getConcurrency 方法来确定结果集是否是可更新的。

例如,假设想提高某些图书的价格,但是在执行 UPDATE 语句时又没有一个简单的提价标准。此时,就可以根据任意设定的条件,迭代遍历所有的图书并更新它们的价格。

```
String query = "SELECT * FROM Books";
ResultSet rs = stat.executeQuery(query);
while (rs.next())
{
   if (. . .)
   {
      double increase = . . .;
      double price = rs.getDouble("Price");
      rs.updateDouble("Price", price + increase);
      rs.updateRow(); // make sure to call updateRow after updating fields
   }
}
```

所有对应于 SQL 类型的数据类型都配有 updateXxx 方法，比如 updateDouble、updateString 等。与 getXxx 方法相同，在使用 updateXxx 方法时必须指定列的名称或序号，然后给该字段设置新的值。

> **注释**：在使用第一个参数为列序号的 updateXxx 方法时，请注意这里的列序号指的是该列在结果集中的序号。它的值可以与数据库中的列序号不同。

updateXxx 方法改变的只是结果集中的行值，而非数据库中的值。当更新完行中的字段值后，必须调用 updateRow 方法，这个方法将当前行中的所有更新信息发送给数据库。如果没有调用 updateRow 方法就将游标移动到其他行上，那么对此行所做的所有更新都将被丢弃，而且永远也不会被传递给数据库。还可以调用 cancelRowUpdates 方法来取消对当前行的更新。

我们在前面的例子中已经介绍过如何修改一个现有的行。如果想在数据库中添加一条新的记录，首先需要使用 moveToInsertRow 方法将游标移动到特定的位置，我们称之为插入行 (insert row)。然后，调用 updateXxx 方法在插入行的位置上创建一个新的行。在上述操作全部完成之后，还需要调用 insertRow 方法将新建的行发送给数据库。完成插入操作后，再调用 moveToCurrentRow 方法将游标移回到调用 moveToInsertRow 方法之前的位置。下面是一段示例程序：

```
rs.moveToInsertRow();
rs.updateString("Title", title);
rs.updateString("ISBN", isbn);
rs.updateString("Publisher_Id", pubid);
rs.updateDouble("Price", price);
rs.insertRow();
rs.moveToCurrentRow();
```

请注意，你无法控制在结果集或数据库中添加新数据的位置。

对于在插入行中没有指定值的列，将被设置为 SQL 的 NULL。但是，如果这个列有 NOT NULL 约束，那么将会抛出异常，而这一行也无法插入。

最后需要说明的是，你可以使用以下方法删除游标所指的行。

```
rs.deleteRow();
```

deleteRow 方法会立即将该行从结果集和数据库中删除。

ResultSet 接口中的 updateRow、insertRow 和 deleteRow 方法的执行效果等同于 SQL 命令中的 UPDATE、INSERT 和 DELETE。不过，习惯于 Java 编程语言的程序员通常会觉得使用结果集来操控数据库要比使用 SQL 语句自然得多。

> **警告**：如果不小心处理的话，就很有可能在使用可更新的结果集时编写出非常低效的代码。执行 UPDATE 语句，要比建立一个查询，然后一边遍历一边修改数据显得高效得多。对于用户能够任意修改数据的交互式程序来说，使用可更新的结果集是非常有意义的。但是相对于大多数通过程序进行修改的情况，使用 SQL 的 UPDATE 语句更合适一些。

> **注释**：JDBC 2 对结果集做了进一步的改进，例如，如果数据被其他的并发数据库连接所修改，那么它可以用最新的数据来更新结果集。JDBC 3 添加了另一种优化，可以指定结果集在事务提交时的行为。但是，这些高级特性超出了本章的范围。我们推荐你参考 Maydene Fisher、Jon Ellis 和 Jonathan Bruce 所著的 *JDBC API Tutorial and Reference*，*Third Edition*（Addison-Wesley 出版社 2003 年出版）和 JDBC 规范，以了解更多的信息。

API *java.sql.Connection* 1.1

- Statement createStatement(int type, int concurrency) 1.2
- PreparedStatement prepareStatement(String command, int type, int concurrency) 1.2

  创建一个语句或预备语句，且该语句可以产生指定类型和并发模式的结果集。type 参数是 ResultSet 接口中的下述常量之一：TYPE_FORWARD_ONLY、TYPE_SCROLL_INSENSITIVE 或者 TYPE_SCROLL_SENSITIVE，concurrency 参数是 ResultSet 接口中的下述常量之一：CONCUR_READ_ONLY 或者 CONCUR_UPDATABLE

API *java.sql.ResultSet* 1.1

- int getType() 1.2

  返回结果集的类型。返回值为以下常量之一：TYPE_FORWARD_ONLY、TYPE_SCROLL_INSENSITIVE 或 TYPE_SCROLL_SENSITIVE。

- int getConcurrency() 1.2

  返回结果集的并发设置。返回值为以下常量之一：CONCUR_READ_ONLY 或 CONCUR_UPDATABLE

- boolean previous() 1.2

  将游标移动到前一行。如果游标位于某一行上，则返回 true；如果游标位于第一行之前的位置，则返回 false。

- int getRow() 1.2

  得到当前行的序号。所有行从 1 开始编号。

- boolean absolute(int r) 1.2

  移动游标到第 r 行。如果游标位于某一行上，则返回 true。

- boolean relative(int d) 1.2

  将游标移动 d 行。如果 d 为负数，则游标向后移动。如果游标位于某一行上，则返回 true。

- boolean first() 1.2
- boolean last() 1.2

  移动游标到第一行或最后一行。如果游标位于某一行上，则返回 true。

- void beforeFirst() 1.2
- void afterLast() 1.2

  移动游标到第一行之前或最后一行之后的位置。

- boolean isFirst() 1.2
- boolean isLast() 1.2

    测试游标是否在第一行或最后一行。
- boolean isBeforeFirst() 1.2
- boolean isAfterLast() 1.2

    测试游标是否在第一行之前或最后一行之后的位置。
- void moveToInsertRow() 1.2

    移动游标到插入行。插入行是一个特殊的行,可以在该行上使用 updateXxx 和 insertRow 方法来插入新数据。
- void moveToCurrentRow() 1.2

    将游标从插入行移回到调用 moveToInsertRow 方法之前它所在的那一行。
- void insertRow() 1.2

    将插入行上的内容插入到数据库和结果集中。
- void deleteRow() 1.2

    从数据库和结果集中删除当前行。
- void update*Xxx*(int column, *Xxx* data) 1.2
- void update*Xxx*(String columnName, *Xxx* data) 1.2

    (*Xxx* 指数据类型,比如 int、double、String、Date 等)更新结果集中当前行上的某个字段值。
- void updateRow() 1.2

    将当前行的更新信息发送到数据库。
- void cancelRowUpdates() 1.2

    撤销对当前行的更新。

API **java.sql.DatabaseMetaData** 1.1

- boolean supportsResultSetType(int type) 1.2

    如果数据库支持给定类型的结果集,则返回 true。type 是 ResultSet 接口中的常量之一: TYPE_FORWARD_ONLY、TYPE_SCROLL_INSENSITIVE 或者 TYPE_SCROLL_SENSITIVE。
- boolean supportsResultSetConcurrency(int type, int concurrency) 1.2

    如果数据库支持给定类型和并发模式的结果集,则返回 true。type 参数是 ResultSet 接口中的下述常量之一: TYPE_FORWARD_ONLY、TYPE_SCROLL_INSENSITIVE 或者 TYPE_SCROLL_SENSITIVE。concurrency 是 ResultSet 接口中的下述常量之一: CONCUR_READ_ONL 或者 CONCUR_UPDATABLE。

5.7 行集

可滚动的结果集虽然功能强大,却有一个重要的缺陷:在与用户的整个交互过程中,必须始终与数据库保持连接。用户也许会离开电脑旁很长一段时间，而在此期间却始终占有着数据库连接。这种方式存在很大的问题，因为数据库连接属于稀有资源。在这种情况下，我们可以使用行集。RowSet 接口扩展自 ResultSet 接口，却无须始终保持与数据库的连接。

行集还适用于将查询结果移动到复杂应用的其他层，或者是诸如手机之类的其他设备中。你可能永远都不会考虑移动一个结果集，因为它的数据结构可能非常庞大，且依赖于数据库连接。

5.7.1 构建行集

以下为 javax.sql.rowset 包提供的接口，它们都扩展了 RowSet 接口：

- CachedRowSet 允许在断开连接的状态下执行相关操作。关于被缓存的行集我们将在下一节中讨论。
- WebRowSet 对象代表了一个被缓存的行集，该行集可以保存为 XML 文件。该文件可以移动到 Web 应用的其他层中，只要在该层中使用另一个 WebRowSet 对象重新打开该文件即可。
- FilteredRowSet 和 JoinRowSet 接口支持对行集的轻量级操作，它们等同于 SQL 中的 SELECT 和 JOIN 操作。这两个接口的操作对象是存储在行集中的数据，因此运行时无须建立数据库连接。
- JdbcRowSet 是 ResultSet 接口的一个瘦包装器。它在 RowSet 接口中添加了一些有用的方法。

在 Java 7 中，有一种获取行集的标准方式：

```
RowSetFactory factory = RowSetProvider.newFactory();
CachedRowSet crs = factory.createCachedRowSet();
```

获取其他行集类型的对象也有类似的方法。

5.7.2 被缓存的行集

一个被缓存的行集中包含了一个结果集中所有的数据。CachedRowSet 是 ResultSet 接口的子接口，所以你完全可以像使用结果集一样来使用被缓存的行集。被缓存的行集有一个非常重要的优点：断开数据库连接后仍然可以使用行集。你将在程序清单 5-4 的示例程序中看到，这种做法大大简化了交互式应用的实现。在执行每个用户命令时，我们只需打开数据库连接、执行查询操作、将查询结果放入被缓存的行集，然后关闭数据库连接即可。

我们甚至可以修改被缓存的行集中的数据。当然，这些修改不会立即反馈到数据库中。相反，必须发起一个显式的请求，以便让数据库真正接受所有修改。此时 CachedRowSet 类会重新连接到数据库，并通过执行 SQL 语句向数据库中写入所有修改后的数据。

可以使用一个结果集来填充 CachedRowSet 对象：

```
ResultSet result = . . .;
RowSetFactory factory = RowSetProvider.newFactory();
CachedRowSet crs = factory.createCachedRowSet();
crs.populate(result);
conn.close(); // now OK to close the database connection
```

或者，也可以让 CachedRowSet 对象自动建立一个数据库连接。首先，设置数据库参数：

```
crs.setURL("jdbc:derby://localhost:1527/COREJAVA");
crs.setUsername("dbuser");
crs.setPassword("secret");
```

然后，设置查询语句和所有参数。

```
crs.setCommand("SELECT * FROM Books WHERE Publisher_ID = ?");
crs.setString(1, publisherId);
```

最后，将查询结果填充到行集中：

```
crs.execute();
```

这个方法调用会建立数据库连接、执行查询操作、填充行集，最后断开连接。

如果查询结果非常大，那我们肯定不想将其全部放入行集中。毕竟，用户可能只是想浏览其中的几行而已。在这种情况下，可以指定每一页的尺寸：

```
CachedRowSet crs = . . .;
crs.setCommand(command);
crs.setPageSize(20);
. . .
crs.execute();
```

现在就能只获得 20 行了。要获取下一批数据，可以调用：

```
crs.nextPage();
```

可以使用与结果集中相同的方法来查看和修改行集中的数据。如果修改了行集中的内容，那么必须调用以下方法将修改写回到数据库中：

```
crs.acceptChanges(conn);
```

或

```
crs.acceptChanges();
```

只有在行集中设置了连接数据库所需的信息（如 URL、用户名和密码）时，上述第二个方法调用才会有效。

在 5.6.2 节中，我们曾经介绍过，并非所有的结果集都是可更新的。同样，如果一个行集包含的是复杂查询的查询结果，那么我们就无法将对该行集数据的修改写回到数据库中。不过，如果行集上的数据都来自同一张数据库表，我们就可以安全地写回数据。

⚠ **警告**：如果是使用结果集来填充行集，那么行集就无从获知需要更新数据的数据库表名。此时，必须调用 setTable 方法来设置表名称。

另一个导致问题复杂化的情况是：在填充了行集之后，数据库中的数据发生了改变，这显然容易产生数据不一致性。为了解决这个问题，参考实现会首先检查行集中的原始值（即，修改前的值）是否与数据库中的当前值一致。如果一致，那么修改后的值将覆盖数据库中的当前值。否则，将抛出 SyncProviderException 异常，且不向数据库写回任何值。在实现行集接口时其他实现也可以采用不同的同步策略。

API *javax.sql.RowSet* 1.4

- String getURL()
- void setURL(String url)

  获取或设置数据库的 URL。

- String getUsername()
- void setUsername(String username)

  获取或设置连接数据库所需的用户名。

- String getPassword()
- void setPassword(String password)

  获取或设置连接数据库所需的密码。

- String getCommand()
- void setCommand(String command)

  获取或设置向行集中填充数据时需要执行的命令。

- void execute()

  通过执行使用 setCommand 方法设置的语句集来填充行集。为了使驱动管理器可以获得连接，必须事先设定 URL、用户名和密码。

API *javax.sql.rowset.CachedRowSet* 5.0

- void execute(Connection conn)

  通过执行使用 setCommand 方法设置的语句集来填充行集。该方法使用给定的连接，并负责关闭它。

- void populate(ResultSet result)

  将指定的结果集中的数据填充到被缓存的行集中。

- String getTableName()
- void setTableName(String tableName)

  获取或设置数据库表名称，填充被缓存的行集时所需的数据来自该表。

- int getPageSize()
- void setPageSize(int size)

  获取和设置页的尺寸。

- boolean nextPage()
- boolean previousPage()

  加载下一页或上一页，如果要加载的页存在，则返回 true。

- void acceptChanges()
- void acceptChanges(Connection conn)

  重新连接数据库，并写回行集中修改过的数据。如果因为数据库中的数据已经被修改而导致无法写回行集中的数据，该方法可能会抛出 SyncProviderException 异常。

API javax.sql.rowset.RowSetProvider 7

- static RowSetFactory newFactory()
  创建一个行集工厂。

API javax.sql.rowset.RowSetFactory 7

- CachedRowSet createCachedRowSet()
- FilteredRowSet createFilteredRowSet()
- JdbcRowSet createJdbcRowSet()
- JoinRowSet createJoinRowSet()
- WebRowSet createWebRowSet()
  创建一个指定类型的行集。

5.8 元数据

在前几节中，我们介绍了如何填充、查询和更新数据库表。其实，JDBC 还可以提供关于数据库及其表结构的详细信息。例如，可以获取某个数据库的所有表的列表，也可以获得某个表中所有列的名称及其数据类型。如果是在开发业务应用时使用事先定义好的数据库，那么数据库结构和表信息就不是非常有用了。毕竟，在设计数据库表时，就已经知道了它们的结构。但是，对于那些编写数据库工具的程序员来说，数据库的结构信息却是极其有用的。

在 SQL 中，描述数据库或其组成部分的数据称为元数据（区别于那些存在数据库中的实际数据）。我们可以获得三类元数据：关于数据库的元数据、关于结果集的元数据以及关于预备语句参数的元数据。

如果要了解数据库的更多信息，可以从数据库连接中获取一个 DatabaseMetaData 对象。

```
DatabaseMetaData meta = conn.getMetaData();
```

现在就可以获取某些元数据了。例如，调用

```
ResultSet mrs = meta.getTables(null, null, null, new String[] { "TABLE" });
```

将返回一个包含所有数据库表信息的结果集（如果要了解该方法的其他参数，请参见本节末尾的 API 说明）。

该结果集中的每一行都包含了数据库中一张表的详细信息，其中，第三列是表的名称。（同样，如果要了解其他列的信息，请参阅 API 说明。）下面的循环可以获取所有的表名：

```
while (mrs.next())
    tableNames.addItem(mrs.getString(3));
```

数据库元数据还有第二个重要应用。数据库是非常复杂的，SQL 标准为数据库的多样性提供了很大的空间。DatabaseMetaData 接口中有上百个方法可以用于查询数据库的相关信息，包括一些使用奇特的名字进行调用的方法，如：

```
meta.supportsCatalogsInPrivilegeDefinitions()
```

和

```
meta.nullPlusNonNullIsNull()
```

显然,这些方法主要是针对有特殊要求的高级用户的,尤其是那些需要编写涉及多个数据库且具有高可移植性的代码的编程人员。

DatabaseMetaData 接口用于提供有关数据库的数据,第二个元数据接口 ResultSetMetaData 则用于提供结果集的相关信息。每当通过查询得到一个结果集时,我们都可以获取该结果集的列数以及每一列的名称、类型和字段宽度。下面是一个典型的循环:

```
ResultSet rs = stat.executeQuery("SELECT * FROM " + tableName);
ResultSetMetaData meta = rs.getMetaData();
for (int i = 1; i <= meta.getColumnCount(); i++)
{
    String columnName = meta.getColumnLabel(i);
    int columnWidth = meta.getColumnDisplaySize(i);
    . . .
}
```

在这一节中,我们将介绍如何编写一个简单的数据库工具,程序清单 5-4 中的程序通过使用元数据来浏览数据库中的所有表,该程序还展示了如何使用缓存的行集。

程序清单 5-4　view/ViewDB.java

```java
 1  package view;
 2
 3  import java.awt.*;
 4  import java.awt.event.*;
 5  import java.io.*;
 6  import java.nio.file.*;
 7  import java.sql.*;
 8  import java.util.*;
 9
10  import javax.sql.*;
11  import javax.sql.rowset.*;
12  import javax.swing.*;
13
14  /**
15   * This program uses metadata to display arbitrary tables in a database.
16   * @version 1.34 2018-05-01
17   * @author Cay Horstmann
18   */
19  public class ViewDB
20  {
21     public static void main(String[] args)
22     {
23        EventQueue.invokeLater(() ->
24           {
25              var frame = new ViewDBFrame();
26              frame.setTitle("ViewDB");
27              frame.setDefaultCloseOperation(JFrame.EXIT_ON_CLOSE);
28              frame.setVisible(true);
```

```java
29              });
30      }
31  }
32
33  /**
34   * The frame that holds the data panel and the navigation buttons.
35   */
36  class ViewDBFrame extends JFrame
37  {
38      private JButton previousButton;
39      private JButton nextButton;
40      private JButton deleteButton;
41      private JButton saveButton;
42      private DataPanel dataPanel;
43      private Component scrollPane;
44      private JComboBox<String> tableNames;
45      private Properties props;
46      private CachedRowSet crs;
47      private Connection conn;
48
49      public ViewDBFrame()
50      {
51          tableNames = new JComboBox<String>();
52
53          try
54          {
55              readDatabaseProperties();
56              conn = getConnection();
57              DatabaseMetaData meta = conn.getMetaData();
58              try (ResultSet mrs = meta.getTables(null, null, null, new String[] { "TABLE" }))
59              {
60                  while (mrs.next())
61                      tableNames.addItem(mrs.getString(3));
62              }
63          }
64          catch (SQLException ex)
65          {
66              for (Throwable t : ex)
67                  t.printStackTrace();
68          }
69          catch (IOException ex)
70          {
71              ex.printStackTrace();
72          }
73
74          tableNames.addActionListener(
75              event -> showTable((String) tableNames.getSelectedItem(), conn));
76          add(tableNames, BorderLayout.NORTH);
77          addWindowListener(new WindowAdapter()
78              {
79                  public void windowClosing(WindowEvent event)
80                  {
81                      try
82                      {
```

```java
                    if (conn != null) conn.close();
                }
                catch (SQLException ex)
                {
                    for (Throwable t : ex)
                        t.printStackTrace();
                }
            }
        });

        var buttonPanel = new JPanel();
        add(buttonPanel, BorderLayout.SOUTH);

        previousButton = new JButton("Previous");
        previousButton.addActionListener(event -> showPreviousRow());
        buttonPanel.add(previousButton);

        nextButton = new JButton("Next");
        nextButton.addActionListener(event -> showNextRow());
        buttonPanel.add(nextButton);

        deleteButton = new JButton("Delete");
        deleteButton.addActionListener(event -> deleteRow());
        buttonPanel.add(deleteButton);

        saveButton = new JButton("Save");
        saveButton.addActionListener(event -> saveChanges());
        buttonPanel.add(saveButton);
        if (tableNames.getItemCount() > 0)
            showTable(tableNames.getItemAt(0), conn);
    }

    /**
     * Prepares the text fields for showing a new table, and shows the first row.
     * @param tableName the name of the table to display
     * @param conn the database connection
     */
    public void showTable(String tableName, Connection conn)
    {
        try (Statement stat = conn.createStatement();
            ResultSet result = stat.executeQuery("SELECT * FROM " + tableName))
        {
            // get result set

            // copy into cached row set
            RowSetFactory factory = RowSetProvider.newFactory();
            crs = factory.createCachedRowSet();
            crs.setTableName(tableName);
            crs.populate(result);

            if (scrollPane != null) remove(scrollPane);
            dataPanel = new DataPanel(crs);
            scrollPane = new JScrollPane(dataPanel);
            add(scrollPane, BorderLayout.CENTER);
```

```java
137          pack();
138          showNextRow();
139       }
140       catch (SQLException ex)
141       {
142          for (Throwable t : ex)
143             t.printStackTrace();
144       }
145    }
146
147    /**
148     * Moves to the previous table row.
149     */
150    public void showPreviousRow()
151    {
152       try
153       {
154          if (crs == null || crs.isFirst()) return;
155          crs.previous();
156          dataPanel.showRow(crs);
157       }
158       catch (SQLException ex)
159       {
160          for (Throwable t : ex)
161             t.printStackTrace();
162       }
163    }
164
165    /**
166     * Moves to the next table row.
167     */
168    public void showNextRow()
169    {
170       try
171       {
172          if (crs == null || crs.isLast()) return;
173          crs.next();
174          dataPanel.showRow(crs);
175       }
176       catch (SQLException ex)
177       {
178          for (Throwable t : ex)
179             t.printStackTrace();
180       }
181    }
182
183    /**
184     * Deletes current table row.
185     */
186    public void deleteRow()
187    {
188       if (crs == null) return;
189       new SwingWorker<Void, Void>()
190       {
```

```java
191        public Void doInBackground() throws SQLException
192        {
193           crs.deleteRow();
194           crs.acceptChanges(conn);
195           if (crs.isAfterLast())
196              if (!crs.last()) crs = null;
197           return null;
198        }
199        public void done()
200        {
201           dataPanel.showRow(crs);
202        }
203     }.execute();
204  }
205  /**
206   * Saves all changes.
207   */
208  public void saveChanges()
209  {
210     if (crs == null) return;
211     new SwingWorker<Void, Void>()
212     {
213        public Void doInBackground() throws SQLException
214        {
215           dataPanel.setRow(crs);
216           crs.acceptChanges(conn);
217           return null;
218        }
219     }.execute();
220  }
221
222  private void readDatabaseProperties() throws IOException
223  {
224     props = new Properties();
225     try (InputStream in = Files.newInputStream(Paths.get("database.properties")))
226     {
227        props.load(in);
228     }
229     String drivers = props.getProperty("jdbc.drivers");
230     if (drivers != null) System.setProperty("jdbc.drivers", drivers);
231  }
232
233  /**
234   * Gets a connection from the properties specified in the file database.properties.
235   * @return the database connection
236   */
237  private Connection getConnection() throws SQLException
238  {
239     String url = props.getProperty("jdbc.url");
240     String username = props.getProperty("jdbc.username");
241     String password = props.getProperty("jdbc.password");
242
243     return DriverManager.getConnection(url, username, password);
244  }
```

```java
245     }
246
247     /**
248      * This panel displays the contents of a result set.
249      */
250     class DataPanel extends JPanel
251     {
252         private java.util.List<JTextField> fields;
253
254         /**
255          * Constructs the data panel.
256          * @param rs the result set whose contents this panel displays
257          */
258         public DataPanel(RowSet rs) throws SQLException
259         {
260             fields = new ArrayList<>();
261             setLayout(new GridBagLayout());
262             var gbc = new GridBagConstraints();
263             gbc.gridwidth = 1;
264             gbc.gridheight = 1;
265
266             ResultSetMetaData rsmd = rs.getMetaData();
267             for (int i = 1; i <= rsmd.getColumnCount(); i++)
268             {
269                 gbc.gridy = i - 1;
270
271                 String columnName = rsmd.getColumnLabel(i);
272                 gbc.gridx = 0;
273                 gbc.anchor = GridBagConstraints.EAST;
274                 add(new JLabel(columnName), gbc);
275
276                 int columnWidth = rsmd.getColumnDisplaySize(i);
277                 var tb = new JTextField(columnWidth);
278                 if (!rsmd.getColumnClassName(i).equals("java.lang.String"))
279                     tb.setEditable(false);
280
281                 fields.add(tb);
282
283                 gbc.gridx = 1;
284                 gbc.anchor = GridBagConstraints.WEST;
285                 add(tb, gbc);
286             }
287         }
288
289         /**
290          * Shows a database row by populating all text fields with the column values.
291          */
292         public void showRow(ResultSet rs)
293         {
294             try
295             {
296                 if (rs == null) return;
297                 for (int i = 1; i <= fields.size(); i++)
298                 {
```

```
299                String field = rs == null ? "" : rs.getString(i);
300                JTextField tb = fields.get(i - 1);
301                tb.setText(field);
302             }
303          }
304          catch (SQLException ex)
305          {
306             for (Throwable t : ex)
307                t.printStackTrace();
308          }
309       }
310
311       /**
312        * Updates changed data into the current row of the row set.
313        */
314       public void setRow(RowSet rs) throws SQLException
315       {
316          for (int i = 1; i <= fields.size(); i++)
317          {
318             String field = rs.getString(i);
319             JTextField tb = fields.get(i - 1);
320             if (!field.equals(tb.getText()))
321                rs.updateString(i, tb.getText());
322          }
323          rs.updateRow();
324       }
325    }
```

顶部的组合框用于显示数据库中的所有表。选中其中一个表，框中央就会显示出该表的所有字段名及其第一条记录的值，见图 5-6。点击 Next 和 Previous 按钮可以滚动遍历表中的所有记录，还可以删除一行或编辑行的值，点击 Save 按钮可以将各种修改保存到数据库中。

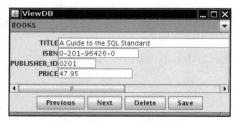

图 5-6　ViewDB 应用程序

> **注释**：许多数据库都配有复杂得多的工具，用于查看和编辑数据库表。如果你使用的数据库没有这样的工具，那么可以求助于 DBeaver（https://dbeaver.io）或者 SQuirreL（http://squirrel-sql.sourceforge.net）。这些程序可以查看任何 JDBC 数据库中的表。我们编写示例程序并非为了取代这些工具，而是为了向你演示如何编写工具来处理任意的数据库表。

API *java.sql.Connection* 1.1

- DatabaseMetaData getMetaData()

  返回一个 DatabaseMetaData 对象，该对象封装了有关数据库连接的元数据。

API *java.sql.DatabaseMetaData* 1.1

- ResultSet getTables(String catalog, String schemaPattern, String tableNamePattern, String types[])

  返回某个目录（catalog）中所有表的描述，该目录必须匹配给定的模式（schema）、表名字模式以及类型标准。（模式用于描述一组相关的表和访问权限，而目录描述的是一组相关的模式，这些概念对组织大型数据库非常重要。）

  catalog 和 schema 参数可以为 ""，用于检索那些没有目录或模式的表。如果不想考虑目录和模式，也可以将上述参数设为 null。

  types 数组包含了所需的表类型的名称，通常表类型有 TABLE、VIEW、SYSTEM TABLE、GLOBAL TEMPORARY、LOCAL TEMPORARY、ALIAS 和 SYNONYM。如果 types 为 null，则返回所有类型的表。返回的结果集共有 5 列，均为 String 类型。

行	名称	解释
1	TABLE_CAT	表目录（可以为 null）
2	TABLE_SCHEM	表模式（可以为 null）
3	TABLE_NAME	表名称
4	TABLE_TYPE	表类型
5	REMARKS	关于表的注释

- int getJDBCMajorVersion() 1.4
- int getJDBCMinorVersion() 1.4

  返回建立数据库连接的 JDBC 驱动程序的主版本号和次版本号。例如，一个 JDBC 3.0 的驱动程序有一个主版本号 3 和一个次版本号 0。

- int getMaxConnections()

  返回可同时连接到数据库的最大并发连接数。

- int getMaxStatements()

  返回单个数据库连接允许同时打开的最大并发语句数。如果对允许打开的语句数目没有限制或者不可知，则返回 0。

API *java.sql.ResultSet* 1.1

- ResultSetMetaData getMetaData()

  返回与当前 ResultSet 对象中的列相关的元数据。

API *java.sql.ResultSetMetaData* 1.1

- int getColumnCount()

返回当前 ResultSet 对象中的列数。
- int getColumnDisplaySize(int column)
  返回给定列序号的列的最大宽度。
- String getColumnLabel(int column)
  返回该列所建议的名称。
- String getColumnName(int column)
  返回指定的列序号所对应的列名。

5.9 事务

我们可以将一组语句构建成一个事务（transaction）。当所有语句都顺利执行之后，事务可以被提交（commit）。否则，如果其中某个语句遇到错误，那么事务将被回滚，就好像没有任何语句被执行过一样。

将多个语句组合成事务的主要原因是为了确保数据库完整性（database integrity）。例如，假设我们需要将钱从一个银行账号转账到另一个账号。此时，一个非常重要的问题就是我们必须同时将钱从一个账号取出并且存入另一个账号。如果在将钱存入其他账号之前系统发生崩溃，那么我们必须撤销取款操作。

如果将更新语句组合成一个事务，那么事务要么成功地执行所有操作并提交，要么在中间某个位置发生失败。在这种情况下，可以执行回滚（rollback）操作，则数据库将自动撤销自上次提交事务以来的所有更新操作产生的影响。

5.9.1 用 JDBC 对事务编程

默认情况下，数据库连接处于自动提交模式（autocommit mode），即每个 SQL 语句一旦被执行便被提交给数据库。一旦命令被提交，就无法对它进行回滚操作。在使用事务时，需要关闭这个默认值：

```
conn.setAutoCommit(false);
```

现在可以按照通常的方式创建一个语句对象：

```
Statement stat = conn.createStatement();
```

然后任意多次地调用 executeUpdate 方法：

```
stat.executeUpdate(command₁);
stat.executeUpdate(command₂);
stat.executeUpdate(command₃);
...
```

如果执行了所有命令之后没有出错，则调用 commit 方法：

```
conn.commit();
```

如果出现错误，则调用：

```
conn.rollback();
```

此时，程序将自动撤销自上次提交以来的所有语句。当事务被 SQLException 异常中断时，典型的办法就是发起回滚操作。

5.9.2 保存点

在使用某些驱动程序时，使用保存点（save point）可以更细粒度地控制回滚操作。创建一个保存点意味着稍后只需返回到这个点，而非放弃整个事务。例如，

```
Statement stat = conn.createStatement(); // start transaction; rollback() goes here
stat.executeUpdate(command1);
Savepoint svpt = conn.setSavepoint(); // set savepoint; rollback(svpt) goes here
stat.executeUpdate(command2);

if (. . .) conn.rollback(svpt); // undo effect of command2
. . .
conn.commit();
```

当不再需要保存点时，应该释放它：

```
conn.releaseSavepoint(svpt);
```

5.9.3 批量更新

假设有一个程序需要执行许多 INSERT 语句，以便将数据填入数据库表中，此时可以使用批量更新的方法来提高程序性能。在使用批量更新（batch update）时，一个语句序列作为一批操作将同时被收集和提交。

> 📖 **注释**：使用 DatabaseMetaData 接口中的 supportsBatchUpdates 方法可以获知数据库是否支持这种特性。

处于同一批中的语句可以是 INSERT、UPDATE 和 DELETE 等操作，也可以是数据库定义语句，如 CREATE TABLE 和 DROP TABLE。但是，在批量处理中添加 SELECT 语句会抛出异常（从概念上讲，批量处理中的 SELECT 语句没有意义，因为它会返回结果集，而并不更新数据库）。

为了执行批量处理，首先必须使用通常的办法创建一个 Statement 对象：

```
Statement stat = conn.createStatement();
```

现在，应该调用 addBatch 方法，而非 executeUpdate 方法：

```
String command = "CREATE TABLE . . ."
stat.addBatch(command);

while (. . .)
{
   command = "INSERT INTO . . . VALUES (" + . . . + ")";
   stat.addBatch(command);
}
```

最后，提交整个批量更新语句：

```
int[] counts = stat.executeBatch();
```

调用 executeBatch 方法将为所有已提交的语句返回一个记录数的数组。

为了在批量模式下正确地处理错误，必须将批量执行的操作视为单个事务。如果批量更新在执行过程中失败，那么必须将它回滚到批量操作开始之前的状态。

首先，关闭自动提交模式，然后收集批量操作，执行并提交该操作，最后恢复最初的自动提交模式：

```
boolean autoCommit = conn.getAutoCommit();
conn.setAutoCommit(false);
Statement stat = conn.getStatement();
. . .
// keep calling stat.addBatch(. . .);
. . .
stat.executeBatch();
conn.commit();
conn.setAutoCommit(autoCommit);
```

API *java.sql.Connection* 1.1

- boolean getAutoCommit()
- void setAutoCommit(boolean b)

  获取该连接中的自动提交模式，或将其设置为 b。如果自动更新为 true，那么所有语句将在执行结束后立刻被提交。

- void commit()

  提交自上次提交以来所有执行过的语句。

- void rollback()

  撤销自上次提交以来所有执行过的语句所产生的影响。

- Savepoint setSavepoint() 1.4
- Savepoint setSavepoint(String name) 1.4

  设置一个匿名或具名的保存点。

- void rollback(Savepoint svpt) 1.4

  回滚到给定保存点。

- void releaseSavepoint(Savepoint svpt) 1.4

  释放给定的保存点。

API *java.sql.Savepoint* 1.4

- int getSavepointId()

  获取该匿名保存点的 ID 号。如果该保存点具有名字，则抛出一个 SQLException 异常。

- String getSavepointName()

  获取该保存点的名称。如果该对象为匿名保存点，则抛出一个 SQLException 异常。

API *java.sql.Statement* 1.1

- void addBatch(String command) 1.2

添加命令到该语句当前的批量命令中。
- int[] executeBatch() 1.2
- long[] executeLargeBatch() 8

执行当前批量更新中的所有命令。返回一个记录数的数组，其中每一个元素都对应一条语句，如果其值非负，则表示受该语句影响的行数；如果其值为 SUCCESS_NO_INFO，则表示该语句成功执行了，但没有提供任何行数；如果其值为 EXECUTE_FAILED，则表示该语句执行失败了。

API *java.sql.DatabaseMetaData* 1.1

- boolean supportsBatchUpdates() 1.2
  如果驱动程序支持批量更新，则返回 true。

5.9.4 高级 SQL 类型

表 5-8 列举了 JDBC 支持的 SQL 数据类型以及它们在 Java 语言中对应的数据类型。

表 5-8 SQL 数据类型及其对应的 Java 类型

SQL 数据类型	Java 数据类型
INTEGER 或 INT	int
SMALLINT	short
NUMERIC(m,n), DECIMAL(m,n) 或 DEC(m,n)	java.math.BigDecimal
FLOAT(n)	double
REAL	float
DOUBLE	double
CHARACTER(n) 或 CHAR(n)	String
VARCHAR(n), LONG VARCHAR	String
BOOLEAN	boolean
DATE	java.sql.Date
TIME	java.sql.Time
TIMESTAMP	java.sql.Timestamp
BLOB	java.sql.Blob
CLOB	java.sql.Clob
ARRAY	java.sql.Array
ROWID	java.sql.RowId
NCHAR(n), NVARCHAR(n), LONG NVARCHAR	String
NCLOB	java.sql.NClob
SQLXML	java.sql.SQLXML

SQL ARRAY（SQL 数组）指的是值的序列。例如，Student 表中通常都会有一个 Scores 列，这个列就应该是 ARRAY OF INTEGER（整数数组）。getArray 方法返回一个接口类型为 java.sql.Array 的对象，该接口中有许多方法可以用来获取数据值。

从数据库中获得一个 LOB 或数组并不等于获取了它的实际内容,只有在访问具体的值时它们才会从数据库中被读取出来。这对改善性能非常有好处,因为通常这些数据的数据量都非常大。

某些数据库支持描述行位置的 ROWID 值,这样就可以非常快捷地获取某一行值。JDBC 4 引入了 java.sql.RowId 接口,并提供了用于在查询中提供行 ID,以及从结果中获取该值的方法。

国家属性字符串(NCHAR 及其变体)按照本地字符编码机制存储字符串,并使用本地排序惯例对这些字符串进行排序。JDBC 4 提供了方法,用于在查询和结果中进行 Java 的 String 对象和国家属性字符串之间的双向转换。

有些数据库可以存储用户自定义的结构化类型。JDBC 3 提供了一种机制用于将 SQL 结构化类型自动映射成 Java 对象。

有些数据库提供用于 XML 数据的本地存储。JDBC 4 引入了 SQLXML 接口,它可以在内部的 XML 表示和 DOM 的 Source/Result 接口或二进制流之间起到中介作用。请查看 SQLXML 类的 API 文档以了解详细信息。

我们不再更深入地讨论这些高级 SQL 类型了,你可以在 *JDBC API Tutorial and Reference* 和 JDBC 4 的规范中找到更多有关这些主题的信息。

5.10 Web 与企业应用中的连接管理

我们在前面几节中曾经介绍过,使用 database.properties 文件可以对数据库连接进行非常简单的设置。这种方法适用于小型的测试程序,但是不适用于规模较大的应用。

在 Web 或企业环境中部署 JDBC 应用时,数据库连接管理与 Java 名字和目录接口(JNDI)是集成在一起的。遍布企业的数据源的属性可以存储在一个目录中,采用这种方式使得我们可以集中管理用户名、密码、数据库名和 JDBC URL。

在这样的环境中,可以使用下列代码创建数据库连接:

```
var jndiContext = new InitialContext();
var source = (DataSource) jndiContext.lookup("java:comp/env/jdbc/corejava");
Connection conn = source.getConnection();
```

请注意,我们不再使用 DriverManager,而是使用 JNDI 服务来定位数据源。数据源是一个能够提供简单的 JDBC 连接和更多高级服务的接口,比如执行涉及多个数据库的分布式事务。javax.sql 标准扩展包定义了 DataSource 接口。

> **注释**:在 Java EE 的容器中,甚至不必编程进行 JNDI 查找,只需在 DataSource 域上使用 Resource 注解,当加载应用时,这个数据源引用将被设置:
>
> ```
> @Resource(name="jdbc/corejava")
> private DataSource source;
> ```

当然,我们必须在某个地方配置数据源。如果你编写的数据库程序将在 Servlet 容器中运行，比如 Apache Tomcat，或在应用服务器中运行，比如 GlassFish，那么必须将数据库配置信息（包括 JNDI 名字、JDBC URL、用户名和密码）放置在配置文件中，或者在管理员 GUI 中进行设置。

用户名管理和登录管理只是众多需要特别关注的问题之一。另一个重要问题则涉及建立数据库连接所需的开销。我们的示例数据库程序使用了两种策略来获取数据库连接：程序清单 5-3 中的 QueryDB 程序在程序的开头建立了到数据库的单个连接，并在程序结尾处关闭了它，而程序清单 5-4 中的 ViewDB 程序在每次需要时都会打开一个新连接。

但是，这两种方式都不令人满意：因为数据库连接是有限的资源，如果用户要离开应用一段时间，那么他占用的连接就不应该保持打开状态；另一方面，每次查询都获取连接并在随后关闭它的代价也是相当高的。

解决上述问题的方法是建立数据库连接池（pool）。这意味着数据库连接在物理上并未被关闭，而是保留在一个队列中并被反复重用。连接池是一种非常重要的服务，JDBC 规范为实现者提供了用以实现连接池服务的手段。不过，JDK 本身并未实现这项服务，数据库供应商提供的 JDBC 驱动程序中通常也不包含这项服务。相反，Web 容器和应用服务器的开发商通常会提供连接池服务的实现。

连接池的使用对程序员来说是完全透明的，可以通过获取数据源并调用 getConnection 方法来得到连接池中的连接。使用完连接后，需要调用 close 方法。该方法并不会在物理上关闭连接，而只是告诉连接池已经使用完该连接。连接池通常还会将池机制作用于预备语句上。

至此，你已经学会了 JDBC 的基本知识，并且已经知道如何实现简单的数据库应用。然而，正如我们在本章的开头所强调的那样，数据库的相关技术非常复杂；本章属于介绍性章节，相当多的高级话题已经超出了本章的范围。如果要全面了解 JDBC 的高级功能，请参阅 *JDBC API Tutorial and Reference* 或 JDBC 规范。

在本章中，我们学习了如何用 Java 操作关系型数据库。下一章将讨论 Java 8 的日期和时间库。

第 6 章 日期和时间 API

- ▲ 时间线
- ▲ 本地日期
- ▲ 日期调整器
- ▲ 本地时间
- ▲ 时区时间
- ▲ 格式化和解析
- ▲ 与遗留代码的互操作

光阴似箭，我们可以很容易地设置一个起点，然后向前和向后以秒来计时。那为什么处理时间还会显得如此之难呢？问题出在人类自身上。如果我们只需告诉对方："1523793600 时来见我，别迟到！"那么一切都会很简单。但是我们希望时间能够和朝夕与季节挂钩，这就使事情变得复杂了。Java 1.0 有一个 Date 类，事后证明它过于简单了，当 Java 1.1 引入 Calendar 类之后，Date 类中的大部分方法就被弃用了。但是，Calendar 的 API 还不够给力，它的实例是可修改的，并且它没有处理诸如闰秒这样的问题。第 3 次升级很吸引人，那就是 Java SE 8 中引入的 java.time API，它修正了过去的缺陷，并且应该会服役相当长的一段时间。在本章中，你将会学习是什么使时间计算变得如此烦人，以及日期和时间 API 是如何解决这些问题的。

6.1 时间线

在历史上，基本的时间单位"秒"是从地球的自转中推导出来的。地球自转一周需要 24 个小时，即 24 × 60 × 60=86 400 秒，因此，看起来这好像只是一个有关如何精确定义 1 秒的天文度量问题。遗憾的是，地球有轻微的摄动，所以需要更加精确的定义。1967 年，人们根据铯 133 原子内在的特性推导出了与历史定义相匹配的秒的新的精确定义。自那以后，一直由一个原子钟网络来维护着官方时间。

官方时间的维护器时常需要将绝对时间与地球自转进行同步。首先，官方的秒需要稍作调整，从 1972 年开始，偶尔需要插入"闰秒"。（在理论上，偶尔也需要移除 1 秒，但是这还从来没发生过。）这又是有关修改系统时间的话题。很明显，闰秒是个痛点，许多计算机系统使用"平滑"方式来人为地在紧邻闰秒之前让时间变慢或变快，以保证每天都是 86 400 秒。这种做法可以奏效，因为计算机上的本地时间并非那么精确，而计算机也惯于将自身时间与外部的时间服务进行同步。

Java 的 Date 和 Time API 规范要求 Java 使用的时间尺度为：

- 每天 86 400 秒
- 每天正午与官方时间精确匹配
- 在其他时间点上，以精确定义的方式与官方时间接近匹配这赋予了 Java 很大的灵活性，使其可以进行调整，以适应官方时间未来的变化。

在 Java 中，Instant 表示时间线上的某个点。被称为"新纪元"的时间线原点被设置为穿过伦敦格林尼治皇家天文台的本初子午线所处时区的 1970 年 1 月 1 日的午夜。这与 UNIX/POSIX 时间中使用的惯例相同。从该原点开始，时间按照每天 86 400 秒向前或向回度量，精确到纳秒。Instant 的值往回可追溯 10 亿年（Instant.MIN）。这对于表示宇宙年龄（大约 135 亿年）来说还差得远，但是对于所有实际应用来说，应该足够了。毕竟，10 亿年前，地球表面还覆盖着冰层，只有当今植物和动物的微生物祖先在繁殖生衍。最大的值 Instant.MAX 是公元 1 000 000 000 年的 12 月 31 日。

静态方法调用 Instant.now() 会给出当前的时刻。你可以按照常用的方式，用 equals 和 compareTo 方法来比较两个 Instant 对象，因此你可以将 Instant 对象用作时间戳。

为了得到两个时刻之间的时间差，可以使用静态方法 Duration.between。例如，下面的代码展示了如何度量算法的运行时间：

```
Instant start = Instant.now();
runAlgorithm();
Instant end = Instant.now();
Duration timeElapsed = Duration.between(start, end);
long millis = timeElapsed.toMillis();
```

Duration 是两个时刻之间的时间量。你可以通过调用 toNanos、toMillis、getSeconds、toMinutes、toHours 和 toDays 来获得 Duration 按照传统单位度量的时间长度。

> **注释**：在 Java 8 中，必须调用 getSeconds 而不是 toSeconds。

如果想要让计算精确到纳秒级，那么就需要当心上溢问题。long 值可以存储大约 300 年时间对应的纳秒数。如果你需要的 Duration 短于这个时间，那么可以直接将其转换为纳秒数。你可以使用更长的 Duration，即让 Duration 对象用一个 long 来存储秒数，用另外一个 int 来存储纳秒数。Duration 接口包含了大量在本节末尾展示的用于执行算术运算的方法。

例如，如果想要检查某个算法是否至少比另一个算法快 10 倍，那么你可以执行如下的计算：

```
Duration timeElapsed2 = Duration.between(start2, end2);
boolean overTenTimesFaster
    = timeElapsed.multipliedBy(10).minus(timeElapsed2).isNegative();
```

这里只展示了语法。因为算法不会运行数百年，所以可以直接使用下面的方法：

```
boolean overTenTimesFaster = timeElapsed.toNanos() * 10 < timeElapsed2.toNanos();
```

> **注释**：Instant 和 Duration 类都是不可修改的类，所以诸如 multipliedBy 和 minus 这样的方法都会返回一个新的实例。

在程序清单 6-1 的示例程序中，可以看到如何使用 Instant 和 Duration 类来对两个算法计时。

程序清单 6-1 timeline/TimeLine.java

```
1  package timeline;
2
3  /**
```

```java
 4   * @version 1.0 2016-05-10
 5   * @author Cay Horstmann
 6   */
 7
 8  import java.time.*;
 9  import java.util.*;
10  import java.util.stream.*;
11
12  public class Timeline
13  {
14     public static void main(String[] args)
15     {
16        Instant start = Instant.now();
17        runAlgorithm();
18        Instant end = Instant.now();
19        Duration timeElapsed = Duration.between(start, end);
20        long millis = timeElapsed.toMillis();
21        System.out.printf("%d milliseconds\n", millis);
22
23        Instant start2 = Instant.now();
24        runAlgorithm2();
25        Instant end2 = Instant.now();
26        Duration timeElapsed2 = Duration.between(start2, end2);
27        System.out.printf("%d milliseconds\n", timeElapsed2.toMillis());
28        boolean overTenTimesFaster = timeElapsed.multipliedBy(10)
29           .minus(timeElapsed2).isNegative();
30        System.out.printf("The first algorithm is %smore than ten times faster",
31           overTenTimesFaster ? "" : "not ");
32     }
33
34     public static void runAlgorithm()
35     {
36        int size = 10;
37        List<Integer> list = new Random().ints().map(i -> i % 100).limit(size)
38           .boxed().collect(Collectors.toList());
39        Collections.sort(list);
40        System.out.println(list);
41     }
42
43     public static void runAlgorithm2()
44     {
45        int size = 10;
46        List<Integer> list = new Random().ints().map(i -> i % 100).limit(size)
47           .boxed().collect(Collectors.toList());
48        while (!IntStream.range(1, list.size())
49           .allMatch(i -> list.get(i - 1).compareTo(list.get(i)) <= 0))
50           Collections.shuffle(list);
51        System.out.println(list);
52     }
53  }
```

API **java.time.Instant** 8

- static Instant now()

从最佳的可用系统时钟中获取当前的时刻。

- Instant plus(TemporalAmount amountToAdd)
- Instant minus(TemporalAmount amountToSubtract)

产生一个时刻，该时刻与当前时刻距离给定的时间量。Duration 和 Period（参阅 6.2 节）实现了 TemporalAmount 接口。

- Instant (plus|minus)(Nanos|Millis|Seconds)(long number)

产生一个时刻，该时刻与当前时刻距离给定数量的纳秒、微秒或秒。

API java.time.Duration 8

- static Duration of(Nanos|Millis|Seconds|Minutes|Hours|Days)(long number)

产生一个给定数量的指定时间单位的时间间隔。

- static Duration between(Temporal startInclusive, Temporal endExclusive)

产生一个在给定时间点之间的 Duration 对象。Instant 类实现了 Temporal 接口，LocalDate/LocalDateTime/LocalTime（参阅 6.4 节）和 ZonedDateTime（参阅 6.5 节）也实现了该接口。

- long toNanos()
- long toMillis()
- long toSeconds() 9
- long toMinutes()
- long toHours()
- long toSeconds()
- long toSeconds()
- long toDays()

获取当前时长按照方法名中的时间单位度量的数量。

- int to(Nanos|Millis|Seconds|Minutes|Hours)Part() 9
- long to(Days|Hours|Minutes|Seconds|Millis|Nanos)Part() 9

当前时长中给定时间单位的部分。例如，在 100 秒的时间间隔中，分钟的部分是 1，秒的部分是 40。

- Instant plus(TemporalAmount amountToAdd)
- Instant minus(TemporalAmount amountToSubtract)

产生一个时刻，该时刻与当前时刻距离给定的时间量。Duration 和 Period 类（参阅 6.2 节）实现了 TemporalAmount 接口。

- Duration multipliedBy(long multiplicand)
- Duration dividedBy(long divisor)
- Duration negated()

产生一个时长，该时长是通过当前时刻乘以或除以给定的量或 –1 得到的。

- boolean isZero()

- boolean isNegative()
  如果当前 Duration 对象是 0 或负数，则返回 true。
- Duration (plus|minus)(Nanos|Millis|Seconds|Minutes|Hours|Days)(long number)
  产生一个时长，该时长是通过当前时刻加上或减去给定的数量的指定时间单位而得到的。

6.2 本地日期

现在，让我们从绝对时间转向人类时间。在 Java API 中有两种人类时间，本地日期/时间和时区时间。本地日期/时间包含日期和当天的时间，但是与时区信息没有任何关联。1903 年 6 月 14 日就是一个本地日期的示例（lambda 演算的发明者 Alonzo Church 在这一天诞生）。因为这个日期既没有当天的时间，也没有时区信息，因此它并不对应精确的时刻。与之相反的是，1969 年 7 月 16 日 09:32:00 EDT（阿波罗 11 号发射的时刻）是一个时区日期/时间，表示的是时间线上的一个精确的时刻。

有许多计算并不需要时区，在某些情况下，时区甚至是一种障碍。假设你安排每周 10:00 开一次会。如果你加 7 天（即 7 × 24 × 60 × 60 秒）到最后一次会议的时区时间上，那么你可能会碰巧跨越了夏令时的时间调整边界，这次会议可能会早一小时或晚一小时！

正是考虑到这个原因，API 的设计者们推荐程序员不要使用时区时间，除非确实想要表示绝对时间的实例。生日、假日、计划时间等通常最好都表示成本地日期和时间。

LocalDate 是带有年、月、日的日期。为了构建 LocalDate 对象，可以使用 now 或 of 静态方法：

```
LocalDate today = LocalDate.now(); // Today's date
LocalDate alonzosBirthday = LocalDate.of(1903, 6, 14);
alonzosBirthday = LocalDate.of(1903, Month.JUNE, 14);
    // Uses the Month enumeration
```

与 UNIX 和 java.util.Date 中使用的月从 0 开始计算而年从 1900 开始计算的不规则的惯用法不同，你需要提供通常使用的月份的数字。或者，你可以使用 Month 枚举。

本节末尾展示了最有用的操作 LocalDate 对象的方法。

例如，程序员日是每年的第 256 天。下面展示了如何很容易地计算出它：

```
LocalDate programmersDay = LocalDate.of(2014, 1, 1).plusDays(255);
    // September 13, but in a leap year it would be September 12
```

回忆一下，两个 Instant 之间的时长是 Duration，而用于本地日期的等价物是 Period，它表示的是流逝的年、月或日的数量。可以调用 birthday.plus(Period.ofYears(1)) 来获取下一年的生日。当然，也可以直接调用 birthday.plusYears(1)。但是 birthday.plus(Duration.ofDays(365)) 在闰年是不会产生正确结果的。

util 方法会产生两个本地日期之间的时长。例如，

```
independenceDay.until(christmas)
```

会产生 5 个月 21 天的一段时长。这实际上并不是很有用，因为每个月的天数不尽相同。为

了确定到底有多少天，可以使用：

```
independenceDay.until(christmas, ChronoUnit.DAYS) // 174 days
```

> **警告**：LocalDate API 中的有些方法可能会创建出并不存在的日期。例如，在 1 月 31 日上加上 1 个月不应该产生 2 月 31 日。这些方法并不会抛出异常，而是会返回该月有效的最后一天。例如，
>
> ```
> LocalDate.of(2016, 1, 31).plusMonths(1)
> ```
>
> 和
>
> ```
> LocalDate.of(2016, 3, 31).minusMonths(1)
> ```
>
> 都将产生 2016 年 2 月 29 日。

getDayOfWeek 会产生星期日期，即 DayOfWeek 枚举的某个值。DayOfWeek.MONDAY 的枚举值为 1，而 DayOfWeek.SUNDAY 的枚举值为 7。例如，

```
LocalDate.of(1900, 1, 1).getDayOfWeek().getValue()
```

会产生 1。DayOfWeek 枚举具有便捷方法 plus 和 minus，以 7 为模计算星期日期。例如，DayOfWeek.SATURDAY.plus(3) 会产生 DayOfWeek.TUESDAY。

> **注释**：周末实际上在每周的末尾。这与 java.util.Calendar 有所差异，在后者中，星期日的值为 1，而星期六的值为 7。

Java 9 添加了两个有用的 datesUntil 方法，它们会产生 LocalDate 对象流。

```
LocalDate start = LocalDate.of(2000, 1, 1);
LocalDate endExclusive = LocalDate.now();
Stream<LocalDate> allDays = start.datesUntil(endExclusive);
Stream<LocalDate> firstDaysInMonth = start.datesUntil(endExclusive, Period.ofMonths(1));
```

除了 LocalDate 之外，还有 MonthDay、YearMonth 和 Year 类可以描述部分日期。例如，12 月 25 日（没有指定年份）可以表示成一个 MonthDay 对象。

程序清单 6-2 中的示例程序展示了如何使用 LocalDate 类。

程序清单 6-2　localdates/LocalDates.java

```java
 1  package localdates;
 2
 3  /**
 4   * @version 1.0 2016-05-10
 5   * @author Cay Horstmann
 6   */
 7  import java.time.*;
 8  import java.time.temporal.*;
 9  import java.util.stream.*;
10
11  public class LocalDates
12  {
13     public static void main(String[] args)
```

```java
14    {
15        LocalDate today = LocalDate.now(); // Today's date
16        System.out.println("today: " + today);
17
18        LocalDate alonzosBirthday = LocalDate.of(1903, 6, 14);
19        alonzosBirthday = LocalDate.of(1903, Month.JUNE, 14);
20        // Uses the Month enumeration
21        System.out.println("alonzosBirthday: " + alonzosBirthday);
22
23        LocalDate programmersDay = LocalDate.of(2018, 1, 1).plusDays(255);
24        // September 13, but in a leap year it would be September 12
25        System.out.println("programmersDay: " + programmersDay);
26
27        LocalDate independenceDay = LocalDate.of(2018, Month.JULY, 4);
28        LocalDate christmas = LocalDate.of(2018, Month.DECEMBER, 25);
29
30        System.out.println("Until christmas: " + independenceDay.until(christmas));
31        System.out.println("Until christmas: "
32            + independenceDay.until(christmas, ChronoUnit.DAYS));
33
34        System.out.println(LocalDate.of(2016, 1, 31).plusMonths(1));
35        System.out.println(LocalDate.of(2016, 3, 31).minusMonths(1));
36
37        DayOfWeek startOfLastMillennium = LocalDate.of(1900, 1, 1).getDayOfWeek();
38        System.out.println("startOfLastMillennium: " + startOfLastMillennium);
39        System.out.println(startOfLastMillennium.getValue());
40        System.out.println(DayOfWeek.SATURDAY.plus(3));
41
42        LocalDate start = LocalDate.of(2000, 1, 1);
43        LocalDate endExclusive = LocalDate.now();
44        Stream<LocalDate> firstDaysInMonth = start.datesUntil(endExclusive, Period.ofMonths(1));
45        System.out.println("firstDaysInMonth: "
46            + firstDaysInMonth.collect(Collectors.toList()));
47    }
48 }
```

API java.time.LocalDate 8

- static LocalDate now()

  获取当前的 LocalDate。

- static LocalDate of(int year, int month, int dayOfMonth)
- static LocalDate of(int year, Month month, int dayOfMonth)

  用给定的年、月（1 到 12 之间的整数或者 Month 枚举的值）和日（1 到 31 之间）产生一个本地日期。

- LocalDate (plus|minus)(Days|Weeks|Months|Years)(long number)

  产生一个 LocalDate，该对象是通过在当前对象上加上或减去给定数量的时间单位获得的。

- LocalDate plus(TemporalAmount amountToAdd)
- LocalDate minus(TemporalAmount amountToSubtract)

  产生一个时刻，该时刻与当前时刻距离给定的时间量。Duration 和 Period 类实现了

TemporalAmount 接口。
- LocalDate withDayOfMonth(int dayOfMonth)
- LocalDate withDayOfYear(int dayOfYear)
- LocalDate withMonth(int month)
- LocalDate withYear(int year)

  返回一个新的 LocalDate，将月份日期、年日期、月或年修改为给定值。

- int getDayOfMonth()

  获取月份日期（1 到 31 之间）。

- int getDayOfYear()

  获取年日期（1 到 366 之间）。

- DayOfWeek getDayOfWeek()

  获取星期日期，返回某个 DayOfWeek 枚举值。

- Month getMonth()
- int getMonthValue()

  获取用 Month 枚举值表示的月份，或者用 1 到 12 之间的数字表示的月份。

- int getYear()

  获取年份，在 –999,999,999 到 999,999,999 之间。

- Period until(ChronoLocalDate endDateExclusive)

  获取直到给定终止日期的 period。LocalDate 和 date 类针对非公历实现了 ChronoLocalDate 接口。

- boolean isBefore(ChronoLocalDate other)
- boolean isAfter(ChronoLocalDate other)

  如果该日期在给定日期之前或之后，则返回 true。

- boolean isLeapYear()

  如果当前是闰年，则返回 true。即，该年份能够被 4 整除，但是不能被 100 整除，或者能够被 400 整除。该算法应该可以应用于所有已经过去的年份，尽管在历史上它并不准确（闰年是在公元前 46 年发明出来的，而涉及整除 100 和 400 的规则是在 1582 年的公历改革中引入的。这场改革经历了 300 年才被广泛接受）。

- Stream<LocalDate> datesUntil(LocalDate endExclusive) 9
- Stream<LocalDate> datesUntil(LocalDate endExclusive, Period step) 9

  产生一个日期流，从当前的 LocalDate 对象直至参数 endExclusive 指定的日期，其中步长尺寸为 1，或者是给定的 period。

API java.time.Period 8

- static Period of(int years, int months, int days)
- Period of(Days|Weeks|Months|Years)(int number)

用给定数量的时间单位产生一个 Period 对象。
- int get(Days|Months|Years)()
  获取当前 Period 对象的日、月或年。
- Period (plus|minus)(Days|Months|Years)(long number)
  产生一个 LocalDate，该对象是通过在当前对象上加上或减去给定数量的时间单位获得的。
- Period plus(TemporalAmount amountToAdd)
- Period minus(TemporalAmount amountToSubtract)
  产生一个时刻，该时刻与当前时刻距离给定的时间量。Duration 和 Period 类实现了 TemporalAmount 接口。
- Period with(Days|Months|Years)(int number)
  返回一个新的 Period，将日、月、年修改为给定值。

6.3 日期调整器

对于日程安排应用来说，经常需要计算诸如"每个月的第一个星期二"这样的日期。TemporalAdjusters 类提供了大量用于常见调整的静态方法。你可以将调整方法的结果传递给 with 方法。例如，某个月的第一个星期二可以像下面这样计算：

```
LocalDate firstTuesday = LocalDate.of(year, month, 1).with(
    TemporalAdjusters.nextOrSame(DayOfWeek.TUESDAY));
```

一如既往，with 方法会返回一个新的 LocalDate 对象，而不会修改原来的对象。本节末尾展示了有关可用的调整器的 API 说明。

还可以通过实现 TemporalAdjuster 接口来创建自己的调整器。下面是用于计算下一个工作日的调整器。

```
TemporalAdjuster NEXT_WORKDAY = w ->
    {
        var result = (LocalDate) w;
        do
        {
            result = result.plusDays(1);
        }
        while (result.getDayOfWeek().getValue() >= 6);
        return result;
    };

LocalDate backToWork = today.with(NEXT_WORKDAY);
```

注意，lambda 表达式的参数类型为 Temporal，它必须被强制转型为 LocalDate。你可以用 ofDateAdjuster 方法来避免这种强制转型，该方法期望得到的参数是类型为 UnaryOperator<LocalDate> 的 lambda 表达式。

```
TemporalAdjuster NEXT_WORKDAY = TemporalAdjusters.ofDateAdjuster(w ->
    {
        LocalDate result = w; // No cast
```

```
        do
        {
            result = result.plusDays(1);
        }
        while (result.getDayOfWeek().getValue() >= 6);
        return result;
    });
```

API java.time.LocalDate 9

- LocalDate with(TemporalAdjuster adjuster)
  返回该日期通过给定的调整器调整后的结果。

API java.time.temporal.TemporalAdjusters 9

- static TemporalAdjuster next(DayOfWeek dayOfWeek)
- static TemporalAdjuster nextOrSame(DayOfWeek dayOfWeek)
- static TemporalAdjuster previous(DayOfWeek dayOfWeek)
- static TemporalAdjuster previousOrSame(DayOfWeek dayOfWeek)
  返回一个调整器，用于将日期调整为给定的星期日期。
- static TemporalAdjuster dayOfWeekInMonth(int n, DayOfWeek dayOfWeek)
- static TemporalAdjuster lastInMonth(DayOfWeek dayOfWeek)
  返回一个调整器，用于将日期调整为月份中第 n 个或最后一个给定的星期日期。
- static TemporalAdjuster firstDayOfMonth()
- static TemporalAdjuster firstDayOfNextMonth()
- static TemporalAdjuster firstDayOfYear()
- static TemporalAdjuster firstDayOfNextYear()
- static TemporalAdjuster lastDayOfMonth()
- static TemporalAdjuster lastDayOfYear()
  返回一个调整器，用于将日期调整为月份或年份中给定的日期。

6.4 本地时间

LocalTime 表示当日时刻，例如 15:30:00。可以用 now 或 of 方法创建其实例：

```
LocalTime rightNow = LocalTime.now();
LocalTime bedtime = LocalTime.of(22, 30); // or LocalTime.of(22, 30, 0)
```

API 说明展示了常见的对本地时间的操作。plus 和 minus 操作是按照一天 24 小时循环操作的。例如，

```
LocalTime wakeup = bedtime.plusHours(8); // wakeup is 6:30:00
```

> **注释**：LocalTime 自身并不关心 AM/PM。这种愚蠢的设计将问题抛给格式器去解决，请参见 6.6 节。

还有一个表示日期和时间的 LocalDateTime 类。这个类适合存储固定时区的时间点，例如，用于排课或排程。但是，如果你的计算需要跨越夏令时，或者需要处理不同时区的用户，那么就应该使用接下来要讨论的 ZonedDateTime 类。

API java.time.LocalTime 8

- static LocalTime now()
  获取当前的 LocalTime。
- static LocalTime of(int hour, int minute)
- static LocalTime of(int hour, int minute, int second)
- static LocalTime of(int hour, int minute, int second, int nanoOfSecond)
  产生一个 LocalTime，它具有给定的小时（0 到 23 之间）、分钟、秒（0 到 59 之间）和纳秒（0 到 999,999,999 之间）。
- LocalTime (plus|minus)(Hours|Minutes|Seconds|Nanos)(long number)
  产生一个 LocalTime，该对象是通过在当前对象上加上或减去给定数量的时间单位获得的。
- LocalTime plus(TemporalAmount amountToAdd)
- LocalTime minus(TemporalAmount amountToSubtract)
  产生一个时刻，该时刻与当前时刻距离给定的时间量。
- LocalTime with(Hour|Minute|Second|Nano)(int value)
  返回一个新的 LocalTime，将小时、分钟、秒或纳秒修改为给定值。
- int getHour()
  获取小时（0 到 23 之间）。
- int getMinute()
- int getSecond()
  获取分钟或秒（0 到 59 之间）。
- int getNano()
  获取纳秒（0 到 999,999,999 之间）。
- int toSecondOfDay()
- long toNanoOfDay()
  产生自午夜到当前 LocalTime 的秒或纳秒数。
- boolean isBefore(LocalTime other)
- boolean isAfter(LocalTime other)
  如果当期日期在给定日期之前或之后，则返回 true。

6.5 时区时间

时区问题比较复杂。在理性的世界中，我们都会遵循格林尼治时间，有些人在 02:00

吃午饭，而有些人却在 22:00 吃午饭。中国横跨了 4 个时区，但是使用了同一个时间。在其他地方，时区显得并不规则，并且还有国际日期变更线，而夏令时则使事情变得更复杂了。

尽管时区显得变化繁多，但这就是无法回避的现实生活。在实现日历应用时，它需要能够为坐飞机在不同国家之间穿梭的人们提供服务。如果你有个 10:00 在纽约召开的电话会议，但是碰巧你人在柏林，那么你肯定希望该应用能够在正确的本地时间点上发出提醒。

互联网编码分配管理机构（Internet Assigned Numbers Authority，IANA）保存着一个数据库，里面存储着世界上所有已知的时区（www.iana.org/time-zones），它每年会更新数次，而批量更新会处理夏令时的变更规则。Java 使用了 IANA 数据库。

每个时区都有一个 ID，例如 America/New_York 和 Europe/Berlin。要想找出所有可用的时区，可以调用 ZoneId.getAvailableZoneIds。在本书撰写之时，有将近 600 个 ID。

给定一个时区 ID，静态方法 ZoneId.of(id) 可以产生一个 ZoneId 对象。可以通过调用 local.atZone(zoneId) 用这个对象将 LocalDateTime 对象转换为 ZonedDateTime 对象，或者可以通过调用静态方法 ZonedDateTime.of(year,month,day,hour,minute,second,nano,zoneId) 来构造一个 ZonedDateTime 对象。例如，

```
ZonedDateTime apollo11launch = ZonedDateTime.of(1969, 7, 16, 9, 32, 0, 0,
    ZoneId.of("America/New_York"));
    // 1969-07-16T09:32-04:00[America/New_York]
```

这是一个具体的时刻，调用 apollo11launch.toInstant 可以获得对应的 Instant 对象。反过来，如果你有一个时刻对象，调用 instant.atZone(ZoneId.of("UTC")) 可以获得格林尼治皇家天文台的 ZonedDateTime 对象，或者使用其他的 ZoneId 获得地球上其他地方的 ZoneId。

> **注释**：UTC 代表"协调世界时"，这是英文"Coordinated Universal Time"和法文"Temps Universel Coordiné"首字母缩写的折中，它与这两种语言中的缩写都不一致。UTC 是不考虑夏令时的格林尼治皇家天文台时间。

ZonedDateTime 的许多方法都与 LocalDateTime 的方法相同（参见本节末尾的 API 说明），它们大多数都很直观，但是夏令时带来了一些复杂性。

当夏令时开始时，时钟要向前拨快一小时。当你构建的时间对象正好落入了这跳过去的一个小时内时，会发生什么？例如，在 2013 年，中欧地区在 3 月 31 日 2:00 切换到夏令时，如果你试图构建的时间是不存在的 3 月 31 日 2:30，那么你实际上得到的是 3:30。

```
ZonedDateTime skipped = ZonedDateTime.of(
    LocalDate.of(2013, 3, 31),
    LocalTime.of(2, 30),
    ZoneId.of("Europe/Berlin"));
    // Constructs March 31 3:30
```

反过来，当夏令时结束时，时钟要向回拨慢一小时，这样同一个本地时间就会有出现两次。当你构建位于这个时间段内的时间对象时，就会得到这两个时刻中较早的一个：

```
ZonedDateTime ambiguous = ZonedDateTime.of(
   LocalDate.of(2013, 10, 27), // End of daylight savings time
   LocalTime.of(2, 30),
   ZoneId.of("Europe/Berlin"));
   // 2013-10-27T02:30+02:00[Europe/Berlin]
ZonedDateTime anHourLater = ambiguous.plusHours(1);
   // 2013-10-27T02:30+01:00[Europe/Berlin]
```

一个小时后的时间会具有相同的小时和分钟，但是时区的偏移量会发生变化。

你还需要在调整跨越夏令时边界的日期时特别注意。例如，如果你将会议设置在下个星期，不要直接加上一个 7 天的 Duration：

```
ZonedDateTime nextMeeting = meeting.plus(Duration.ofDays(7));
   // Caution! Won't work with daylight savings time
```

而是应该使用 Period 类。

```
ZonedDateTime nextMeeting = meeting.plus(Period.ofDays(7)); // OK
```

> **警告**：还有一个 OffsetDateTime 类，它表示与 UTC 具有偏移量的时间，但是没有时区规则的束缚。这个类被设计用于专用应用，这些应用特别需要剔除这些规则的约束，例如某些网络协议。对于人类时间，还是应该使用 ZonedDateTime。

程序清单 6-3 中的示例程序演示了 ZonedDateTime 类的用法。

程序清单 6-3   zonedtimes/ZonedTimes.java

```
 1  package zonedtimes;
 2
 3  /**
 4   * @version 1.0 2016-05-10
 5   * @author Cay Horstmann
 6   */
 7
 8  import java.time.*;
 9
10  public class ZonedTimes
11  {
12     public static void main(String[] args)
13     {
14        ZonedDateTime apollo11launch = ZonedDateTime.of(1969, 7, 16, 9, 32, 0, 0,
15           ZoneId.of("America/New_York")); // 1969-07-16T09:32-04:00[America/New_York]
16        System.out.println("apollo11launch: " + apollo11launch);
17
18        Instant instant = apollo11launch.toInstant();
19        System.out.println("instant: " + instant);
20
21        ZonedDateTime zonedDateTime = instant.atZone(ZoneId.of("UTC"));
22        System.out.println("zonedDateTime: " + zonedDateTime);
23
24        ZonedDateTime skipped = ZonedDateTime.of(LocalDate.of(2013, 3, 31),
25           LocalTime.of(2, 30), ZoneId.of("Europe/Berlin")); // Constructs March 31 3:30
26        System.out.println("skipped: " + skipped);
```

```
27
28      ZonedDateTime ambiguous = ZonedDateTime.of(
29         LocalDate.of(2013, 10, 27), // End of daylight savings time
30         LocalTime.of(2, 30), ZoneId.of("Europe/Berlin"));
31         // 2013-10-27T02:30+02:00[Europe/Berlin]
32      ZonedDateTime anHourLater = ambiguous.plusHours(1);
33         // 2013-10-27T02:30+01:00[Europe/Berlin]
34      System.out.println("ambiguous: " + ambiguous);
35      System.out.println("anHourLater: " + anHourLater);
36
37      ZonedDateTime meeting = ZonedDateTime.of(LocalDate.of(2013, 10, 31),
38         LocalTime.of(14, 30), ZoneId.of("America/Los_Angeles"));
39      System.out.println("meeting: " + meeting);
40      ZonedDateTime nextMeeting = meeting.plus(Duration.ofDays(7));
41         // Caution! Won't work with daylight savings time
42      System.out.println("nextMeeting: " + nextMeeting);
43      nextMeeting = meeting.plus(Period.ofDays(7)); // OK
44      System.out.println("nextMeeting: " + nextMeeting);
45   }
46 }
```

API java.time.ZonedDateTime 8

- static ZonedDateTime now()
  获取当前的 ZonedDateTime。

- static ZonedDateTime of(int year, int month, int dayOfMonth, int hour, int minute, int second, int nanoOfSecond, ZoneId zone)

- static ZonedDateTime of(LocalDate date, LocalTime time, ZoneId zone)

- static ZonedDateTime of(LocalDateTime localDateTime, ZoneId zone)

- static ZonedDateTime ofInstant(Instant instant, ZoneId zone)
  用给定的参数和时区产生一个 ZonedDateTime。

- ZonedDateTime (plus|minus)(Days|Weeks|Months|Years|Hours|Minutes|Seconds|Nanos)(long number)
  产生一个 ZonedDateTime，该对象是通过在当前对象上加上或减去给定数量的时间单位获得的。

- ZonedDateTime plus(TemporalAmount amountToAdd)

- ZonedDateTime minus(TemporalAmount amountToSubtract)
  产生一个时刻，该时刻与当前时刻距离给定的时间量。

- ZonedDateTime with(DayOfMonth|DayOfYear|Month|Year|Hour|Minute|Second|Nano)(int value)
  返回一个新的 ZonedDateTime，用给定的值替换给定的时间单位。

- ZonedDateTime withZoneSameInstant(ZoneId zone)

- ZonedDateTime withZoneSameLocal(ZoneId zone)
  返回一个新的 ZonedDateTime，位于给定的时区，它与当前对象要么表示相同的时刻，要么表示相同的本地时间。

- int getDayOfMonth()

  获取月份日期（1 到 31 之间）。
- int getDayOfYear()

  获取年份日期（1 到 366 之间）。
- DayOfWeek getDayOfWeek()

  获取星期日期，返回 DayOfWeek 枚举的值。
- Month getMonth()
- int getMonthValue()

  获取用 Month 枚举值表示的月份，或者用 1 到 12 之间的数字表示的月份。
- int getYear()

  获取年份，在 –999 999 999 到 999 999999 之间。
- int getHour()

  获取小时（0 到 23 之间）。
- int getMinute()
- int getSecond()

  获取分钟到秒（0 到 59 之间）。
- int getNano()

  获取纳秒（0 到 999 999 999 之间）。
- public ZoneOffset getOffset()

  获取与 UTC 的时间差距。差距可在 –12:00 ～ +14:00 变化。有些时区还有小数时间差。时间差会随着夏令时变化。
- LocalDate toLocalDate()
- LocalTime toLocalTime()
- LocalDateTime toLocalDateTime()
- Instant toInstant()

  生成当地日期、时间，或日期/时间，或相应的瞬间。
- boolean isBefore(ChronoZonedDateTime other)
- boolean isAfter(ChronoZonedDateTime other)

  如果这个时区日期/时间在给定的时区日期/时间之前或之后，则返回 true。

6.6 格式化和解析

DateTimeFormatter 类提供了三种用于打印日期/时间值的格式器：
- 预定义的格式器（参见表 6-1）
- locale 相关的格式器
- 带有定制模式的格式器

表 6-1　预定义的格式器

格式器	描述	示例
BASIC_ISO_DATE	年、月、日、时区偏移量，中间没有分隔符	19690716-0500
ISO_LOCAL_DATE, ISO_LOCAL_TIME, ISO_LOCAL_DATE_TIME	分隔符为 -、：、T	1969-07-16, 09:32:00, 1969-07-16T09:32:00
ISO_OFFSET_DATE, ISO_OFFSET_TIME, ISO_OFFSET_DATE_TIME	类似 ISO_LOCAL_XXX，但是有时区偏移量	1969-07-16-05:00, 09:32:00-05:00, 1969-07-16T09:32:00-05:00
ISO_ZONED_DATE_TIME	有时区偏移量和时区 ID	1969-07-16T09:32:00-05:00[America/New_York]
ISO_INSTANT	在 UTC 中，用 Z 时区 ID 来表示	1969-07-16T14:32:00Z
ISO_DATE, ISO_TIME, ISO_DATE_TIME	类似 ISO_OFFSET_DATE、ISO_OFFSET_TIME 和 ISO_ZONED_DATE_TIME，但是时区信息是可选的	1969-07-16-05:00, 09:32:00-05:00, 1969-07-16T09:32:00-05:00[America/New_York]
ISO_ORDINAL_DATE	LocalDate 的年和年日期	1969-197
ISO_WEEK_DATE	LocalDate 的年、星期和星期日期	1969-W29-3
RFC_1123_DATE_TIME	用于邮件时间戳的标准，编纂于 RFC822，并在 RFC1123 中将年份更新到 4 位	Wed, 16 Jul 1969 09:32:00 -0500

要使用标准的格式器，可以直接调用其 format 方法：

```
String formatted = DateTimeFormatter.ISO_OFFSET_DATE_TIME.format(apollo11launch);
    // 1969-07-16T09:32:00-04:00"
```

标准格式器主要是为了机器可读的时间戳而设计的。为了向人类读者表示日期和时间，可以使用 locale 相关的格式器。对于日期和时间而言，有 4 种与 locale 相关的格式化风格，即 SHORT、MEDIUM、LONG 和 FULL，参见表 6-2。

表 6-2　locale 相关的格式化风格

风格	日期	时间
SHORT	7/16/69	9:32 AM
MEDIUM	Jul 16, 1969	9:32:00 AM
LONG	July 16, 1969	9:32:00 AM EDT
FULL	Wednesday, July 16, 1969	9:32:00 AM EDT

静态方法 ofLocalizedDate、ofLocalizedTime 和 ofLocalizedDateTime 可以创建这种格式器。例如：

```
DateTimeFormatter formatter = DateTimeFormatter.ofLocalizedDateTime(FormatStyle.LONG);
String formatted = formatter.format(apollo11launch);
    // July 16, 1969 9:32:00 AM EDT
```

这些方法使用了默认的 locale。为了切换到不同的 locale，可以直接使用 withLocale 方法。

```
formatted = formatter.withLocale(Locale.FRENCH).format(apollo11launch);
    // 16 juillet 1969 09:32:00 EDT
```

DayOfWeek 和 Month 枚举都有 getDisplayName 方法，可以按照不同的 locale 和格式给出星期日期和月份的名字。

```
for (DayOfWeek w : DayOfWeek.values())
    System.out.print(w.getDisplayName(TextStyle.SHORT, Locale.ENGLISH) + " ");
    // Prints Mon Tue Wed Thu Fri Sat Sun
```

请查看第 7 章以了解更多有关 locale 的信息。

> **注释**：java.time.format.DateTimeFormatter 类被设计用来替代 java.util.DateFormat。如果你为了向后兼容性而需要后者的实例，那么可以调用 formatter.toFormat()。

最后，可以通过指定模式来定制自己的日期格式。例如，

```
formatter = DateTimeFormatter.ofPattern("E yyyy-MM-dd HH:mm");
```

会将日期格式化为 Wed 1969-07-16 09:32 的形式。按照人们日积月累而制定的显得有些晦涩的规则，每个字母都表示一个不同的时间域，而字母重复的次数对应于所选择的特定格式。表 6-3 展示了最有用的模式元素。

表 6-3　常用的日期 / 时间格式的格式化符号

时间域或目的	示例
ERA	G: AD, GGGG: Anno Domini, GGGGG: A
YEAR_OF_ERA	yy: 69, yyyy: 1969
MONTH_OF_YEAR	M: 7, MM: 07, MMM: Jul, MMMM: July, MMMMM: J
DAY_OF_MONTH	d: 6, dd: 06
DAY_OF_WEEK	e: 3, E: Wed, EEEE: Wednesday, EEEEE: W
HOUR_OF_DAY	H: 9, HH: 09
CLOCK_HOUR_OF_AM_PM	K: 9, KK: 09
AMPM_OF_DAY	a: AM
MINUTE_OF_HOUR	mm: 02
SECOND_OF_MINUTE	ss: 00
NANO_OF_SECOND	nnnnnn: 000000
时区 ID	VV: America/New_York
时区名	z: EDT, zzzz: Eastern Daylight Time V:ET, VVVV:Eastern time
时区偏移量	x: -04, xx: -0400, xxx: -04:00, XXX: 与 xxx 相同，但是 Z 表示 0
本地化的时区偏移量	O: GMT-4, OOOO: GMT-04:00
修改后的儒略日	g:58243

为了解析字符串中的日期 / 时间值，可以使用众多的静态 parse 方法之一。例如，

```
LocalDate churchsBirthday = LocalDate.parse("1903-06-14");
ZonedDateTime apollo11launch =
    ZonedDateTime.parse("1969-07-16 03:32:00-0400",
        DateTimeFormatter.ofPattern("yyyy-MM-dd HH:mm:ssxx"));
```

第一个调用使用了标准的 ISO_LOCAL_DATE 格式器，而第二个调用使用的是一个定制的格式器。

程序清单 6-4 中的程序展示了如何格式化和解析日期与时间。

程序清单 6-4 formatting/Formatting.java

```java
 1  package formatting;
 2
 3  /**
 4   * @version 1.0 2016-05-10
 5   * @author Cay Horstmann
 6   */
 7
 8  import java.time.*;
 9  import java.time.format.*;
10  import java.util.*;
11
12  public class Formatting
13  {
14     public static void main(String[] args)
15     {
16        ZonedDateTime apollo11launch = ZonedDateTime.of(1969, 7, 16, 9, 32, 0, 0,
17           ZoneId.of("America/New_York"));
18
19        String formatted = DateTimeFormatter.ISO_OFFSET_DATE_TIME.format(apollo11launch);
20        // 1969-07-16T09:32:00-04:00
21        System.out.println(formatted);
22
23        DateTimeFormatter formatter = DateTimeFormatter.ofLocalizedDateTime(FormatStyle.LONG);
24        formatted = formatter.format(apollo11launch);
25        // July 16, 1969 9:32:00 AM EDT
26        System.out.println(formatted);
27        formatted = formatter.withLocale(Locale.FRENCH).format(apollo11launch);
28        // 16 juillet 1969 09:32:00 EDT
29        System.out.println(formatted);
30
31        formatter = DateTimeFormatter.ofPattern("E yyyy-MM-dd HH:mm");
32        formatted = formatter.format(apollo11launch);
33        System.out.println(formatted);
34
35        LocalDate churchsBirthday = LocalDate.parse("1903-06-14");
36        System.out.println("churchsBirthday: " + churchsBirthday);
37        apollo11launch = ZonedDateTime.parse("1969-07-16 03:32:00-0400",
38           DateTimeFormatter.ofPattern("yyyy-MM-dd HH:mm:ssxx"));
39        System.out.println("apollo11launch: " + apollo11launch);
40
41        for (DayOfWeek w : DayOfWeek.values())
42           System.out.print(w.getDisplayName(TextStyle.SHORT, Locale.ENGLISH) + " ");
43     }
44  }
```

API java.time.format.DateTimeFormatter 8

- String format(TemporalAccessor temporal)

  格式化给定值。Instant、LocalDate、LocalTime、LocalDateTime 和 ZonedDateTime，以及许多其他类，都实现了 TemporalAccessor 接口。

- static DateTimeFormatter ofLocalizedDate(FormatStyle dateStyle)
- static DateTimeFormatter ofLocalizedTime(FormatStyle timeStyle)
- static DateTimeFormatter ofLocalizedDateTime(FormatStyle dateTimeStyle)
- static DateTimeFormatter ofLocalizedDateTime(FormatStyle dateStyle, FormatStyle timeStyle)

  产生一个用于给定风格的格式器。FormatStyle 枚举的值包括 SHORT、MEDIUM、LONG 和 FULL。

- DateTimeFormatter withLocale(Locale locale)

  用给定的地点产生一个等价于当前格式器的格式器。

- static DateTimeFormatter ofPattern(String pattern)
- static DateTimeFormatter ofPattern(String pattern, Locale locale)

  用给定的模式和地点产生一个格式器。参阅表 6-3 有关模式的语法。

API java.time.LocalDate 8

- static LocalDate parse(CharSequence text)
- static LocalDate parse(CharSequence text, DateTimeFormatter formatter)

  用默认的格式器或给定的格式器产生一个 LocalDate。

API java.time.ZonedDateTime 8

- static ZonedDateTime parse(CharSequence text)
- static ZonedDateTime parse(CharSequence text, DateTimeFormatter formatter)

  用默认的格式器或给定的格式器产生一个 ZonedDateTime。

6.7 与遗留代码的互操作

作为全新的创造，Java Date 和 Time API 必须能够与已有类之间进行互操作，特别是无处不在的 java.util.Date、java.util.GregorianCalendar 和 java.sql.Date/Time/Timestamp。

Instant 类近似于 java.util.Date。在 Java 8 中，这个类有两个额外的方法：将 Date 转换为 Instant 的 toInstant 方法，以及反方向转换的静态的 from 方法。

类似地，ZonedDateTime 近似于 java.util.GregorianCalendar，在 Java 8 中，这个类有细粒度的转换方法。toZonedDateTime 方法可以将 GregorianCalendar 转换为 ZonedDateTime，而静态的 from 方法可以执行反方向的转换。

另一个可用于日期和时间类的转换集位于 java.sql 包中。你还可以传递一个 DateTimeFormatter 给使用 java.text.Format 的遗留代码。表 6-4 对这些转换进行了总结。

表 6-4 java.time 类与遗留类之间的转换

类	转换到遗留类	转换自遗留类
Instant ↔ java.util.Date	Date.from(instant)	date.toInstant()

（续）

类	转换到遗留类	转换自遗留类
ZonedDateTime ↔ java.util.GregorianCalendar	GregorianCalendar.from(zonedDateTime)	cal.toZonedDateTime()
Instant ↔ java.sql.Timestamp	TimeStamp.from(instant)	timestamp.toInstant()
LocalDateTime ↔ java.sql.Timestamp	Timestamp.valueOf(localDateTime)	timeStamp.toLocalDateTime()
LocalDate ↔ java.sql.Date	Date.valueOf(localDate)	date.toLocalDate()
LocalTime ↔ java.sql.Time	Time.valueOf(localTime)	time.toLocalTime()
DateTimeFormatter ↔ java.text.DateFormat	formatter.toFormat()	无
java.util.TimeZone ↔ ZoneId	Timezone.getTimeZone(id)	timeZone.toZoneId()
java.nio.file.attribute.FileTime ↔ Instant	FileTime.from(instant)	fileTime.toInstant()

你现在知道如何使用 Java 8 的日期和时间库来操作全世界的日期和时间值了。下一章将进一步讨论如何为国际受众编程。你将会看到如何以对客户而言有意义的方式来格式化程序的消息、数字和货币，无论这些客户身处世界的何处。

第 7 章 国 际 化

- ▲ locale
- ▲ 数字格式
- ▲ 日期和时间
- ▲ 排序和规范化
- ▲ 消息格式化
- ▲ 文本输入和输出
- ▲ 资源包
- ▲ 一个完整的例子

世界丰富多彩，我们希望大部分居民都能对你的软件感兴趣。一方面，因特网早已为我们打破了国家之间的界限。另一方面，如果你不去关注国际用户，你的产品的应用情况就会受到限制。

Java 编程语言是第一种设计成为全面支持国际化的语言。从一开始，它就具备了进行有效的国际化所必需的一个重要特性：使用 Unicode 来处理所有字符串。由于支持 Unicode，在 Java 编程语言中编写程序来操作多种语言的字符串变得异常方便。

多数程序员认为将程序进行国际化需要做的所有事情就是支持 Unicode 并在用户接口中对消息进行翻译。但是，在本章你将会看到，国际化一个程序所要做的事情绝不仅仅是提供 Unicode 支持。在世界的不同地方，日期、时间、货币甚至数字的格式都不相同。你需要用一种简单的方法来为不同的语言配置菜单与按钮的名字、消息字符串和快捷键。

在本章中，我们将演示如何编写国际化的 Java 应用程序以及如何将日期、时间、数字、文本和图形用户界面本地化，还将演示 Java 提供的编写国际化程序的工具。最后以一个完整的例子来作为本章的结束，它是一个退休金计算器，带有英语、德语和中文用户界面。

7.1 locale

当你看到一个面向国际市场的应用软件时，它与其他软件最明显的区别就是语言。其实如果以这种外在的不同来判断是不是真正的国际化就太片面了：不同的国家可以使用相同的语言，但是为了使两个国家的用户都满意，你还有很多工作要做。就像 Oscar Wilde 所说的那样，"我们现在真的是每件东西都和美国一样，当然，语言除外。"

7.1.1 为什么需要 locale

当你提供程序的国际化版本时，所有程序消息都需要转换为本地语言。当然，直接翻译用户界面的文本是不够的，还有许多更细微的差异，例如，数字在英语和德语中格式很不相同。对于德国用户，数字

　　123,456.78

应该显示为

123.456,78

小数点和十进制数的逗号分隔符的角色是相反的！在日期的显示上也有相似的变化。在美国，日期显示为月／日／年，这有些不合理。德国使用的是更合理的顺序，即日／月／年，而在中国，则使用年／月／日。因此，对于德国用户，日期

3/22/61

应该被表示为

22.03.1961

当然，如果月份的名称被显式地写了出来，那么语言之间的不同就显而易见了。英语

March 22, 1961

在德国应该被表示成

22. März 1961

在中国则是

1961年3月22日

locale 捕获了像上面这类偏好特征。无论何时，只要你表示数字、日期、货币值以及其他格式会随语言或地点发生变化的项，都需要使用 locale 感知的 API。

7.1.2 指定 locale

locale 由多达 5 个部分构成：

1. 一种语言，由 2 个或 3 个小写字母表示，例如 en（英语）、de（德语）和 zh（中文）。表 7-1 展示了常用的代码。

表 7-1　常见的 ISO-639-1 语言代码

语言	代码	语言	代码
Chinese	zh	Italian	it
Danish	da	Japanese	ja
Dutch	nl	Korean	ko
English	en	Norwegian	no
French	fr	Portuguese	pt
Finnish	fi	Spanish	es
German	de	Swedish	sv
Greek	el	Turkish	tr

2. 可选的一段脚本，由首字母大写的四个字母表示，例如 Latn（拉丁文）、Cyrl（西里尔文）和 Hant（繁体中文）。这个部分很有用，因为有些语言，例如塞尔维亚语，可以用拉丁文或西里尔文书写，而有些中文读者更喜欢阅读繁体中文而不是简体中文。

3. 可选的一个国家或地区，由 2 个大写字母或 3 个数字表示，例如 US（美国）和 CH（瑞士）。表 7-2 展示了常用的代码。

表 7-2 常见的 ISO-3166-1 国家代码

国家	代码	国家	代码
Austria	AT	Japan	JP
Belgium	BE	Korea	KR
Canada	CA	The Netherlands	NL
China	CN	Norway	NO
Denmark	DK	Portugal	PT
Finland	FI	Spain	ES
Germany	DE	Sweden	SE
Great Britain	GB	Switzerland	CH
Ireland	IE	Turkey	TR
Italy	IT	United States	US

4. 可选的一个变体，用于指定各种杂项特性，例如方言和拼写规则。变体现在已经很少使用了。过去曾经有一种挪威语的变体"尼诺斯克语"，但是它现在已经用另一种不同的代码 nn 来表示了。过去曾经用于日本帝国历和泰语数字的变体现在也都被表示成了扩展（请参见下一条）。

5. 可选的一个扩展。扩展描述了日历（例如日本历）和数字（替代西方数字的泰语数字）等内容的本地偏好。Unicode 标准规范了其中的某些扩展，这些扩展应该以 u- 和两个字母的代码开头，这两个字母的代码指定了该扩展处理的是日历（ca）还是数字（nu），或者是其他内容。例如，扩展 u-nu-thai 表示使用泰语数字。其他扩展是完全任意的，并且以 x- 开头，例如 x-java。

locale 的规则在 Internet Engineering Task Force 的 "Best Current Practices" 备忘录 BCP 47 (http://tools.ietf.org/html/bcp47) 中进行了明确阐述。你可以在 www.w3.org/International/articles/language-tags 处找到更容易理解的总结。

语言和国家的代码看起来有点乱，因为它们中的有些是从本地语言导出的。德语在德语中是 Deutsch，中文在中文里是 zhongwen，因此它们分别是 de 和 zh。瑞士是 CH，这是从瑞士联邦的拉丁语 Confoederatio Helvetica 中导出的。

locale 是用标签描述的，标签是由 locale 的各个元素通过连字符连接起来的字符串，例如 en-US。

在德国，你可以使用 de-DE。瑞士有 4 种官方语言（德语、法语、意大利语和里托罗曼斯语）。在瑞士讲德语的人希望使用的 locale 是 de-CH。这个 locale 会使用德语的规则，但是货币值会表示成瑞士法郎而不是欧元。

如果只指定了语言，例如 de，那么该 locale 就不能用于与国家相关的场景，例如货币。

我们可以像下面这样用标签字符串来构建 Locale 对象：

```
Locale usEnglish = Locale.forLanguageTag("en-US");
```

toLanguageTag 方法可以生成给定 locale 的语言标签。例如，Local.US.toLanguageTag() 生成的字符串是 "en-US"。

为方便起见，有许多为各个国家预定义的 Locale 对象：

```
Locale.CANADA
Locale.CANADA_FRENCH
Locale.CHINA
Locale.FRANCE
Locale.GERMANY
Locale.ITALY
Locale.JAPAN
Locale.KOREA
Locale.PRC
Locale.UK
Locale.US
```

还有许多预定义的语言 Locale，它们只设定了语言而没有设定位置：

```
Locale.CHINESE
Locale.ENGLISH
Locale.FRENCH
Locale.GERMAN
Locale.ITALIAN
Locale.JAPANESE
Locale.KOREAN
Locale.SIMPLIFIED_CHINESE
Locale.TRADITIONAL_CHINESE
```

最后，静态的 getAvailableLocales 方法会返回由 Java 虚拟机能够识别的所有 locale 构成的数组。

> **注释**：可以用 Locale.getISOLanguages() 获取所有语言代码，用 Locale.getISOCountries() 获取所有国家代码。

7.1.3 默认 locale

Locale 类的静态 getDefault 方法可以获得作为本地操作系统的一部分而存放的默认 locale。可以调用 setDefault 来改变默认的 Java locale，但是，这种改变只对你的程序有效，不会对操作系统产生影响。

有些操作系统允许用户为显示消息和格式化指定不同的 locale。例如，生活在美国的说法语的人菜单是法语的，但是货币值是用美元来表示的。

要想获取这些偏好，可以调用

```
Locale displayLocale = Locale.getDefault(Locale.Category.DISPLAY);
Locale formatLocale = Locale.getDefault(Locale.Category.FORMAT);
```

> **注释**：在 UNIX 中，可以为数字、货币和日期分别设置 LC_NUMERIC、LC_MONETARY 和 LC_TIME 环境变量来指定不同的 locale。但是 Java 并不会关注这些设置。

> **提示**：为了测试，你也许希望改变你的程序的默认 locale，可以在启动程序时提供语言和地域特性。比如，下面的语句将默认的 locale 设为 de-CH：
> ```
> java -Duser.language=de -Duser.region=CH MyProgram
> ```

7.1.4 显示名字

一旦有了一个 locale，你能用它做什么呢？答案是它所能做的事情很有限。Locale 类中唯一有用的是那些识别语言和国家代码的方法，其中最重要的一个是 getDisplayName，它返回一个描述 locale 的字符串。这个字符串并不包含前面所说的由两个字母组成的代码，而是以一种面向用户的形式来表现，比如

```
German (Switzerland)
```

事实上，这里有一个问题，显示的名字是以默认的 locale 来表示的，这可能不太恰当。如果你的用户已经选择了德语作为首选的语言，那么你可能希望将字符串显示成德语。通过将 German locale 作为参数传递就可以做到这一点：代码

```
var loc = new Locale("de", "CH");
System.out.println(loc.getDisplayName(Locale.GERMAN));
```

将打印出

```
Deutsch (Schweiz)
```

这个例子说明了为什么需要 Locale 对象。你把它传给 locale 感知的那些方法，这些方法将根据不同的地域产生不同形式的文本。在后面各节中你可以见到大量的例子。

> ⚠️ **警告**：即使是像把字符串中的字母全部转换为小写或大写这样简单的操作，也可能是与 locale 相关的。例如，在土耳其 locale 中，字母 I 的小写是不带点的 ı。那些试图通过将字符串存储为小写格式来正则化字符串的程序对于土耳其客户来说就会显得很失败，因为 I 和带点的 i 没有相同的小写格式。一种好的做法是总是使用 toUpperCase 和 toLowerCase 的变体，这种变体会接受一个 Locale 参数。例如，试试下面的代码：
>
> ```
> String cmd = "QUIT".toLowerCase(Locale.forLanguageTag("tr"));
>     // "quit" with a dotless i
> ```
>
> 当然，在土耳其，Locale.getDefault() 产生的就是那里的 locale，"QUIT".toLowerCase() 与 "quit" 不同。
>
> 如果想要将英语字符串正则化为小写形式，那么就应该将英语的 locale 传递给 toLowerCase 方法。

> 📋 **注释**：你可以显式地指定输入/输出操作的 locale。
> - 当从 Scanner 读入数字时，可以用 useLocale 方法设置它的 locale。
> - String.format 和 PrintWriter.printf 方法也可以接受一个 Locale 参数。

API `java.util.Locale` 1.1

- Locale(String language)
- Locale(String language, String country)
- Locale(String language, String country, String variant)

  用给定的语言、国家和变量创建一个 locale。在新代码中不要使用变体，应该使用

IETF BCP 47 语言标签。

- static Locale forLanguageTag(String languageTag)　7

  构建与给定的语言标签相对应的 locale。

- static Locale getDefault()

  返回默认的 locale。

- static void setDefault(Locale loc)

  设定默认的 locale。

- String getDisplayName()

  返回一个在当前的 locale 中所表示的用来描述 locale 的名字。

- String getDisplayName(Locale loc)

  返回一个在给定的 locale 中所表示的用来描述 locale 的名字。

- String getLanguage()

  返回语言代码，它是两个小写字母组成的 ISO 639 代码。

- String getDisplayLanguage()

  返回在当前 locale 中所表示的语言名称。

- String getDisplayLanguage(Locale loc)

  返回在给定 locale 中所表示的语言名称。

- String getCountry()

  返回国家代码，它是由两个大写字母组成的 ISO 3166 代码。

- static String[] getISOCountries()

- static Set<String> getISOCountries(Locale.IsoCountryCode type)　9

  获取所有两字母的国家代码，或者所有 2、3、4 个字母的国家代码。type 参数是枚举常量 PART1_ALPHA2、PART1_ALPHA3 和 PART3 之一。

- String getDisplayCountry()

  返回在当前 locale 中所表示的国家名。

- String getDisplayCountry(Locale loc)

  返回在给定 locale 中所表示的国家名。

- String toLanguageTag()　7

  返回该 locale 的语言标签，例如 "de-CH"。

- String toString()

  返回 locale 的描述，包括语言和国家，用下划线分隔（比如，"de_CH"）。应该只在调试时使用该方法。

7.2　数字格式

我们已经提到了数字和货币的格式是高度依赖于 locale 的。Java 类库提供了一个格式器

（formatter）对象的集合，它可以对 java.text 包中的数字值进行格式化和解析。

7.2.1 格式化数字值

可以通过下面的步骤对特定 locale 的数字进行格式化：
1. 使用上一节的方法，得到 Locale 对象。
2. 使用一个"工厂方法"得到一个格式器对象。
3. 使用这个格式器对象来完成格式化和解析工作。

工厂方法是 NumberFormat 类的静态方法，它们接受一个 Locale 类型的参数。总共有 3 个工厂方法 getNumberInstance、getCurrencyInstance 和 getPercentInstance，这些方法返回的对象可以分别对数字、货币量和百分比进行格式化和解析。例如，下面显示了如何对德语中的货币值进行格式化。

```
Locale loc = Locale.GERMAN;
NumberFormat currFmt = NumberFormat.getCurrencyInstance(loc);
double amt = 123456.78;
String result = currFmt.format(amt);
```

结果是

123.456,78 €

请注意，货币符号是€，而且位于字符串的最后。同时还要注意到小数点和十进制分隔符与其他语言中的情况是相反的。

相反，如果想读取一个按照某个 locale 的惯用法而输入或存储的数字，那么就需要使用 parse 方法。比如，下面的代码解析了用户输入到文本框中的值。parse 方法能够处理小数点和分隔符以及其他语言中的数字。

```
TextField inputField;
...
NumberFormat fmt = NumberFormat.getNumberInstance();
// get the number formatter for default locale
Number input = fmt.parse(inputField.getText().trim());
double x = input.doubleValue();
```

parse 的返回类型是抽象类型 Number。返回的对象是一个 Double 或 Long 的包装器对象，这取决于被解析的数字是否是浮点数。如果不关心两者的差异，可以直接使用 Number 类的 doubleValue 方法来读取被包装的数字。

⚠️ **警告**：Number 类型的对象并不能自动转换成相关的基本类型，因此，不能直接将一个 Number 对象赋给一个基本类型，而应该使用 doubleValue 或 intValue 方法。

如果数字文本的格式不正确，该方法会抛出一个 ParseException 异常。例如，字符串以空白字符开头是不允许的（可以调用 trim 方法来去掉它）。但是，任何跟在数字之后的字符都将被忽略，所以这些跟在后面的字符是不会抛出异常的。

请注意，由 get*Xxx*Instance 工厂方法返回的类并非是 NumberFormat 类型的。NumberFormat 类

型是一个抽象类，而我们实际上得到的格式器是它的一个子类。工厂方法只知道如何定位属于特定 locale 的对象。

可以用静态的 getAvailableLocales 方法得到一个当前支持的 locale 列表。这个方法返回一个 locale 数组，从中可以获得针对它们的数字格式器对象。

本节的示例程序让你体会到了数字格式器的用法（参见图 7-1）。图上方的组合框包含所有带数字格式器的 locale，可以在数字、货币和百分率格式器之间进行选择。每次改变选择，文本框中的数字就会被重新格式化。在尝试了几种 locale 后，你就会对有这么多种方式来格式化数字和货币值而感到吃惊。也可以输入不同的数字并点击 Parse 按钮来调用

图 7-1　NumberFormatTest 程序

parse 方法，这个方法会尝试解析你输入的内容。如果解析成功，format 方法就会将结果显示出来。如果解析失败，文本框中会显示"Parse error"消息。

程序清单 7-1 给出了它的代码，非常直观。在构造器中，我们调用 NumberFormat.getAvailableLocales。对每一个 locale，我们调用 getDisplayName，并把返回的结果字符串填入组合框（字符串没有被排序，在 7.4 节中我们将深入研究排序问题）。一旦用户选择了另一个 locale 或点击了单选按钮，就创建一个新的格式器对象并更新文本框。当用户点击 Parse 按钮后，调用 Parse 方法来基于选中的 locale 进行实际的解析操作。

> **注释**：可以使用 Scanner 来读取本地化的整数和浮点数。可以调用 useLocale 方法来设置 locale。

程序清单 7-1　numberFormat/NumberFormatTest.java

```java
 1  package numberFormat;
 2
 3  import java.awt.*;
 4  import java.awt.event.*;
 5  import java.text.*;
 6  import java.util.*;
 7
 8  import javax.swing.*;
 9
10  /**
11   * This program demonstrates formatting numbers under various locales.
12   * @version 1.15 2018-05-01
13   * @author Cay Horstmann
14   */
15  public class NumberFormatTest
16  {
17     public static void main(String[] args)
18     {
19        EventQueue.invokeLater(() ->
20           {
21              var frame = new NumberFormatFrame();
22              frame.setTitle("NumberFormatTest");
```

```java
23                  frame.setDefaultCloseOperation(JFrame.EXIT_ON_CLOSE);
24                  frame.setVisible(true);
25              });
26      }
27  }
28
29  /**
30   * This frame contains radio buttons to select a number format, a combo box to pick a locale,
31   * a text field to display a formatted number, and a button to parse the text field contents.
32   */
33  class NumberFormatFrame extends JFrame
34  {
35      private Locale[] locales;
36      private double currentNumber;
37      private JComboBox<String> localeCombo = new JComboBox<>();
38      private JButton parseButton = new JButton("Parse");
39      private JTextField numberText = new JTextField(30);
40      private JRadioButton numberRadioButton = new JRadioButton("Number");
41      private JRadioButton currencyRadioButton = new JRadioButton("Currency");
42      private JRadioButton percentRadioButton = new JRadioButton("Percent");
43      private ButtonGroup rbGroup = new ButtonGroup();
44      private NumberFormat currentNumberFormat;
45
46      public NumberFormatFrame()
47      {
48          setLayout(new GridBagLayout());
49
50          ActionListener listener = event -> updateDisplay();
51
52          var p = new JPanel();
53          addRadioButton(p, numberRadioButton, rbGroup, listener);
54          addRadioButton(p, currencyRadioButton, rbGroup, listener);
55          addRadioButton(p, percentRadioButton, rbGroup, listener);
56
57          add(new JLabel("Locale:"), new GBC(0, 0).setAnchor(GBC.EAST));
58          add(p, new GBC(1, 1));
59          add(parseButton, new GBC(0, 2).setInsets(2));
60          add(localeCombo, new GBC(1, 0).setAnchor(GBC.WEST));
61          add(numberText, new GBC(1, 2).setFill(GBC.HORIZONTAL));
62          locales = (Locale[]) NumberFormat.getAvailableLocales().clone();
63          Arrays.sort(locales, Comparator.comparing(Locale::getDisplayName));
64          for (Locale loc : locales)
65              localeCombo.addItem(loc.getDisplayName());
66          localeCombo.setSelectedItem(Locale.getDefault().getDisplayName());
67          currentNumber = 123456.78;
68          updateDisplay();
69
70          localeCombo.addActionListener(listener);
71
72          parseButton.addActionListener(event ->
73              {
74                  String s = numberText.getText().trim();
75                  try
76                  {
```

```java
77                    Number n = currentNumberFormat.parse(s);
78                    currentNumber = n.doubleValue();
79                    updateDisplay();
80                 }
81                 catch (ParseException e)
82                 {
83                    numberText.setText(e.getMessage());
84                 }
85           });
86       pack();
87    }
88
89    /**
90     * Adds a radio button to a container.
91     * @param p the container into which to place the button
92     * @param b the button
93     * @param g the button group
94     * @param listener the button listener
95     */
96    public void addRadioButton(Container p, JRadioButton b, ButtonGroup g,
97          ActionListener listener)
98    {
99       b.setSelected(g.getButtonCount() == 0);
100      b.addActionListener(listener);
101      g.add(b);
102      p.add(b);
103   }
104
105   /**
106    * Updates the display and formats the number according to the user settings.
107    */
108   public void updateDisplay()
109   {
110      Locale currentLocale = locales[localeCombo.getSelectedIndex()];
111      currentNumberFormat = null;
112      if (numberRadioButton.isSelected())
113         currentNumberFormat = NumberFormat.getNumberInstance(currentLocale);
114      else if (currencyRadioButton.isSelected())
115         currentNumberFormat = NumberFormat.getCurrencyInstance(currentLocale);
116      else if (percentRadioButton.isSelected())
117         currentNumberFormat = NumberFormat.getPercentInstance(currentLocale);
118      String formatted = currentNumberFormat.format(currentNumber);
119      numberText.setText(formatted);
120   }
121 }
```

API java.text.NumberFormat 1.1

- static Locale[] getAvailableLocales()

  返回一个 Locale 对象的数组，其成员为可用的 NumberFormat 格式器。

- static NumberFormat getNumberInstance()
- static NumberFormat getNumberInstance(Locale l)

- static NumberFormat getCurrencyInstance()
- static NumberFormat getCurrencyInstance(Locale l)
- static NumberFormat getPercentInstance()
- static NumberFormat getPercentInstance(Locale l)

  为当前或给定的 locale 提供处理数字、货币量或百分比的格式器。

- String format(double x)
- String format(long x)

  对给定的浮点数或整数进行格式化并以字符串的形式返回结果。

- Number parse(String s)

  解析给定的字符串并返回数字值，如果输入字符串描述了一个浮点数，返回类型就是 Double，否则返回类型就是 Long。字符串必须以一个数字开头，以空白字符开头是不允许的。数字之后可以跟随其他字符，但它们都将被忽略。解析失败时抛出 ParseException 异常。

- void setParseIntegerOnly(boolean b)
- boolean isParseIntegerOnly()

  设置或获取一个标志，该标志指示这个格式器是否应该只解析整数值。

- void setGroupingUsed(boolean b)
- boolean isGroupingUsed()

  设置或获取一个标志，该标志指示这个格式器是否会添加和识别十进制分隔符（比如，100,000）。

- void setMinimumIntegerDigits(int n)
- int getMinimumIntegerDigits()
- void setMaximumIntegerDigits(int n)
- int getMaximumIntegerDigits()
- void setMinimumFractionDigits(int n)
- int getMinimumFractionDigits()
- void setMaximumFractionDigits(int n)
- int getMaximumFractionDigits()

  设置或获取整数或小数部分所允许的最大或最小位数。

7.2.2 货币

为了格式化货币值，可以使用 NumberFormat.getCurrencyInstance 方法。但是，这个方法的灵活性不好，它返回的是一个只针对一种货币的格式器。假设你为一个美国客户准备了一张货物单，货物单中有些货物的金额是用美元表示的，有些是用欧元表示的，此时，你不能只是使用两种格式器：

```
NumberFormat dollarFormatter = NumberFormat.getCurrencyInstance(Locale.US);
NumberFormat euroFormatter = NumberFormat.getCurrencyInstance(Locale.GERMANY);
```

这是因为，这样一来，你的发票看起来非常奇怪，有些金额的格式像 $100,000，另一些则像 100.000 €（注意，欧元值使用小数点而不是逗号作为分隔符）。

处理这样的情况，应该使用 Currency 类来控制被格式器处理的货币。可以通过将一个货币标识符传给静态的 Currency.getInstance 方法来得到一个 Currency 对象，然后对每一个格式器都调用 setCurrency 方法。下面展示了如何为你的美国客户设置欧元的格式：

```
NumberFormat euroFormatter = NumberFormat.getCurrencyInstance(Locale.US);
euroFormatter.setCurrency(Currency.getInstance("EUR"));
```

货币标识符由 ISO 4217 定义，可参考 https://www.iso.org/iso-4217-currency-codes.html。表 7-3 提供了其中的一部分。

表 7-3 货币标识符

货币值	标识符	货币代号	货币值	标识符	货币代号
U.S. Dollar	USD	840	Chinese Renminbi (Yuan)	CNY	156
Euro	EUR	978	Indian Rupee	INR	356
British Pound	GBP	826	Russian Ruble	RUB	643
Japanese Yen	JPY	392			

API java.util.Currency 1.4

- static Currency getInstance(String currencyCode)
- static Currency getInstance(Locale locale)

  返回与给定的 ISO 4217 货币代号或给定的 locale 中的国家相对应的 Currency 对象。

- String toString()
- String getCurrencyCode()
- String getNumericCode() 7
- String getNumericCodeAsString() 9

  获取该货币的 ISO 4217 代码。

- String getSymbol()
- String getSymbol(Locale locale)

  根据默认或给定的 locale 得到该货币的格式化符号。比如美元的格式化符号可能是 "$" 或 "US$"，具体是哪种形式取决于 locale。

- int getDefaultFractionDigits()

  获取该货币小数点后的默认位数。

- static Set<Currency> getAvailableCurrencies() 7

  获取所有可用的货币。

7.3 日期和时间

当格式化日期和时间时，需要考虑 4 个与 locale 相关的问题：

- 月份和星期应该用本地语言来表示。
- 年、月、日的顺序要符合本地习惯。
- 公历可能不是本地首选的日期表示方法。
- 必须要考虑本地的时区。

java.time 包中的 DateTimeFormatter 类可以处理这些问题。首先挑选表 7-4 中所示的一种格式风格，然后获取一个格式器：

```
FormatStyle style = . . .; // One of FormatStyle.SHORT, FormatStyle.MEDIUM, . . .
DateTimeFormatter dateFormatter = DateTimeFormatter.ofLocalizedDate(style);
DateTimeFormatter timeFormatter = DateTimeFormatter.ofLocalizedTime(style);
DateTimeFormatter dateTimeFormatter = DateTimeFormatter.ofLocalizedDateTime(style);
   // or DateTimeFormatter.ofLocalizedDateTime(style1, style2)
```

表 7-4 日期和时间的格式化风格

风格	日期	时间
SHORT	7/16/69	9:32 AM
MEDIUM	Jul 16, 1969	9:32:00 AM
LONG	July 16, 1969	9:32:00 AM EDT in en-US, 9:32:00 MSZ in de-DE（只用于 ZonedDateTime）
FULL	Wednesday, July 16, 1969	9:32:00 AM EDT in en-US, 9:32 Uhr MSZ in de-DE（只用于 ZonedDateTime）

这些格式器都会使用当前的 locale。为了使用不同的 locale，需要使用 withLocale 方法：

```
DateTimeFormatter dateFormatter =
   DateTimeFormatter.ofLocalizedDate(style).withLocale(locale);
```

现在你可以格式化 LocalDate、LocalDateTime、LocalTime 和 ZonedDateTime 了：

```
ZonedDateTime appointment = . . .;
String formatted = formatter.format(appointment);
```

> 注释：这里我们使用的是 java.time 包中的 DateTimeFormatter。还有一种来自 Java 1.1 的遗留的 java.text.DateFormatter 类，它可以操作 Date 和 Calendar 对象。

可以使用 LocalDate、LocalDateTime、LocalTime 和 ZonedDateTime 的静态 parse 方法之一来解析字符串中的日期和时间：

```
LocalTime time = LocalTime.parse("9:32 AM", formatter);
```

这些方法不适合解析人类的输入，至少不适合解析未做预处理的人类输入。例如，用于美国的短时间格式器可以解析 "9:32 AM"，但是解析不了 "9:32AM" 和 "9:32 am"。

> 警告：日期格式器可以解析不存在的日期，例如 November 31，它会将这种日期调整为给定月份的最后一天。

有时，你需要显示星期和月份的名字，例如在日历应用中。此时可以调用 DayOfWeek 和 Month 枚举的 getDisplayName 方法：

```
for (Month m : Month.values())
    System.out.println(m.getDisplayName(textStyle, locale) + " ");
```

表 7-5 展示了文本风格，其中 STANDALONE 版本用于格式化日期之外的显示。例如，在芬兰语中，一月在日期中是"tammikuuta"，但是单独显示时是"tammikuu"。

表 7-5　java.time.format.TextStyle 枚举

风格	示例
FULL / FULL_STANDALONE	January
SHORT / SHORT_STANDALONE	Jan
NARROW / NARROW_STANDALONE	J

注释：星期的第一天可以是星期六、星期日或星期一，这取决于 locale。你可以像下面这样获取星期的第一天：

```
DayOfWeek first = WeekFields.of(locale).getFirstDayOfWeek();
```

程序清单 7-2 展示了如何在实际中使用 DateFormat 类，用户可以选择一个 locale 并看看日期和时间在世界上的不同地区是如何格式化的。

程序清单 7-2　dateFormat/DateTimeFormatterTest.java

```
 1  package dateFormat;
 2
 3  import java.awt.*;
 4  import java.awt.event.*;
 5  import java.time.*;
 6  import java.time.format.*;
 7  import java.util.*;
 8
 9  import javax.swing.*;
10
11  /**
12   * This program demonstrates formatting dates under various locales.
13   * @version 1.01 2018-05-01
14   * @author Cay Horstmann
15   */
16  public class DateTimeFormatterTest
17  {
18     public static void main(String[] args)
19     {
20        EventQueue.invokeLater(() ->
21           {
22              var frame = new DateTimeFormatterFrame();
23              frame.setTitle("DateFormatTest");
24              frame.setDefaultCloseOperation(JFrame.EXIT_ON_CLOSE);
25              frame.setVisible(true);
26           });
27     }
28  }
29
```

```java
30  /**
31   * This frame contains combo boxes to pick a locale, date and time formats, text fields to
32   * display formatted date and time, buttons to parse the text field contents, and a "lenient"
33   * check box.
34   */
35  class DateTimeFormatterFrame extends JFrame
36  {
37     private Locale[] locales;
38     private LocalDate currentDate;
39     private LocalTime currentTime;
40     private ZonedDateTime currentDateTime;
41     private DateTimeFormatter currentDateFormat;
42     private DateTimeFormatter currentTimeFormat;
43     private DateTimeFormatter currentDateTimeFormat;
44     private JComboBox<String> localeCombo = new JComboBox<>();
45     private JButton dateParseButton = new JButton("Parse");
46     private JButton timeParseButton = new JButton("Parse");
47     private JButton dateTimeParseButton = new JButton("Parse");
48     private JTextField dateText = new JTextField(30);
49     private JTextField timeText = new JTextField(30);
50     private JTextField dateTimeText = new JTextField(30);
51     private EnumCombo<FormatStyle> dateStyleCombo = new EnumCombo<>(FormatStyle.class,
52        "Short", "Medium", "Long", "Full");
53     private EnumCombo<FormatStyle> timeStyleCombo = new EnumCombo<>(FormatStyle.class,
54        "Short", "Medium");
55     private EnumCombo<FormatStyle> dateTimeStyleCombo = new EnumCombo<>(FormatStyle.class,
56        "Short", "Medium", "Long", "Full");
57  
58     public DateTimeFormatterFrame()
59     {
60        setLayout(new GridBagLayout());
61        add(new JLabel("Locale"), new GBC(0, 0).setAnchor(GBC.EAST));
62        add(localeCombo, new GBC(1, 0, 2, 1).setAnchor(GBC.WEST));
63  
64        add(new JLabel("Date"), new GBC(0, 1).setAnchor(GBC.EAST));
65        add(dateStyleCombo, new GBC(1, 1).setAnchor(GBC.WEST));
66        add(dateText, new GBC(2, 1, 2, 1).setFill(GBC.HORIZONTAL));
67        add(dateParseButton, new GBC(4, 1).setAnchor(GBC.WEST));
68  
69        add(new JLabel("Time"), new GBC(0, 2).setAnchor(GBC.EAST));
70        add(timeStyleCombo, new GBC(1, 2).setAnchor(GBC.WEST));
71        add(timeText, new GBC(2, 2, 2, 1).setFill(GBC.HORIZONTAL));
72        add(timeParseButton, new GBC(4, 2).setAnchor(GBC.WEST));
73  
74        add(new JLabel("Date and time"), new GBC(0, 3).setAnchor(GBC.EAST));
75        add(dateTimeStyleCombo, new GBC(1, 3).setAnchor(GBC.WEST));
76        add(dateTimeText, new GBC(2, 3, 2, 1).setFill(GBC.HORIZONTAL));
77        add(dateTimeParseButton, new GBC(4, 3).setAnchor(GBC.WEST));
78  
79        locales = (Locale[]) Locale.getAvailableLocales().clone();
80        Arrays.sort(locales, Comparator.comparing(Locale::getDisplayName));
81        for (Locale loc : locales)
82           localeCombo.addItem(loc.getDisplayName());
83        localeCombo.setSelectedItem(Locale.getDefault().getDisplayName());
```

```java
         currentDate = LocalDate.now();
         currentTime = LocalTime.now();
         currentDateTime = ZonedDateTime.now();
         updateDisplay();

         ActionListener listener = event -> updateDisplay();
         localeCombo.addActionListener(listener);
         dateStyleCombo.addActionListener(listener);
         timeStyleCombo.addActionListener(listener);
         dateTimeStyleCombo.addActionListener(listener);

         addAction(dateParseButton, () ->
            {
               currentDate = LocalDate.parse(dateText.getText().trim(), currentDateFormat);
            });
         addAction(timeParseButton, () ->
            {
               currentTime = LocalTime.parse(timeText.getText().trim(), currentTimeFormat);
            });
         addAction(dateTimeParseButton, () ->
            {
               currentDateTime = ZonedDateTime.parse(
                  dateTimeText.getText().trim(), currentDateTimeFormat);
            });

         pack();
      }

      /**
       * Adds the given action to the button and updates the display upon completion.
       * @param button the button to which to add the action
       * @param action the action to carry out when the button is clicked
       */
      public void addAction(JButton button, Runnable action)
      {
         button.addActionListener(event ->
            {
               try
               {
                  action.run();
                  updateDisplay();
               }
               catch (Exception e)
               {
                  JOptionPane.showMessageDialog(null, e.getMessage());
               }
            });
      }

      /**
       * Updates the display and formats the date according to the user settings.
       */
      public void updateDisplay()
      {
```

```
138            Locale currentLocale = locales[localeCombo.getSelectedIndex()];
139            FormatStyle dateStyle = dateStyleCombo.getValue();
140            currentDateFormat = DateTimeFormatter.ofLocalizedDate(
141               dateStyle).withLocale(currentLocale);
142            dateText.setText(currentDateFormat.format(currentDate));
143            FormatStyle timeStyle = timeStyleCombo.getValue();
144            currentTimeFormat = DateTimeFormatter.ofLocalizedTime(
145               timeStyle).withLocale(currentLocale);
146            timeText.setText(currentTimeFormat.format(currentTime));
147            FormatStyle dateTimeStyle = dateTimeStyleCombo.getValue();
148            currentDateTimeFormat = DateTimeFormatter.ofLocalizedDateTime(
149               dateTimeStyle).withLocale(currentLocale);
150            dateTimeText.setText(currentDateTimeFormat.format(currentDateTime));
151         }
152      }
```

图 7-2 显示了该程序（已安装中文字体）。就像你看到的那样，输出能够正确显示。

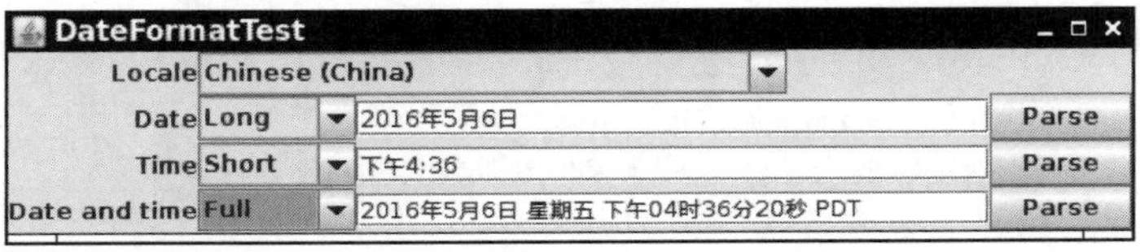

图 7-2　DateFormatTest 程序

也可以对解析进行试验。输入一个日期或时间，或同时输入日期和时间，然后点击 Parse 按钮。

我们使用了辅助类 EnumCombo 来解决一个技术问题（参见程序清单 7-3）。我们想用 Short、Medium 和 Long 等值来填充一个组合框（combo），然后自动将用户的选择转换成整数值 FormatStyle.SHORT、FormatStyle.MEDIUM 和 FormatStyle.LONG。我们并没有编写重复的代码，而是使用了反射：我们将用户的选择转换成大写字母，所有空格都用下划线替换，然后找到使用这个名字的静态域的值。（更多关于反射的内容参见卷 I 第 5 章。）

程序清单 7-3　dateFormat/EnumCombo.java

```
 1  package dateFormat;
 2
 3  import java.util.*;
 4  import javax.swing.*;
 5
 6  /**
 7   * A combo box that lets users choose from among static field
 8   * values whose names are given in the constructor.
 9   * @version 1.15 2016-05-06
10   * @author Cay Horstmann
11   */
12  public class EnumCombo<T> extends JComboBox<String>
13  {
14     private Map<String, T> table = new TreeMap<>();
```

```java
15
16      /**
17       * Constructs an EnumCombo yielding values of type T.
18       * @param cl a class
19       * @param labels an array of strings describing static field names
20       * of cl that have type T
21       */
22      public EnumCombo(Class<?> cl, String... labels)
23      {
24         for (String label : labels)
25         {
26            String name = label.toUpperCase().replace(' ', '_');
27            try
28            {
29               java.lang.reflect.Field f = cl.getField(name);
30               @SuppressWarnings("unchecked") T value = (T) f.get(cl);
31               table.put(label, value);
32            }
33            catch (Exception e)
34            {
35               label = "(" + label + ")";
36               table.put(label, null);
37            }
38            addItem(label);
39         }
40         setSelectedItem(labels[0]);
41      }
42
43      /**
44       * Returns the value of the field that the user selected.
45       * @return the static field value
46       */
47      public T getValue()
48      {
49         return table.get(getSelectedItem());
50      }
51   }
```

API java.time.format.DateTimeFormatter 8

- static DateTimeFormatter ofLocalizedDate(FormatStyle dateStyle)
- static DateTimeFormatter ofLocalizedTime(FormatStyle dateStyle)
- static DateTimeFormatter ofLocalizedDateTime(FormatStyle dateTimeStyle)
- static DateTimeFormatter ofLocalizedDate(FormatStyle dateStyle, FormatStyle timeStyle)

  返回用指定的风格格式化日期、时间或日期和时间的 DateTimeFormatter 实例。

- DateTimeFormatter withLocale(Locale locale)

  返回当前格式器的具有给定 locale 的副本。

- String format(TemporalAccessor temporal)

  返回格式化给定日期/时间所产生的字符串。

API java.time.LocalDate 8
　　java.time.LocalTime 8
　　java.time.LocalDateTime 8
　　java.time.ZonedDateTime 8

- static *Xxx* parse(CharSequence text, DateTimeFormatter formatter)
  解析给定的字符串并返回其中描述的 LocalDate、LocalTime、LocalDateTime 或 ZonedDateTime。如果解析不成功，则抛出 DateTimeParseException 异常。

7.4 排序和规范化

大多数程序员都知道如何使用 String 类中的 compareTo 方法对字符串进行比较。但是，当与人类用户交互时，这个方法就不是很有用了。compareTo 方法使用的是字符串的 UTF-16 编码值，这会导致很荒唐的结果，即使在英文比较中也是如此。比如，下面的 5 个字符串进行排序的结果为：

```
America
Zulu
able
zebra
Ångström
```

按照字典中的顺序，你希望将大写和小写看作是等价的。对于一个说英语的读者来说，期望的排序结果应该是：

```
able
America
Ångström
zebra
Zulu
```

但是，这种顺序对于瑞典用户是不可接受的。在瑞典语中，字母 Å 和字母 A 是不同的，它应该排在字母 Z 之后！就是说，瑞典用户希望排序的结果是：

```
able
America
zebra
Zulu
Ångström
```

为了获得 locale 敏感的比较器，可以调用静态的 Collator.getInstance 方法：

```
Collator coll = Collator.getInstance(locale);
words.sort(coll); // Collator implements Comparator<Object>
```

因为 Collator 类实现了 Comparator 接口，因此，可以传递一个 Collator 对象给 list.sort (Comparator) 方法来对一组字符串进行排序。

排序器有几个高级设置项。你可以设置排序器的强度来选择不同的排序行为。字符间的差别可以被分为首要的（primary）、其次的（secondary）和再次的（tertiary）。比如，在英语中，"A"和"Z"之间的差别被归为首要的，而"A"和"Å"之间的差别是其次的，"A"和"a"之间的差别是再次的。

如果将排序器的强度设置成 Collator.PRIMARY，那么排序器将只关注 primary 级的差别。如果设置成 Collator.SECONDARY，排序器将把 secondary 级的差别也考虑进去。就是说，两个字符串在"secondary"或"tertiary"强度下更容易被区分开来，如表 7-6 所示。

表 7-6　不同强度下的排序（英语 locale）

首要	其次	再次
Angstrom = Ångström	Angstrom ≠ Ångström	Angstrom ≠ Ångström
Able = able	Able = able	Able ≠ able

如果强度被设置为 Collator.IDENTICAL，则不允许有任何差别。这种设置在与排序器的另一种具有相当技术性的设置即分解模式（decomposition mode）联合使用时，显得非常有用。我们接下来将讨论分解模式。

偶尔我们会碰到一个字符或字符序列在被描述成 Unicode 时有多种方式的情况。例如，"Å"可以是 Unicode 字符 U+00C5，或者可以表示成普通的 A（U+0065）后跟°（"上方组合环"，U+030A）。也许让你吃惊的是，字母序列"ffi"可以用代码 U+FB03 描述成单个字符"拉丁小连字 ffi"。(有人会说这是表示方法的不同，不应该因此产生不同的 Unicode 字符，但规则不是我们定的。)

Unicode 标准对字符串定义了四种规范化形式（normalization form）：D、KD、C 和 KC。请查看 http://www.unicode.org/unicode/reports/tr15/tr15-23.html 以了解详细信息。在规范化形式 C 中，重音符号总是组合的。例如，A 和上方组合环°被组合成了单个字符 Å。在规范化形式 D 中，重音字符被分解为基字符和组合重音。例如，Å 就被转换成由字母 A 和上方组合环°构成的序列。规范化形式 KC 和 KD 也会分解字符，例如连字或商标符号。

我们可以选择排序器所使用的规范化程度：Collator.NO_DECOMPOSITION 表示不对字符串做任何规范化，这个选项处理速度较快，但是对于以多种形式表示字符的文本就不适用了；默认值 Collator.CANONICAL_DECOMPOSITION 使用规范化形式 D，这对于包含重音但不包含连字的文本是非常有用的形式；最后是使用规范化形式 KD 的"完全分解"。请参见表 7-7 中的示例。

表 7-7　分解模式之间的差异

不分解	规范分解	完全分解
Å ≠ A°	Å=A°	Å=A°
™ ≠ TM	™ ≠ TM	™=TM

让排序器去多次分解一个字符串是很浪费的。如果一个字符串要和其他字符串进行多次比较，可以将分解的结果保存在一个排序键对象中。getCollationKey 方法返回一个 CollationKey 对象，可以用它来进行更进一步、更快速的比较操作。下面是一个例子：

```
String a = . . .;
CollationKey aKey = coll.getCollationKey(a);
if(aKey.compareTo(coll.getCollationKey(b)) == 0) // fast comparison
    . . .
```

最后，有可能在你不需要进行排序时，也希望将字符串转换成其规范化形式。例如，在将字符串存储到数据库中，或与其他程序进行通信时。java.text.Normalizer 类实现了对规范化的处理。例如：

```
String name = "Ångström";
String normalized = Normalizer.normalize(name, Normalizer.Form.NFD); // uses normalization
                                                                    // form D
```

上面的字符串规范化后包含 10 个字符，其中"Å"和"ö"被替换成了"A°"和"o¨"序列。

但是，这种形式通常并不是用于存储或传输的最佳形式。规范化形式 C 首先进行分解，然后将重音按照标准化的顺序组合在后面。根据 W3C 的标准，这是用于在因特网上进行数据传输的推荐模式。

程序清单 7-4 中的程序让你体验了一下比较排序。你可以向文本框中输入一个词然后点击 Add 按钮把它添加到一个单词列表中。每当添加一个单词，或选择 locale、强度或分解模式时，列表中的单词就会被重新排列。＝号表示这两个词被认为是等同的（参见图 7-3）。

组合框中 locale 名字的显示顺序，是用默认 locale 的排序器进行排序而产生的顺序。如果用美国英语 locale 运行这个程序，即使逗号的 Unicode 值比右括号的 Unicode 值大，"Norwegian (Norway, Nynorsk)"也会显示在"Norwegian (Norway)"的前面。

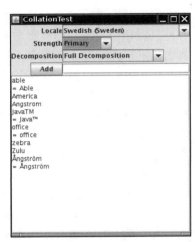

图 7-3 CollationTest 程序

程序清单 7-4 collation/CollationTest.java

```
 1  package collation;
 2
 3  import java.awt.*;
 4  import java.awt.event.*;
 5  import java.text.*;
 6  import java.util.*;
 7  import java.util.List;
 8
 9  import javax.swing.*;
10
11  /**
12   * This program demonstrates collating strings under various locales.
13   * @version 1.16 2018-05-01
14   * @author Cay Horstmann
15   */
```

```java
16   public class CollationTest
17   {
18      public static void main(String[] args)
19      {
20         EventQueue.invokeLater(() ->
21            {
22               var frame = new CollationFrame();
23               frame.setTitle("CollationTest");
24               frame.setDefaultCloseOperation(JFrame.EXIT_ON_CLOSE);
25               frame.setVisible(true);
26            });
27      }
28   }
29
30   /**
31    * This frame contains combo boxes to pick a locale, collation strength and decomposition
32    * rules, a text field and button to add new strings, and a text area to list the collated
33    * strings.
34    */
35   class CollationFrame extends JFrame
36   {
37      private Collator collator = Collator.getInstance(Locale.getDefault());
38      private List<String> strings = new ArrayList<>();
39      private Collator currentCollator;
40      private Locale[] locales;
41      private JComboBox<String> localeCombo = new JComboBox<>();
42      private JTextField newWord = new JTextField(20);
43      private JTextArea sortedWords = new JTextArea(20, 20);
44      private JButton addButton = new JButton("Add");
45      private EnumCombo<Integer> strengthCombo = new EnumCombo<>(Collator.class, "Primary",
46         "Secondary", "Tertiary", "Identical");
47      private EnumCombo<Integer> decompositionCombo = new EnumCombo<>(Collator.class,
48         "Canonical Decomposition", "Full Decomposition", "No Decomposition");
49
50      public CollationFrame()
51      {
52         setLayout(new GridBagLayout());
53         add(new JLabel("Locale"), new GBC(0, 0).setAnchor(GBC.EAST));
54         add(new JLabel("Strength"), new GBC(0, 1).setAnchor(GBC.EAST));
55         add(new JLabel("Decomposition"), new GBC(0, 2).setAnchor(GBC.EAST));
56         add(addButton, new GBC(0, 3).setAnchor(GBC.EAST));
57         add(localeCombo, new GBC(1, 0).setAnchor(GBC.WEST));
58         add(strengthCombo, new GBC(1, 1).setAnchor(GBC.WEST));
59         add(decompositionCombo, new GBC(1, 2).setAnchor(GBC.WEST));
60         add(newWord, new GBC(1, 3).setFill(GBC.HORIZONTAL));
61         add(new JScrollPane(sortedWords), new GBC(0, 4, 2, 1).setFill(GBC.BOTH));
62
63         locales = (Locale[]) Collator.getAvailableLocales().clone();
64         Arrays.sort(locales,
65            (l1, l2) -> collator.compare(l1.getDisplayName(), l2.getDisplayName()));
66         for (Locale loc : locales)
67            localeCombo.addItem(loc.getDisplayName());
68         localeCombo.setSelectedItem(Locale.getDefault().getDisplayName());
69
```

```java
70        strings.add("America");
71        strings.add("able");
72        strings.add("Zulu");
73        strings.add("zebra");
74        strings.add("\u00C5ngstr\u00F6m");
75        strings.add("A\u030angstro\u0308m");
76        strings.add("Angstrom");
77        strings.add("Able");
78        strings.add("office");
79        strings.add("o\uFB03ce");
80        strings.add("Java\u2122");
81        strings.add("JavaTM");
82        updateDisplay();
83
84        addButton.addActionListener(event ->
85           {
86              strings.add(newWord.getText());
87              updateDisplay();
88           });
89
90        ActionListener listener = event -> updateDisplay();
91
92        localeCombo.addActionListener(listener);
93        strengthCombo.addActionListener(listener);
94        decompositionCombo.addActionListener(listener);
95        pack();
96     }
97
98     /**
99      * Updates the display and collates the strings according to the user settings.
100     */
101    public void updateDisplay()
102    {
103       Locale currentLocale = locales[localeCombo.getSelectedIndex()];
104       localeCombo.setLocale(currentLocale);
105
106       currentCollator = Collator.getInstance(currentLocale);
107       currentCollator.setStrength(strengthCombo.getValue());
108       currentCollator.setDecomposition(decompositionCombo.getValue());
109
110       strings.sort(currentCollator);
111
112       sortedWords.setText("");
113       for (int i = 0; i < strings.size(); i++)
114       {
115          String s = strings.get(i);
116          if (i > 0 && currentCollator.compare(s, strings.get(i - 1)) == 0)
117             sortedWords.append("= ");
118          sortedWords.append(s + "\n");
119       }
120       pack();
121    }
122 }
```

API java.text.Collator 1.1

- static Locale[] getAvailableLocales()
  返回 Locale 对象的一个数组，该 Collator 对象可用于这些对象。
- static Collator getInstance()
- static Collator getInstance(Locale l)
  为默认或给定的 locale 返回一个排序器。
- int compare(String a, String b)
  如果 a 在 b 之前，则返回负值；如果它们相等，则返回 0；否则返回正值。
- boolean equals(String a, String b)
  如果 a 和 b 相等，则返回 true，否则返回 false。
- void setStrength(int strength)
- int getStrength()
  设置或获取排序器的强度。更强的排序器可以区分更多的词。强度的值可以是 Collator.PRIMARY、Collator.SECONDARY 和 Collator.TERTIARY。
- void setDecomposition(int decomp)
- int getDecompositon()
  设置或获取排序器的分解模式。分解越细，判断两个字符串是否相等时就越严格。分解的等级值可以是 Collator.NO_DECOMPOSITION、Collator.CANONICAL_DECOMPOSITION 和 Collator.FULL_DECOMPOSITION。
- CollationKey getCollationKey(String a)
  返回一个排序器键，这个键包含一个对一组字符按特定格式分解的结果，可以快速地和其他排序器键进行比较。

API java.text.CollationKey 1.1

- int compareTo(CollationKey b)
  如果这个键在 b 之前，则返回一个负值；如果两者相等，则返回 0，否则返回正值。

API java.text.Normalizer 6

- static String normalize(CharSequence str, Normalizer.Form form)
  返回 str 的规范化形式，form 的值是 ND、NKD、NC 或 NKC 之一。

7.5 消息格式化

Java 类库中有一个用来对包含变量部分的文本进行格式化的 MessageFormat 类，它的格式化方式与用 printf 方法进行格式化很类似，但是它支持 locale，并且可以对数字和日期进行格式化。我们将在以下各节中审视这种机制。

7.5.1 格式化数字和日期

下面是一个典型的消息格式化字符串：

```
"On {2}, a {0} destroyed {1} houses and caused {3} of damage."
```

括号中的数字是占位符，可以用实际的名字和值来替换它们。使用静态方法 MessageFormat.format 可以用实际的值来替换这些占位符。它是一个"varargs"方法，所以可以通过下面的方法提供参数：

```
String msg
    = MessageFormat.format("On {2}, a {0} destroyed {1} houses and caused {3} of damage.",
        "hurricane", 99, new GregorianCalendar(1999, 0, 1).getTime(), 10.0E8);
```

在这个例子中，占位符 {0} 被 "hurricane" 替换，{1} 被 99 替换，等等。

上述例子的结果是下面的字符串：

```
On 1/1/99 12:00 AM, a hurricane destroyed 99 houses and caused 100,000,000 of damage.
```

这只是开始，离完美还有距离。我们不想将时间显示为"12:00 AM"，而且我们想将造成的损失量打印成货币值。通过为占位符提供可选的格式，就可以做到这一点：

```
"On {2,date,long}, a {0} destroyed {1} houses and caused {3,number,currency} of damage."
```

这段示例代码将打印出：

```
On January 1, 1999, a hurricane destroyed 99 houses and caused $100,000,000 of damage.
```

一般来说，占位符索引后面可以跟一个类型（type）和一个风格（style），它们之间用逗号隔开。类型可以是：

```
number
time
date
choice
```

如果类型是 number，那么风格可以是

```
integer
currency
percent
```

或者是数字格式模式，如 $,##0。（关于格式的更多信息，可参见 DecimalFormat 类的文档。）

如果类型是 time 或 date，那么风格可以是

```
short
medium
long
full
```

或者是一个日期格式模式，如 yyyy-MM-dd。（关于格式的更多信息，可参见 SimpleDateFormat 类的文档。）

> ⚠️ **警告**：静态的 MessageFormat.format 方法使用当前的 locale 对值进行格式化。要想用任

意的 locale 进行格式化，还有一些工作要做，因为这个类还没有提供任何可以使用的 "varargs" 方法。你需要把将要格式化的值置于 Object[] 数组中，就像下面这样：

```
var mf = new MessageFormat(pattern, loc);
String msg = mf.format(new Object[] { values });
```

API java.text.MessageFormat 1.1

- MessageFormat(String pattern)
- MessageFormat(String pattern, Locale loc)
  用给定的模式和 locale 构建一个消息格式对象。
- void applyPattern(String pattern)
  给消息格式对象设置特定的模式。
- void setLocale(Locale loc)
- Locale getLocale()
  设置或获取消息中占位符所使用的 locale。这个 locale 仅仅被通过调用 applyPattern 方法所设置的后续模式使用。
- static String format(String pattern, Object... args)
  通过使用 args[i] 作为占位符 {i} 的输入来格式化 pattern 字符串。
- StringBuffer format(Object args, StringBuffer result, FieldPosition pos)
  格式化 MessageFormat 的模式。args 参数必须是一个对象数组。被格式化的字符串会被附加到 result 末尾，并返回 result。如果 pos 等于 new FieldPosition(MessageFormat.Field.ARGUMENT)，就用它的 beginIndex 和 endIndex 属性值来设置替换占位符 {1} 的文本位置。如果不关心位置信息，可以将它设为 null。

API java.text.Format 1.1

- String format(Object obj)
  按照格式器的规则格式化给定的对象，这个方法将调用 format(obj,new StringBuffer(), new FieldPosition(1)).toString()。

7.5.2 选择格式

让我们仔细地看看前面一节所提到的模式：

"On {2}, a {0} destroyed {1} houses and caused {3} of damage."

如果我们用 "earthquake" 来替换代表灾难的占位符 {0}，那么，在英语中，这句话的语法就不正确了。

On January 1, 1999, a earthquake destroyed . . .

这说明，我们真正希望的是将冠词 "a" 集成到占位符中去：

"On {2}, {0} destroyed {1} houses and caused {3} of damage."

这样我们就应该用 "a hurricane" 或 "an earthquake" 来替换 {0}。当消息需要被翻译成某种语言，而该语言中的词会随词性的变化而变化时，这种替换方式特别适用。比如，在德语中，模式可能是：

"{0} zerstörte am {2} {1} Häuser und richtete einen Schaden von {3} an."

这样，占位符将被正确地替换成冠词和名词的组合，比如 "Ein Wirbelsturm" 或 "Eine Naturkatastrophe"。

让我们来看看参数 {1}。如果灾难的后果不严重，{1} 的替换值可能是数字 1，消息就变成：

On January 1, 1999, a mudslide destroyed 1 houses and . . .

我们当然希望消息能够随占位符的值而变化，这样就能根据具体的值形成

```
no houses
one house
2 houses
. . .
```

choice 格式化选项就是为了这个目的而设计的。

一个选择格式是由一个序列对构成的，每一个对包括：

- 一个下限（lower limit）
- 一个格式字符串（format string）

下限和格式字符串由一个 # 符号分隔，对与对之间由符号 | 分隔。

例如，

`{1,choice,0#no houses|1#one house|2#{1} houses}`

表 7-8 显示了格式字符串对 {1} 的不同值产生的作用。

表 7-8 由选择格式进行格式化的字符串

{1}	结果	{1}	结果
0	"no houses"	3	"3 houses"
1	"one house"	-1	"no houses"

为什么在格式字符串中两次用到了 {1}？当消息格式将选择格式应用于占位符 {1} 而且替换值是 2 时，选择格式会返回 "{1} houses"。这个字符串由消息格式再次格式化，并将这次的结果和上一次的叠加。

> **注释**：这个例子说明选择格式的设计者有些糊涂了。如果你有 3 个格式字符串，就需要两个下限来分隔它们。一般来说，你需要的下限数目比格式字符串数目少 1。就像你在表 7-8 中见到的，MessageFormat 类将忽略第一个下限。
> 如果这个类的设计者意识到下限只在两个选择之间出现，那么语法就要清楚得多，比如，
>
> `no houses|1|one house|2|{1} houses // not the actual format`

可以使用 < 符号来表示如果替换值严格小于下限，则选中这个选择项。

也可以使用 ≤（Unicode 中的代码是 \u2264）来实现和 # 相同的效果。如果愿意的话，甚至可以将第一个下限的值定义为 –∞（Unicode 代码是 -\u221E）。

例如，

```
-∞<no houses|0<one house|2≤{1} houses
```

或者使用 Unicode 转义字符，

```
-\u221E<no houses|0<one house|2\u2264{1} houses
```

让我们来结束自然灾害的场景。如果我们将选择字符串放到原始消息字符串中，那么会得到下面的格式化指令：

```
String pattern = "On {2,date,long}, {0} destroyed {1,choice,0#no houses|1#one house|2#{1} houses}" + "and caused {3,number,currency} of damage.";
```

在德语中，即

```
String pattern
    = "{0} zerstörte am {2,date,long} {1,choice,0#kein Haus|1#ein Haus|2#{1} Häuser}"
      + "und richtete einen Schaden von {3,number,currency} an.";
```

请注意，在德语中词的顺序和英语中是不同的，但是你传给 format 方法的对象数组是相同的。可以用格式字符串中占位符的顺序来处理单词顺序的改变。

7.6 文本输入和输出

众所周知，Java 编程语言自身是完全基于 Unicode 的。但是，Windows 和 Mac OS X 仍旧支持遗留的字符编码机制，例如西欧国家的 Windows-1252 和 Mac Roman。因此，与用户通过文本沟通并非看上去那么简单。下面各节将讨论你可能会碰到的各种复杂情况。

7.6.1 文本文件

当今最好是使用 UTF-8 来存储和加载文本文件，但是你可能需要操作遗留文件。如果你知道遗留文件所希望使用的字符编码机制，那么可以在读写文本文件时指定它：

```
var out = new PrintWriter(filename, "Windows-1252");
```

如果想要获得可用的最佳编码机制，可以通过下面的调用来获得"平台的编码机制"：

```
Charset platformEncoding = Charset.defaultCharset();
```

7.6.2 行结束符

这不是 locale 的问题，而是平台的问题。在 Windows 中，文本文件希望在每行末尾使用 \r\n，而基于 UNIX 的系统只需要一个 \n 字符。当今，大多数 Windows 程序都可以处理只有一个 \n 的情况，一个重要的例外是记事本。如果"用户可以在你的应用所产生的文本文件上双击并在记事本中浏览它"对你来说非常重要,那么你就要确保该文本文件使用了正确的行结束符。

任何用 println 方法写入的行都将是被正确终止的。唯一的问题是你是否打印了包含 \n 字符的行。它们不会被自动修改为平台的行结束符。

与在字符串中使用 \n 不同,可以使用 printf 和 %n 格式说明符来产生平台相关的行结束符。例如,

```
out.printf("Hello%nWorld%n");
```

会在 Windows 上产生

```
Hello\r\nWorld\r\n
```

而在其他所有平台上产生

```
Hello\nWorld\n
```

7.6.3 控制台

如果你编写的程序是通过 System.in/System.out 或 System.console() 与用户交互的,那么就不得不面对控制台使用的字符编码机制与 Charset.defaultCharset() 报告的平台编码机制有差异的可能性。当使用 Windows 上的 cmd 工具时,这个问题尤其需要注意。在美国版本的 Windows 10 中,命令行 Shell 使用的是陈旧的 IBM437 编码机制,它源自 1982 年 IBM 的个人计算机。没有任何官方的 API 可以揭示该信息。Charset.defaultCharset() 方法将返回 Windows-1252 字符集,它与 IBM437 完全不同。例如,在 Windows-1252 中有欧元符号€,但是在 IBM437 中没有。如果调用

```
System.out.println("100 €");
```

控制台会显示

```
100 ?
```

你可以建议用户切换控制台的字符编码机制。在 Windows 中,这可以通过 chcp 命令实现。例如:

```
chcp 1252
```

会将控制台变换为 Windows-1252 编码页。

当然,理想情况下你的用户应该将控制台切换到 UTF-8。在 Windows 中,该命令为

```
chcp 65001
```

遗憾的是,这种命令还不足以让 Java 在控制台中使用 UTF-8,我们还必须使用非官方的 file.encoding 系统属性来设置平台的编码机制:

```
java -Dfile.encoding=UTF-8 MyProg
```

7.6.4 日志文件

当来自 java.util.logging 库的日志消息被发送到控制台时,它们会用控制台的编码机制来

书写。在上一节中你看到了如何进行控制。但是，文件中的日志消息会使用 FileHandler 来处理，它在默认情况下使用平台的编码机制。

要想将编码机制修改为 UTF-8，需要修改日志管理器的设置。具体做法是在日志配置文件中做如下设置：

```
java.util.logging.FileHandler.encoding=UTF-8
```

7.6.5 UTF-8 字节顺序标志

正如我们已经提到的，尽可能地让文本文件使用 UTF-8 是一个好的做法。如果你的应用必须读取其他程序创建的 UTF-8 文本文件，那么你可能会碰到另一个问题。在文件中添加一个"字节顺序标志"字符 U+FEFF 作为文件的第一个字符，是一种完全合法的做法。在 UTF-16 编码机制中，每个码元都是一个两字节的数字，字节顺序标志可以告诉读入器该文件使用的是"高字节在前"还是"低字节在前"的字节顺序。UTF-8 是一种单字节编码机制，因此不需要指定字节的顺序。但是如果一个文件以字节 0xEF 0xBB 0xBF（U+FEFF 的 UTF-8 编码）开头，那么这就是一个强烈暗示，表示该文件使用了 UTF-8。正是这个原因，Unicode 标准鼓励这种实践方式。任何读入器都被认为会丢弃最前面的字节顺序标志。

还有一个美中不足的瑕疵。Oracle 的 Java 实现很固执地因潜在的兼容性问题而拒绝遵循 Unicode 标准。作为程序员，这对你而言意味着必须去执行平台并不会执行的操作。在读入文本文件时，如果开头碰到了 U+FEFF，那就需要忽略它。

> ⚠️ **警告**：遗憾的是，JDK 的实现没有遵循这项建议。在向 javac 编译器传递有效的以字节顺序标志开头的 UTF-8 源文件时，编译会以产生错误消息"illegal character: \65279"而失败。

7.6.6 源文件的字符编码

作为程序员，要牢记你需要与 Java 编译器交互，这种交互需要通过本地系统的工具来完成。例如，可以使用中文版的记事本来写你的 Java 源代码文件。但这样写出来的源码不是随处可用的，因为它们使用的是本地的字符编码。只有编译后的 class 文件才能随处使用，因为它们会自动地使用"modified UTF-8"编码来处理标识符和字符串。这意味着即使在程序编译和运行时，也涉及 3 种字符编码：

- 源文件：平台编码
- 类文件：modified UTF-8
- 虚拟机：UTF-16

关于 modified UTF-8 和 UTF-16 格式的定义，参见第 1 章。

> ✅ **提示**：可以用 -encoding 标记来设定源文件的字符编码，例如：
> ```
> javac -encoding UTF-8 Myfile.java
> ```

7.7 资源包

当本地化一个应用时，可能会有大量的消息字符串、按钮标签和其他的东西需要被翻译。为了能灵活地完成这项任务，你肯定希望在外部定义消息字符串，这些消息字符串通常被称为资源（resource）。这样，翻译人员不需要接触程序源代码就可以很容易地编辑资源文件。

在 Java 中，要使用属性文件来设定字符串资源，并为其他类型的资源实现相应的类。

> **注释**：Java 技术资源与 Windows 或 Macintosh 资源不同。Macintosh 或 Windows 可执行文件在程序代码以外的地方存储类似菜单、对话框、图标和消息这样的资源。资源编辑器能够在不影响程序代码的情况下检查并更新这些资源。

> **注释**：卷 I 第 5 章描述了 JAR 文件资源的概念，以及为何数据文件、声音和图片可以存放在 JAR 文件中。Class 类的 getResource 方法可以找到相应的文件，打开它并返回资源的 URL。通过将文件放到 JAR 文件中，将查找这些资源文件的工作留给了类的加载器去处理，加载器知道如何定位 JAR 文件中的项。但是，这种机制不支持 locale。

7.7.1 定位资源包

当本地化一个应用时，会产生很多资源包（resource bundle）。每一个包都是一个属性文件或者是一个描述了与 locale 相关的项的类（比如消息、标签等）。对于每一个包，都要为所有你想要支持的 locale 提供相应的版本。

需要对这些包使用一种统一的命名规则。例如，为德国定义的资源放在一个名为"*baseName_de_DE*"的文件中，而所有说德语的国家所共享的资源则放在名为"*baseName_de*"的文件中。一般来说，使用

baseName_language_country

来命名所有和国家相关的资源，使用

baseName_language

来命名所有和语言相关的资源。最后，作为后备，可以把默认资源放到一个没有后缀的文件中。

可以用下面的命令加载一个包：

```
ResourceBundle currentResources = ResourceBundle.getBundle(baseName, currentLocale);
```

getBundle 方法试图加载匹配当前 locale 定义的语言和国家的包。如果失败，通过依次放弃国家和语言来继续进行查找，然后同样的查找被应用于默认的 locale，最后，如果还不行的话就去查看默认的包文件，如果这也失败了，则抛出一个 MissingResourceException 异常。

这就是说，getBundle 方法会试图加载以下包：

```
baseName_currentLocaleLanguage_currentLocaleCountry
baseName_currentLocaleLanguage
baseName_currentLocaleLanguage_defaultLocaleCountry
baseName_defaultLocaleLanguage
baseName
```

一旦 getBundle 方法定位了一个包，比如，*baseName_de_DE*，它还会继续查找 *baseName_de* 和 *baseName* 这两个包。如果这些包也存在，它们在资源层次中就成为 *baseName_de_DE* 的父包。以后，当查找一个资源时，如果在当前包中没有找到，就去查找其父包。就是说，如果一个特定的资源在当前包中没有找到，比如，某个特定资源在 *baseName_de_DE* 中没有找到，那么就会去查找 *baseName_de* 和 *baseName*。

这是一项非常有用的服务，如果手工来编写将会非常麻烦。Java 编程语言的资源包机制会自动定位与给定的 locale 匹配得最好的项。可以很容易地把越来越多的本地化信息加到已有的程序中：你需要做的只是增加额外的资源包。

> 📝 **注释**：我们简化了对资源包查找的讨论。如果 locale 中包含脚本或变体，那么查找就会复杂得多。可以查看 ResourceBundle.Control.getCandidateLocales 方法的文档以了解其细节。

> ✅ **提示**：不需要把你的程序的所有资源都放到同一个包中。可以用一个包来存放按钮标签，用另一个包存放错误消息等。

7.7.2 属性文件

对字符串进行国际化是很直接的，可以把所有字符串放到一个属性文件中，比如 MyProgramStrings.properties，这是一个每行存放一个键-值对的文本文件。典型的属性文件看起来像下面这样：

```
computeButton=Rechnen
colorName=black
defaultPaperSize=210×297
```

然后像上一节描述的那样命名属性文件，例如，

```
MyProgramStrings.properties
MyProgramStrings_en.properties
MyProgramStrings_de_DE.properties
```

可以加载包，例如：

```
ResourceBundle bundle = ResourceBundle.getBundle("MyProgramStrings", locale);
```

要查找一个具体的字符串，可以调用

```
String computeButtonLabel = bundle.getString("computeButton");
```

> ⚠️ **警告**：在 Java 9 之前，存储属性的文件都是 ASCII 文件。如果你使用的是旧版本的 Java，并且需要将 Unicode 字符放到属性文件中，那么请用 \uxxxx 编码方式对它们进

行编码。比如，要设定 "colorName=Grün"，可以使用

```
colorName=Gr\u00FCn
```

你可以使用 native2ascii 工具来产生这些文件。

7.7.3 包类

为了提供字符串以外的资源，需要定义类，它必须扩展自 ResourceBundle 类。应该使用标准的命名规则来命名你的类，比如

```
MyProgramResources.java
MyProgramResources_en.java
MyProgramResources_de_DE.java
```

可以使用与加载属性文件相同的 getBundle 方法来加载这个类：

```
ResourceBundle bundle = ResourceBundle.getBundle("MyProgramResources", locale);
```

⚠️ **警告**：当搜索包时，如果类中的包和属性文件中的包都存在匹配，则优先选择类中的包。

每一个资源包类都实现了一个查询表。你需要为每一个你想定位的设置提供一个关键字字符串，使用这个字符串来提取相应的设置。例如，

```
var backgroundColor = (Color) bundle.getObject("backgroundColor");
double[] paperSize = (double[]) bundle.getObject("defaultPaperSize");
```

实现资源包类的最简单方法就是继承 ListResourceBundle 类。ListResourceBundle 类让你把所有资源都放到一个对象数组中并提供查找功能。要遵循以下的代码框架：

```
public class baseName_language_country extends ListResourceBundle
{
    private static final Object[][] contents =
    {
        { key1, value1 },
        { key2, value2 },
        ...
    }
    public Object[][] getContents() { return contents; }
}
```

例如，

```
public class ProgramResources_de extends ListResourceBundle
{
    private static final Object[][] contents =
    {
        { "backgroundColor", Color.black },
        { "defaultPaperSize", new double[] { 210, 297 } }
    }
    public Object[][] getContents() { return contents; }
}
```

```java
public class ProgramResources_en_US extends ListResourceBundle
{
    private static final Object[][] contents =
    {
        { "backgroundColor", Color.blue },
        { "defaultPaperSize", new double[] { 216, 279 } }
    }
    public Object[][] getContents() { return contents; }
}
```

> **注释**：纸的尺寸是以毫米为单位给出的。在世界上，除了加拿大和美国，其他地区都使用 ISO 216 规格的纸。更多信息见 http://www.cl.cam.ac.uk/~mgk25/iso-paper.html。

或者，你的资源包类可以继承 ResourceBundle 类。然后需要实现两个方法，一是枚举所有键，二是用给定的键查找相应的值：

```
Enumeration<String> getKeys()
Object handleGetObject(String key)
```

ResourceBundle 类的 getObject 方法会调用你提供的 handleGetObject 方法。

API java.util.ResourceBundle 1.1

- static ResourceBundle getBundle(String baseName, Locale loc)
- static ResourceBundle getBundle(String baseName)

  在给定或默认的 locale 下以给定的名字加载资源包类和它的父类。如果资源包类位于一个 Java 包中，那么类的名字必须包含完整的包名，例如 "intl.ProgramResources"。资源包类必须是 public 的，这样 getBundle 方法才能访问它们。

- Object getObject(String name)

  从资源包或它的父包中查找一个对象。

- String getString(String name)

  从资源包或它的父包中查找一个对象并把它转型成字符串。

- String[] getStringArray(String name)

  从资源包或它的父包中查找一个对象并把它转型成字符串数组。

- Enumeration<String> getKeys()

  返回一个枚举对象，枚举出资源包中的所有键，也包括父包中的键。

- Object handleGetObject(String key)

  如果你要定义自己的资源查找机制，那么这个方法就需要被覆写，用来查找与给定的键相关联的资源的值。

7.8 一个完整的例子

在这一节中，我们使用本章中的内容来对退休金计算器小程序进行本地化，这个小程序可以计算你是否为退休存够了钱。你需要输入年龄，每个月存多少钱等信息（参见图 7-4）。

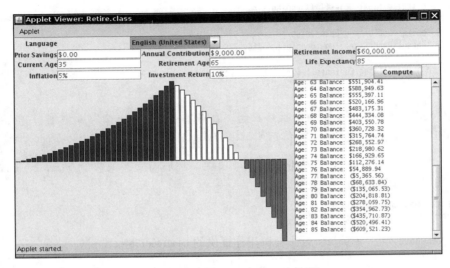

图 7-4　使用英语的退休金计算器

文本域和图表显示每年退休金账户中的余额。如果你后半生的退休金余额变成负数，并且表中的数据条在 x 轴以下，你就需要做些什么了。例如，存更多的钱、推迟退休等。

这个退休金计算器可以在三种 locale（英语、德语和中文）下工作。下面是进行国际化时的一些要点：

- 标签、按钮和消息被翻译成德语和中文。你可以在 RetireResources_de 和 RetireResources_zh 中找到它们。英语作为后备，见 RetireResources 文件。
- 当 locale 改变时，我们重置标签并格式化文本域中的内容。
- 文本域以本地格式处理数字、货币值和百分数。
- 计算域使用了 MessageFormat。格式字符串被存储在每种语言的资源包中。
- 为了使展示的确可行，我们按照用户选择的语言为条形图使用不同的颜色。

程序清单 7-5 到程序清单 7-8 展示了代码，而程序清单 7-9 到程序清单 7-11 是本地化的字符串的属性文件。图 7-5 和图 7-6 分别显示了在德语和中文下的输出。为了显示中文，请确认你已经在 Java 运行环境中安装并配置了中文字体。否则，所有的中文字符将会显示"missing character"图标。

程序清单 7-5　retire/Retire.java

```
1  package retire;
2
3  import java.awt.*;
4  import java.awt.geom.*;
5  import java.text.*;
```

```java
 6 import java.util.*;
 7
 8 import javax.swing.*;
 9
10 /**
11  * This program shows a retirement calculator. The UI is displayed in English, German, and
12  * Chinese.
13  * @version 1.25 2018-05-01
14  * @author Cay Horstmann
15  */
16 public class Retire
17 {
18    public static void main(String[] args)
19    {
20       EventQueue.invokeLater(() ->
21          {
22             var frame = new RetireFrame();
23             frame.setDefaultCloseOperation(JFrame.EXIT_ON_CLOSE);
24             frame.setVisible(true);
25          });
26    }
27 }
28
29 class RetireFrame extends JFrame
30 {
31    private JTextField savingsField = new JTextField(10);
32    private JTextField contribField = new JTextField(10);
33    private JTextField incomeField = new JTextField(10);
34    private JTextField currentAgeField = new JTextField(4);
35    private JTextField retireAgeField = new JTextField(4);
36    private JTextField deathAgeField = new JTextField(4);
37    private JTextField inflationPercentField = new JTextField(6);
38    private JTextField investPercentField = new JTextField(6);
39    private JTextArea retireText = new JTextArea(10, 25);
40    private RetireComponent retireCanvas = new RetireComponent();
41    private JButton computeButton = new JButton();
42    private JLabel languageLabel = new JLabel();
43    private JLabel savingsLabel = new JLabel();
44    private JLabel contribLabel = new JLabel();
45    private JLabel incomeLabel = new JLabel();
46    private JLabel currentAgeLabel = new JLabel();
47    private JLabel retireAgeLabel = new JLabel();
48    private JLabel deathAgeLabel = new JLabel();
49    private JLabel inflationPercentLabel = new JLabel();
50    private JLabel investPercentLabel = new JLabel();
51    private RetireInfo info = new RetireInfo();
52    private Locale[] locales = { Locale.US, Locale.CHINA, Locale.GERMANY };
53    private Locale currentLocale;
54    private JComboBox<Locale> localeCombo = new LocaleCombo(locales);
55    private ResourceBundle res;
56    private ResourceBundle resStrings;
57    private NumberFormat currencyFmt;
58    private NumberFormat numberFmt;
59    private NumberFormat percentFmt;
```

```java
60   public RetireFrame()
61   {
62      setLayout(new GridBagLayout());
63      add(languageLabel, new GBC(0, 0).setAnchor(GBC.EAST));
64      add(savingsLabel, new GBC(0, 1).setAnchor(GBC.EAST));
65      add(contribLabel, new GBC(2, 1).setAnchor(GBC.EAST));
66      add(incomeLabel, new GBC(4, 1).setAnchor(GBC.EAST));
67      add(currentAgeLabel, new GBC(0, 2).setAnchor(GBC.EAST));
68      add(retireAgeLabel, new GBC(2, 2).setAnchor(GBC.EAST));
69      add(deathAgeLabel, new GBC(4, 2).setAnchor(GBC.EAST));
70      add(inflationPercentLabel, new GBC(0, 3).setAnchor(GBC.EAST));
71      add(investPercentLabel, new GBC(2, 3).setAnchor(GBC.EAST));
72      add(localeCombo, new GBC(1, 0, 3, 1));
73      add(savingsField, new GBC(1, 1).setWeight(100, 0).setFill(GBC.HORIZONTAL));
74      add(contribField, new GBC(3, 1).setWeight(100, 0).setFill(GBC.HORIZONTAL));
75      add(incomeField, new GBC(5, 1).setWeight(100, 0).setFill(GBC.HORIZONTAL));
76      add(currentAgeField, new GBC(1, 2).setWeight(100, 0).setFill(GBC.HORIZONTAL));
77      add(retireAgeField, new GBC(3, 2).setWeight(100, 0).setFill(GBC.HORIZONTAL));
78      add(deathAgeField, new GBC(5, 2).setWeight(100, 0).setFill(GBC.HORIZONTAL));
79      add(inflationPercentField, new GBC(1, 3).setWeight(100, 0).setFill(GBC.HORIZONTAL));
80      add(investPercentField, new GBC(3, 3).setWeight(100, 0).setFill(GBC.HORIZONTAL));
81      add(retireCanvas, new GBC(0, 4, 4, 1).setWeight(100, 100).setFill(GBC.BOTH));
82      add(new JScrollPane(retireText),
83         new GBC(4, 4, 2, 1).setWeight(0, 100).setFill(GBC.BOTH));
84
85      computeButton.setName("computeButton");
86      computeButton.addActionListener(event ->
87         {
88            getInfo();
89            updateData();
90            updateGraph();
91         });
92      add(computeButton, new GBC(5, 3));
93
94      retireText.setEditable(false);
95      retireText.setFont(new Font("Monospaced", Font.PLAIN, 10));
96
97      info.setSavings(0);
98      info.setContrib(9000);
99      info.setIncome(60000);
100     info.setCurrentAge(35);
101     info.setRetireAge(65);
102     info.setDeathAge(85);
103     info.setInvestPercent(0.1);
104     info.setInflationPercent(0.05);
105
106     int localeIndex = 0; // US locale is default selection
107     for (int i = 0; i < locales.length; i++)
108        // if current locale one of the choices, select it
109        if (getLocale().equals(locales[i])) localeIndex = i;
110     setCurrentLocale(locales[localeIndex]);
111
112     localeCombo.addActionListener(event ->
113        {
```

```java
115                 setCurrentLocale((Locale) localeCombo.getSelectedItem());
116                 validate();
117             });
118         pack();
119     }
120
121     /**
122      * Sets the current locale.
123      * @param locale the desired locale
124      */
125     public void setCurrentLocale(Locale locale)
126     {
127         currentLocale = locale;
128         localeCombo.setLocale(currentLocale);
129         localeCombo.setSelectedItem(currentLocale);
130
131         res = ResourceBundle.getBundle("retire.RetireResources", currentLocale);
132         resStrings = ResourceBundle.getBundle("retire.RetireStrings", currentLocale);
133         currencyFmt = NumberFormat.getCurrencyInstance(currentLocale);
134         numberFmt = NumberFormat.getNumberInstance(currentLocale);
135         percentFmt = NumberFormat.getPercentInstance(currentLocale);
136
137         updateDisplay();
138         updateInfo();
139         updateData();
140         updateGraph();
141     }
142
143     /**
144      * Updates all labels in the display.
145      */
146     public void updateDisplay()
147     {
148         languageLabel.setText(resStrings.getString("language"));
149         savingsLabel.setText(resStrings.getString("savings"));
150         contribLabel.setText(resStrings.getString("contrib"));
151         incomeLabel.setText(resStrings.getString("income"));
152         currentAgeLabel.setText(resStrings.getString("currentAge"));
153         retireAgeLabel.setText(resStrings.getString("retireAge"));
154         deathAgeLabel.setText(resStrings.getString("deathAge"));
155         inflationPercentLabel.setText(resStrings.getString("inflationPercent"));
156         investPercentLabel.setText(resStrings.getString("investPercent"));
157         computeButton.setText(resStrings.getString("computeButton"));
158     }
159
160     /**
161      * Updates the information in the text fields.
162      */
163     public void updateInfo()
164     {
165         savingsField.setText(currencyFmt.format(info.getSavings()));
166         contribField.setText(currencyFmt.format(info.getContrib()));
167         incomeField.setText(currencyFmt.format(info.getIncome()));
168         currentAgeField.setText(numberFmt.format(info.getCurrentAge()));
```

```java
169        retireAgeField.setText(numberFmt.format(info.getRetireAge()));
170        deathAgeField.setText(numberFmt.format(info.getDeathAge()));
171        investPercentField.setText(percentFmt.format(info.getInvestPercent()));
172        inflationPercentField.setText(percentFmt.format(info.getInflationPercent()));
173     }
174
175     /**
176      * Updates the data displayed in the text area.
177      */
178     public void updateData()
179     {
180        retireText.setText("");
181        var retireMsg = new MessageFormat("");
182        retireMsg.setLocale(currentLocale);
183        retireMsg.applyPattern(resStrings.getString("retire"));
184
185        for (int i = info.getCurrentAge(); i <= info.getDeathAge(); i++)
186        {
187           Object[] args = { i, info.getBalance(i) };
188           retireText.append(retireMsg.format(args) + "\n");
189        }
190     }
191
192     /**
193      * Updates the graph.
194      */
195     public void updateGraph()
196     {
197        retireCanvas.setColorPre((Color) res.getObject("colorPre"));
198        retireCanvas.setColorGain((Color) res.getObject("colorGain"));
199        retireCanvas.setColorLoss((Color) res.getObject("colorLoss"));
200        retireCanvas.setInfo(info);
201        repaint();
202     }
203
204     /**
205      * Reads the user input from the text fields.
206      */
207     public void getInfo()
208     {
209        try
210        {
211           info.setSavings(currencyFmt.parse(savingsField.getText()).doubleValue());
212           info.setContrib(currencyFmt.parse(contribField.getText()).doubleValue());
213           info.setIncome(currencyFmt.parse(incomeField.getText()).doubleValue());
214           info.setCurrentAge(numberFmt.parse(currentAgeField.getText()).intValue());
215           info.setRetireAge(numberFmt.parse(retireAgeField.getText()).intValue());
216           info.setDeathAge(numberFmt.parse(deathAgeField.getText()).intValue());
217           info.setInvestPercent(percentFmt.parse(investPercentField.getText()).doubleValue());
218           info.setInflationPercent(
219              percentFmt.parse(inflationPercentField.getText()).doubleValue());
220        }
221        catch (ParseException ex)
222        {
```

```
223            ex.printStackTrace();
224         }
225      }
226 }
227
228 /**
229  * The information required to compute retirement income data.
230  */
231 class RetireInfo
232 {
233    private double savings;
234    private double contrib;
235    private double income;
236    private int currentAge;
237    private int retireAge;
238    private int deathAge;
239    private double inflationPercent;
240    private double investPercent;
241    private int age;
242    private double balance;
243
244    /**
245     * Gets the available balance for a given year.
246     * @param year the year for which to compute the balance
247     * @return the amount of money available (or required) in that year
248     */
249    public double getBalance(int year)
250    {
251       if (year < currentAge) return 0;
252       else if (year == currentAge)
253       {
254          age = year;
255          balance = savings;
256          return balance;
257       }
258       else if (year == age) return balance;
259       if (year != age + 1) getBalance(year - 1);
260       age = year;
261       if (age < retireAge) balance += contrib;
262       else balance -= income;
263       balance = balance * (1 + (investPercent - inflationPercent));
264       return balance;
265    }
266
267    /**
268     * Gets the amount of prior savings.
269     * @return the savings amount
270     */
271    public double getSavings()
272    {
273       return savings;
274    }
275
276    /**
```

```java
277      * Sets the amount of prior savings.
278      * @param newValue the savings amount
279      */
280     public void setSavings(double newValue)
281     {
282        savings = newValue;
283     }
284
285     /**
286      * Gets the annual contribution to the retirement account.
287      * @return the contribution amount
288      */
289     public double getContrib()
290     {
291        return contrib;
292     }
293
294     /**
295      * Sets the annual contribution to the retirement account.
296      * @param newValue the contribution amount
297      */
298     public void setContrib(double newValue)
299     {
300        contrib = newValue;
301     }
302
303     /**
304      * Gets the annual income.
305      * @return the income amount
306      */
307     public double getIncome()
308     {
309        return income;
310     }
311
312     /**
313      * Sets the annual income.
314      * @param newValue the income amount
315      */
316     public void setIncome(double newValue)
317     {
318        income = newValue;
319     }
320
321     /**
322      * Gets the current age.
323      * @return the age
324      */
325     public int getCurrentAge()
326     {
327        return currentAge;
328     }
329
330     /**
```

```java
331       * Sets the current age.
332       * @param newValue the age
333       */
334      public void setCurrentAge(int newValue)
335      {
336         currentAge = newValue;
337      }
338
339      /**
340       * Gets the desired retirement age.
341       * @return the age
342       */
343      public int getRetireAge()
344      {
345         return retireAge;
346      }
347
348      /**
349       * Sets the desired retirement age.
350       * @param newValue the age
351       */
352      public void setRetireAge(int newValue)
353      {
354         retireAge = newValue;
355      }
356
357      /**
358       * Gets the expected age of death.
359       * @return the age
360       */
361      public int getDeathAge()
362      {
363         return deathAge;
364      }
365
366      /**
367       * Sets the expected age of death.
368       * @param newValue the age
369       */
370      public void setDeathAge(int newValue)
371      {
372         deathAge = newValue;
373      }
374
375      /**
376       * Gets the estimated percentage of inflation.
377       * @return the percentage
378       */
379      public double getInflationPercent()
380      {
381         return inflationPercent;
382      }
383
384      /**
385       * Sets the estimated percentage of inflation.
```

```java
386      * @param newValue the percentage
387      */
388     public void setInflationPercent(double newValue)
389     {
390        inflationPercent = newValue;
391     }
392
393     /**
394      * Gets the estimated yield of the investment.
395      * @return the percentage
396      */
397     public double getInvestPercent()
398     {
399        return investPercent;
400     }
401
402     /**
403      * Sets the estimated yield of the investment.
404      * @param newValue the percentage
405      */
406     public void setInvestPercent(double newValue)
407     {
408        investPercent = newValue;
409     }
410  }
411
412  /**
413   * This component draws a graph of the investment result.
414   */
415  class RetireComponent extends JComponent
416  {
417     private static final int PANEL_WIDTH = 400;
418     private static final int PANEL_HEIGHT = 200;
419     private static final Dimension PREFERRED_SIZE = new Dimension(800, 600);
420     private RetireInfo info = null;
421     private Color colorPre;
422     private Color colorGain;
423     private Color colorLoss;
424
425     public RetireComponent()
426     {
427        setSize(PANEL_WIDTH, PANEL_HEIGHT);
428     }
429
430     /**
431      * Sets the retirement information to be plotted.
432      * @param newInfo the new retirement info.
433      */
434     public void setInfo(RetireInfo newInfo)
435     {
436        info = newInfo;
437        repaint();
438     }
439
440     public void paintComponent(Graphics g)
```

```java
   {
      var g2 = (Graphics2D) g;
      if (info == null) return;

      double minValue = 0;
      double maxValue = 0;
      int i;
      for (i = info.getCurrentAge(); i <= info.getDeathAge(); i++)
      {
         double v = info.getBalance(i);
         if (minValue > v) minValue = v;
         if (maxValue < v) maxValue = v;
      }
      if (maxValue == minValue) return;

      int barWidth = getWidth() / (info.getDeathAge() - info.getCurrentAge() + 1);
      double scale = getHeight() / (maxValue - minValue);

      for (i = info.getCurrentAge(); i <= info.getDeathAge(); i++)
      {
         int x1 = (i - info.getCurrentAge()) * barWidth + 1;
         int y1;
         double v = info.getBalance(i);
         int height;
         int yOrigin = (int) (maxValue * scale);

         if (v >= 0)
         {
            y1 = (int) ((maxValue - v) * scale);
            height = yOrigin - y1;
         }
         else
         {
            y1 = yOrigin;
            height = (int) (-v * scale);
         }

         if (i < info.getRetireAge()) g2.setPaint(colorPre);
         else if (v >= 0) g2.setPaint(colorGain);
         else g2.setPaint(colorLoss);
         var bar = new Rectangle2D.Double(x1, y1, barWidth - 2, height);
         g2.fill(bar);
         g2.setPaint(Color.black);
         g2.draw(bar);
      }
   }

   /**
    * Sets the color to be used before retirement.
    * @param color the desired color
    */
   public void setColorPre(Color color)
   {
      colorPre = color;
      repaint();
```

```
496      }
497
498      /**
499       * Sets the color to be used after retirement while the account balance is positive.
500       * @param color the desired color
501       */
502      public void setColorGain(Color color)
503      {
504         colorGain = color;
505         repaint();
506      }
507
508      /**
509       * Sets the color to be used after retirement when the account balance is negative.
510       * @param color the desired color
511       */
512      public void setColorLoss(Color color)
513      {
514         colorLoss = color;
515         repaint();
516      }
517
518      public Dimension getPreferredSize() { return PREFERRED_SIZE; }
519   }
```

程序清单 7-6   retire/RetireResources.java

```
1  package retire;
2
3  import java.awt.*;
4
5  /**
6   * These are the English non-string resources for the retirement calculator.
7   * @version 1.21 2001-08-27
8   * @author Cay Horstmann
9   */
10 public class RetireResources extends java.util.ListResourceBundle
11 {
12    private static final Object[][] contents = {
13    // BEGIN LOCALIZE
14       { "colorPre", Color.blue }, { "colorGain", Color.white }, { "colorLoss", Color.red }
15    // END LOCALIZE
16    };
17
18    public Object[][] getContents()
19    {
20       return contents;
21    }
22 }
```

程序清单 7-7   retire/RetireResources_de.java

```
1  package retire;
2
```

```
 3  import java.awt.*;
 4
 5  /**
 6   * These are the German non-string resources for the retirement calculator.
 7   * @version 1.21 2001-08-27
 8   * @author Cay Horstmann
 9   */
10  public class RetireResources_de extends java.util.ListResourceBundle
11  {
12     private static final Object[][] contents = {
13     // BEGIN LOCALIZE
14         { "colorPre", Color.yellow }, { "colorGain", Color.black }, { "colorLoss", Color.red }
15     // END LOCALIZE
16     };
17
18     public Object[][] getContents()
19     {
20        return contents;
21     }
22  }
```

程序清单 7-8　retire/RetireResources_zh.java

```
 1  package retire;
 2
 3  import java.awt.*;
 4
 5  /**
 6   * These are the Chinese non-string resources for the retirement calculator.
 7   * @version 1.21 2001-08-27
 8   * @author Cay Horstmann
 9   */
10  public class RetireResources_zh extends java.util.ListResourceBundle
11  {
12     private static final Object[][] contents = {
13     // BEGIN LOCALIZE
14         { "colorPre", Color.red }, { "colorGain", Color.blue }, { "colorLoss", Color.yellow }
15     // END LOCALIZE
16     };
17
18     public Object[][] getContents()
19     {
20        return contents;
21     }
22  }
```

程序清单 7-9　retire/RetireStrings.properties

```
1  language=Language
2  computeButton=Compute
3  savings=Prior Savings
4  contrib=Annual Contribution
5  income=Retirement Income
6  currentAge=Current Age
```

```
 7  retireAge=Retirement Age
 8  deathAge=Life Expectancy
 9  inflationPercent=Inflation
10  investPercent=Investment Return
11  retire=Age: {0,number} Balance: {1,number,currency}
```

程序清单 7-10　retire/RetireStrings_de.properties

```
 1  language=Sprache
 2  computeButton=Rechnen
 3  savings=Vorherige Ersparnisse
 4  contrib=Jährliche Einzahlung
 5  income=Einkommen nach Ruhestand
 6  currentAge=Jetziges Alter
 7  retireAge=Ruhestandsalter
 8  deathAge=Lebenserwartung
 9  inflationPercent=Inflation
10  investPercent=Investitionsgewinn
11  retire=Alter: {0,number} Guthaben: {1,number,currency}
```

程序清单 7-11　retire/RetireStrings_zh.properties

```
 1  language=语言
 2  computeButton=计算
 3  savings=既存
 4  contrib=每年存金
 5  income=退休收入
 6  currentAge=现龄
 7  retireAge=退休年龄
 8  deathAge=预期寿命
 9  inflationPercent=通货膨胀
10  investPercent=投资报酬
11  retire=年龄: {0,number} 总结: {1,number,currency}
```

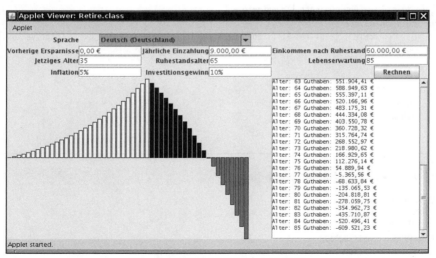

图 7-5　使用德语的退休金计算器

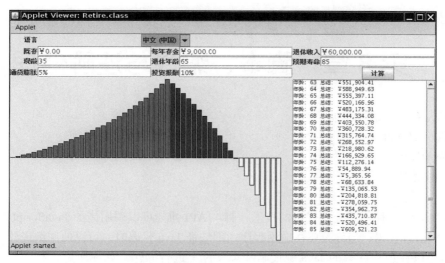

图 7-6 使用中文的退休金计算器

本章进述了如何运用 Java 语言的国际化特性。现在你可以使用资源包来提供多种语言的转换，也可以使用格式器和排序器来处理与 locale 相关的文本了。

下一章将研究脚本编写、编译和注解处理。

第 8 章 脚本、编译与注解处理

- ▲ Java 平台的脚本机制
- ▲ 编译器 API
- ▲ 使用注解
- ▲ 注解语法
- ▲ 标准注解
- ▲ 源码级注解处理
- ▲ 字节码工程

本章将介绍三种用于处理代码的技术：脚本 API 使你可以调用诸如 JavaScript 和 Groovy 这样的脚本语言代码；当你希望在应用程序内部编译 Java 代码时，可以使用编译器 API；注解处理器可以在包含注解的 Java 源代码和类文件上进行操作。如你所见，有许多应用程序都可以用来处理注解，从简单的诊断到"字节码工程"，后者可以将字节码插入到类文件中，甚至可以插入到运行程序中。

8.1 Java 平台的脚本机制

脚本语言是一种通过在运行时解释程序文本，从而避免使用通常的编辑/编译/链接/运行循环的语言。脚本语言有许多优势：
- 便于快速变更，鼓励不断试验。
- 可以修改运行着的程序的行为。
- 支持程序用户的定制化。

另一方面，大多数脚本语言都缺乏可以使编写复杂应用受益的特性，例如强类型、封装和模块化。

因此人们在尝试将脚本语言和传统语言的优势相结合。脚本 API 使你可以在 Java 平台上实现这个目的，它支持在 Java 程序中对用 JavaScript、Groovy、Ruby，甚至是更奇异的诸如 Scheme 和 Haskell 等语言编写的脚本进行调用。例如，Renjin 项目（www.renjin.org）就提供了一个 R 编程语言的 Java 实现和相应的脚本 API 的"引擎"，R 语言被广泛应用于统计编程中。

在下面的小节中，我们将向你展示如何为某种特定的语言选择一个引擎，如何执行脚本，以及如何利用某些脚本引擎提供的先进特性。

8.1.1 获取脚本引擎

脚本引擎是一个可以执行用某种特定语言编写的脚本的类库。当虚拟机启动时，它会发现可用的脚本引擎。为了枚举这些引擎，需要构造一个 `ScriptEngineManager`，并调用 `getEngineFactories` 方法。可以向每个引擎工厂询问它们所支持的引擎名、MIME 类型和文件扩展名。表 8-1 显示了这些内容的典型值。

表 8-1 脚本引擎工厂的属性

引擎	名字	MIME 类型	文件扩展
Nashorn（包含在 JDK 中）	nashorn, Nashorn, js, JS, JavaScript, javascript, ECMAScript, ecmascript	application/javascript, application/ecmascript, text/javascript, text/ecmascript	js
Groovy	groovy	无	groovy
Renjin	Renjin	text/x-R	R, r, S, s

通常，你知道所需要的引擎，因此可以直接通过名字、MIME 类型或文件扩展来请求它，例如：

```
ScriptEngine engine = manager.getEngineByName("nashorn");
```

Java 8 引入了 Nashorn，这是由 Oracle 开发的一个 JavaScript 解释器。可以通过在类路径中提供必要的 JAR 文件来添加对更多语言的支持。

API *javax.script.ScriptEngineManager* 6

- List<ScriptEngineFactory> getEngineFactories()
  获取所有发现的引擎工厂的列表。
- ScriptEngine getEngineByName(String name)
- ScriptEngine getEngineByExtension(String extension)
- ScriptEngine getEngineByMimeType(String mimeType)
  获取给定名字、脚本文件扩展名或 MIME 类型的脚本引擎。

API *javax.script.ScriptEngineFactory* 6

- List<String> getNames()
- List<String> getExtensions()
- List<String> getMimeTypes()
  获取该工厂所了解的名字、脚本文件扩展名和 MIME 类型。

8.1.2 脚本计算与绑定

一旦拥有了引擎，就可以通过下面的调用来直接调用脚本：

```
Object result = engine.eval(scriptString);
```

如果脚本存储在文件中，那么需要先打开一个 Reader，然后调用：

```
Object result = engine.eval(reader);
```

可以在同一个引擎上调用多个脚本。如果一个脚本定义了变量、函数或类，那么大多数引擎都会保留这些定义，以供将来使用。例如：

```
engine.eval("n = 1728");
Object result = engine.eval("n + 1");
```

将返回 1729。

> **注释**：要想知道在多个线程中并发执行脚本是否安全，可以调用
> ```
> Object param = factory.getParameter("THREADING");
> ```
> 其返回的是下列值之一：
> - null：并发执行不安全。
> - "MULTITHREADED"：并发执行安全。一个线程的执行效果对另外的线程有可能是可视的。
> - "THREAD-ISOLATED"：除了 "MULTITHREADED"，还会为每个线程维护不同的变量绑定。
> - "STATELESS"：除了 "THREAD-ISOLATED"，脚本还不会改变变量绑定。

我们经常希望能够向引擎中添加新的变量绑定。绑定由名字及其关联的 Java 对象构成。例如，考虑下面的语句：

```
engine.put("k", 1728);
Object result = engine.eval("k + 1");
```

脚本代码从"引擎作用域"中的绑定里读取 k 的定义。这一点非常重要，因为大多数脚本语言都可以访问 Java 对象，通常使用的是比 Java 语法更简单的语法。例如，

```
engine.put("b", new JButton());
engine.eval("b.text = 'Ok'");
```

反过来，也可以获取由脚本语句绑定的变量：

```
engine.eval("n = 1728");
Object result = engine.get("n");
```

除了引擎作用域之外，还有全局作用域。任何添加到 ScriptEngineManager 中的绑定对所有引擎都是可视的。

除了向引擎或全局作用域添加绑定之外，还可以将绑定收集到一个类型为 Bindings 的对象中，然后将其传递给 eval 方法：

```
Bindings scope = engine.createBindings();
scope.put("b", new JButton());
engine.eval(scriptString, scope);
```

如果绑定集不应该为了将来对 eval 方法的调用而持久化，那么这么做就很有用。

> **注释**：你可能希望除了引擎作用域和全局作用域之外还有其他的作用域。例如，Web 容器可能需要请求作用域或会话作用域。但是，这需要你自己去解决。你需要实现一个类，它实现了 ScriptContext 接口，并管理着一个作用域集合。每个作用域都是由一个整数标识的，而且越小的数字应该越先被搜索。（标准类库提供了 SimpleScriptContext 类，但是它只能持有全局作用域和引擎作用域。）

API *javax.script.ScriptEngine* 6

- Object eval(String script)
- Object eval(Reader reader)

- Object eval(String script, Bindings bindings)
- Object eval(Reader reader, Bindings bindings)

  对由字符串或读取器给定的脚本进行计算,并服从给定的绑定。
- Object get(String key)
- void put(String key, Object value)

  在引擎作用域内获取或放置一个绑定。
- Bindings createBindings()

  创建一个适合该引擎的空 Bindings 对象。

API *javax.script.ScriptEngineManager* 6

- Object get(String key)
- void put(String key, Object value)

  在全局作用域内获取或放置一个绑定。

API *javax.script.Bindings* 6

- Object get(String key)
- void put(String key, Object value)

  在由该 Bindings 对象表示的作用域内获取或放置一个绑定。

8.1.3 重定向输入和输出

可以通过调用脚本上下文的 setReader 和 setWriter 方法来重定向脚本的标准输入和输出。例如,

```
var writer = new StringWriter();
engine.getContext().setWriter(new PrintWriter(writer, true));
```

在上例中,任何用 JavaScript 的 print 和 println 函数产生的输出都会被发送到 writer。

setReader 和 setWriter 方法只会影响脚本引擎的标准输入和输出源。例如,如果执行下面的 JavaScript 代码:

```
println("Hello");
java.lang.System.out.println("World");
```

则只有第一个输出会被重定向。

Nashorn 引擎没有标准输入源的概念,因此调用 setReader 没有任何效果。

API *javax.script.ScriptEngine* 6

- ScriptContext getContext()

  获得该引擎的默认的脚本上下文。

API *javax.script.ScriptContext* 6

- Reader getReader()

- void setReader(Reader reader)
- Writer getWriter()
- void setWriter(Writer writer)
- Writer getErrorWriter()
- void setErrorWriter(Writer writer)

获取或设置用于输入的读入器或用于正常与错误输出的写出器。

8.1.4 调用脚本的函数和方法

对于许多脚本引擎而言,我们都可以调用脚本语言的函数,而不必对实际的脚本代码进行计算。如果允许用户用他们所选择的脚本语言来实现服务,那么这种机制就很有用了。

提供这种功能的脚本引擎实现了 Invocable 接口。特别是,Nashorn 引擎就是实现了 Invocable 接口。

要调用一个函数,需要用函数名来调用 invokeFunction 方法,函数名后面是函数的参数:

```
// Define greet function in JavaScript
engine.eval("function greet(how, whom) { return how + ', ' + whom + '!' }");

// Call the function with arguments "Hello", "World"
result = ((Invocable) engine).invokeFunction("greet", "Hello", "World");
```

如果脚本语言是面向对象的,那就可以调用 invokeMethod:

```
// Define Greeter class in JavaScript
engine.eval("function Greeter(how) { this.how = how }");
engine.eval("Greeter.prototype.welcome = "
  + " function(whom) { return this.how + ', ' + whom + '!' }");

// Construct an instance
Object yo = engine.eval("new Greeter('Yo')");

// Call the welcome method on the instance
result = ((Invocable) engine).invokeMethod(yo, "welcome", "World");
```

> **注释**:关于如何用 JavaScript 定义类的更多细节,可以参阅 *JavaScript: The Good Parts*,Douglas Grockford 著(O'Reilly, 2008)。

> **注释**:即使脚本引擎没有实现 Invocable 接口,你也可能仍旧可以以一种独立于语言的方式来调用某个方法。ScriptEngineFactory 类的 getMethodCallSyntax 方法可以产生一个字符串,你可以将其传递给 eval 方法。但是,所有的方法参数必须都与名字绑定,而 invokeMethod 方法是可以用任意值调用的。

我们可以更进一步,让脚本引擎去实现一个 Java 接口,然后就可以用 Java 方法调用的语法来调用脚本函数。

其细节依赖于脚本引擎,但是典型情况是我们需要为该接口中的每个方法都提供一个函数。例如,考虑下面的 Java 接口:

```
public interface Greeter
{
    String welcome(String whom);
}
```

如果在 Nashorn 中定义了具有相同名字的函数，那么可通过这个接口来调用它：

```
// Define welcome function in JavaScript
engine.eval("function welcome(whom) { return 'Hello, ' + whom + '!' }");

// Get a Java object and call a Java method
Greeter g = ((Invocable) engine).getInterface(Greeter.class);
result = g.welcome("World");
```

在面向对象的脚本语言中，可以通过相匹配的 Java 接口来访问一个脚本类。例如，下面的代码展示了如何使用 Java 的语法来调用 JavaScript 的 SimpleGreeter 类：

```
Greeter g = ((Invocable) engine).getInterface(yo, Greeter.class);
result = g.welcome("World");
```

总之，如果你希望从 Java 中调用脚本代码，同时又不想因这种脚本语言的语法而受到困扰，那么 Invocable 接口就很有用。

API *javax.script.Invocable* 6

- Object invokeFunction(String name, Object... parameters)
- Object invokeMethod(Object implicitParameter, String name, Object... explicitParameters)
  用给定的名字调用函数或方法，并传递给定的参数。
- <T> T getInterface(Class<T> iface)
  返回给定接口的实现，该实现用脚本引擎中的函数实现了接口中的方法。
- <T> T getInterface(Object implicitParameter, Class<T> iface)
  返回给定接口的实现，该实现用给定对象的方法实现了接口中的方法。

8.1.5 编译脚本

某些脚本引擎出于对执行效率的考虑，可以将脚本代码编译为某种中间格式。这些引擎实现了 Compilable 接口。下面的示例展示了如何编译和计算包含在脚本文件中的代码：

```
var reader = new FileReader("myscript.js");
CompiledScript script = null;
if (engine implements Compilable)
    script = ((Compilable) engine).compile(reader);
```

一旦该脚本被编译，就可以执行它。下面的代码将会在编译成功的情况下执行编译后的脚本，如果引擎不支持编译，则执行原始的脚本。

```
if (script != null)
    script.eval();
else
    engine.eval(reader);
```

当然，只有需要重复执行时，我们才希望编译脚本。

API `javax.script.Compilable` 6

- CompiledScript compile(String script)
- CompiledScript compile(Reader reader)
  编译由字符串或读入器给定的脚本。

API `javax.script.CompiledScript` 6

- Object eval()
- Object eval(Bindings bindings)
  对该脚本计算。

8.1.6 示例：用脚本处理 GUI 事件

为了演示脚本 API，我们将开发一个样例程序，它允许用户指定使用他们所选择的脚本语言编写的事件处理器。

让我们看看程序清单 8-1 中的程序，它可以将脚本添加到任意的框体类中。默认情况下，它会读取程序清单 8-2 中的 ButtonFrame 类，ButtonFrame 类与卷 I 中介绍的事件处理演示程序类似，但是有两个差异：

- 每个构件都有其自己的 name 属性集。
- 没有任何事件处理器。

程序清单 8-1   script/ScriptTest.java

```
 1  package script;
 2
 3  import java.awt.*;
 4  import java.beans.*;
 5  import java.io.*;
 6  import java.lang.reflect.*;
 7  import java.util.*;
 8  import javax.script.*;
 9  import javax.swing.*;
10
11  /**
12   * @version 1.03 2018-05-01
13   * @author Cay Horstmann
14   */
15  public class ScriptTest
16  {
17     public static void main(String[] args)
18     {
19        EventQueue.invokeLater(() ->
20           {
21              try
22              {
23                 var manager = new ScriptEngineManager();
24                 String language;
25                 if (args.length == 0)
```

```java
            {
               System.out.println("Available factories: ");
               for (ScriptEngineFactory factory : manager.getEngineFactories())
                  System.out.println(factory.getEngineName());

               language = "nashorn";
            }
            else language = args[0];

            final ScriptEngine engine = manager.getEngineByName(language);
            if (engine == null)
            {
               System.err.println("No engine for " + language);
               System.exit(1);
            }

            final String frameClassName
               = args.length < 2 ? "buttons1.ButtonFrame" : args[1];

            var frame
               = (JFrame) Class.forName(frameClassName).getConstructor().newInstance();
            InputStream in = frame.getClass().getResourceAsStream("init." + language);
            if (in != null) engine.eval(new InputStreamReader(in));
            var components = new HashMap<String, Component>();
            getComponentBindings(frame, components);
            components.forEach((name, c) -> engine.put(name, c));

            var events = new Properties();
            in = frame.getClass().getResourceAsStream(language + ".properties");
            events.load(in);

            for (Object e : events.keySet())
            {
               String[] s = ((String) e).split("\\.");
               addListener(s[0], s[1], (String) events.get(e), engine, components);
            }
            frame.setTitle("ScriptTest");
            frame.setDefaultCloseOperation(JFrame.EXIT_ON_CLOSE);
            frame.setVisible(true);
         }
         catch (ReflectiveOperationException | IOException
               | ScriptException | IntrospectionException ex)
         {
            ex.printStackTrace();
         }
      });
   }

   /**
    * Gathers all named components in a container.
    * @param c the component
    * @param namedComponents a map into which to enter the component names and components
    */
   private static void getComponentBindings(Component c,
```

```java
 80            Map<String, Component> namedComponents)
 81     {
 82        String name = c.getName();
 83        if (name != null) { namedComponents.put(name, c); }
 84        if (c instanceof Container)
 85        {
 86           for (Component child : ((Container) c).getComponents())
 87              getComponentBindings(child, namedComponents);
 88        }
 89     }
 90
 91     /**
 92      * Adds a listener to an object whose listener method executes a script.
 93      * @param beanName the name of the bean to which the listener should be added
 94      * @param eventName the name of the listener type, such as "action" or "change"
 95      * @param scriptCode the script code to be executed
 96      * @param engine the engine that executes the code
 97      * @param bindings the bindings for the execution
 98      * @throws IntrospectionException
 99      */
100     private static void addListener(String beanName, String eventName, final String scriptCode,
101           ScriptEngine engine, Map<String, Component> components)
102           throws ReflectiveOperationException, IntrospectionException
103     {
104        Object bean = components.get(beanName);
105        EventSetDescriptor descriptor = getEventSetDescriptor(bean, eventName);
106        if (descriptor == null) return;
107        descriptor.getAddListenerMethod().invoke(bean,
108           Proxy.newProxyInstance(null, new Class[] { descriptor.getListenerType() },
109              (proxy, method, args) ->
110                 {
111                    engine.eval(scriptCode);
112                    return null;
113                 }));
114     }
115
116     private static EventSetDescriptor getEventSetDescriptor(Object bean, String eventName)
117           throws IntrospectionException
118     {
119        for (EventSetDescriptor descriptor : Introspector.getBeanInfo(bean.getClass())
120              .getEventSetDescriptors())
121           if (descriptor.getName().equals(eventName)) return descriptor;
122        return null;
123     }
124 }
```

程序清单 8-2　buttons1/ButtonFrame.java

```java
1 package buttons1;
2
3 import javax.swing.*;
4
5 /**
6  * A frame with a button panel.
```

```
 7     * @version 1.00 2007-11-02
 8     * @author Cay Horstmann
 9     */
10    public class ButtonFrame extends JFrame
11    {
12       private static final int DEFAULT_WIDTH = 300;
13       private static final int DEFAULT_HEIGHT = 200;
14
15       private JPanel panel;
16       private JButton yellowButton;
17       private JButton blueButton;
18       private JButton redButton;
19
20       public ButtonFrame()
21       {
22          setSize(DEFAULT_WIDTH, DEFAULT_HEIGHT);
23
24          panel = new JPanel();
25          panel.setName("panel");
26          add(panel);
27
28          yellowButton = new JButton("Yellow");
29          yellowButton.setName("yellowButton");
30          blueButton = new JButton("Blue");
31          blueButton.setName("blueButton");
32          redButton = new JButton("Red");
33          redButton.setName("redButton");
34
35          panel.add(yellowButton);
36          panel.add(blueButton);
37          panel.add(redButton);
38       }
39    }
```

事件处理器是在属性文件中定义的。每个属性定义都具有下面的形式：

componentName.eventName = scriptCode

例如，如果选择使用 JavaScript，那就要在 js.properties 文件中提供事件处理器：

```
yellowButton.action=panel.background = java.awt.Color.YELLOW
blueButton.action=panel.background = java.awt.Color.BLUE
redButton.action=panel.background = java.awt.Color.RED
```

本书附带的代码还包括用于 Groovy、R 和 SISC Scheme 的文件。

该程序以加载在命令行中指定的语言所需的引擎开始，如果未指定语言，则使用 JavaScript。然后，我们处理 init.*language* 脚本，如果该文件存在的话。这对 R 语言和 Scheme 语言而言很有用，因为这些语言需要某些麻烦的初始化工作，我们不希望在每个事件处理器的脚本中都包括这部分工作。

接下来，我们递归地遍历所有的子构件，并在构件映射表中添加（名字，对象）绑定，然后，将它们添加到引擎中。

然后，我们读入 language.properties 文件。对于每一个属性，都合成其事件处理器代理，使得脚本代码得以执行。其细节有些技术性，如果你希望了解实现的细节，请参阅卷 I 第 6 章有关代理的小节。但是，其精髓部分是每个事件处理器都会调用下面的方法：

```
engine.eval(scriptCode);
```

让我们详细看看 yellowButton。当下面一行被处理时，

```
yellowButton.action=panel.background = java.awt.Color.YELLOW
```

我们找到了具有"yellowButton"名字的 JButton 构件，然后附着一个 ActionListener，它拥有 actionPerformed 方法，该方法将执行下面的脚本，如果该脚本是用 Nashorn 执行的：

```
panel.background = java.awt.Color.YELLOW
```

引擎包含一个将名字"panel"与这个 JPanel 对象绑定在一起的绑定。当事件发生时，该面板的 setBackground 方法就会执行，并且其颜色也会改变。

只需要执行下面的命令，就可以运行这个带有 JavaScript 事件处理器的程序：

```
java ScriptTest
```

对于 Groovy 处理器，需要使用

```
java -classpath .:groovy/lib/\* ScriptTest groovy
```

这里，groovy 是 Groovy 的安装目录。

对于 R 的 Renjin 实现，要在类路径中包含 Renjin Studio 的 JAR 文件以及 Renjin 脚本引擎。它们都可以在 www.renjin.org/downloads.html 处获得。

这个应用演示了如何在 Java GUI 编程中使用脚本机制。大家可以更进一步，用 XML 文件来描述 GUI，就像在第 3 章中看到的那样。然后我们的程序就会变成解释器，去解释那些由 XML 文件定义可视化表示以及用脚本语言定义行为的 GUI。请注意这与动态 HTML 页面或动态服务器端脚本环境之间的相似性。

8.2 编译器 API

有许多工具都需要编译 Java 代码。很明显，教授 Java 编程的开发环境和程序就位于其列，测试和自动化构建工具也属于这类工具。另一个例子是 JavaServer Pages 的处理工具，JSP 是一种嵌入了 Java 语句的网页。

8.2.1 调用编译器

调用编译器非常简单，下面是一个示范调用：

```java
JavaCompiler compiler = ToolProvider.getSystemJavaCompiler();
OutputStream outStream = ...;
OutputStream errStream = ...;
int result = compiler.run(null, outStream, errStream,
    "-sourcepath", "src", "Test.java");
```

返回值为 0 表示编译成功。

编译器会向提供给它的流发送输出和错误消息。如果将这些参数设置为 null，编译器就会使用 System.out 和 System.err。run 方法的第一个参数是输入流，由于编译器不会接受任何控制台输入，因此总是应该让其保持为 null。（run 方法是从泛化的 Tool 接口继承而来的，它考虑到某些工具需要读取输入。）

如果在命令行调用 javac，那么 run 方法其余的参数就会作为变量传递给 javac。这些变量是一些选项或文件名。

8.2.2 发起编译任务

可以通过使用 CompilationTask 对象来对编译过程进行更多的控制。如果要从字符串中提供源码，在内存中捕获类文件，或者处理错误和警告消息，这样做就会显得很有用。

要想获取 CompilationTask 对象，需要以前一节中描述的 compiler 对象开始，然后按照下面的方式调用：

```
JavaCompiler.CompilationTask task = compiler.getTask(
    errorWriter, // Uses System.err if null
    fileManager, // Uses the standard file manager if null
    diagnostics, // Uses System.err if null
    options, // null if no options
    classes, // For annotation processing; null if none
    sources);
```

最后三个参数是 Iterable 的实例。例如，选项序列可以像下面这样指定：

```
Iterable<String> options = List.of("-d", "bin");
```

sources 参数是 JavaFileObject 实例的 Iterable。如果想要编译磁盘文件，需要获取一个 StandardJavaFileManager 对象，并调用其 getJavaFileObjects 方法：

```
StandardJavaFileManager fileManager = compiler.getStandardFileManager(null, null, null);
Iterable<JavaFileObject> sources
    = fileManager.getJavaFileObjectsFromStrings(List.of("File1.java", "File2.java"));
JavaCompiler.CompilationTask task = compiler.getTask(
    null, null, null, options, null, sources);
```

> 注释：classes 参数只用于注解处理。在这种情况下，还需要用一个 Processor 对象的列表来调用 task.processors(annotationProcessors)。请参见 8.6 节中有关注解处理的示例。

getTask 方法会返回任务对象，但是并不会启动编译过程。CompilationTask 类扩展了 Callable<Boolean>，我们可以将其对象传递给 ExecutorService 以并行执行，或者只是做出如下的同步调用：

```
Boolean success = task.call();
```

8.2.3 捕获诊断消息

为了监听错误消息，需要安装一个 DiagnosticListener。这个监听器在编译器报告警告或错误消息时会收到一个 Diagnostic 对象。DiagnosticCollector 类实现了这个接口，它将收集所有的诊断信息，使得你可以在编译完成之后遍历这些信息。

```
DiagnosticCollector<JavaFileObject> collector = new DiagnosticCollector<>();
compiler.getTask(null, fileManager, collector, null, null, sources).call();
for (Diagnostic<? extends JavaFileObject> d : collector.getDiagnostics())
{
    System.out.println(d);
}
```

Diagnostic 对象包含有关问题位置的信息（包括文件名、行号和列号）以及人类可阅读的描述。

还可以在标准的文件管理器上安装一个 DiagnosticListener 对象，这样就可以捕获到有关文件缺失的消息：

```
StandardJavaFileManager fileManager
    = compiler.getStandardFileManager(diagnostics, null, null);
```

8.2.4 从内存中读取源文件

如果动态地生成了源代码，那么就可以从内存中获取它来进行编译，而无须在磁盘上保存文件。可以使用下面的类来持有代码：

```java
public class StringSource extends SimpleJavaFileObject
{
    private String code;

    StringSource(String name, String code)
    {
        super(URI.create("string:///" + name.replace('.','/') + ".java"), Kind.SOURCE);
        this.code = code;
    }

    public CharSequence getCharContent(boolean ignoreEncodingErrors)
    {
        return code;
    }
}
```

然后，生成类的代码，并提交给编译器一个 StringSource 对象的列表：

```java
List<StringSource> sources = List.of(
    new StringSource(className1, class1CodeString), . . .);
task = compiler.getTask(null, fileManager, diagnostics, null, null, sources);
```

8.2.5 将字节码写出到内存中

如果动态地编译类，那么就无须将类文件写出到硬盘上。可以将它们存储在内存中，并立即加载它们。

首先，要有一个类来持有这些字节：

```java
public class ByteArrayClass extends SimpleJavaFileObject
{
```

```java
   private ByteArrayOutputStream out;

   ByteArrayClass(String name)
   {
      super(URI.create("bytes:///" + name.replace('.','/') + ".class"), Kind.CLASS);
   }

   public byte[] getCode()
   {
      return out.toByteArray();
   }

   public OutputStream openOutputStream() throws IOException
   {
      out = new ByteArrayOutputStream();
      return out;
   }
}
```

接下来,需要将文件管理器配置为使用这些类作为输出:

```java
List<ByteArrayClass> classes = new ArrayList<>();
StandardJavaFileManager stdFileManager
   = compiler.getStandardFileManager(null, null, null);
JavaFileManager fileManager
   = new ForwardingJavaFileManager<JavaFileManager>(stdFileManager)
      {
         public JavaFileObject getJavaFileForOutput(Location location,
            String className, Kind kind, FileObject sibling)
            throws IOException
         {
            if (kind == Kind.CLASS)
            {
               ByteArrayClass outfile = new ByteArrayClass(className);
               classes.add(outfile);
               return outfile;
            }
            else
               return super.getJavaFileForOutput(location, className, kind, sibling);
         }
      };
```

为了加载这些类,需要使用类加载器(参见第 10 章):

```java
public class ByteArrayClassLoader extends ClassLoader
{
   private Iterable<ByteArrayClass> classes;

   public ByteArrayClassLoader(Iterable<ByteArrayClass> classes)
   {
      this.classes = classes;
   }

   public Class<?> findClass(String name) throws ClassNotFoundException
   {
      for (ByteArrayClass cl : classes)
```

```java
        {
            if (cl.getName().equals("/" + name.replace('.','/') + ".class"))
            {
                byte[] bytes = cl.getCode();
                return defineClass(name, bytes, 0, bytes.length);
            }
        }
        throw new ClassNotFoundException(name);
    }
}
```

编译完成后，用上面的类加载器调用 Class.forName 方法：

```
ByteArrayClassLoader loader = new ByteArrayClassLoader(classes);
Class<?> cl = Class.forName(className, true, loader);
```

8.2.6 示例：动态 Java 代码生成

在用于动态 Web 页面的 JSP 技术中，可以在 HTML 中混杂 Java 代码，例如：

```
<p>The current date and time is <b><%= new java.util.Date() %></b>.</p>
```

JSP 引擎动态地将 Java 代码编译到 Servlet 中。在示例应用中，我们使用了一个更简单的示例，它可以动态生成 Swing 代码。其基本思想是使用 GUI 构建器在窗体中放置构件，并在一个外部文件中指定构件的行为。程序清单 8-4 展示了一个非常简单的窗体类实例，而程序清单 8-5 展示了按钮动作的代码。请注意，窗体类的构造器调用了抽象方法 addEventHandlers。我们的代码生成器将产生一个实现了 addEventHandlers 方法的子类，并且对 action.properties 文件中的每一行都添加了动作监听器。（我们给读者留下了一个典型的练习，即扩展代码的生成功能，使其支持其他事件类型。）

我们将这个子类置于名字为 x 的包中，因为我们不希望在程序的其他地方用到它。所生成的代码有如下形式：

```
package x;
public class Frame extends SuperclassName
{
    protected void addEventHandlers()
    {
        componentName₁.addActionListener(event ->
            {
                code for event handler₁
            });
        // repeat for the other event handlers . . .
    }
}
```

程序清单 8-3 的程序中的 buildSource 方法构建了这些代码，并将它们放到了 StringBuilder-JavaSource 对象中。该对象会传递给 Java 编译器。

如前一节所述，我们使用了一个 ForwardingJavaFileManager 对象，它会为每一个编译过的类都构造一个 ByteArrayClass 对象，这些对象会捕获 x.Frame 类被编译时所生成的类文件。该方法将每个文件对象都添加到了一个列表中，然后将其返回，以使得我们稍后可以定位这些字节码。

编译完成后，我们使用前一节中描述的类加载器来加载存储在这个列表中的所有类。然后，我们构造并显示应用程序的窗体类。

```
var loader = new ByteArrayClassLoader(classFileObjects);
var frame = (JFrame) loader.loadClass("x.Frame").getConstructor().newInstance();
frame.setVisible(true);
```

当点击按钮时，背景色会按照常规方式进行修改。为了查看这些动作是动态编译的，可以更改 action.properties 文件中一行，例如，修改成下面这样：

```
yellowButton=panel.setBackground(java.awt.Color.YELLOW); yellowButton.setEnabled(false);
```

再次运行这个程序，现在，黄色按钮在点击之后就变得禁用了。再看看代码目录，你不会发现 x 包中的类的任何源文件和类文件。这个示例向你演示了如何通过内存中的源文件和类文件来使用动态编译。

程序清单 8-3 compiler/CompilerTest.java

```java
 1  package compiler;
 2
 3  import java.awt.*;
 4  import java.io.*;
 5  import java.nio.file.*;
 6  import java.util.*;
 7  import java.util.List;
 8
 9  import javax.swing.*;
10  import javax.tools.*;
11  import javax.tools.JavaFileObject.*;
12
13  /**
14   * @version 1.10 2018-05-01
15   * @author Cay Horstmann
16   */
17  public class CompilerTest
18  {
19     public static void main(final String[] args)
20           throws IOException, ReflectiveOperationException
21     {
22        JavaCompiler compiler = ToolProvider.getSystemJavaCompiler();
23
24        var classFileObjects = new ArrayList<ByteArrayClass>();
25        var diagnostics = new DiagnosticCollector<JavaFileObject>();
26
27        JavaFileManager fileManager = compiler.getStandardFileManager(diagnostics, null, null);
28        fileManager = new ForwardingJavaFileManager<JavaFileManager>(fileManager)
29           {
30              public JavaFileObject getJavaFileForOutput(Location location,
31                 String className, Kind kind, FileObject sibling) throws IOException
32              {
33                 if (kind == Kind.CLASS)
34                 {
35                    var fileObject = new ByteArrayClass(className);
36                    classFileObjects.add(fileObject);
```

```java
                    return fileObject;
                }
                else return super.getJavaFileForOutput(location, className, kind, sibling);
            }
        };

        String frameClassName = args.length == 0 ? "buttons2.ButtonFrame" : args[0];
        //compiler.run(null, null, null, frameClassName.replace(".", "/") + ".java");

        StandardJavaFileManager fileManager2 = compiler.getStandardFileManager(null, null, null);
        var sources = new ArrayList<JavaFileObject>();
        for (JavaFileObject o : fileManager2.getJavaFileObjectsFromStrings(
            List.of(frameClassName.replace(".", "/") + ".java")))
          sources.add(o);

        JavaFileObject source = buildSource(frameClassName);
        JavaCompiler.CompilationTask task = compiler.getTask(null, fileManager, diagnostics,
            null, null, List.of(source));
        Boolean result = task.call();

        for (Diagnostic<? extends JavaFileObject> d : diagnostics.getDiagnostics())
            System.out.println(d.getKind() + ": " + d.getMessage(null));
        fileManager.close();
        if (!result)
        {
            System.out.println("Compilation failed.");
            System.exit(1);
        }

        var loader = new ByteArrayClassLoader(classFileObjects);
        var frame = (JFrame) loader.loadClass("x.Frame").getConstructor().newInstance();

        EventQueue.invokeLater(() ->
            {
                frame.setDefaultCloseOperation(JFrame.EXIT_ON_CLOSE);
                frame.setTitle("CompilerTest");
                frame.setVisible(true);
            });
    }

    /*
     * Builds the source for the subclass that implements the addEventHandlers method.
     * @return a file object containing the source in a string builder
     */
    static JavaFileObject buildSource(String superclassName)
        throws IOException, ClassNotFoundException
    {
        var builder = new StringBuilder();
        builder.append("package x;\n\n");
        builder.append("public class Frame extends " + superclassName + " {\n");
        builder.append("protected void addEventHandlers() {\n");
        var props = new Properties();
        props.load(Files.newInputStream(Paths.get(
            superclassName.replace(".", "/")).getParent().resolve("action.properties")));
```

```java
 91        for (Map.Entry<Object, Object> e : props.entrySet())
 92        {
 93           var beanName = (String) e.getKey();
 94           var eventCode = (String) e.getValue();
 95           builder.append(beanName + ".addActionListener(event -> {\n");
 96           builder.append(eventCode);
 97           builder.append("\n} );\n");
 98        }
 99        builder.append("} \n");
100        return new StringSource("x.Frame", builder.toString());
101     }
102 }
```

程序清单 8-4　buttons2/ButtonFrame.java

```java
 1  package buttons2;
 2  import javax.swing.*;
 3
 4  /**
 5   * A frame with a button panel.
 6   * @version 1.00 2007-11-02
 7   * @author Cay Horstmann
 8   */
 9  public abstract class ButtonFrame extends JFrame
10  {
11     public static final int DEFAULT_WIDTH = 300;
12     public static final int DEFAULT_HEIGHT = 200;
13
14     protected JPanel panel;
15     protected JButton yellowButton;
16     protected JButton blueButton;
17     protected JButton redButton;
18
19     protected abstract void addEventHandlers();
20
21     public ButtonFrame()
22     {
23        setSize(DEFAULT_WIDTH, DEFAULT_HEIGHT);
24
25        panel = new JPanel();
26        add(panel);
27
28        yellowButton = new JButton("Yellow");
29        blueButton = new JButton("Blue");
30        redButton = new JButton("Red");
31
32        panel.add(yellowButton);
33        panel.add(blueButton);
34        panel.add(redButton);
35
36        addEventHandlers();
37     }
38  }
```

程序清单 8-5 buttons2/action.properties

```
1 yellowButton=panel.setBackground(java.awt.Color.YELLOW);
2 blueButton=panel.setBackground(java.awt.Color.BLUE);
```

API *javax.tools.Tool* 6

- int run(InputStream in, OutputStream out, OutputStream err, String... arguments)
  用给定的输入、输出、错误流，以及给定的参数来运行该工具。返回值为 0 表示成功，非 0 值表示失败。

API *javax.tools.JavaCompiler* 6

- StandardJavaFileManager getStandardFileManager(DiagnosticListener<? super JavaFileObject> diagnosticListener, Locale locale, Charset charset)
  获取该编译器的标准文件管理器。如果要使用默认的错误报告机制、locale 和字符集等参数，则可以提供 null。

- JavaCompiler.CompilationTask getTask(Writer out, JavaFileManager fileManager, DiagnosticListener<? super JavaFileObject> diagnosticListener, Iterable<String> options, Iterable<String> classesForAnnotationProcessing, Iterable<? extends JavaFileObject> sourceFiles)
  获取编译任务，在被调用时，该任务将编译给定的源文件。参见前一节中有关这部分内容的详细讨论。

API *javax.tools.StandardJavaFileManager* 6

- Iterable<? extends JavaFileObject> getJavaFileObjectsFromStrings(Iterable<String> fileNames)
- Iterable<? extends JavaFileObject> getJavaFileObjectsFromFiles(Iterable<? extends File> files)
  将文件名或文件序列转译成一个 JavaFileObject 实例序列。

API *javax.tools.JavaCompiler.CompilationTask* 6

- Boolean call()
  执行编译任务。

API *javax.tools.DiagnosticCollector<S>* 6

- DiagnosticCollector()
  构造一个空收集器。
- List<Diagnostic<? extends S>> getDiagnostics()
  获取收集到的诊断信息。

API *javax.tools.Diagnostic<S>* 6

- S getSource()

获取与该诊断信息相关联的源对象。
- Diagnostic.Kind getKind()
  获取该诊断信息的类型，返回值为 ERROR，WARNING，MANDATORY_WARNING，NOTE 或 OTHER 之一。
- String getMessage(Locale locale)
  获取一条消息，这条消息描述了由该诊断信息所揭示的问题。如果要使用默认的 locale，则传递 null。
- long getLineNumber()
- long getColumnNumber()
  获取由该诊断信息所揭示的问题的位置。

API javax.tools.SimpleJavaFileObject 6

- CharSequence getCharContent(boolean ignoreEncodingErrors)
  对于表示源文件并产生源代码的文件对象，需要覆盖该方法。
- OutputStream openOutputStream()
  对于表示类文件并产生字节码可写入其中的流的文件对象，需要覆盖该方法。

API javax.tools.ForwardingJavaFileManager<M extends JavaFileManager> 6

- protected ForwardingJavaFileManager(M fileManager)
  构造一个 JavaFileManager，它将所有的调用都代理给指定的文件管理器。
- FileObject   getFileForOutput(JavaFileManager.Location   location,   String   className, JavaFileObject.Kind kind, FileObject sibling)
  如果希望替换用于写出类文件的文件对象，则需要拦截该调用。kind 的值是 SOURCE，CLASS，HTML 或 OTHER 之一。

8.3 使用注解

注解是那些插入到源代码中使用其他工具可以对其进行处理的标签。这些工具可以在源码层次上进行操作，或者可以处理编译器在其中放置了注解的类文件。

注解不会改变程序的编译方式。Java 编译器对于包含注解和不包含注解的代码会生成相同的虚拟机指令。

为了能够受益于注解，你需要选择一个处理工具，然后向你的处理工具可以理解的代码中插入注解，之后运用该处理工具处理代码。

注解的使用范围还是很广泛的，并且这种广泛性让人乍一看会觉得有些杂乱无章。下面是关于注解的一些可能的用法：

- 附属文件的自动生成，例如部署描述符或者 bean 信息类。
- 测试、日志、事务语义等代码的自动生成。

8.3.1 注解简介

我们首先介绍基本概念,然后将这些概念运用到一个具体示例中:我们将某些方法标注为 AWT 构件的事件监听器,然后向你展示一个能够分析注解和连接监听器的注解处理器。然后,我们对其语法规则进行详细讨论。最后我们以两个注解处理的高级示例结束本章。其中一个可以处理源代码级别的注解。另外一个使用了 Apache 的字节码工程类库,可以向注解过的方法中添加额外的字节码。

下面是一个简单注解的示例:

```
public class MyClass
{
    ...
    @Test public void checkRandomInsertions()
}
```

注解 @Test 用于注解 checkRandomInsertions 方法。

在 Java 中,注解是当作一个修饰符来使用的,它被置于被注解项之前,中间没有分号。(修饰符就是诸如 public 和 static 之类的关键词。)每一个注解的名称前面都加上了 @ 符号,这有点类似于 Javadoc 的注释。然而,Javadoc 注释出现在 /**...*/ 定界符的内部,而注解是代码的一部分。

@Test 注解自身并不会做任何事情,它需要工具支持才会有用。例如,当测试一个类的时候,JUnit4 测试工具(可以从 http://junit.org 处获得)可能会调用所有标识为 @Test 的方法。另一个工具可能会删除一个类文件中的所有测试方法,以便在对这个类测试完毕后,不会将这些测试方法与程序装载在一起。

注解可以定义成包含元素的形式,例如:

```
@Test(timeout="10000")
```

这些元素可以被读取这些注解的工具去处理。其他形式的元素也是有可能的;我们将会在本章的随后部分进行讨论。

除了方法外,还可以注解类、成员以及局部变量,这些注解可以存在于任何可以放置一个像 public 或者 static 这样的修饰符的地方。另外,正如在 8.4 节中看到的,你还可以注解包、参数变量、类型参数和类型用法。

每个注解都必须通过一个注解接口进行定义。这些接口中的方法与注解中的元素相对应。例如,JUnit 的注解 Test 可以用下面这个接口进行定义:

```
@Target(ElementType.METHOD)
@Retention(RetentionPolicy.RUNTIME)
public @interface Test
{
    long timeout() default 0L;
    ...
}
```

@interface 声明创建了一个真正的 Java 接口。处理注解的工具将接收那些实现了这个注解接口的对象。这类工具可以调用 timeout 方法来获取某个特定 Test 注解的 timeout 元素。

注解 Target 和 Retention 是元注解。它们注解了 Test 注解，即将 Test 注解标识成一个只能运用到方法上的注解，并且当类文件载入到虚拟机的时候，它仍可以保留下来。我们将在 8.5.3 节详细讨论这些元注解。

你现在已经清楚了程序的元数据和注解这两个概念。在接下来的小节中，我们将深入到一个注解处理的具体示例中继续探讨。

> **注释**：对于注解引人入胜的用法，可以查看 JCommander (http:// jcommander.org) 和 picocli (http://picocli.info)。这些类库将注解用于命令行参数的处理。

8.3.2 示例：注解事件处理器

在用户界面编程中，一件更令人讨厌的事情就是组装事件源上的监听器。很多监听器是下面这种形式的：

```
myButton.addActionListener(() -> doSomething());
```

在本节，我们设计了一个注解来免除这种苦差事。该注解是在程序清单 8-8 中定义的，其使用方式如下：

```
@ActionListenerFor(source="myButton") void doSomething() { ... }
```

程序员不再需要去调用 addActionListener 了。相反地，每个方法直接用一个注解标记起来。程序清单 8-7 展示了卷 I 第 10 章的 ButtonFrame 程序，但是使用上述这类注解重新实现了一遍。

我们还需要定义一个注解接口，代码在程序清单 8-8 中。

程序清单 8-6 runtimeAnnotations/ActionListenerInstaller.java

```java
1  package runtimeAnnotations;
2
3  import java.awt.event.*;
4  import java.lang.reflect.*;
5
6  /**
7   * @version 1.00 2004-08-17
8   * @author Cay Horstmann
9   */
10 public class ActionListenerInstaller
11 {
12    /**
13     * Processes all ActionListenerFor annotations in the given object.
14     * @param obj an object whose methods may have ActionListenerFor annotations
15     */
16    public static void processAnnotations(Object obj)
17    {
18       try
19       {
20          Class<?> cl = obj.getClass();
```

```java
21              for (Method m : cl.getDeclaredMethods())
22              {
23                 ActionListenerFor a = m.getAnnotation(ActionListenerFor.class);
24                 if (a != null)
25                 {
26                    Field f = cl.getDeclaredField(a.source());
27                    f.setAccessible(true);
28                    addListener(f.get(obj), obj, m);
29                 }
30              }
31           }
32           catch (ReflectiveOperationException e)
33           {
34              e.printStackTrace();
35           }
36        }
37
38        /**
39         * Adds an action listener that calls a given method.
40         * @param source the event source to which an action listener is added
41         * @param param the implicit parameter of the method that the listener calls
42         * @param m the method that the listener calls
43         */
44        public static void addListener(Object source, final Object param, final Method m)
45              throws ReflectiveOperationException
46        {
47           var handler = new InvocationHandler()
48              {
49                 public Object invoke(Object proxy, Method mm, Object[] args) throws Throwable
50                 {
51                    return m.invoke(param);
52                 }
53              };
54
55           Object listener = Proxy.newProxyInstance(null,
56              new Class[] { java.awt.event.ActionListener.class }, handler);
57           Method adder = source.getClass().getMethod("addActionListener", ActionListener.class);
58           adder.invoke(source, listener);
59        }
60  }
```

程序清单 8-7　buttons3/ButtonFrame.java

```java
1  package buttons3;
2
3  import java.awt.*;
4  import javax.swing.*;
5  import runtimeAnnotations.*;
6
7  /**
8   * A frame with a button panel.
9   * @version 1.00 2004-08-17
10  * @author Cay Horstmann
11  */
```

```java
12  public class ButtonFrame extends JFrame
13  {
14     private static final int DEFAULT_WIDTH = 300;
15     private static final int DEFAULT_HEIGHT = 200;
16
17     private JPanel panel;
18     private JButton yellowButton;
19     private JButton blueButton;
20     private JButton redButton;
21
22     public ButtonFrame()
23     {
24        setSize(DEFAULT_WIDTH, DEFAULT_HEIGHT);
25
26        panel = new JPanel();
27        add(panel);
28
29        yellowButton = new JButton("Yellow");
30        blueButton = new JButton("Blue");
31        redButton = new JButton("Red");
32
33        panel.add(yellowButton);
34        panel.add(blueButton);
35        panel.add(redButton);
36
37        ActionListenerInstaller.processAnnotations(this);
38     }
39
40     @ActionListenerFor(source = "yellowButton")
41     public void yellowBackground()
42     {
43        panel.setBackground(Color.YELLOW);
44     }
45
46     @ActionListenerFor(source = "blueButton")
47     public void blueBackground()
48     {
49        panel.setBackground(Color.BLUE);
50     }
51
52     @ActionListenerFor(source = "redButton")
53     public void redBackground()
54     {
55        panel.setBackground(Color.RED);
56     }
57  }
```

程序清单 8-8　runtimeAnnotations/ActionListenerFor.java

```java
1  package runtimeAnnotations;
2
3  import java.lang.annotation.*;
4
5  /**
```

```
 6    * @version 1.00 2004-08-17
 7    * @author Cay Horstmann
 8    */
 9   @Target(ElementType.METHOD)
10   @Retention(RetentionPolicy.RUNTIME)
11   public @interface ActionListenerFor
12   {
13      String source();
14   }
```

当然,这些注解本身不会做任何事情,它们只是存在于源文件中。编译器将它们置于类文件中,并且虚拟机会将它们载入。我们现在需要的是一个分析注解以及安装行为监听器的机制。这也是类 ActionListenerInstaller 的职责所在。ButtonFrame 构造器将调用下面的方法:

```
ActionListenerInstaller.processAnnotations(this);
```

静态的 processAnnotations 方法可以枚举出某个对象接收到的所有方法。对于每一个方法,它先获取 ActionListenerFor 注解对象,然后再对它进行处理。

```
Class<?> cl = obj.getClass();
for (Method m : cl.getDeclaredMethods())
{
    ActionListenerFor a = m.getAnnotation(ActionListenerFor.class);
    if (a != null) . . .
}
```

这里,我们使用了定义在 AnnotatedElement 接口中的 getAnnotation 方法。Method、Constructor、Field、Class 和 Package 这些类都实现了这个接口。

源成员域的名字是存储在注解对象中的。我们可以通过调用 source 方法对它进行检索,然后查找匹配的成员域。

```
String fieldName = a.source();
Field f = cl.getDeclaredField(fieldName);
```

这表明我们的注解有点局限。源元素必须是一个成员域的名字,而不能是局部变量。

代码的剩余部分相当具有技术性。对于每一个被注解的方法,我们构造了一个实现了 ActionListener 接口的代理对象,其 actionPerformed 方法将调用这个被注解过的方法。(关于代理的更多信息见卷 I 第 6 章。)细节并不重要,关键要知道注解的功能是通过 processAnnotations 方法建立起来的。

图 8-1 展示了在本例中注解是如何被处理的。

在这个示例中,注解是在运行时进行处理的。另外也可以在源码级别上对它们进行处理,这样,源代码生成器将产生用于添加监听器的代码。注解也可以在字节码级别上进行处理,字节码编辑器可以将对 addActionListener 的调用注入框体构造器中。听起来似乎很复杂,不过可以利用一些类库相对直截了当地实现这项任务。

对于用户界面程序员来说,我们这个示例并不能看作是一个严格意义上的工具。因为,用于添加监听器的实用方法对于程序员来说和添加一条注解一样方便。(实际上,java.beans.EventHandler 类试图实现的就是这样。通过在这个类中提供一个可以添加事件处理器的方法,而不只是构建它,就可以很容易地对它进行改进。)

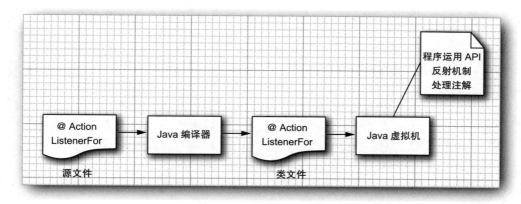

图 8-1 在运行时处理注解

不过,这个示例展示了对一个程序进行注解以及对这些注解进行分析的机制。既然你已经领会了这个具体示例,那么,现在可能已经为后续小节详述注解语法做好了更充分的准备(这也是我们所希望的)。

API *java.lang.reflect.AnnotatedElement* 5.0

- boolean isAnnotationPresent(Class<? extends Annotation> annotationType)
  如果该项具有给定类型的注解,则返回 true。
- <T extends Annotation> T getAnnotation(Class<T> annotationType)
  获得给定类型的注解,如果该项不具有这样的注解,则返回 null。
- <T extends Annotation> T[] getAnnotationsByType(Class<T> annotationType) 8
  获得某个可重复注解类型的所有注解(查阅 8.5.3 节),或者返回长度为 0 的数组。
- Annotation[] getAnnotations()
  获得作用于该项的所有注解,包括继承而来的注解。如果没有出现任何注解,那么将返回一个长度为 0 的数组。
- Annotation[] getDeclaredAnnotations()
  获得为该项声明的所有注解,不包含继承而来的注解。如果没有出现任何注解,那么将返回一个长度为 0 的数组。

8.4 注解语法

本节将介绍你必须了解的注解语法。

8.4.1 注解接口

注解是由注解接口来定义的:

```
modifiers @interface AnnotationName
{
    elementDeclaration₁
    elementDeclaration₂
    ...
}
```

每个元素声明都具有下面这种形式：

type elementName();

或者

type elementName() default *value*;

例如，下面这个注解具有两个元素：assignedTo 和 severity。

```
public @interface BugReport
{
    String assignedTo() default "[none]";
    int severity();
}
```

所有的注解接口都隐式地扩展自 java.lang.annotation.Annotation 接口。这个接口是一个常规接口，不是一个注解接口。请查看本节最后为该接口提供的一些方法所做的 API 注解。

你无法扩展注解接口。换句话说，所有的注解接口都直接扩展自 java.lang.annotation.Annotation。

你从来不用为注解接口提供实现类。

注解元素的类型为下列之一：

- 基本类型（int、short、long、byte、char、double、float 或者 boolean）。
- String。
- Class（具有一个可选的类型参数，例如 Class<? extends MyClass>）。
- enum 类型。
- 注解类型。
- 由前面所述类型组成的数组（由数组组成的数组不是合法的元素类型）。

下面是一些合法的元素声明的例子：

```
public @interface BugReport
{
    enum Status { UNCONFIRMED, CONFIRMED, FIXED, NOTABUG };
    boolean showStopper() default false;
    String assignedTo() default "[none]";
    Class<?> testCase() default Void.class;
    Status status() default Status.UNCONFIRMED;
    Reference ref() default @Reference(); // an annotation type
    String[] reportedBy();
}
```

API *java.lang.annotation.Annotation* 5.0

- Class<? extends Annotation> annotationType()

返回 Class 对象，它用于描述该注解对象的注解接口。注意：调用注解对象上的 getClass 方法可以返回真正的类，而不是接口。
- boolean equals(Object other)

  如果 other 是一个实现了与该注解对象相同的注解接口的对象，并且如果该对象和 other 的所有元素彼此相等，那么返回 True。
- int hashCode()

  返回一个与 equals 方法兼容、由注解接口名以及元素值衍生而来的散列码。
- String toString()

  返回一个包含注解接口名以及元素值的字符串表示，例如，@BugReport (assignedTo= [none], severity=0)。

8.4.2 注解

每个注解都具有下面这种格式：

@*AnnotationName*(*elementName*$_1$=*value*$_1$, *elementName*$_2$=*value*$_2$, . . .)

例如，

@BugReport(assignedTo="Harry", severity=10)

元素的顺序无关紧要。下面这个注解和前面那个一样。

@BugReport(severity=10, assignedTo="Harry")

如果某个元素的值并未指定，那么就使用声明的默认值。例如，考虑一下下面这个注解：

@BugReport(severity=10)

元素 assignedTo 的值是字符串 "[none]"。

> ⚠️ **警告**：默认值并不是和注解存储在一起的；相反地，它们是动态计算而来的。例如，如果你将元素 assignedTo 的默认值更改为 "[]"，然后重新编译 BugReport 接口，那么注解 @BugReport(severity=10) 将使用这个新的默认值，甚至在那些在默认值修改之前就已经编译过的类文件中也是如此。

有两个特殊的快捷方式可以用来简化注解。

如果没有指定元素，要么是因为注解中没有任何元素，要么是因为所有元素都使用默认值，那么你就不需要使用圆括号了。例如，

@BugReport

和下面这个注解是一样的

@BugReport(assignedTo="[none]", severity=0)

这样的注解又称为标记注解。

另外一种快捷方式是单值注解。如果一个元素具有特殊的名字 value，并且没有指定其他元素，那么你就可以忽略掉这个元素名以及等号。例如，既然我们已经在前面将 Action

ListenerFor 注解接口定义为如下形式：

```
public @interface ActionListenerFor
{
    String value();
}
```

那么，我们可以将这个注解书写成如下形式：

`@ActionListenerFor("yellowButton")`

而不是

`@ActionListenerFor(value="yellowButton")`

一个项可以有多个注解：

```
@Test
@BugReport(showStopper=true, reportedBy="Joe")
public void checkRandomInsertions()
```

如果注解的作者将其声明为可重复的，那么你就可以多次重复使用同一个注解：

```
@BugReport(showStopper=true, reportedBy="Joe")
@BugReport(reportedBy={"Harry", "Carl"})
public void checkRandomInsertions()
```

> 注释：因为注解是由编译器计算而来的，因此，所有元素值必须是编译期常量。例如，
> `@BugReport(showStopper=true, assignedTo="Harry", testCase=MyTestCase.class, status=BugReport.Status.CONFIRMED, ...)`

> 警告：一个注解元素永远不能设置为 null，甚至不允许其默认值为 null。这样在实际应用中会相当不方便。你必须使用其他的默认值，例如 "" 或者 Void.class。

如果元素值是一个数组，那么要将它的值用括号括起来，像下面这样：

`@BugReport(..., reportedBy={"Harry", "Carl"})`

如果该元素具有单值，那么可以忽略这些括号：

`@BugReport(..., reportedBy="Joe") // OK, same as {"Joe"}`

既然一个注解元素可以是另一个注解，那么就可以创建出任意复杂的注解。例如，

`@BugReport(ref=@Reference(id="3352627"), ...)`

> 注释：在注解中引入循环依赖是一种错误。例如，因为 BugReport 具有一个注解类型为 Reference 的元素，所以 Reference 就不能再拥有一个类型为 BugReport 的元素。

8.4.3 注解各类声明

注解可以出现在许多地方，这些地方可以分为两类：声明和类型用法声明注解可以出现在下列声明处：

- 包

- 类（包括 enum）
- 接口（包括注解接口）
- 方法
- 构造器
- 实例域（包含 enum 常量）
- 局部变量
- 参数变量
- 类型参数

对于类和接口，需要将注解放置在 class 和 interface 关键词的前面：

```
@Entity public class User { . . . }
```

对于变量，需要将它们放置在类型的前面：

```
@SuppressWarnings("unchecked") List<User> users = . . .;
public User getUser(@Param("id") String userId)
```

泛化类或方法中的类型参数可以像下面这样被注解：

```
public class Cache<@Immutable V> { . . . }
```

包是在文件 package-info.java 中注解的，该文件只包含以注解先导的包语句。

```
/**
    Package-level Javadoc
*/
@GPL(version="3")
package com.horstmann.corejava;
import org.gnu.GPL;
```

> 注释：对局部变量的注解只能在源码级别上进行处理。类文件并不描述局部变量。因此，所有的局部变量注解在编译完一个类的时候就会被遗弃掉。同样地，对包的注解不能在源码级别之外存在。

8.4.4 注解类型用法

声明注解提供了正在被声明的项的相关信息。例如，在下面的声明中

```
public User getUser(@NonNull String userId)
```

就断言 userId 参数不为空。

> 注 释：@NonNull 注 解 是 Checker Framework 的 一 部 分（http://types.cs.washington.edu/checker-framework）。通过使用这个框架，可以在程序中包含断言，例如某个参数不为空，或者某个 String 包含一个正则表达式。然后，静态分析工具将检查在给定的源代码段中这些断言是否有效。

现在，假设我们有一个类型为 List<String> 的参数，并且想要表示其中所有的字符串都不为 null。这就是类型用法注解大显身手之处，可以将该注解放置到类型参数之前：List<@

NonNull String>。

类型用法注解可以出现在下面的位置：
- 与泛化类型参数一起使用：List<@NonNull String>, Comparator.<@NonNull String> reverseOrder()。
- 数组中的任何位置：@NonNull String[][] words（words[i][j] 不为 null），String @NonNull [][] words（words 不为 null），String[] @NonNull [] words（words[i] 不为 null）。
- 与超类和实现接口一起使用：class Warning extends @Localized Message。
- 与构造器调用一起使用：new @Localized String(...)。
- 与强制转型和 instanceof 检查一起使用：(@Localized String) text, if (text instanceof @Localized String)。（这些注解只供外部工具使用，它们对强制转型和 instanceof 检查不会产生任何影响。）
- 与异常规约一起使用：public String read() throws @Localized IOException。
- 与通配符和类型边界一起使用：List<@Localized ? extends Message>, List<? extends @Localized Message>。
- 与方法和构造器引用一起使用：@Localized Message::getText。

有多种类型位置是不能被注解的：

```
@NonNull String.class // ERROR: Cannot annotate class literal
import java.lang.@NonNull String; // ERROR: Cannot annotate import
```

可以将注解放置到诸如 private 和 static 这样的其他修饰符的前面或后面。习惯（但不是必需）的做法，是将类型用法注解放置到其他修饰符的后面和将声明注解放置到其他修饰符的前面。例如，

```
private @NonNull String text; // Annotates the type use
@Id private String userId; // Annotates the variable
```

> **注释**：注解的作者需要指定特定的注解可以出现在哪里。如果一个注解可以同时应用于变量和类型用法，并且它确实被应用到了某个变量声明上，那么该变量和类型用法就都被注解了。例如，请考虑
>
> ```
> public User getUser(@NonNull String userId)
> ```
>
> 如果 @NonNull 可以同时应用于参数和类型用法，那么 userId 参数就被注解了，而其参数类型是 @NonNull String。

8.4.5 注解 this

假设想要将参数注解为在方法中不会被修改。

```
public class Point
{
    public boolean equals(@ReadOnly Object other) { ... }
}
```

那么，处理这个注解的工具在看到下面的调用时

```
p.equals(q)
```

就会推理出 q 没有被修改过。

但是 p 呢？

当该方法被调用时，this 变量是绑定到 p 的。但是 this 从来都没有被声明过，因此你无法注解它。

实际上，你可以用一种很少使用的语法变体来声明它，这样你就可以添加注解了：

```
public class Point
{
   public boolean equals(@ReadOnly Point this, @ReadOnly Object other) { ... }
}
```

第一个参数被称为接收器参数，它必须被命名为 this，而它的类型就是要构建的类。

> **注释**：你只能为方法而不能为构造器提供接收器参数。从概念上讲，构造器中的 this 引用在构造器没有执行完之前还不是给定类型的对象。所以，放置在构造器上的注解描述的是被构建的对象的属性。

传递给内部类构造器的是另一个不同的隐藏参数，即对其外围类对象的引用。你也可以让这个参数显式化：

```
public class Sequence
{
   private int from;
   private int to;

   class Iterator implements java.util.Iterator<Integer>
   {
      private int current;
      public Iterator(@ReadOnly Sequence Sequence.this)
      {
         this.current = Sequence.this.from;
      }
      ...
   }
   ...
}
```

这个参数的名字必须像引用它时那样，叫做 EnclosingClass.this，其类型为外围类。

8.5 标准注解

Java SE 在 java.lang、java.lang.annotation 和 javax.annotation 包中定义了大量的注解接口。其中四个是元注解，用于描述注解接口的行为属性，其他的三个是规则接口，可以用它们来注解你的源代码中的项。表 8-2 列出了这些注解。我们将会在随后的两个小节中给予详细介绍。

表 8-2 标准注解

注解接口	应用场合	目的
Deprecated	全部	将项标记为过时的
SuppressWarnings	除了包和注解之外的所有情况	阻止某个给定类型的警告信息
SafeVarargs	方法和构造器	断言 varargs 参数可安全使用
Override	方法	检查该方法是否覆盖了某一个超类方法
FunctionalInterface	接口	将接口标记为只有一个抽象方法的函数式接口
PostConstruct PreDestroy	方法	被标记的方法应该在构造之后或移除之前立即被调用
Resource	类、接口、方法、域	在类或接口上:标记为在其他地方要用到的资源。在方法或域上:为"注入"而标记
Resources	类、接口	一个资源数组
Generated	全部	
Target	注解	指明可以应用这个注解的那些项
Retention	注解	指明这个注解可以保留多久
Documented	注解	指明这个注解应该包含在注解项的文档中
Inherited	注解	指明当这个注解应用于一个类的时候,能够自动被它的子类继承
Repeatable	注解	指明这个注解可以在同一个项上应用多次

8.5.1 用于编译的注解

@Deprecated 注解可以被添加到任何不再鼓励使用的项上。所以,当你使用一个已过时的项时,编译器将会发出警告。这个注解与 Javadoc 标签 @deprecated 具有同等功效。但是,该注解会一直持久化到运行时。

> **注释**: jdeprscan 工具可以扫描 JAR 文件集中的过时元素,它是 JDK 的组成部分。

@SuppressWarnings 注解会告知编译器阻止特定类型的警告信息,例如,

@SuppressWarnings("unchecked")

@Override 这种注解只能应用到方法上。编译器会检查具有这种注解的方法是否真正覆盖了一个来自于超类的方法。例如,如果你声明:

```
public MyClass
{
    @Override public boolean equals(MyClass other);
    ...
}
```

那么编译器会报告一个错误。毕竟,这个 equals 方法没有覆盖 Object 类的 equals 方法。因为那个方法有一个类型为 Object 而不是 MyClass 的参数。

@Generated 注解的目的是供代码生成工具来使用。任何生成的源代码都可以被注解,从而与程序员提供的代码区分开。例如,代码编辑器可以隐藏生成的代码,或者代码生成器可以

移除生成代码的旧版本。每个注解都必须包含一个表示代码生成器的唯一标识符，而日期字符串（ISO8601 格式）和注释字符串是可选的。例如，

```
@Generated("com.horstmann.beanproperty", "2008-01-04T12:08:56.235-0700");
```

8.5.2 用于管理资源的注解

@PostConstruct 和 @PreDestroy 注解用于控制对象生命周期的环境中，例如 Web 容器和应用服务器。标记了这些注解的方法应该在对象被构建之后，或者在对象被移除之前，紧接着调用。

@Resource 注解用于资源注入。例如，考虑一下访问数据库的 Web 应用。当然，数据库访问信息不应该被硬编码到 Web 应用中。而是应该让 Web 容器提供某种用户接口，以便设置连接参数和数据库资源的 JNDI 名字。在这个 Web 应用中，可以像下面这样引用数据源：

```
@Resource(name="jdbc/mydb")
private DataSource source;
```

当包含这个域的对象被构造时，容器会"注入"一个对该数据源的引用。

8.5.3 元注解

@Target 元注解可以应用于一个注解，以限制该注解可以应用到哪些项上。例如，

```
@Target({ElementType.TYPE, ElementType.METHOD})
public @interface BugReport
```

表 8-3 显示了所有可能的取值情况，它们属于枚举类型 ElementType。可以指定任意数量的元素类型，用括号括起来。

表 8-3 @Target 注解的元素类型

元素类型	注解适用场合	元素类型	注解适用场合
ANNOTATION_TYPE	注解类型声明	FIELD	成员域（包括 enum 常量）
PACKAGE	包	PARAMETER	方法或构造器参数
TYPE	类（包括 enum）及接口（包括注解类型）	LOCAL_VARIABLE	局部变量
METHOD	方法	TYPE_PARAMETER	类型参数
CONSTRUCTOR	构造器	TYPE_USE	类型用法

一条没有 @Target 限制的注解可以应用于任何项上。编译器将检查你是否将一条注解只应用到了某个允许的项上。例如，如果将 @BugReport 应用于一个成员域上，则会导致一个编译器错误。

@Retention 元注解用于指定一条注解应该保留多长时间。只能将其指定为表 8-4 中的任意值，其默认值是 RetentionPolicy.CLASS。

表 8-4 用于 @Retention 注解的保留策略

保留规则	描述
SOURCE	不包括在类文件中的注解
CLASS	包括在类文件中的注解,但是虚拟机不需要将它们载入
RUNTIME	包括在类文件中的注解,并由虚拟机载入。通过反射 API 可获得它们

在程序清单 8-8 中,@ActionListenerFor 注解声明为具有 RetentionPolicy.RUNTIME,因为我们是使用反射机制进行注解处理的。在随后的两个小节里,你将会看到一些在源码级别和类文件级别上怎样对注解进行处理的示例。

@Documented 元注解为像 Javadoc 这样的归档工具提供了一些提示。应该像处理其他修饰符(例如 protected 和 static)一样来处理归档注解,以实现其归档目的。其他注解的使用并不会纳入归档的范畴。例如,假定我们将 @ActionListenerFor 作为一个归档注解来声明:

```
@Documented
@Target(ElementType.METHOD)
@Retention(RetentionPolicy.RUNTIME)
public @interface ActionListenerFor
```

现在每一个被该注解标注过的方法的归档就会含有这条注解,如图 8-2 所示。

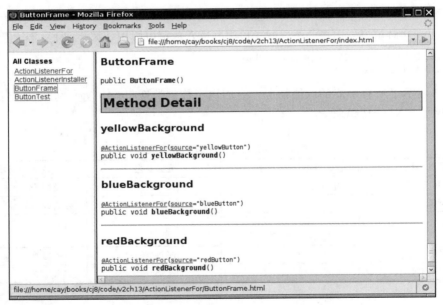

图 8-2 归档注解

如果某个注解是暂时性的(例如 @BugReport),那么就不应该对它们的用法进行归档。

> **注释**:将一个注解应用到它自身上是合法的。例如,@Documented 注解被它自身注解为 @Documented。因此,针对注解的 Javadoc 文档可以表明它们是否可被归档。

@Inherited 元注解只能应用于对类的注解。如果一个类具有继承注解，那么它的所有子类都自动具有同样的注解。这使得创建一个与 Serializable 这样的标记接口具有相同运行方式的注解变得很容易。

实际上，@Serializable 注解应该比没有任何方法的 Serializable 标记接口更适用。一个类之所以可以被序列化，是因为存在着对它的成员域进行读写的运行期支持，而不是因为任何面向对象的设计原则。注解比接口继承更擅长描述这一事实。当然，可序列化接口是在 JDK1.1 中产生的，远比注解出现得早。

假设定义了一个继承注解 @Persistent 来指明一个类的对象可以存储到数据库中，那么该持久类的子类就会自动被注解为是持久性的。

```
@Inherited @interface Persistent { }
@Persistent class Employee { ... }
class Manager extends Employee { ... } // also @Persistent
```

在持久化机制去查找存储在数据库中的对象时，它就会同时探测到 Employee 对象以及 Manager 对象。

对于 Java SE 8 来说，将同种类型的注解多次应用于某一项是合法的。为了向后兼容，可重复注解的实现者需要提供一个容器注解，它可以将这些重复注解存储到一个数组中。

下面是如何定义 @TestCase 注解以及它的容器的代码：

```
@Repeatable(TestCases.class)
@interface TestCase
{
    String params();
    String expected();
}

@interface TestCases
{
    TestCase[] value();
}
```

无论何时，只要用户提供了两个或更多个 @TestCase 注解，那么它们就会自动地被包装到一个 @TestCases 注解中。

> **警告**：在处理可重复注解时必须非常仔细。如果调用 getAnnotation 来查找某个可重复注解，而该注解又确实重复了，那么就会得到 null。这是因为重复注解被包装到了容器注解中。在这种情况下，应该调用 getAnnotationsByType。这个调用会"遍历"容器，并给出一个重复注解的数组。如果只有一条注解，那么该数组的长度就为 1。通过使用这个方法，你就不用操心如何处理容器注解了。

8.6 源码级注解处理

在上一节中，你看到了如何分析正在运行的程序中的注解。注解的另一种用法是自动处理源代码以产生更多的源代码、配置文件、脚本或其他任何我们想要生成的东西。

8.6.1 注解处理器

注解处理已经被集成到了 Java 编译器中。在编译过程中，你可以通过运行下面的命令来调用注解处理器。

```
javac -processor ProcessorClassName₁,ProcessorClassName₂,... sourceFiles
```

编译器会定位源文件中的注解。每个注解处理器会依次执行，并得到它表示感兴趣的注解。如果某个注解处理器创建了一个新的源文件，那么上述过程将重复执行。如果某次处理循环没有再产生任何新的源文件，那么就编译所有的源文件。

> **注释**：注解处理器只能产生新的源文件，它无法修改已有的源文件。

注解处理器通常通过扩展 AbstractProcessor 类而实现 Processor 接口。你需要指定你的处理器支持的注解，我们的案例如下：

```java
@SupportedAnnotationTypes("com.horstmann.annotations.ToString")
@SupportedSourceVersion(SourceVersion.RELEASE_8)
public class ToStringAnnotationProcessor extends AbstractProcessor
{
    public boolean process(Set<? extends TypeElement> annotations,
        RoundEnvironment currentRound)
    {
        ...
    }
}
```

处理器可以声明具体的注解类型或诸如 "com.horstmann*" 这样的通配符（com.horstmann 包及其所有子包中的注解），甚至是 "*"（所有注解）。

在每一轮中，process 方法都会被调用一次，调用时会传递给由这一轮在所有文件中发现的所有注解构成的集，以及包含了有关当前处理轮次的信息的 RoundEnvironment 引用。

8.6.2 语言模型 API

应该使用语言模型 API 来分析源码级的注解。与用来呈现类和方法的虚拟机表示形式的反射 API 不同，语言模型 API 让我们可以根据 Java 语言的规则去分析 Java 程序。

编译器会产生一棵树，其节点是实现了 javax.lang.model.element.Element 接口及其 TypeElement、VariableElement、ExecutableElement 等子接口的类的实例。这些节点可以类比于编译时的 Class、Field/Parament 和 Method/Constructor 反射类。

本书并不会详细讨论该 API，但我们要强调的是，你需要知道它是如何处理注解的。

- RoundEnvironment 通过调用下面的方法交给你一个由特定注解标注过的所有元素构成的集。

    ```
    Set<? extends Element> getElementsAnnotatedWith(Class<? extends Annotation> a)
    ```

- 在源码级别上等价于 AnnotatedElement 接口的是 AnnotatedConstruct。使用下面的方法就可以获得属于给定注解类的单条注解或重复的注解。

    A getAnnotation(Class<A> annotationType)
    A[] getAnnotationsByType(Class<A> annotationType)

- TypeElement 表示一个类或接口，而 getEnclosedElements 方法会产生一个由它的域和方法构成的列表。
- 在 Element 上调用 getSimpleName 或在 TypeElement 上调用 getQualifiedName 会产生一个 Name 对象，它可以用 toString 方法转换为一个字符串。

8.6.3 使用注解来生成源码

作为示例，我们将使用注解来减少实现 toString 方法时枯燥的编程工作量。我们不能将这些方法放到原来的类中，因为注解处理器只能产生新的类，而不能修改已有的类。

因此，我们将所有方法添加到工具类 ToStrings 中：

```
public class ToStrings
{
    public static String toString(Point obj)
    {
        Generated code
    }
    public static String toString(Rectangle obj)
    {
        Generated code
    }
    . . .
    public static String toString(Object obj)
    {
        return Objects.toString(obj);
    }
}
```

我们不想使用反射，因此对访问器方法而不是域进行注解：

```
@ToString
public class Rectangle
{
    . . .
    @ToString(includeName=false) public Point getTopLeft() { return topLeft; }
    @ToString public int getWidth() { return width; }
    @ToString public int getHeight() { return height; }
}
```

然后，注解处理器应该生成下面的源码：

```
public static String toString(Rectangle obj)
{
    var result = new StringBuilder();
    result.append("Rectangle");
    result.append("[");
    result.append(toString(obj.getTopLeft()));
    result.append(",");
```

```
    result.append("width=");
    result.append(toString(obj.getWidth()));
    result.append(",");
    result.append("height=");
    result.append(toString(obj.getHeight()));
    result.append("]");
    return result.toString();
}
```

其中,灰色的是"模板"代码。下面的框架所描述的方法可以为具有给定的 TypeElement 的类产生 toString 方法:

```
private void writeToStringMethod(PrintWriter out, TypeElement te)
{
    String className = te.getQualifiedName().toString();
    Print method header and declaration of string builder
    ToString ann = te.getAnnotation(ToString.class);
    if (ann.includeName())
        Print code to add class name
    for (Element c : te.getEnclosedElements())
    {
        ann = c.getAnnotation(ToString.class);
        if (ann != null)
        {
            if (ann.includeName()) Print code to add field name
            Print code to append toString(obj.methodName())
        }
    }
    Print code to return string
}
```

而下面给出的是注解处理器的 process 方法的框架。它会创建助手类的源文件,并为每个被注解标注的类编写类头和一个 toString 方法。

```
public boolean process(Set<? extends TypeElement> annotations,
    RoundEnvironment currentRound)
{
    if (annotations.size() == 0) return true;
    try
    {
        JavaFileObject sourceFile = processingEnv.getFiler().createSourceFile(
            "com.horstmann.annotations.ToStrings");
        try (var out = new PrintWriter(sourceFile.openWriter()))
        {
            Print code for package and class
            for (Element e : currentRound.getElementsAnnotatedWith(ToString.class))
            {
                if (e instanceof TypeElement)
                {
                    TypeElement te = (TypeElement) e;
                    writeToStringMethod(out, te);
                }
            }
            Print code for toString(Object)
        }
```

```
        catch (IOException ex)
        {
            processingEnv.getMessager().printMessage(
                Kind.ERROR, ex.getMessage());
        }
    }
    return true;
}
```

对于具体的那些显得有些冗长的代码,可以去查看本书附带的代码。

注意,process 方法在后续轮次中是用空的注解列表调用的,然后,它会立即返回,因此它并不会多次创建源文件。

首先,编译注解处理器,然后编译并运行测试程序,就像下面这样:

```
javac sourceAnnotations/ToStringAnnotationProcessor.java
javac -processor sourceAnnotations.ToStringAnnotationProcessor rect/*.java
java rect.SourceLevelAnnotationDemo
```

> **提示**:要想查看轮次,可以用 -XprintRounds 标记来运行 javac 命令:
> ```
> Round 1:
>     input files: {rect.Point, rect.Rectangle,
>         rect.SourceLevelAnnotationDemo}
>     annotations: [sourceAnnotations.ToString]
>     last round: false
> Round 2:
>     input files: {sourceAnnotations.ToStrings}
>     annotations: []
>     last round: false
> Round 3:
>     input files: {}
>     annotations: []
>     last round: true
> ```

这个示例演示了工具可以如何获取源文件注解以产生其他文件。生成的文件并非一定要是源文件。注解处理器可以选择生成 XML 描述符、属性文件、Shell 脚本、HTML 文档等。

> **注释**:有些人建议使用注解来完成一项更繁重的体力活。如果琐碎的获取器和设置器可以自动生成,那岂不是很好?例如,用下面的注解:
>
>   @Property private String title;
>
> 来产生下面的方法:
>
>   public String getTitle() { return title; }
>   public void setTitle(String title) { this = title; }
>
> 但是,这些方法需要被添加到同一个类中。这需要编辑源文件而不是产生另一个文件,而这超出了注解处理器的能力范围。我们可以为实现此目的而构建另一个工具,但是这种工具超出了注解的职责范围。注解被设计为对代码项的描述,而不是添加或修改代码的指令。

8.7 字节码工程

你已经看到了我们是怎样在运行期或者在源码级别上对注解进行处理的。还有第 3 种可能：在字节码级别上进行处理。除非将注解在源码级别上删除，否则它们会一直存在于类文件中。类文件格式是归过档的（参阅 http://docs.oracle.com/javase/specs/jvms/se10/html），这种格式相当复杂，并且在没有特殊类库支持的情况下，处理类文件具有很大的挑战性。ASM 库就是这样的特殊类库之一，可以从网站 http://asm.ow2.org 上获得。

8.7.1 修改类文件

在本小节，我们使用 ASM 向已注解方法中添加日志信息。如果一个方法被这样注解过：

```
@LogEntry(logger=loggerName)
```

那么，在方法的开头部分，我们将添加下面这条语句的字节码：

```
Logger.getLogger(loggerName).entering(className, methodName);
```

例如，如果对 Item 类的 hashCode 方法做了如下注解：

```
@LogEntry(logger="global") public int hashCode()
```

那么，在任何时候调用该方法，都会报告一条与下面打印出来的消息相似的消息：

```
May 17, 2016 10:57:59 AM Item hashCode
FINER: ENTRY
```

为了实现这项任务，我们需要遵循下面几点：
1. 加载类文件中的字节码。
2. 定位所有的方法。
3. 对于每个方法，检查它是不是有一个 LogEntry 注解。
4. 如果有，在方法开头部分添加下面所列指令的字节码：

```
ldc loggerName
invokestatic
    java/util/logging/Logger.getLogger:(Ljava/lang/String;)Ljava/util/logging/Logger;
ldc className
ldc methodName
invokevirtual
    java/util/logging/Logger.entering:(Ljava/lang/String;Ljava/lang/String;)V
```

插入这些字节码看起来相当棘手，不过 ASM 却使它变得相当简单。我们不会详细描述和分析插入字节码的过程。关键之处是程序清单 8-9 中的程序可以编辑一个类文件，并且在已经用 LogEntry 注解标注过的方法的开头部分插入日志调用。

例如，下面展示了应该怎样向程序清单 8-10 中的 Item.java 文件添加记录日志指令，其中 asm 是安装 ASM 库的目录。

```
javac set/Item.java
javac -classpath .:asm/lib/\* bytecodeAnnotations/EntryLogger.java
java -classpath .:asm/lib/\* bytecodeAnnotations.EntryLogger set.Item
```

在对 Item 类文件进行修改之前和之后分别试运行一下：

```
javap -c set.Item
```

就可以看到在 hashCode、equals 以及 compareTo 方法的开头部分插入的那些指令。

```
public int hashCode();
  Code:
   0: ldc     #85; // String global
   2: invokestatic  #80;
      // Method
      // java/util/logging/Logger.getLogger:(Ljava/lang/String;)Ljava/util/logging/Logger;
   5: ldc     #86; //String Item
   7: ldc     #88; //String hashCode
   9: invokevirtual  #84;
      // Method java/util/logging/Logger.entering:(Ljava/lang/String;Ljava/lang/String;)V
  12: bipush  13
  14: aload_0
  15: getfield    #2; // Field description:Ljava/lang/String;
  18: invokevirtual  #15; // Method java/lang/String.hashCode:()I
  21: imul
  22: bipush  17
  24: aload_0
  25: getfield    #3; // Field partNumber:I
  28: imul
  29: iadd
  30: ireturn
```

程序清单 8-11 中的 SetTest 程序会将 Item 对象插入到一个散列集中。当你用修改过的类文件来运行该程序时，会看到下面的日志记录信息：

```
May 17, 2016 10:57:59 AM Item hashCode
FINER: ENTRY
May 17, 2016 10:57:59 AM Item hashCode
FINER: ENTRY
May 17, 2016 10:57:59 AM Item hashCode
FINER: ENTRY
May 17, 2016 10:57:59 AM Item equals
FINER: ENTRY
[[description=Toaster, partNumber=1729], [description=Microwave, partNumber=4104]]
```

当将同一项插入两次时，请注意一下对 equals 的调用。

这个示例显示了字节码工程的强大之处：注解可以用来向程序中添加一些指示，而字节码编辑工具则可以提取这些指示，然后修改虚拟机指令。

程序清单 8-9 bytecodeAnnotations/EntryLogger.java

```
1 package bytecodeAnnotations;
2
3 import java.io.*;
4 import java.nio.file.*;
5
6 import org.objectweb.asm.*;
7 import org.objectweb.asm.commons.*;
8
```

```java
  9  /**
 10   * Adds "entering" logs to all methods of a class that have the LogEntry annotation.
 11   * @version 1.21 2018-05-01
 12   * @author Cay Horstmann
 13   */
 14  public class EntryLogger extends ClassVisitor
 15  {
 16     private String className;
 17
 18     /**
 19      * Constructs an EntryLogger that inserts logging into annotated methods of a given class.
 20      * @param cg the class
 21      */
 22     public EntryLogger(ClassWriter writer, String className)
 23     {
 24        super(Opcodes.ASM5, writer);
 25        this.className = className;
 26     }
 27
 28     public MethodVisitor visitMethod(int access, String methodName, String desc,
 29           String signature, String[] exceptions)
 30     {
 31        MethodVisitor mv = cv.visitMethod(access, methodName, desc, signature, exceptions);
 32        return new AdviceAdapter(Opcodes.ASM5, mv, access, methodName, desc)
 33           {
 34              private String loggerName;
 35
 36              public AnnotationVisitor visitAnnotation(String desc, boolean visible)
 37              {
 38                 return new AnnotationVisitor(Opcodes.ASM5)
 39                    {
 40                       public void visit(String name, Object value)
 41                       {
 42                          if (desc.equals("LbytecodeAnnotations/LogEntry;")
 43                                && name.equals("logger"))
 44                             loggerName = value.toString();
 45                       }
 46                    };
 47              }
 48
 49              public void onMethodEnter()
 50              {
 51                 if (loggerName != null)
 52                 {
 53                    visitLdcInsn(loggerName);
 54                    visitMethodInsn(INVOKESTATIC, "java/util/logging/Logger", "getLogger",
 55                          "(Ljava/lang/String;)Ljava/util/logging/Logger;", false);
 56                    visitLdcInsn(className);
 57                    visitLdcInsn(methodName);
 58                    visitMethodInsn(INVOKEVIRTUAL, "java/util/logging/Logger", "entering",
 59                          "(Ljava/lang/String;Ljava/lang/String;)V", false);
 60                    loggerName = null;
 61                 }
 62              }
```

```
 63          };
 64       }
 65
 66       /**
 67        * Adds entry logging code to the given class.
 68        * @param args the name of the class file to patch
 69        */
 70       public static void main(String[] args) throws IOException
 71       {
 72          if (args.length == 0)
 73          {
 74             System.out.println("USAGE: java bytecodeAnnotations.EntryLogger classfile");
 75             System.exit(1);
 76          }
 77          Path path = Paths.get(args[0]);
 78          var reader = new ClassReader(Files.newInputStream(path));
 79          var writer = new ClassWriter(
 80                ClassWriter.COMPUTE_MAXS | ClassWriter.COMPUTE_FRAMES);
 81          var entryLogger = new EntryLogger(writer,
 82                path.toString().replace(".class", "").replaceAll("[/\\\\]", "."));
 83          reader.accept(entryLogger, ClassReader.EXPAND_FRAMES);
 84          Files.write(Paths.get(args[0]), writer.toByteArray());
 85       }
 86    }
```

程序清单 8-10 set/Item.java

```
 1  package set;
 2
 3  import java.util.*;
 4  import bytecodeAnnotations.*;
 5
 6  /**
 7   * An item with a description and a part number.
 8   * @version 1.01 2012-01-26
 9   * @author Cay Horstmann
10   */
11  public class Item
12  {
13     private String description;
14     private int partNumber;
15
16     /**
17      * Constructs an item.
18      * @param aDescription the item's description
19      * @param aPartNumber the item's part number
20      */
21     public Item(String aDescription, int aPartNumber)
22     {
23        description = aDescription;
24        partNumber = aPartNumber;
25     }
26
27     /**
```

```java
28      * Gets the description of this item.
29      * @return the description
30      */
31     public String getDescription()
32     {
33        return description;
34     }
35
36     public String toString()
37     {
38        return "[description=" + description + ", partNumber=" + partNumber + "]";
39     }
40
41     @LogEntry(logger = "com.horstmann")
42     public boolean equals(Object otherObject)
43     {
44        if (this == otherObject) return true;
45        if (otherObject == null) return false;
46        if (getClass() != otherObject.getClass()) return false;
47        var other = (Item) otherObject;
48        return Objects.equals(description, other.description) && partNumber == other.partNumber;
49     }
50
51     @LogEntry(logger = "com.horstmann")
52     public int hashCode()
53     {
54        return Objects.hash(description, partNumber);
55     }
56  }
```

程序清单 8-11 set/SetTest.java

```java
1  package set;
2
3  import java.util.*;
4  import java.util.logging.*;
5
6  /**
7   * @version 1.03 2018-05-01
8   * @author Cay Horstmann
9   */
10 public class SetTest
11 {
12    public static void main(String[] args)
13    {
14       Logger.getLogger("com.horstmann").setLevel(Level.FINEST);
15       var handler = new ConsoleHandler();
16       handler.setLevel(Level.FINEST);
17       Logger.getLogger("com.horstmann").addHandler(handler);
18
19       var parts = new HashSet<Item>();
20       parts.add(new Item("Toaster", 1279));
21       parts.add(new Item("Microwave", 4104));
22       parts.add(new Item("Toaster", 1279));
```

```
23        System.out.println(parts);
24    }
25 }
```

8.7.2 在加载时修改字节码

在前一节中，已经看到了一个用于编辑类文件的工具。但是，在把另一个工具添加到程序的构建过程中时，会显得笨重不堪。更吸引人的做法是将字节码工程延迟到载入时，即类加载器加载类的时候。

设备（*instrumentation*）API 提供了一个安装字节码转换器的挂钩。不过，必须在程序的 main 方法调用之前安装这个转换器。通过定义一个代理，即被加载用来按照某种方式监视程序的一个类库，就可以处理这个需求。代理代码可以在 premain 方法中执行初始化。

下面是构建一个代理所需的步骤：

1. 实现一个具有下面这个方法的类：

   `public static void premain(String arg, Instrumentation instr)`

   当加载代理时，此方法会被调用。代理可以获取一个单一的命令行参数，该参数是通过 arg 参数传递进来的。instr 参数可以用来安装各种各样的挂钩。

2. 制作一个清单文件 EntryLoggingAgent.mf 来设置 Premain-Class 属性。例如：

   `Premain-Class: bytecodeAnnotations.EntryLoggingAgent`

3. 将代理代码打包，并生成一个 JAR 文件，例如：

   ```
   javac -classpath .:asm/lib/\* bytecodeAnnotations/EntryLoggingAgent.java
   jar cvfm EntryLoggingAgent.jar bytecodeAnnotations/EntryLoggingAgent.mf \
      bytecodeAnnotations/Entry*.class
   ```

为了运行一个具有该代理的 Java 程序，需要使用下面这个命令行选项：

`java -javaagent:AgentJARFile=agentArgument ...`

例如，运行具有日志代理的 SetTest 程序需调用：

```
javac set/SetTest.java
java -javaagent:EntryLoggingAgent.jar=set.Item -classpath .:asm/lib/\* set.SetTest
```

Item 参数是代理应该修改的类的名称。

程序清单 8-12 展示了这个代理的代码。该代理安装了一个类文件转换器，这个转换器首先检验类名是否与代理参数相匹配。如果匹配，那么它会利用上一节那个 EntryLogger 类修改字节码。不过，修改过的字节码并不保存成文件。相反地，转换器只是将它们返回，以加载到虚拟机中（参见图 8-3）。换句话说，这项技术实现的是 "即时"（just in time）字节码修改。

程序清单 8-12 bytecodeAnnotations/EntryLoggingAgent.java

```
1  package bytecodeAnnotations;
2
3  import java.lang.instrument.*;
4
```

```
 5  import org.objectweb.asm.*;
 6
 7  /**
 8   * @version 1.11 2018-05-01
 9   * @author Cay Horstmann
10   */
11  public class EntryLoggingAgent
12  {
13     public static void premain(final String arg, Instrumentation instr)
14     {
15        instr.addTransformer((loader, className, cl, pd, data) ->
16           {
17              if (!className.replace("/", ".").equals(arg)) return null;
18              var reader = new ClassReader(data);
19              var writer = new ClassWriter(
20                 ClassWriter.COMPUTE_MAXS | ClassWriter.COMPUTE_FRAMES);
21              var el = new EntryLogger(writer, className);
22              reader.accept(el, ClassReader.EXPAND_FRAMES);
23              return writer.toByteArray();
24           });
25     }
26  }
```

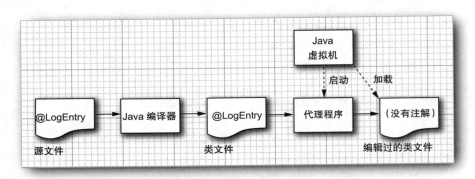

图 8-3 在加载时修改类

在本章，你已经学习到了以下的知识：
- 怎样向 Java 程序中添加注解。
- 怎样设计你自己的注解接口。
- 怎样实现可以利用注解的工具。

你已经看到了三种处理代码的技术：编写脚本、编译 Java 程序和处理注解。前两种技术十分简单。而另一方面，构建注解工具可能会很复杂，但这并非是大多数开发者都需要解决的问题。本章向你介绍了一些背景知识，有助于你去理解可能会碰到的注解工具内部工作机制，但这些背景知识可能会挫伤你自行开发工具的积极性。

下一章将讨论 Java 平台模块系统，它是 Java 9 的关键特性，是促进 Java 平台向前发展的重要动力。

第 9 章 Java 平台模块系统

- ▲ 模块的概念
- ▲ 对模块命名
- ▲ 模块化的"Hello, World！"程序
- ▲ 对模块的需求
- ▲ 导出包
- ▲ 模块的 JAR
- ▲ 模块和反射式访问
- ▲ 自动模块
- ▲ 不具名模块
- ▲ 用于迁移的命令行标志
- ▲ 传递的需求和静态的需求
- ▲ 限定导出和开放
- ▲ 服务加载
- ▲ 操作模块的工具

封装是面向对象编程的一个重要特性。类的声明由公有接口和私有实现构成，类可以通过只修改实现而不影响其用户的方式而得以演化。模块系统为编程带来了大致相同的益处。模块使类和包可以有选择性地获取，从而使得模块的演化可以受控。

多个现有的 Java 模块系统都依赖于类加载器来实现类之间的隔离。但是，Java 9 引入了一个由 Java 编译器和虚拟机支持的新系统，称为 Java 平台模块系统。它被设计用来模块化基于 Java 平台的大型代码基。如果愿意，也可以使用这个系统来模块化我们自己的应用程序。

无论是否在自己的应用程序中使用 Java 平台模块，都可能会受到模块化的 Java 平台的影响。本章将展示如何声明和使用 Java 平台模块。你还会学习到如何迁移你的应用程序，使其能够与模块化的 Java 平台和第三方模块一起工作。

9.1 模块的概念

在面向对象编程中，基础的构建要素就是类。类提供了封装，私有特性只能被具有明确访问权限的代码访问，即，只能被其所属类中的方法访问，这使得对访问权限的推断成为可能。如果某个私有变量发生了变化，那么我们就会发现一系列可能出错的方法。如果需要修改私有表示，那么就需要知道哪些方法会受到影响。

在 Java 中，包提供了更高一级的组织方式，包是类的集合。包也提供了一种封装级别，具有包访问权限的所有特性（无论是公有的还是私有的）都只能被同一个包中的方法访问。

但是，在大型系统中，这种级别的访问控制仍显不足。所有公有特性（即在包的外部也可以访问的特性）可以从任何地方访问。假设我们想要修改或剔除一个很少使用的特性，如果它是公有的，那么就没有办法推断这个变化所产生的影响。

Java 平台的设计者们面对的就是这种情况。过去 20 年中，JDK 呈跨越式发展，但是有

些特性现在明显过时了。有大家喜欢提到的例子，即 CORBA。你最后一次使用它是什么时候？但是 org.omg.corba 包仍旧打包在每一个 JDK 中，直至 Java 10。到了 Java 11，仍旧需要这个包的那些极少量的人就必须将所需的 JAR 文件自己添加到他们的项目中了。

java.awt 的情况又如何呢？服务器端的应用程序并不需要它，对吗？但是，java.awt.DataFlavor 类在 SOAP 的实现中仍在使用，这是一种基于 XML 的 Web 服务协议。

Java 平台的设计者们在面对规模超大且盘根错节的代码时，认为他们需要一种能够提供更多控制能力的构建机制。他们研究了现有的模块系统（例如 OSGi），发现它们都不适用于他们的问题。于是，他们设计了一个新的系统，称为 Java 平台模块系统，现在成了 Java 语言和虚拟机的一部分。这个系统已经成功地用于 Java API 的模块化，如果愿意，也可以使用这个系统来模块化我们自己的应用程序。

一个 Java 平台模块包含：
- 一个包集合
- 可选地包含资源文件和像本地库这样的其他文件
- 一个有关模块中可访问的包的列表
- 一个有关这个模块依赖的所有其他模块的列表

Java 平台在编译时和在虚拟机中都强制执行封装和依赖。

为什么在我们自己的程序中要考虑使用 Java 平台模块系统而不是传统的使用类路径上的 JAR 文件呢？因为这样做有以下两个优点。

1. 强封装：我们可以控制哪些包是可访问的，并且无须操心去维护那些我们不想开放给公众去访问的代码。

2. 可靠的配置：我们可以避免诸如类重复或丢失这类常见的类路径问题。

还有一些有关 Java 平台模块系统的话题我们没有涉及，例如模块的版本管理。当前还不支持指定要求使用模块的具体版本，或者在同一个程序中使用某个模块的多个版本。这些特性可能正是人们所期望的，但是如果需要用到它们，就必须使用 Java 平台模块系统之外的机制。

9.2 对模块命名

模块是包的集合。模块中的包名无须彼此相关。例如，java.sql 模块中就包含了 java.sql、javax.sql 和 javax.transaction.xa 这几个包。并且，正如这个例子所示，模块名和包名相同是完全可行的。

就像路径名一样，模块名是由字母、数字、下划线和句点构成的。而且，和路径名一样，模块之间没有任何层次关系。如果有一个模块是 com.horstmann，另一个模块是 com.horstmann.corejava，那么就模块系统而言，它们是无关的。

当创建供他人使用的模块时，重要的是要确保它的名字是全局唯一的。我们期望大多数的模块名都遵循"反向域名"惯例，就像包名一样。

命名模块最简单的方式就是按照模块提供的顶级包来命名。例如，SLF4J 日志记录外观有一个 org.slf4j 模块，其中包含的包为 org.slf4j、org.slf4j.spi、org.slf4j.event 和 org.slf4j.helpers。

这个惯例可以防止模块中产生包名冲突，因为任何给定的模块都只能被放到一个模块中。如果模块名是唯一的，并且包名以模块名开头，那么包名也就是唯一的。

我们可以使用更短的模块名来命名不打算给其他程序员使用的模块，例如包含某个应用程序的模块。只是为了展示这样做可行，本章就使用了这种方式，那些貌似应该成为库代码的模块都具有像 com.horstmann.util 这样的名字，而包含程序（具有 main 方法的类）的模块都具有像 v2ch09.hellomod 这样很容易记忆的名字。

> **注释**：模块名只用于模块声明中。在 Java 类的源文件中，永远都不应该引用模块名，而是应该按照一如既往的方式去使用包名。

9.3 模块化的"Hello, World!"程序

让我们把传统的"Hello, World!"程序转换为一个模块。首先，我们需要将这个类放到一个包中，"不具名的"包是不能包含在模块中的。下面是代码：

```java
package com.horstmann.hello;

public class HelloWorld
{
    public static void main(String[] args)
    {
        System.out.println("Hello, Modular World!");
    }
}
```

到目前为止，还没有任何东西有变化。为了创建包含这个包的 v2ch09.hellomod 模块，需要添加一个模块声明，可以将其置于名为 module.info.java 的文件中，该文件位于基目录中（即，与包含 com 目录的目录相同）。按照惯例，基目录的名字与模块名相同。

```
v2ch09.hellomod/
└ module-info.java
  com/
  └ horstmann/
    └ hello/
      └ HelloWorld.java
```

module-info.java 文件包含模块声明：

```java
module v2ch09.hellomod
{
}
```

这个模块声明之所以为空，是因为该模块没有任何可以向其他人提供的内容，它也不需要依赖任何东西。

现在，按照往常一样编译它：

```
javac v2ch09.hellomod/module-info.java v2ch09.hellomod/com/horstmann/hello/HelloWorld.java
```

module.info.java 这个文件看起来与 Java 资源文件不同，当然，也不可能存在名为 module-info 的类，因为类名不能包含连字符。关键词 module 和在下一节将会看到的 requires、exports 等关键词都是"限定关键词"，即只在模块声明中具有特殊含义。这个文件会以二进制形式编译到包含该模块定义的类文件 module-info.class 中。

为了让这个程序作为模块化应用程序来运行，需要指定模块路径，它与类路径相似，但是包含的是模块。还需要以模块名 / 类名的形式指定主类：

```
java --module-path v2ch09.hellomod --module v2ch09.hellomod/com.horstmann.hello.HelloWorld
```

也可以不使用 --module-path 和 -module，而是使用单字母选项 -p 和 -m：

```
java -p v2ch09.hellomod -m v2ch09.hellomod/com.horstmann.hello.HelloWorld
```

无论哪种方式，都会显示问候语 "Hello, Module World!"，证明我们成功地模块化了第一个应用程序。

> **注释**：在编译这个模块时，会获得一条警告消息：
>
> ```
> warning: [module] module name component v2ch09 should avoid terminal digits
> ```
>
> 这条警告意在建议程序员不要给模块名添加版本号。你可以忽略这个警告，或者用注解来抑制它：
>
> ```
> @SuppressWarnings("module")
> module v2ch09.hellomod
> {
> }
> ```
>
> 在这一点上，module 声明就像类声明一样：可以对其进行注解。（注解类型必须具有值为 ElementType.MODULE 的 target。）

9.4 对模块的需求

让我们创建一个新的模块 v2ch09.requiremod，其中使用一个 JOptionPane 对象展示了消息 "Hello, Modular World"：

```java
package com.horstmann.hello;

import javax.swing.JOptionPane;

public class HelloWorld
{
    public static void main(String[] args)
    {
        JOptionPane.showMessageDialog(null, "Hello, Modular World!");
    }
}
```

现在，编译会失败并报下面的消息：

```
error: package javax.swing is not visible
  (package javax.swing is declared in module java.desktop,
  but module v2ch09.requiremod does not read it)
```

JDK 已经被模块化了，并且 javax.swing 包现在包含在 java.desktop 模块中。我们的模块需要声明它依赖于这个模块：

```
module v2ch09.requiremod
{
    requires java.desktop;
}
```

模块系统的设计目标之一就是模块需要明确它们的需求，使得虚拟机可以确保在启动程序之前所有的需求都得以满足。

在前一节中，并没有产生明确的需求，因为我们只用到了 java.lang 和 java.io 包。这些包都包含在默认需要的 java.base 模块中。

注意，我们的 v2ch09.requiremod 模块只列出了它自己的模块需求。它需要 java.desktop 模块，这样它才能使用 javax.swing 包。java.desktop 模块自身声明了它需要其他三个包，即 java.datatransfer、java.prefs 和 java.xml。

图 9-1 展示了一张模块图，图中的节点是模块，而图中的边，也就是连接节点的箭头，要么声明了需求，要么在没有声明任何需求时表示需要 java.base。

在模块图中不能有环，即，一个模块不能直接或间接地对自己产生依赖。

模块不会自动地将访问权限传递给其他模块。在我们的示例中，java.desktop 模块声明它需要 java.prefs，而 java.prefs 模块声明它需要 java.xml，但是这并不会赋予 java.desktop 使用来自 java.xml 模块中的包的权力。按照数学术语描述，require 不是"传递性"的。通常，这种行为正是我们想要的，因为它使得需求必须明确化，

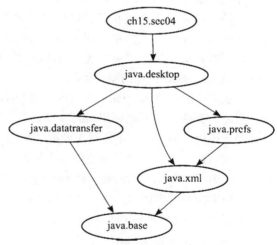

图 9-1　Swing 应用程序 "Hello, Modular World!" 的模块图

但是正如你将会在 9.11 节中看到的，在某些情况下，可以放松这条限制。

> **注释**：本节开头部分给出的错误消息声明我们的 v2ch99.requiremod 模块没有 "读入" java.desktop 模块。按照 Java 模块系统的用语，模块 M 会在下列情况下**读入**模块 N：
> 1. M 需要 N
> 2. M 需要某个模块，而该模块传递性地需要 N（参阅 9.11 节）
> 3. N 是 M 或 java.base

9.5 导出包

在前一节中,我们看到一个模块如果想要使用其他模块中的包,就必须声明需要该模块。但是,这并不会自动使得所需模块中所有的包都可用。模块可以用 exports 关键词来声明它的哪些包可用。例如,下面是 java.xml 模块的模块声明中的一部分:

```
module java.xml
{
    exports javax.xml;
    exports javax.xml.catalog;
    exports javax.xml.datatype;
    exports javax.xml.namespace;
    exports javax.xml.parsers;
    ...
}
```

这个模块让许多包都可用,但是通过不导出其他的包而隐藏了它们(例如 jdk.xml.internal)。当包被导出时,它的 public 和 protected 的类和接口,以及 public 和 protected 的成员,在模块的外部也是可以访问的(如往常一样,protected 的类型和成员只有在子类中才是可访问的)。

但是,没有导出的包在其自己的模块之外是不可访问的,这与 Java 模块化之前很不相同。在过去,我们可以使用任何包中公有的类,尽管它可能并非公有 API 的一部分。例如,当公有 API 没有提供相对应的适合的功能时,通常会推荐使用像 sun.misc.BASE64Encoder 或 com.sun.rowset.CachedRowSetImpl 这样的类。

现在,不能再访问 Java 平台 API 中未导出的包了,因为所有的这些包都包含在模块的内部。因此,有些程序不能再用 Java 9 来运行了。当然,从来没有人承诺过会让非公有的 API 一直保持可用,因此大家不应该对此感到震惊。

让我们在一个简单场景中使用导出机制。我们将准备一个 com.horstmann.greet 模块,它会导出一个名字也是 com.horstmann.greet 的包,这遵循了向他人提供代码的模块应该按照其内部的顶层包来命名的惯例。还有一个名为 com.horstmann.greet.internal 的包,我们并不会导出它。

公有的 Greeter 接口在第一个包中:

```
package com.horstmann.greet;

public interface Greeter
{
    static Greeter newInstance()
    {
        return new com.horstmann.greet.internal.GreeterImpl();
    }

    String greet(String subject);
}
```

第二个包有一个实现了该接口的类。这个类是公有的,因为它需要在第一个包中是可访问的:

```
package com.horstmann.greet.internal;
```

```
import com.horstmann.greet.Greeter;

public class GreeterImpl implements Greeter
{
   public String greet(String subject)
   {
      return "Hello, " + subject + "!";
   }
}
```

com.horstmann.greet 模块包含这两个包,但是只会导出第一个包:

```
module com.horstmann.greet
{
   exports com.horstmann.greet;
}
```

第二个包在模块外部是不可访问的。

我们将应用程序放到第二个包中,它需要用到第一个模块:

```
module v2ch09.exportedpkg
{
   requires com.horstmann.greet;
}
```

> 注释:exports 语句跟在包名后面,而 requires 语句跟在模块名后面。

现在,我们的应用程序将使用 Greeter 来获取问候语:

```
package com.horstmann.hello;

import com.horstmann.greet.Greeter;

public class HelloWorld
{
   public static void main(String[] args)
   {
      Greeter greeter = Greeter.newInstance();
      System.out.println(greeter.greet("Modular World"));
   }
}
```

下面是这两个模块的源文件结构:

```
com.horstmann.greet
├ module-info.java
└ com
   └ horstmann
      └ greet
         ├ Greeter.java
         └ internal
            └ GreeterImpl.java
v2ch09.exportedpkg
├ module-info.java
└ com
   └ horstmann
      └ hello
         └ HelloWorld.java
```

为了构建这个应用程序，首先要编译 com.horstmann.greet 模块：

```
javac com.horstmann.greet/module-info.java \
    com.horstmann.greet/com/horstmann/greet/Greeter.java \
    com.horstmann.greet/com/horstmann/greet/internal/GreeterImpl.java
```

然后，用模块路径上的第一个模块来编译这个应用程序模块：

```
javac -p com.horstmann.greet v2ch09.exportedpkg/module-info.java \
    v2ch09.exportedpkg/com/horstmann/hello/HelloWorld.java
```

最后，用模块路径上的这两个模块来运行这个程序：

```
java -p v2ch09.exportedpkg:com.horstmann.greet \
    -m v2ch09.exportedpkg/com.horstmann.hello.HelloWorld
```

☑ 提示：如果要用 Eclipse 来构建这个应用程序，需要为每一个模块建立一个单独的工程。在 v2ch09.exportedpkg 项目中，编辑项目属性。在 Projects 配置页上，添加 com.horstmann.greet 模块到模块路径中，参阅图 9-2。

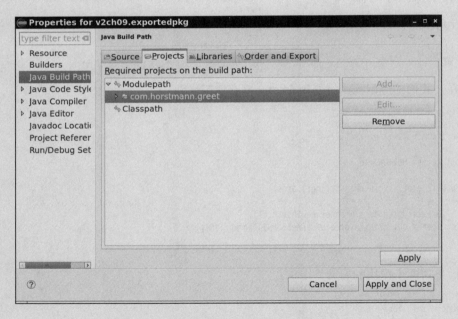

图 9-2　添加依赖的模块到 Eclipse 项目中

现在，你已经看到了构成 Java 平台模块系统基础的 requires 和 exports 语句。正如你所见，模块系统在概念上很简单。模块指定了它们需要哪些模块，以及它们可以向其他模块提供哪些包。9.12 节将展示 exports 语句的一个次要变体。

❗ 警告：模块没有作用域的概念。不能在不同的模块中放置两个具有相同名字的包。即使是隐藏的包（即，不会导出的包），情况也是如此。

9.6 模块化的 JAR

到目前为止，我们直接将模块编译到了源代码的目录树中。很明显，这无法满足部署的要求。模块可以通过将其所有的类都置于一个 JAR 文件中而得以部署，其中 module-info.class 在 JAR 文件的根部。这样的 JAR 文件被称为模块的 JAR。

要想创建模块化的 JAR 文件，只需以通常的方式使用 jar 工具。如果有多个包，那么最好是用 -d 选项来编译，这样可以将类文件置于单独的目录中，如果该目录不存在，则会创建该目录。然后，在收集这些类文件时使用 -c 选项的 jar 命令来修改该目录。

```
javac -d modules/com.horstmann.greet $(find com.horstmann.greet -name *.java)
jar -cvf com.horstmann.greet.jar -C modules/com.horstmann.greet .
```

如果你使用的是像 Maven、Ant 或 Gradle 这样的构建工具，那么只需按照你惯用的方式来构建 JAR 文件。只要 module-info.class 包含在内，就可以得到该模块的 JAR 文件。

然后，在模块路径中包含该模块化的 JAR，该模块就会被加载。

> ⚠️ **警告**：在过去，包中的类有时会分布在多个 JAR 文件中。（这种包被称为"分离包"。）这可能从来就不是一个好注意，对于模块来说也不可能是个好主意。

就像常规的 JAR 文件一样，可以指定模块化的 JAR 中的主类：

```
javac -p com.horstmann.greet.jar \
    -d modules/v2ch09.exportedpkg $(find v2ch09.exportedpkg -name *.java)
jar -c -v -f v2ch09.exportedpkg.jar -e com.horstmann.hello.HelloWorld \
    -C modules/v2ch09.exportedpkg .
```

当启动该程序时，可以指定包含主类的模块：

```
java -p com.horstmann.greet.jar:v2ch09.exportedpkg.jar -m v2ch09.exportedpkg
```

在创建 JAR 文件时，可以选择指定版本号。使用 --module-version 选项，以及在 JAR 文件名上添加 @ 和版本号：

```
jar -c -v -f com.horstmann.greet@1.0.jar --module-version 1.0 -C com.horstmann.greet .
```

正如已经讨论过的，Java 平台模块系统并不会使用版本号来解析模块，但是可以通过其他工具和框架来查询版本号。

> 📝 **注释**：可以通过反射 API 找到版本号。在我们的示例中：
> ```
> Optional<String> version = Greeter.class.getModule().getDescriptor().rawVersion();
> ```
> 将产生一个包含版本号字符串 "1.0" 的 Optional。

> 📝 **注释**：等价于类加载器的模块是一个层。Java 平台模块系统会将 JDK 模块和应用程序模块加载到**启动层**（boot layer）。程序还可以使用分层 API 加载其他模块（本书不会讨论该 API）。这种程序可以选择考虑模块的版本。Java 期望像 Java EE 应用服务器这样的程序的开发者会利用分层 API 来提供对模块的支持。

> **提示：** 如果想要加载模块到 JShell 中，需要将 JAR 包含在模块路径中，并使用 --add-modules 选项：
>
> ```
> jshell --module-path com.horstmann.greet@1.0.jar --add-modules com.horstmann.greet
> ```

9.7 模块和反射式访问

在前面的章节中，我们看到了模块系统是如何强制执行封装的。模块只能访问显式地由其他包导出的包。在过去，总是可以通过使用反射来克服令人讨厌的访问权限问题。正如在卷 I 第 5 章中看到的，反射可以访问任何类的私有成员。

但是，在模块化的世界中，这条路再也行不通了。如果一个类在某个模块中，那么对非公有成员的反射式访问将失败。特别是，回忆一下我们是如何访问私有域的：

```
Field f = obj.getClass().getDeclaredField("salary");
f.setAccessible(true);
double value = f.getDouble(obj);
f.setDouble(obj, value * 1.1);
```

`f.setAccessible(true)` 调用会成功，除非安全管理器不允许对私有域的访问。但是，使用安全管理器来运行 Java 应用程序并不常见，并且有许多使用反射式访问的库。典型的例子包括像 JPA 这样的对象-关系映射器，它们会自动地将对象持久化到数据库中，以及在对象和 XML 或 JSON 之间转换的库中，例如 JAXB 和 JSON-B。

如果使用这种库，并且还想使用模块，那么就必须格外小心。为了演示这个问题，让我们将卷 I 第 5 章中的 ObjectAnalyzer 类放到 com.horstmann.util 模块中。这个类有一个 toString 方法，可以使用反射机制来打印出对象的域。

单独的 v2ch09.openpkg 模块包含一个简单 Country 类：

```
package com.horstmann.places;

public class Country
{
   private String name;
   private double area;

   public Country(String name, double area)
   {
      this.name = name;
      this.area = area;
   }
   // . . .
}
```

下面的短程序演示了如何分析 Country 对象：

```
package com.horstmann.places;

import com.horstmann.util.*;
```

```
public class Demo
{
   public static void main(String[] args) throws ReflectiveOperationException
   {
      var belgium = new Country("Belgium", 30510);
      var analyzer = new ObjectAnalyzer();
      System.out.println(analyzer.toString(belgium));
   }
}
```

现在编译模块和 Demo 程序：

```
javac com.horstmann.util/module-info.java \
   com.horstmann.util/com/horstmann/util/ObjectAnalyzer.java
javac -p com.horstmann.util v2ch09.openpkg/module-info.java \
   v2ch09.openpkg/com/horstmann/places/*.java
java -p v2ch09.openpkg:com.horstmann.util -m v2ch09.openpkg/com.horstmann.places.Demo
```

该程序会以下面的异常而失败：

```
Exception in thread "main" java.lang.reflect.InaccessibleObjectException:
   Unable to make field private java.lang.String com.horstmann.places.Country.name
   accessible: module v2ch09.openpkg does not "opens com.horstmann.places" to module
   com.horstmann.util
```

当然，按照纯理论来说，破坏对象的封装并窥视其私有成员是错误的。但是像对象-关系映射或 XML/JSON 绑定这样的机制应用非常广泛，使得模块系统必须接纳它们。

通过使用 opens 关键词，模块就可以打开包，从而启动对给定包中的类的所有实例进行反射式访问。下面是我们的模块必须执行的操作：

```
module v2ch09.openpkg
{
   requires com.horstmann.util;
   opens com.horstmann.places;
}
```

有了这样的变化，`ObjectAnalyzer` 就可以正确地工作了。

模块可以像下面这样声明为 Open（开放的）：

```
open module v2ch09.openpkg
{
   requires com.horstmann.util;
}
```

开放的模块可以授权对其所有包的运行时访问，就像所有的包都用 exports 和 opens 声明过一样。但是，在运行时只有显式导出的包是可访问的。开放模块将模块系统编译时的安全性和经典的授权许可的运行时行为结合在一起。

回忆一下卷 I 第 5 章，JAR 文件除了类文件和清单外，还可以包含文件资源，它们可以被 `Class.getResourceAsStream` 方法加载，现在还可以被 `Module.getResourceAsStream` 加载。如果资源存储在匹配模块的某个包的目录中，那么这个包必须对调用者是开放的。在其他目录中的资源，以及类文件和清单，可以被任何人读取。

> **注释：** 作为更贴近实际的例子，我们把 Country 对象转换为 XML 或 JSON。Java 9 和 10 中包含用于转换为 XML 的 java.xml.bind 模块。该模块已经从 Java 11 中移除了（同时被移除的模块还有 java.activation、java.corba、java.transaction、java.xml.ws 和 java.xml.ws.annotation）。这些模块包含的包也是 Jakarta EE（之前的 Java EE）规范的一部分，其中的 API 比 Java SE 内涵更广。如果 JDK 中包含有与其冲突的包，那么企业应用服务器不能被模块化。遗憾的是，到本书撰写之时，还没有用于 XML 绑定的模块化替代物出现。
>
> 但是，对于 JSON-B 的实现，如果我们从源码构建它，它就会提供模块化的 JAR 文件。可以期望的是，在你阅读这部分内容时，这些 JAR 文件已经进入了 Maven Central。将这些 JAR 文件放到模块路径上，然后运行 com.horstmann.places.Demo2 程序，当 com.horstmann.places 包开放时，向 JSON 的转换就会成功。

> **注释：** 未来的库可能会使用**变量句柄**而不是反射来读写域。VarHandle 类似于 Field。我们可以使用它来读写具体类的任何实例的具体域。但是，为了获得 VarHandle 对象，库代码需要一个 Lookup 对象：
>
> ```
> public Object getFieldValue(Object obj, String fieldName, Lookup lookup)
>         throws NoSuchFieldException, IllegalAccessException
> {
>     Class<?> cl = obj.getClass();
>     Field field = cl.getDeclaredField(fieldName);
>     VarHandle handle = MethodHandles.privateLookupIn(cl, lookup)
>         .unreflectVarHandle(field);
>     return handle.get(obj);
> }
> ```
>
> 只要该模块中生成的 Lookup 对象拥有对该域的访问权，这段代码就可以工作。在模块中的某些方法可以直接调用 MethodHandles.lookup()，它会产生一个封装了调用者访问权限的对象。在这种方式下，一个模块可以赋予另一个模块访问私有成员的权限。在实践中，需要解决如何以麻烦最少的方式赋予这些权限的问题。

9.8 自动模块

现在你知道了如何使用 Java 平台模块系统。如果从全新的项目开始，其中所有的代码都由我们自己编写，那么就可以设计模块、声明模块依赖关系，并将应用程序打包成模块化的 JAR 文件。

但是，这是一种非常罕见的场景，几乎所有的项目都依赖于第三方的库。当然，我们可以等到所有库的提供商都将库演化成模块，然后再模块化我们自己的代码。

但是如果等不及怎么办呢？ Java 平台模块系统提供了两种机制来填补将当今的前模块化世界与完全模块化应用程序割裂开来的鸿沟：自动化模块和不具名模块。

如果是为了迁移，我们可以通过把任何 JAR 文件置于模块路径的目录而不是类路径的目录中，实现将其转换成一个模块。模块路径上没有 module-info.class 文件的 JAR 被称为自动模块。自动模块具有下面的属性：

1. 模块隐式地包含对其他所有模块的 requires 子句。
2. 其所有包都被导出，且是开放的。
3. 如果在 JAR 文件清单 META-INF/MANIFEST.MF 中具有键为 Automatic-Module-Name 的项，那么它的值会变为模块名。
4. 否则，模块名将从 JAR 文件文件名中获得，将文件名中尾部的版本号删除，并将非字母数字的字符替换为句点。

前两条规则表明自动模块中的包的行为和在类路径上一样。使用模块路径的原因是为了让其他模块受益，使得它们可以表示对这个模块的依赖关系。

例如，假设我们正在实现一个处理 CSV 文件的模块，并使用了 Apache Commons CSV 库。我们想要在 module-info.java 文件中表示模块需要依赖 Apache Commons CSV。

如果在模块路径中添加 commons-csv-1.5.jar，那么我们的模块就可以引用这个模块了。它的名字是 commons.csv，因为去掉了尾部版本号 -1.5，而非字母数字字符 - 被替换成了句点。

这个名字也许算是一个可接受的模块名，因为 Commons CSV 人们耳熟能详，其他人也不太可能会用这个名字来命名其他的模块。但是，如果这个 JAR 文件的维护者同意保留反向域名，使用更好的顶级包名 org.apache.commons.csv 作为模块名，那会显得更好。他们只需在 JAR 中的 META-INF/MANIFEST.MF 文件里添加一行：

    Automatic-Module-Name: org.apache.commons.csv

最终，我们期望他们能够在 module-info.java 中添加保留的模块名将这个 JAR 文件转换成一个真正的模块，而每个用该模块名引用了这个 CSV 模块的模块也都能够继续工作。

> **注释**：模块的迁移计划是一项伟大的社会实验，没有人知道它是否能够顺利实施。在将第三方的 JAR 放到模块路径之前，请检查它们是否是模块化的。如果不是，那它们的清单是否有模块名；如果没有，仍旧需要将这样的 JAR 转换成自动模块，但是要准备好以后更新该模块名。

在撰写本书时，Commons CSV JAR 文件的 1.5 版本还没有模块描述符或自动模块名。尽管如此，它在模块路径上工作良好。我们可以从 https://commons.apache.org/proper/commons-csv 处下载这个库，解压并将 commons-csv.1.5.jar 放到 v2ch09.automod 模块的目录中。这个模块包含了一个很简单的从 CSV 文件中读取国家数据的程序：

```
package com.horstmann.places;

import java.io.*;
import org.apache.commons.csv.*;

public class CSVDemo
{
```

```java
public static void main(String[] args) throws IOException
{
   var in = new FileReader("countries.csv");
   Iterable<CSVRecord> records = CSVFormat.EXCEL.withDelimiter(';')
         .withHeader().parse(in);
   for (CSVRecord record : records)
   {
      String name = record.get("Name");
      double area = Double.parseDouble(record.get("Area"));
      System.out.println(name + " has area " + area);
   }
}
```

因为我们将 commons-csv-1.5.jar 用作自动模块, 所以我们要声明需要它:

```
@SuppressWarnings("module")
module v2ch09.automod
{
   requires commons.csv;
}
```

下面是编译和运行该程序的命令:

```
javac -p v2ch09.automod:commons-csv-1.5.jar \
   v2ch09.automod/com/horstmann/places/CSVDemo.java \
   v2ch09.automod/module-info.java
java -p v2ch09.automod:commons-csv-1.5.jar \
   -m v2ch09.automod/com.horstmann.places.CSVDemo
```

9.9 不具名模块

任何不在模块路径中的类都是不具名模块的一部分。从技术上说, 可能会有多个不具名模块, 但是它们合起来看就像是单个不具名的模块。与自动模块一样, 不具名模块可以访问所有其他的模块, 它的所有包都会被导出, 并且都是开放的。

但是, 没有任何明确模块可以访问不具名的模块。(明确模块是指既不是自动模块也不是不具名模块的模块, 即, module-info.class 在模块路径上的模块。) 换句话说, 明确模块总是可以避免 "类路径的坑"。

例如, 考虑前一节的程序, 假设将 commons-csv.1.5.jar 放到类路径而不是模块路径上:

```
java --module-path v2ch09.automod \
   --class-path commons-csv-1.5.jar \
   -m v2ch09.automod/com.horstmann.places.CSVDemo
```

现在, 这个程序将无法启动:

```
Error occurred during initialization of boot layer
java.lang.module.FindException: Module commons.csv not found, required by v2ch09.automod
```

因此, 迁移到 Java 平台模块系统必须按照自底向上的方式处理:

1. Java 平台自身被模块化。
2. 接下来, 库被模块化, 要么通过使用自动模块, 要么将它们转换为明确模块。

3. 一旦应用程序使用的所有库都被模块化，就可以将应用程序的代码转换为一个模块。

> **注释**：自动模块**可以**读取不具名模块，因此它们的依赖关系放在类路径中。

9.10 用于迁移的命令行标识

即使我们的程序没有使用模块，在使用 Java 9 或更新的版本时，我们也无法逃离模块化的世界。即使应用程序的代码位于不具名模块的类路径上，并且所有的包都被导出且开放，它也需要与模块化的 Java 平台交互。

到了 Java 11，默认行为是允许非法的模块访问，但是会在每种违规行为第一次出现时在控制台上显示一条警告消息。在 Java 未来的版本中，默认行为会发生变化，非法访问会被拒绝。为了未雨绸缪地应对这种变化，我们应该用 --illegal-access 标志来测试我们的应用程序。下面是 4 种可能的设置：

1. --illegal-access=permit 是 Java 9 默认的行为，它会在每一种非法访问第一次出现时打印一条消息。
2. --illegal-access=warn 对每次非法访问都打印一条消息。
3. --illegal-access=debug 对每次非法访问都打印一条消息和栈轨迹。
4. --illegal-access=deny 是未来的默认行为，直接拒绝所有非法访问。

现在是时候用 --illegal-access=deny 来测试了，这样我们就可以为这种行为变成默认行为时做好准备。

考虑这样的一个应用程序，它使用了一个不再能继续访问的内部 API，例如 com.sun.rowset.CachedRowSetImpl。最好的解决方案就是修改这个实现。（在 Java 7 中，可以从 RowSetProvider 中获取一个缓冲的行集。）但是，假设我们不能访问源代码。

在这种情况下，用 --add-exports 标志启动该应用程序，指定希望导出的模块和包，以及将包导出到的模块，在我们所举的例子中，包会导出到不具名模块中。

```
java --illegal-access=deny --add-exports java.sql.rowset/com.sun.rowset=ALL_UNNAMED \
   -jar MyApp.jar
```

现在，假设我们的应用程序使用反射来访问私有域或方法，那么在不具名模块内的反射是可行的，但是对 Java 平台类的非公有成员的反射式访问就再也不可行了。例如，有些动态生成 Java 类的库会通过反射来调用受保护的 ClassLoader.defineClass。如果某个应用程序使用了这样的库，那么需要添加下面的标志

```
--add-opens java.base/java.lang=ALL-UNNAMED
```

当添加这些命令行选项来让遗留应用程序工作时，你可能最终会被这些吓人的命令行吓倒。为了更好地管理多个选项，可以将它们放到一个或多个用 @ 前缀指定的文件中。例如

```
java @options1 @options2 -jar MyProg.java
```

其中文件 options1 和 options2 包含 java 命令的选项。

对于选项文件,有多条相关的语法规则:
- 用空格、制表符和换行符将各个选项分离
- 用双引号将包括空格在内的参数括起来,例如 "Program Files"
- 在一行的末尾用一个 \ 来合并下一行
- 反斜杠必须转义,例如 C:\\Users\\Fred
- 注释以 # 开头

9.11 传递的需求和静态的需求

在 9.4 节中,你已经看到了 requires 语句的基本形式。在本节中,你将看到偶尔会用到的它的两种变体。

在某些情况下,对于给定模块的用户而言,声明所有需要的模块会显得很冗长。例如,考虑一下包含像按钮这样的 JavaFX 用户界面元素的 javafx.controls 模块。javafx.controls 需要 javafx.base 模块,因此每个使用 javafx.controls 的程序也都需要 javafx.base 模块。(如果没有获取 javafx.base 模块中的包,那么我们就无法用像按钮这样的用户界面控件做太多的事情。)因为这个原因,javafx.controls 模块声明了需要使用 transitive 修饰符:

```
module javafx.controls
{
    requires transitive javafx.base;
    ...
}
```

任何声明需要 javafx.controls 的模块现在都自动地需要 javafx.base。

> 注释:有些程序员推荐在来自另一个模块的包会在公有 API 中用到时,应该总是使用 requires transitive。但是,这并不是 Java 语言的规则。例如,考虑 java.sql 模块:
>
> ```
> module java.sql
> {
>     requires transitive java.logging;
>     ...
> }
> ```
>
> 在整个 java.sql API 中,唯一用到 java.logging 模块中的包的地方,就是 java.sql.Driver.parentLogger 方法,它会返回一个 java.util.logging.Logger 对象。此时,最可接受的方式是不要将这个模块需求声明成传递性的。然后,那些真正使用这个方法的模块,并且也只有那些模块,需要声明它们需要 java.logging。

requires transitive 语句的一种很有吸引力的用法是聚集模块,即没有任何包,只有传递性需求的模块。java.se 模块就是这样的模块,它被声明成下面的样子:

```
module java.se
{
    requires transitive java.compiler;
    requires transitive java.datatransfer;
```

```
requires transitive java.desktop;
...
requires transitive java.sql;
requires transitive java.sql.rowset;
requires transitive java.xml;
requires transitive java.xml.crypto;
}
```

对细粒度模块依赖不感兴趣的程序员可以直接声明需要 java.se，然后获取 Java SE 平台的所有模块。

最终，还有一种不常见的 requires static 变体，它声明一个模块必须在编译时出现，而在运行时是可选的。下面是两个用例：

1. 访问在编译时进行处理的注解，而该注解是在不同的模块中声明的。
2. 对于位于不同模块中的类，如果它可用，就使用它，否则就执行其他操作，例如：

```
try
{
   new oracle.jdbc.driver.OracleDriver();
   ...
}
catch (NoClassDefFoundError er)
{
   Do something else
}
```

9.12 限定导出和开放

在本节中，你将会看到 exports 和 opens 语句的一种变体，将它们的作用域窄化到指定的模块集。例如，javafx.base 模块包含下面的语句：

```
exports com.sun.javafx.collections to
    javafx.controls, javafx.graphics, javafx.fxml, javafx.swing;
```

这样的语句被称为限定导出，所列的模块可以访问这个包，但是其他模块不行。

过多地使用限定导出表明模块化结构比较糟糕。尽管如此，在模块化现有代码基时，这种情况还是会发生。这里，Java 平台的设计者们将 JavaFX 的代码分布到了多个模块中，这是个好主意，因为并非所有的 JavaFX 实现都需要用到 FXML 和 Swing 的交互性。但是，JavaFX 的实现者们可以在他们的代码中不受限制地使用像 com.sun.javafx.collections.ListListener-Helper 这样的内部类。在新创建的项目中，人们可以设计更健壮的公有 API。

类似地，可以将 opens 语句限制到具体的模块。例如，在 9.7 节中，我们使用了下面这样的限定 opens 语句：

```
module v2ch09.openpkg
{
   requires com.horstmann.util;
   opens com.horstmann.places to com.horstmann.util;
}
```

现在，com.horstmann.places 包就只对 com.horstmann.util 模块开放了。

9.13 服务加载

ServiceLoader 类（卷 I 第 6 章）提供了一种轻量级机制，用于将服务接口与实现匹配起来。Java 平台模块系统使得这种机制更易于使用。

下面是对服务加载的一个快速回顾。服务拥有一个接口和一个或多个可能的实现。下面是一个简单的接口示例：

```
public interface GreeterService
{
    String greet(String subject);
    Locale getLocale();
}
```

有一个或多个模块提供了实现，例如

```
public class FrenchGreeter implements GreeterService
{
    public String greet(String subject) { return "Bonjour " + subject; }
    public Locale getLocale() { return Locale.FRENCH; }
}
```

服务消费者必须基于其认为适合的标准在提供的所有实现中选择一个。

```
ServiceLoader<GreeterService> greeterLoader = ServiceLoader.load(GreeterService.class);
GreeterService chosenGreeter;
for (GreeterService greeter : greeterLoader)
{
    if (...)
    {
        chosenGreeter = greeter;
    }
}
```

在过去，实现是通过将文本文件放置到包含实现类的 JAR 文件的 META-INF/services 目录中而提供给服务消费者的。模块系统提供了一种更好的方式，与提供文本文件不同，可以添加语句到模块描述符中。

提供服务实现的模块可以添加一条 provides 语句，它列出了服务接口（可能定义在任何模块中），以及实现类（必须是该模块的一部分）。下面是来自 jdk.security.auth 模块的一个例子：

```
module jdk.security.auth
{
    . . .
    provides javax.security.auth.spi.LoginModule with
        com.sun.security.auth.module.Krb5LoginModule,
        com.sun.security.auth.module.UnixLoginModule,
        com.sun.security.auth.module.JndiLoginModule,
        com.sun.security.auth.module.KeyStoreLoginModule,
        com.sun.security.auth.module.LdapLoginModule,
        com.sun.security.auth.module.NTLoginModule;
}
```

这与 META-INF/services 文件等价。

使用它的消费模块包含一条 uses 语句：

```
module java.base
{
   ...
   uses javax.security.auth.spi.LoginModule;
}
```

当消费模块中的代码调用 ServiceLoader.load(*ServiceInterface*.class) 时，匹配的提供者类将被加载，尽管它们可能不在可访问的包中。

在我们的代码示例中，我们为 com.horstmann.greetsvc.internal 包中的德语和法语问候者提供了相关的实现。该服务模块导出了 com.horstmann.greetsvc 包，但是没有导出包含实现的包。provides 语句声明了在未导出包中的服务及其实现类：

```
module com.horstmann.greetsvc
{
   exports com.horstmann.greetsvc;

   provides com.horstmann.greetsvc.GreeterService with
      com.horstmann.greetsvc.internal.FrenchGreeter,
      com.horstmann.greetsvc.internal.GermanGreeterFactory;
}
```

v2ch09.useservice 模块会消费该服务。通过使用 ServiceLoader 工具，我们会迭代提供的所有服务，并挑选出匹配所期望语言的服务：

```
package com.horstmann.hello;

import java.util.*;
import com.horstmann.greetsvc.*;

public class HelloWorld
{
   public static void main(String[] args)
   {
      ServiceLoader<GreeterService> greeterLoader
         = ServiceLoader.load(GreeterService.class);
      String desiredLanguage = args.length > 0 ? args[0] : "de";
      GreeterService chosenGreeter = null;
      for (GreeterService greeter : greeterLoader)
      {
         if (greeter.getLocale().getLanguage().equals(desiredLanguage))
            chosenGreeter = greeter;
      }
      if (chosenGreeter == null)
         System.out.println("No suitable greeter.");
      else
         System.out.println(chosenGreeter.greet("Modular World"));
   }
}
```

该模块声明需要服务模块，并声明 GreeterService 正在被使用。

```
module v2ch09.useservice
{
```

```
    requires com.horstmann.greetsvc;
    uses com.horstmann.greetsvc.GreeterService;
}
```

provides 和 uses 声明的效果，是使得消费该服务的模块允许访问私有实现类。

为了构建并运行该程序，首先要编译服务：

```
javac com.horstmann.greetsvc/module-info.java \
    com.horstmann.greetsvc/com/horstmann/greetsvc/GreeterService.java \
    com.horstmann.greetsvc/com/horstmann/greetsvc/internal/*.java
```

然后，编译并运行消费模块：

```
javac -p com.horstmann.greetsvc \
    v2ch09.useservice/com/horstmann/hello/HelloWorld.java \
    v2ch09.useservice/module-info.java
java -p com.horstmann.greetsvc:v2ch09.useservice \
    -m v2ch09.useservice/com.horstmann.hello.HelloWorld
```

9.14 操作模块的工具

jdeps 工具可以分析给定的 JAR 文件集之间的依赖关系。例如，假设我们想要模块化 Junit 4。运行

```
jdeps -s junit-4.12.jar hamcrest-core-1.3.jar
```

-s 标志会产生总结性的输出：

```
hamcrest-core-1.3.jar -> java.base
junit-4.12.jar -> hamcrest-core-1.3.jar
junit-4.12.jar -> java.base
junit-4.12.jar -> java.management
```

它告知了我们下面的模块图：

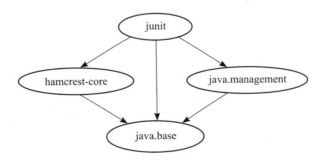

如果删除 -s 标志，那么我们得到的是模块的总结，后面跟着一个映射表，将包映射到所需要的包和模块上。如果添加 -v 标志，那么列出的清单会将类映射到所需要的包和模块上。

--generate-module-info 选项会对每个分析过的模块产生 module-info 文件：

```
jdeps --generate-module-info /tmp/junit junit-4.12.jar hamcrest-core-1.3.jar
```

> **注释：** 还有一个选项，可以用"dot"语言生成用于描述图的图形化输出。假设我们已经安装了 dot 工具，那么运行下面的命令：
>
>   jdeps -s -dotoutput /tmp/junit junit-4.12.jar hamcrest-core-1.3.jar
>   dot -Tpng /tmp/junit/summary.dot > /tmp/junit/summary.png
>
> 就会得到下面的 summary.png 图：
>
>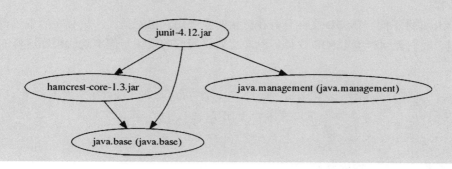

使用 jlink 工具可以产生执行时无须单独的 Java 运行时环境的应用程序。所产生的镜像比整个 JDK 要小很多。我们可以指定想要包含的模块和输出目录：

```
jlink --module-path com.horstmann.greet.jar:v2ch09.exportedpkg.jar:$JAVA_HOME/jmods \
    --add-modules v2ch09.exportedpkg --output /tmp/hello
```

输出目录有一个包含 java 可执行文件的子目录 bin。如果运行

```
bin/java -m v2ch09.exportedpkg
```

那么该模块的主类的 main 方法就会被调用。

jlink 的关键是它将运行应用程序所需的最小的模块集打包在一起。我们可以列出其中包含的所有模块：

```
bin/java --list-modules
```

在这个示例中，输出是

```
v2ch09.exportedpkg
com.horstmann.greet
java.base@9
```

所有模块都包含在运行时镜像文件 lib/modules 中。在我的计算机上，这个文件有 23MB，而所有 JDK 模块的运行时镜像会占据 181MB。整个应用占据 45MB，比 486MB 的 JDK 的 10% 还少。

这可以成为用于打包应用程序的实用工具的基础。我们仍旧需要产生针对多平台的文件集和针对应用程序的脚本。

> **注释：** 我们可以用 jimage 命令来审视运行时镜像。但是，其格式对 JVM 来说是内部的，并且运行时镜像并不是为其他工具而生成并供其他工具所使用的。

最后，jmod 工具可以构建并审视包含在 JDK 中的模块文件。当查看 JDK 内的 jmods 目录时，会发现针对每个模块都有一个扩展名为 jmod 的文件。要注意的是，现在再也没有 rt.jar 文件了。

与 JAR 文件一样，这些文件也包括类文件。此外，它们还可以包括本地代码库、命令、头文件、配置文件和合法的通知。JMOD 文件使用 ZIP 文件格式，可以用任意 ZIP 工具查看它们的内容。

与 JAR 文件不同，JMOD 文件只有在链接时才有用，也就是说，只有在产生运行时镜像时才有用。我们无须产生 JMOD 文件，除非想要将我们的模块与像本地代码库这样的二进制文件绑定。

> **注释**：因为 rt.jar 和 tools.jar 文件不再包含在 Java 9 中，所以需要更新所有对它们的引用。例如，如果在安全策略文件中引用了 tools.jar，那么需要将它修改为对下面的模块的引用：
>
> ```
> grant codeBase "jrt:/jdk.compiler"
> {
>     permission java.security.AllPermission;
> };
> ```
>
> jrt: 语法表示 Java 运行时文件。

现在到了该结束 Java 平台模块系统这一章的时候了。下一章将讨论另一个重要的主题：安全。安全已经成为 Java 平台的核心特性之一。我们生活和计算的世界正在变得越来越危险，彻底理解 Java 的安全对许多开发者来说重要性与日俱增。

第 10 章 安　　全

- ▲ 类加载器
- ▲ 安全管理器与访问权限
- ▲ 用户认证
- ▲ 数字签名
- ▲ 加密

当 Java 技术刚刚问世时，令人激动的并不是因为它是一种设计完美的编程语言，而是因为它能够安全地运行通过因特网传播的各种 applet。很显然，只有当用户确信 applet 的代码不会破坏他的计算机时，用户才会接受在网上传播的可执行的 applet。因此，安全是 Java 技术的设计人员和使用者所关心的一个重大问题。这就意味着，Java 与其他的语言和系统有所不同，在那些语言和系统中安全是在事后才想到要去实现的，或者是对破坏的一种应对措施，而对 Java 来说，安全机制是一个不可分割的组成部分。

Java 技术提供了以下三种确保安全的机制：

- 语言设计特性（对数组的边界进行检查，无不受检查的类型转换，无指针算法等）。
- 访问控制机制，用于控制代码能够执行的操作（比如文件访问，网络访问等）。
- 代码签名，利用该特性，代码的作者就能够用标准的加密算法来认证 Java 代码。这样，该代码的使用者就能够准确地知道谁创建了该代码，以及代码签名后是否被修改过。

首先，我们来讨论类加载器，它可以在将类加载到虚拟机中的时候检查类的完整性。我们将展示这种机制是如何探测类文件中的损坏的。

为了获得最大的安全性，无论是加载类的默认机制，还是自定义的类加载器，都需要与负责控制代码运行的安全管理器类协同工作。后面我们还要详细介绍如何配置 Java 平台的安全性。

最后，我们要介绍 java.security 包提供的加密算法，用来进行代码的签名和用户认证。

与我们的一贯宗旨一样，我们将重点介绍应用程序编程人员最感兴趣的话题。如果要深入研究，推荐阅读 Li Gong、Gary Ellison 和 Mary Dageforde 撰写的 *Inside Java 2 Platform Security: Architecture, API Design, and Implementation* 一书，该书由 Prentice Hall 出版社于 2003 年出版。

10.1 类加载器

Java 编译器会为虚拟机转换源指令。虚拟机代码存储在以 .class 为扩展名的类文件中，每个类文件都包含某个类或者接口的定义和实现代码。在以下各节中，你将会看到虚拟机是如何加载这些类文件的。

10.1.1 类加载过程

请注意，虚拟机只加载程序执行时所需要的类文件。例如，假设程序从 `MyProgram.class` 开始运行，下面是虚拟机执行的步骤：

1. 虚拟机有一个用于加载类文件的机制，例如，从磁盘上读取文件或者请求 Web 上的文件，它使用该机制来加载 `MyProgram` 类文件中的内容。
2. 如果 `MyProgram` 类拥有类型为另一个类的域，或者是拥有超类，那么这些类文件也会被加载。（加载某个类所依赖的所有类的过程称为类的解析。）
3. 接着，虚拟机执行 `MyProgram` 中的 main 方法（它是静态的，无须创建类的实例）。
4. 如果 main 方法或者 main 调用的方法要用到更多的类，那么接下来就会加载这些类。

然而，类加载机制并非只使用单个的类加载器。每个 Java 程序至少拥有三个类加载器：

- 引导类加载器
- 平台类加载器
- 系统类加载器（有时也称为应用类加载器）

引导类加载器负责加载包含在下列模块以及大量的 JDK 内部模块中的平台类：

```
java.base
java.datatransfer
java.desktop
java.instrument
java.logging
java.management
java.management.rmi
java.naming
java.prefs
java.rmi
java.security.sasl
java.xml
```

引导类加载器没有对应的 `ClassLoader` 对象，例如，方法

```
StringBuilder.class.getClassLoader()
```

将返回 null。

在 Java 9 之前，Java 平台类位于 rt.jar 中。如今，Java 平台是模块化的，每个平台模块都包含一个 JMOD 文件（参见第 9 章）。平台类加载器会加载引导类加载器没有加载的 Java 平台中的所有类。

系统类加载器会从模块路径和类路径中加载应用类。

> **注释**：在 Java 9 之前，"扩展类加载器"会加载 jre/lib/ext 目录中的"标准扩展"，而"授权标准覆盖"机制提供了一种方式，可以用更新的版本覆盖某些平台类（包括 CORBA 和 XML 的实现）。这两种机制都被移除了。

10.1.2 类加载器的层次结构

类加载器有一种父/子关系。除了引导类加载器外，每个类加载器都有一个父类加载器。根据规定，类加载器会为它的父类加载器提供一个机会，以便加载任何给定的类，并且只有在其父类加载器加载失败时，它才会加载该给定类。例如，当要求系统类加载器加载一个系统类（比如，`java.lang.StringBuilder`）时，它首先要求平台类加载器进行加载，该加载器则首先要求引导类加载器进行加载。引导类加载器会找到并加载这个类，而无须其他类加载器做更多的搜索。

某些程序具有插件架构，其中代码的某些部分是作为可选的插件打包的。如果插件被打包为 JAR 文件，那就可以直接用 `URLClassLoader` 类的实例去加载插件类。

```
var url = new URL("file:///path/to/plugin.jar");
var pluginLoader = new URLClassLoader(new URL[] { url });
Class<?> cl = pluginLoader.loadClass("mypackage.MyClass");
```

由于在 `URLClassLoader` 构造器中没有指定父类加载器，因此 `pluginLoader` 的父亲就是系统类加载器。图 10-1 展示了这种层次结构。

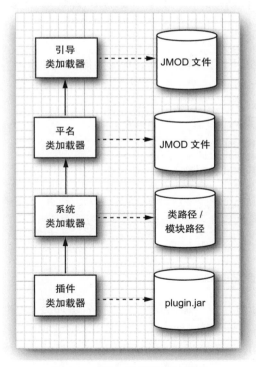

图 10-1 类加载器的层次结构

> ⚠ **警告**：在 Java 9 之前，系统类加载器是 `URLClassLoaser` 类的实例。有些程序员会使用强制转型来访问其 `getURLs` 方法，或者通过反射机制调用受保护的 `addURLs` 方法将 JAR 文件添加到类路径中。现在无法这样操作了。

大多数时候，你不必操心类加载的层次结构。通常，类是由于其他的类需要它而被加载的，而这个过程对你是透明的。

偶尔，你也会需要干涉和指定类加载器。考虑下面的例子：

- 你的应用的代码包含一个助手方法，它要调用 Class.forName(classNameString)。
- 这个方法是从一个插件类中被调用的。
- classNameString 指定的正是一个包含在这个插件的 JAR 中的类。

插件的作者期望这个类会被加载。但是，助手方法的类是由系统类加载器加载的，这正是 Class.forName 所使用的类加载器。而对于它来说，插件 JAR 中的类是不可见的，这种现象称为类加载器倒置。

要解决这个问题，助手方法需要使用恰当的类加载器，它可以要求类加载器作为其一个参数传递给它。或者，它可以要求将恰当的类加载器设置成为当前线程的上下文类加载器，这种策略在许多框架中都得到了应用（例如 JAXP 和 JNDI）。

每个线程都有一个对类加载器的引用，称为上下文类加载器。主线程的上下文类加载器是系统类加载器。当新线程创建时，它的上下文类加载器会被设置成为创建该线程的上下文类加载器。因此，如果不做任何特殊的操作，那么所有线程就都会将它们的上下文类加载器设置为系统类加载器。

但是，我们也可以通过下面的调用将其设置成为任何类加载器。

```
Thread t = Thread.currentThread();
t.setContextClassLoader(loader);
```

然后助手方法可以获取这个上下文类加载器：

```
Thread t = Thread.currentThread();
ClassLoader loader = t.getContextClassLoader();
Class<?> cl = loader.loadClass(className);
```

> 提示：如果你编写了一个按名字来加载类的方法，那么让调用者在传递显式的类加载器和使用上下文类加载器之间进行选择是一种好的做法。不要直接使用该方法所属的类的类加载器。

10.1.3 将类加载器用作命名空间

每个 Java 程序员都知道，包的命名是为了消除名字冲突。在标准类库中，有两个名为 Date 的类，它们的实际名字分别为 java.util.Date 和 java.sql.Date。使用简单的名字只是为了方便程序员，它们要求程序包含恰当的 import 语句。在一个正在执行的程序中，所有的类名都包含它们的包名。

然而，令人惊奇的是，在同一个虚拟机中，可以有两个类，它们的类名和包名都是相同的。类是由它的全名和类加载器来确定的。这项技术在加载来自多处的代码时很有用。例如，应用服务器会为每一个应用使用单独的类加载器，这使得虚拟机可以区分来自不同应用的类，而无论它们是怎样命名的。图 10-2 展示了一个示例。假设一个应用服务器加载了两个不同的应用，它们都有一个名为 Util 的类。因为每个类都是由单独的类加载器加载的，所以这些类可以彻底地区分开而不会产生任何冲突。

10.1.4 编写你自己的类加载器

我们可以编写自己的用于特殊目的的类加载器，这使得我们可以在向虚拟机传递字节码之前执行定制的检查。例如，我们可以编写一个类加载器，它可以拒绝加载没有标记为"paid for"的类。

如果要编写自己的类加载器，只需要继承 ClassLoader 类，然后覆盖下面这个方法：

findClass(String className)

ClassLoader 超类的 loadClass 方法用于将类的加载操作委托给其父类加载器去进行，只有当该类尚未加载并且父类加载器也无法加载该类时，才调用 findClass 方法。

如果要实现该方法，必须做到以下几点：

1. 为来自本地文件系统或者其他来源的类加载其字节码。

2. 调用 ClassLoader 超类的 defineClass 方法，向虚拟机提供字节码。

在程序清单 10-1 中，我们实现了一个类加载器，用于加载加密过的类文件。该程序要求用户输入第一个要加载的类的名字（即包含 main 方法的类）和密钥。然后，使用一个专门的类加载器来加载指定的类并调用 main 方法。该类加载器对指定的类和

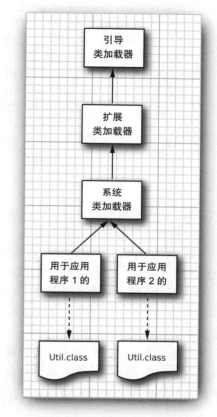

图 10-2 两个类加载器分别加载具有相同名字的两个类

所有被其引用的非系统类进行解密。最后，该程序会调用已加载类的 main 方法（参见图 10-3）。

为了简单起见，我们忽略了密码学领域 2000 年来所取得的技术进展，而是采用了传统的 Caesar 密码对类文件进行加密。

> **注释**：David Kahn 的佳作 *The Codebreakers*（纽约 Macmillan 出版社 1967 年出版）第 84 页中称 Suetonius 是 Caesar 密码的发明人。Caesar 将罗马字母表的 24 个字母移动了 3 个字母的位置，在那个时代这可以迷惑对手。
>
> 第一次撰写本章时，美国政府限制高强度加密方法的出口。因此，我们在实例中使用的是 Caesar 的加密方法，因为该方法的出口显然是合法的。

我们的 Caesar 密码版本使用的密钥是 1 到 255 之间的一个数字，解密时，只需将密钥与每个字节相加，然后对 256 取余。程序清单 10-2 的 Caesar.java 程序就实现了这种加密行为。

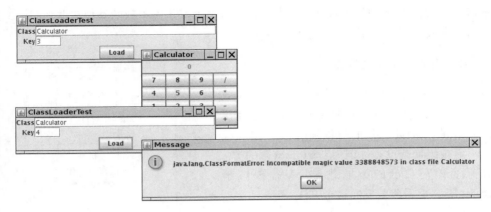

图 10-3 ClassLoaderTest 程序

为了不与常规的类加载器相混淆，我们对加密的类文件使用了不同的扩展名 .caesar。

解密时，类加载器只需要将每个字节减去该密钥即可。在本书的程序代码中，可以找到 4 个类文件，它们都是用 "3" 这个传统的密钥值进行加密的。为了运行加密程序，需要使用在我们的 ClassLoaderTest 程序中定义的定制类加载器。

对类文件进行加密有很大的用途（当然，使用的密码的强度应该高于 Caesar 密码的）。如果没有加密密钥，类文件就毫无用处。它们既不能由标准虚拟机来执行，也不能轻易地被反汇编。

这就是说，可以使用定制的类加载器来认证类用户的身份，或者确保程序在运行之前已经支付了软件费用。当然，加密只是定制类加载器的应用之一。可以使用其他类型的加载器来解决别的问题，例如，将类文件存储到数据库中。

程序清单 10-1　classLoader/ClassLoaderTest.java

```
1  package classLoader;
2
3  import java.io.*;
4  import java.lang.reflect.*;
5  import java.nio.file.*;
6  import java.awt.*;
7  import java.awt.event.*;
8  import javax.swing.*;
9
10 /**
11  * This program demonstrates a custom class loader that decrypts class files.
12  * @version 1.25 2018-05-01
13  * @author Cay Horstmann
14  */
15 public class ClassLoaderTest
16 {
17    public static void main(String[] args)
18    {
19       EventQueue.invokeLater(() ->
```

```java
20          {
21             var frame = new ClassLoaderFrame();
22             frame.setTitle("ClassLoaderTest");
23             frame.setDefaultCloseOperation(JFrame.EXIT_ON_CLOSE);
24             frame.setVisible(true);
25          });
26       }
27    }
28
29    /**
30     * This frame contains two text fields for the name of the class to load and the decryption
31     * key.
32     */
33    class ClassLoaderFrame extends JFrame
34    {
35       private JTextField keyField = new JTextField("3", 4);
36       private JTextField nameField = new JTextField("Calculator", 30);
37       private static final int DEFAULT_WIDTH = 300;
38       private static final int DEFAULT_HEIGHT = 200;
39
40       public ClassLoaderFrame()
41       {
42          setSize(DEFAULT_WIDTH, DEFAULT_HEIGHT);
43          setLayout(new GridBagLayout());
44          add(new JLabel("Class"), new GBC(0, 0).setAnchor(GBC.EAST));
45          add(nameField, new GBC(1, 0).setWeight(100, 0).setAnchor(GBC.WEST));
46          add(new JLabel("Key"), new GBC(0, 1).setAnchor(GBC.EAST));
47          add(keyField, new GBC(1, 1).setWeight(100, 0).setAnchor(GBC.WEST));
48          var loadButton = new JButton("Load");
49          add(loadButton, new GBC(0, 2, 2, 1));
50          loadButton.addActionListener(event -> runClass(nameField.getText(), keyField.getText()));
51          pack();
52       }
53
54       /**
55        * Runs the main method of a given class.
56        * @param name the class name
57        * @param key the decryption key for the class files
58        */
59       public void runClass(String name, String key)
60       {
61          try
62          {
63             var loader = new CryptoClassLoader(Integer.parseInt(key));
64             Class<?> c = loader.loadClass(name);
65             Method m = c.getMethod("main", String[].class);
66             m.invoke(null, (Object) new String[] {});
67          }
68          catch (Throwable t)
69          {
70             JOptionPane.showMessageDialog(this, t);
71          }
72       }
73    }
```

```java
/**
 * This class loader loads encrypted class files.
 */
class CryptoClassLoader extends ClassLoader
{
   private int key;

   /**
    * Constructs a crypto class loader.
    * @param k the decryption key
    */
   public CryptoClassLoader(int k)
   {
      key = k;
   }

   protected Class<?> findClass(String name) throws ClassNotFoundException
   {
      try
      {
         byte[] classBytes = null;
         classBytes = loadClassBytes(name);
         Class<?> cl = defineClass(name, classBytes, 0, classBytes.length);
         if (cl == null) throw new ClassNotFoundException(name);
         return cl;
      }
      catch (IOException e)
      {
         throw new ClassNotFoundException(name);
      }
   }

   /**
    * Loads and decrypt the class file bytes.
    * @param name the class name
    * @return an array with the class file bytes
    */
   private byte[] loadClassBytes(String name) throws IOException
   {
      String cname = name.replace('.', '/') + ".caesar";
      byte[] bytes = Files.readAllBytes(Paths.get(cname));
      for (int i = 0; i < bytes.length; i++)
         bytes[i] = (byte) (bytes[i] - key);
      return bytes;
   }
}
```

程序清单 10-2　classLoader/Caesar.java

```java
package classLoader;

import java.io.*;

```

```java
 5  /**
 6   * Encrypts a file using the Caesar cipher.
 7   * @version 1.02 2018-05-01
 8   * @author Cay Horstmann
 9   */
10  public class Caesar
11  {
12     public static void main(String[] args) throws Exception
13     {
14        if (args.length != 3)
15        {
16           System.out.println("USAGE: java classLoader.Caesar in out key");
17           return;
18        }
19
20        try (var in = new FileInputStream(args[0]);
21             var out = new FileOutputStream(args[1]))
22        {
23           int key = Integer.parseInt(args[2]);
24           int ch;
25           while ((ch = in.read()) != -1)
26           {
27              byte c = (byte) (ch + key);
28              out.write(c);
29           }
30        }
31     }
32  }
```

API java.lang.Class 1.0

- ClassLoader getClassLoader()
  获取加载该类的类加载器。

API java.lang.ClassLoader 1.0

- ClassLoader getParent() 1.2
  返回父类加载器，如果父类加载器是引导类加载器，则返回 null。
- static ClassLoader getSystemClassLoader() 1.2
  获取系统类加载器，即用于加载第一个应用类的类加载器。
- protected Class findClass(String name) 1.2
  类加载器应该覆盖该方法，以查找类的字节码，并通过调用 defineClass 方法将字节码传给虚拟机。在类的名字中，使用 . 作为包名分隔符，并且不使用 .class 后缀。
- Class defineClass(String name, byte[] byteCodeData, int offset, int length)
  将一个新的类添加到虚拟机中，其字节码在给定的数据范围中。

API java.net.URLClassLoader 1.2

- URLClassLoader(URL[] urls)

- URLClassLoader(URL[] urls, ClassLoader parent)

  构建一个类加载器，它可以从给定的 URL 处加载类。如果 URL 以 / 结尾，那么它表示的是一个目录，否则，它表示的是一个 JAR 文件。

API java.lang.Thread 1.0

- ClassLoader getContextClassLoader() 1.2

  获取类加载器，该线程的创建者将其指定为执行该线程时最适合使用的类加载器。

- void setContextClassLoader(ClassLoader loader) 1.2

  为该线程中的代码设置一个类加载器，以获取要加载的类。如果在启动一个线程时没有显式地设置上下文类加载器，则使用父线程的上下文类加载器。

10.1.5 字节码校验

当类加载器将新加载的 Java 平台类的字节码传递给虚拟机时，这些字节码首先要接受校验器（verifier）的校验。校验器负责检查那些指令无法执行的明显有破坏性的操作。除了系统类外，所有的类都要被校验。

下面是校验器执行的一些检查：

- 变量要在使用之前进行初始化。
- 方法调用与对象引用类型之间要匹配。
- 访问私有数据和方法的规则没有被违反。
- 对本地变量的访问都落在运行时堆栈内。
- 运行时堆栈没有溢出。

如果以上这些检查中任何一条没有通过，那么该类就被认为遭到了破坏，并且不予加载。

> **注释**：如果熟悉 Gödel 定理，那么你可能想知道校验器究竟是如何证明某个类文件不存在类型不匹配、变量没有初始化和堆栈溢出等问题的。根据 Gödel 定理，不可能设计出这样的算法：它能够处理程序，确定其是否具有特定的属性（比如不出现堆栈溢出问题）。这是否属于 Oracle 公司的公共关系部门和逻辑法则之间的矛盾呢？不——事实上，校验器并非是一个 Gödel 意义上的决策算法。如果校验器接受了一个程序，那么该程序就确实是安全的。然而，也有许多程序尽管是安全的，但却被校验器拒绝了。（在强制用哑元值来初始化一个变量时，你就会碰到这个问题，因为编译器无法了解这个变量是否可以被正确地初始化。）

这种严格的校验是出于安全上的考虑，有一些偶然性的错误，比如变量没有初始化，如果没有被捕获，就很容易对系统造成严重的破坏。更为重要的是，在因特网这样开放的环境中，你必须保护自己以防恶意的程序员对你实施攻击，因为他们的目的就是要造成恶劣的影响。例如，通过修改运行时堆栈中的值，或者向系统对象的私有数据字段写入数据，某个程序就会突破浏览器的安全防线。

当然，你可能想知道为什么要有一个专门的校验器来检查这些特性。毕竟，编译器绝不会允许你生成一个这样的类文件：该类文件中有未初始化的变量或者可以通过另一个类来访问该类的某个私有数据字段。实际上，用 Java 语言编译器生成的类文件总是可以通过校验的。然而，类文件中使用的字节码格式是文档记录良好的，对于具有汇编程序设计经验并且拥有十六进制编辑器的人来说，要手工地创建一个对 Java 虚拟机来说由合法但是不安全的指令构成的类文件，是一件非常容易的事情。再次提醒你，要记住，校验器总是在防范被故意篡改的类文件，而不只是检查编译器产生的类文件。

下面的例子将展示如何创建一个变动过的类文件。我们从程序清单 10-3 中的程序 VerifierTest.java 开始。这是一个简单的程序，它调用一个方法，并且显示方法的运行结果。该程序既可以在控制台运行，也可以作为一个 applet 程序来运行。其中的 fun 方法本身只是负责计算 1+2。

```java
static int fun()
{
   int m;
   int n;
   m = 1;
   n = 2;
   int r = m + n;
   return r;
}
```

程序清单 10-3 verifier/VerifierTest.java

```java
 1  package verifier;
 2
 3  import java.awt.*;
 4
 5  /**
 6   * This application demonstrates the bytecode verifier of the virtual machine. If you use a
 7   * hex editor to modify the class file, then the virtual machine should detect the tampering.
 8   * @version 1.10 2018-05-05
 9   * @author Cay Horstmann
10   */
11  public class VerifierTest
12  {
13     public static void main(String[] args)
14     {
15        System.out.println("1 + 2 == " + fun());
16     }
17
18     /**
19      * A function that computes 1 + 2.
20      * @return 3, if the code has not been corrupted
21      */
22     public static int fun()
23     {
24        int m;
25        int n;
26        m = 1;
```

```
27        n = 2;
28        // use hex editor to change to "m = 2" in class file
29        int r = m + n;
30        return r;
31    }
32 }
```

作为一次实验，请尝试编译下面这个对该程序进行修改后的文件。

```
static int fun()
{
   int m = 1;
   int n;
   m = 1;
   m = 2;
   int r = m + n;
   return r;
}
```

在这种情况下，n 没有被初始化，它可以是任何随机值。当然，编译器能够检测到这个问题并拒绝编译该程序。如果要建立一个不良的类文件，我们必须得多花点工夫。首先，运行 javap 程序，以便知晓编译器是如何翻译 fun 方法的。命令

```
javap -c verifier.VerifierTest
```

用助记（mnemonic）格式显示了类文件中的字节码。

```
Method int fun()
   0 iconst_1
   1 istore_0
   2 iconst_2
   3 istore_1
   4 iload_0
   5 iload_1
   6 iadd
   7 istore_2
   8 iload_2
   9 ireturn
```

我们使用一个十六进制编辑器将指令 3 从 istore_1 改为 istore_0，也就是说，局部变量 0（即 m）被初始化了两次，而局部变量 1（即 n）则根本没有初始化。我们必须知道这些指令的十六进制值，这些值可以从 Java 虚拟机规范中获知：https://docs.oracle.com/javase/specs/jvms/se11/html/index.html。

```
0 iconst_1  04
1 istore_0  3B
2 iconst_2  05
3 istore_1  3C
4 iload_0   1A
5 iload_1   1B
6 iadd      60
7 istore_2  3D
8 iload_2   1C
9 ireturn   AC
```

可以使用任何十六进制编辑器来执行这种修改。在图 10-4 中，你可以看到类文件 VerifierTest.class 被加载到了 Gnome 编辑器中，fun 方法的字节码已经被选定。

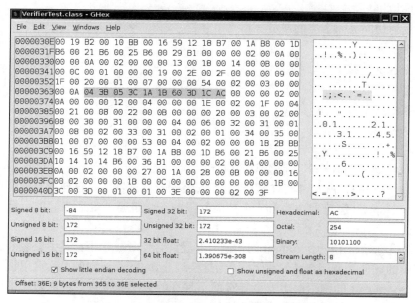

图 10-4　使用十六进制编辑器修改字节码

将 3C 改为 3B 并保存类文件。然后尝试运行 VerifierTest 程序，将会看到下面的出错信息：

```
Exception in thread "main" java.lang.VerifyError: (class: VerifierTest, method:fun signature:
()I) Accessing value from uninitialized register 1
```

这很好——虚拟机发现了我们所做的修改。

现在用 -noverify 选项（或者 -Xverify:none）来运行程序：

```
java -noverify verifier.VerifierTest
```

从表面上看，fun 方法似乎返回了一个随机值。但实际上，该值是 2 与存储在尚未初始化的变量 n 中的值相加得到的结果。下面是典型的输出结果：

```
1 + 2 == 15102330
```

10.2　安全管理器与访问权限

一旦某个类被加载到虚拟机中，并由检验器检查过之后，Java 平台的第二种安全机制就会启动，这个机制就是安全管理器。下面几小节将讨论这种机制。

10.2.1　权限检查

安全管理器是一个负责控制具体操作是否允许执行的类。安全管理器负责检查的操作包括以下内容：
- 创建一个新的类加载器
- 退出虚拟机
- 使用反射访问另一个类的成员
- 访问本地文件
- 打开 socket 连接
- 启动打印作业
- 访问系统剪贴板
- 访问 AWT 事件队列
- 打开一个顶层窗口

整个 Java 类库中还有许多其他类似的检查。

在运行 Java 应用程序时，默认的设置是不安装安全管理器的，这样所有的操作都是允许的。另一方面，applet 浏览器会执行一个功能受限的安全策略。更严格的安全性对其他情况也具有意义。

例如，假设你运行了一个 Tomcat 的实例，并允许合作者或学生在其中安装 Servlet。你并不想让他们中的任何人调用 System.exit，因为这会终止该 Tomcat 实例。你可以设置一个安全策略，让对 System.exit 的调用抛出安全异常而不是真的关闭虚拟机。下面将详细说明这种情况。Runtime 类的 exit 方法会调用安全管理器的 checkExit 方法，下面是 exit 方法的全部代码：

```java
public void exit(int status)
{
    SecurityManager security = System.getSecurityManager();
    if (security != null)
        security.checkExit(status);
    exitInternal(status);
}
```

这时安全管理器要检查退出请求是来自浏览器还是单个的 applet 程序。如果安全管理器同意了退出请求，那么 checkExit 便直接返回并继续处理下面正常的操作。但是，如果安全管理器不同意退出请求，那么 checkExit 方法就会抛出一个 SecurityException 异常。

只有当没有任何异常发生时，exit 方法才能继续执行。然后它调用本地私有的 exitInternal 方法，以真正终止虚拟机的运行。没有其他的方法可以终止虚拟机的运行，因为 exitInternal 方法是私有的，任何其他类都不能调用它。因此，任何试图退出虚拟机的代码都必须通过 exit 方法，从而在不触发安全异常的情况下，通过 checkExit 安全检查。

显然，安全策略的完整性依赖于谨慎的编码。标准类库中系统服务的提供者，在试图继续任何敏感的操作之前，都必须与安全管理器进行协商。

Java 平台的安全管理器，不仅允许系统管理员，而且允许程序员对各个安全访问权限实施细致的控制。我们将在下一节介绍这些特性。首先，我们将介绍 Java 2 平台的安全模型的概况，然后介绍如何使用策略文件对各个权限实施控制。最后，我们要介绍如何来定义你自己的权限类型。

10.2.2 Java 平台安全性

JDK 1.0 具有一个非常简单的安全模型，即本地类拥有所有的权限，而远程类只能在沙盒里运行。就像儿童只能在沙盒里玩沙子一样，远程代码只被允许打印屏幕和与用户进行交互。applet 的安全管理器拒绝了远程代码对本地资源的所有访问。JDK 1.1 对此进行了微小的修改，如果远程代码带有可信赖的实体的签名，将被赋予和本地类相同的访问权限。不过，JDK 1.0 和 1.1 这两个版本提供的都是一种"要么都有，要么都没有"的权限赋予方法。程序要么拥有所有的访问权限，要么必须在沙盒里运行。

从 Java 1.2 开始，Java 平台拥有了更灵活的安全机制，它的安全策略建立了代码来源和访问权限集之间的映射关系（参见图 10-5）。

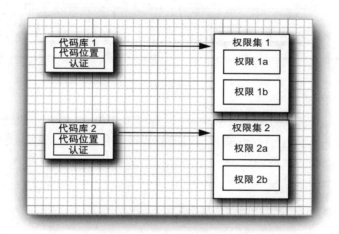

图 10-5　一个安全策略

代码来源（code source）是由一个代码位置和一个证书集指定的。代码位置指定了代码的来源。例如，远程 applet 代码的代码位置是下载 applet 的 HTTP URL，位于 JAR 文件中的代码的代码位置是该文件的 URL。证书的目的是要由某一方来保障代码没有被篡改过。我们将在本章的后面部分讨论证书。

权限（permission）是指由安全管理器负责检查的任何属性。Java 平台支持许多访问权限类，每个类都封装了特定权限的详细信息。例如，下面这个 FilePermission 类的实例表示：允许在 /tmp 目录下读取和写入任何文件。

```
var p = new FilePermission("/tmp/*", "read,write");
```

更为重要的是，Policy 类的默认实现可从访问权限文件中读取权限。在权限文件中，同样的读权限表示为：

```
permission java.io.FilePermission "/tmp/*", "read,write";
```

我们将在下一节介绍权限文件。

图 10-6 显示了 Java 1.2 中提供的权限类的层次结构。JDK 的后续版本添加了更多的权限类。

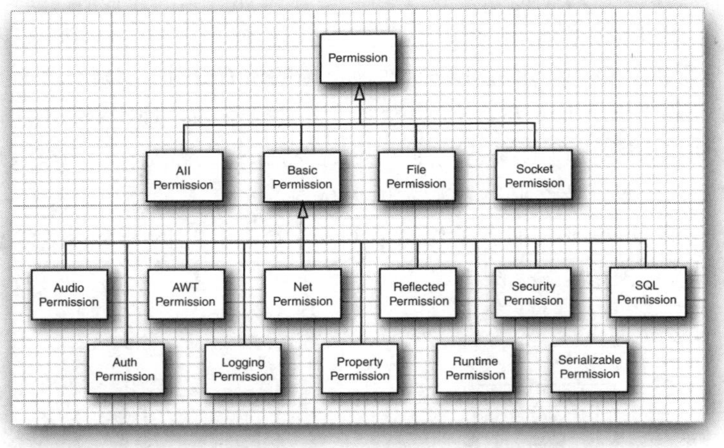

图 10-6　权限类的层次结构

在上一节中，我们看到了 SecurityManager 类有许多诸如 checkExit 的安全检查方法，这些方法的存在，只是为了程序员的方便和向后的兼容性，它们都已被映射为标准的权限检查，例如，下面是 checkExit 方法的源代码：

```
public void checkExit()
{
    checkPermission(new RuntimePermission("exitVM"));
}
```

每个类都有一个保护域，它是一个用于封装类的代码来源和权限集合的对象。当 Security-Manager 类需要检查某个权限时，它要查看当前位于调用堆栈上的所有方法的类，然后它要获得所有类的保护域，并且询问每个保护域，其权限集合是否允许执行当前正在被检查的操作。如果所有的域都同意，那么检查得以通过。否则，就会抛出一个 SecurityException 异常。

为什么在调用堆栈上的所有方法都必须允许某个特定的操作呢？让我们通过一个实例来说明这个问题。假设一个 Servlet 的 init 方法想要打开一个文件，它可能会调用下面的语句：

```
var in = new FileReader(name);
```

FileReader 构造器调用 FileInputStream 构造器，而 FileInputStream 构造器调用安全管理器的 checkRead 方法，安全管理器最后用 FilePermission(name, "read") 对象调用 checkPermission。表 10-1 显示了该调用堆栈。

表 10-1　权限检查期间的调用堆栈

类	方法	代码来源	权限
SecurityManager	checkPermission	null	AllPermission
SecurityManager	checkRead	null	AllPermission

（续）

类	方法	代码来源	权限
FileInputStream	Constructor	null	AllPermission
FileReader	Constructor	null	AllPermission
Servlet	init	Servlet 代码来源	TomcatWeb 应用权限

FileInputStream 和 SecurityManager 类都属于系统类，它们的 CodeSource 为 null，它们的权限都是由 AllPermission 类的一个实例组成的，AllPermission 类允许执行所有的操作。显然地，仅仅根据它们的权限是无法确定检查结果的。正如我们所看到的那样，checkPermission 方法必须考虑 applet 类的受限制的权限问题。通过检查整个调用堆栈，安全机制就能够确保一个类决不会要求另一个类代表自己去执行某个敏感的操作。

> **注释**：上面关于如何进行权限检查的简要介绍，向你展示了这方面的基本概念。不过我们在这里省略了对许多技术细节的说明。对于安全性的细节问题，我们建议你阅读 Li Gong 撰写的著作，以便了解更多的内容。有关 Java 平台安全模型的更多重要信息，请查阅 Gary McGraw 和 Ed Felten 撰写的 *Securing Java: Getting Down to Business with Mobile Code* 第 2 版一书，该书由 Wiley 出版社于 1999 年出版。你可以在下面的网站上找到该书的在线版本：http://www.securing java.com。

API java.lang.SecurityManager 1.0

- void checkPermission(Permission p) 1.2
  检查当前的安全管理器是否授予给定的权限。如果没有授予该权限，本方法抛出一个 SecurityException 异常。

API java.lang.Class 1.0

- ProtectionDomain getProtectionDomain() 1.2
  获取该类的保护域，如果该类被加载时没有保护域，则返回 null。

API java.security.ProtectionDomain 1.2

- ProtectionDomain(CodeSource source, PermissionCollection permissions)
  用给定的代码来源和权限构建一个保护域。
- CodeSource getCodeSource()
  获取该保护域的代码来源。
- boolean implies(Permission p)
  如果该保护域允许给定的权限，则返回 true。

API java.security.CodeSource 1.2

- Certificate[] getCertificates()

获取与该代码来源相关联的用于类文件签名的证书链。
- URL getLocation()
    获取与该代码来源相关联的类文件代码位置。

10.2.3 安全策略文件

策略管理器要读取相应的策略文件，这些文件包含了将代码来源映射为权限的指令。下面是一个典型的策略文件：

```
grant codeBase "http://www.horstmann.com/classes"
{
    permission java.io.FilePermission "/tmp/*", "read,write";
};
```

该文件给所有下载自 http://www.horstmann.com/classes 的代码授予在 /tmp 目录下读取和写入文件的权限。

可以将策略文件安装在标准位置上。默认情况下，有两个位置可以安装策略文件：
- Java 平台主目录的 java.policy 文件。
- 用户主目录的 .java.policy 文件（注意文件名前面的圆点）。

> 注释：可以在 jdk/Conf/Security 目录下 java.security 配置文件中修改这些文件的位置，默认位置设定为：
>
> ```
> policy.url.1=file:${java.home}/lib/security/java.policy
> policy.url.2=file:${user.home}/.java.policy
> ```
>
> 系统管理员可以修改 java.security 文件，并可以指定驻留在另外一台服务器上并且用户无法修改的策略 URL。策略文件中允许存放任意数量的策略 URL（这些 URL 带有连续的编号）。所有文件的权限都被组合了在一起。
>
> 如果想将策略文件存储到文件系统之外，那么可以去实现 Policy 类的一个子类，让其去收集所允许的权限。然后在 java.security 配置文件中更改下面这行：
>
> ```
> policy.provider=sun.security.provider.PolicyFile
> ```

在测试期间，我们不喜欢经常地修改这些标准文件。因此，我们更愿意为每一个应用程序单独命名策略文件，这样将权限写入一个独立的文件（比如 MyApp.policy）中即可。要应用这个策略文件，可以有两个选择。一种是在应用程序的 main 方法内部设置系统属性：

```
System.setProperty("java.security.policy", "MyApp.policy");
```

或者，可以像下面这样启动虚拟机：

```
java -Djava.security.policy=MyApp.policy MyApp
```

在这些例子中，MyApp.policy 文件被添加到了其他有效的策略中。如果在命令行中添加了第二个等号，比如：

```
java -Djava.security.policy==MyApp.policy MyApp
```

那么应用程序就只使用指定的策略文件，而标准策略文件将被忽略。

> **警告**：在测试期间，一个容易犯的错误是在当前目录中留下了一个 .java.policy 文件，该文件授予了许许多多的权限，甚至可能授予了 AllPermission。如果发现你的应用程序似乎没有应用策略文件中的规定，就应该检查当前目录下是否留有 .java.policy 文件。如果使用的是 UNIX 系统，就更容易犯这样的错误，因为在 UNIX 中，文件名以圆点开头的文件默认是不显示的。

正如前面所说，在默认情况下，Java 应用程序是不安装安全管理器的。因此，在安装安全管理器之前，看不到策略文件的作用。当然，可以将这行代码：

```
System.setSecurityManager(new SecurityManager());
```

添加到 main 方法中，或者在启动虚拟机的时候添加命令行选项 -Djava.security.manager。

```
java -Djava.security.manager -Djava.security.policy=MyApp.policy MyApp
```

在本节的剩余部分，我们将要详细介绍如何描述策略文件的权限。我们将介绍整个策略文件的格式，不过不包括代码证书部分，代码证书将在本章的后面部分介绍。

一个策略文件包含一系列 grant 项。每一项都具有以下的形式：

```
grant codesource
{
    permission₁;
    permission₂;
    . . .
};
```

代码来源包含一个代码基（如果某一项适用于所有来源的代码，则代码基可以省略）和值得信赖的用户特征（principal）与证书签名者的名字（如果不要求对该项签名，则可以省略）。

代码基可以设定为：

```
codeBase "url"
```

如果 URL 以 "/" 结束，那么它是一个目录。否则，它将被视为一个 JAR 文件的名字。例如：

```
grant codeBase "www.horstmann.com/classes/" { . . . };
grant codeBase "www.horstmann.com/classes/MyApp.jar" { . . . };
```

代码基是一个 URL 并且总是以斜杠作为文件分隔符，即使是 Windows 中的文件 URL，也是如此。例如：

```
grant codeBase "file:C:/myapps/classes/" { . . . };
```

> **注释**：大家都知道 http 格式的 URL 都以双斜杠（http://）开头的，但是它很容易与 file 格式的 URL 搞混淆，策略文件阅读器接受两种格式的 file URL，即 file://localFile 和 file:localFile。此外，Windows 驱动器名前面的斜杠是可有可无的。也就是说，下面的各种表示都是可以接受的：

```
file:C:/dir/filename.ext
file:/C:/dir/filename.ext
file://C:/dir/filename.ext
file:///C:/dir/filename.ext
```

实际上，我们的测试结果是 file:////C:/dir/filename.ext 也是允许的，对此我们无法解释。

> **注释**：请考虑编译 Java 代码的应用程序，它需要大量的权限。在 JDK 9 之前，你可以被授权获得对 tools.jar 中的代码的所有权限。这个 JAR 文件现在已经不存在了。因此，需要像下面这样授予对适合的模块进行访问的权限：
> ```
> grant codeBase "jrt:/jdk.compiler"
> {
>     permission java.security.AllPermission;
> };
> ```

权限采用下面的结构：

```
permission className targetName, actionList;
```

类名是权限类的全称类名（比如 java.io.FilePermission）。目标名是个与权限相关的值，例如，文件权限中的目录名或者文件名，或者是 socket 权限中的主机和端口。操作列表同样是与权限相关的，它是一个操作方式的列表，比如 read 或者 connect 等操作，用逗号分隔。有些权限类并不需要目标名和操作列表。表 10-2 列出了标准的权限和它们执行的操作。

表 10-2 权限及其相关的目标和操作

权限	目标	操作
java.io.FilePermission	文件目标（见正文）	read, write, execute, delete
java.net.SocketPermission	Socket 目标（见正文）	accept, connect, listen, resolve
java.util.PropertyPermission	属性目标（见正文）	read, write
java.lang.RuntimePermission	createClassLoader getClassLoader setContextClassLoader enableContextClassLoaderOverride createSecurityManager setSecurityManager exitVM getenv.variableName shutdownHooks setFactory setIO modifyThread stopThread modifyThreadGroup getProtectionDomain readFileDescriptor writeFileDescriptor	无

（续）

权限	目标	操作
java.lang.RuntimePermission	loadLibrary.libraryName accessClassInPackage.packageName defineClassInPackage.packageName accessDeclaredMembers.className queuePrintJob getStackTrace setDefaultUncaughtExceptionHandler preferences usePolicy	无
java.awt.AWTPermission	showWindowWithoutWarningBanner accessClipboard accessEventQueue createRobot fullScreenExclusive listenToAllAWTEvents readDisplayPixels replaceKeyboardFocusManager watchMousePointer setWindowAlwaysOnTop setAppletStub	无
java.net.NetPermission	setDefaultAuthenticator specifyStreamHandler requestPasswordAuthentication setProxySelector getProxySelector setCookieHandler getCookieHandler setResponseCache getResponseCache	无
java.lang.reflect.ReflectPermission	suppressAccessChecks	无
java.io.SerializablePermission	enableSubclassImplementation enableSubstitution	无
java.security.SecurityPermission	createAccessControlContext getDomainCombiner getPolicy setPolicy getProperty.keyName setProperty.keyName insertProvider.providerName removeProvider.providerName setSystemScope setIdentityPublicKey setIdentityInfo	无

（续）

权限	目标	操作
java.security.SecurityPermission	addIdentityCertificate removeIdentityCertificate printIdentity clearProviderProperties.providerName putProviderProperty.providerName removeProviderProperty.providerName getSignerPrivateKey setSignerKeyPair	无
java.security.AllPermission	无	无
javax.audio.AudioPermission	播放录音	无
javax.security.auth.AuthPermission	doAs doAsPrivileged getSubject getSubjectFromDomainCombiner setReadOnly modifyPrincipals modifyPublicCredentials modifyPrivateCredentials refreshCredential destroyCredential createLoginContext.contextName getLoginConfiguration setLoginConfiguration refreshLoginConfiguration	无
java.util.logging.LoggingPermission	control	无
java.sql.SQLPermission	setLog	无

正如表 10-2 中所示，大部分权限只允许执行某种特定的行为。可以将这些行为视为带有一个隐含操作"permit"的目标。这些权限类都继承自 BasicPermission 类（参见本章图 10-6）。然而，文件、socket 和属性权限的目标都比较复杂，我们必须对它们进行详细介绍。

文件权限的目标可以有下面几种形式：

file	文件
directory/	目录
directory/*	目录中的所有文件
*	当前目录中的所有文件
directory/-	目录和其子目录中的所有文件
-	当然目录和其子目录中所有文件
<<ALL FILES>>	文件系统中的所有文件

例如，下面的权限项赋予对 /myapp 目录和它的子目录中的所有文件的访问权限。

```
permission java.io.FilePermission "/myapp/-", "read,write,delete";
```

必须使用 \\ 转义字符序列来表示 Window 文件名中的反斜杠。

```
permission java.io.FilePermission "c:\\myapp\\-", "read,write,delete";
```

Socket 权限的目标由主机和端口范围组成。对主机的描述具有下面几种形式：

hostname 或 *IPaddress*	单个主机
localhost 或空字符串	本地主机
*.domainSuffix	以给定后缀结尾的域中所有的主机
*	所有主机

端口范围是可选的，具有下面几种形式：

:n	单个端口
:n-	编号大于等于 *n* 的所有端口
:-n	编号小于等于 *n* 的所有端口
:n1-n2	位于给定范围内的所有端口

下面是一个权限的实例：

```
permission java.net.SocketPermission "*.horstmann.com:8000-8999", "connect";
```

最后，属性权限的目标可以采用下面两种形式之一：

property	一个具体的属性
propertyPrefix.*	带有给定前缀的所有属性

"java.home" 和 "java.vm.*" 就是这样的例子。

例如，下面的权限项允许程序读取以 java.vm 开头的所有属性。

```
permission java.util.PropertyPermission "java.vm.*", "read";
```

可以在策略文件中使用系统属性，其中的 ${property} 标记会被属性值替代，例如，${user.home} 会被用户主目录替代。下面是在访问权限项中使用系统属性的典型应用。

```
permission java.io.FilePermission "${user.home}", "read,write";
```

为了创建平台无关的策略文件，使用 file.separator 属性而不是使用显式的 / 或者 \\ 分隔符绝对是个好主意。如果要使它更加简单，可以使用符号 ${/} 作为 ${file.separator} 的缩写。例如，

```
permission java.io.FilePermission "${user.home}${/}-", "read,write";
```

是一个可在平台之间移植的项，用于授予对在用户的主目录及其子目录中的文件进行读写的权限。

10.2.4　定制权限

在本节中，我们将要介绍如何把自己的权限类提供给用户，以使得他们可以在策略文件中引用这些权限类。

如果要实现自己的权限类，可以继承 Permission 类，并提供以下方法：

- 带有两个 String 参数的构造器，这两个参数分别是目标和操作列表
- String getActions()
- boolean equals(Object other)
- int hashCode()
- boolean implies(Permission other)

最后一个方法是最重要的。权限有一个排序，其中更加泛化的权限隐含了更加具体的权限。请考虑下面的文件权限：

```
p1 = new FilePermission("/tmp/-", "read, write");
```

该权限允许读写 /tmp 目录以及子目录中的任何文件。

该权限隐含了其他更加具体的权限：

```
p2 = new FilePermission("/tmp/-", "read");
p3 = new FilePermission("/tmp/aFile", "read, write");
p4 = new FilePermission("/tmp/aDirectory/-", "write");
```

换句话说，如果

1. p1 的目标文件集包含 p2 的目标文件集。
2. p1 的操作集包含 p2 的操作集。

那么，文件访问权限 p1 就隐含了另一个文件访问权限 p2。

请考虑下面关于 implies 方法的用法举例。当 FileInputStream 构造器想要打开一个文件，以读取该文件时，要检查它是否拥有操作权限。如果要执行这种检查，就应将一个具体的文件权限对象传递给 checkPermission 方法：

```
checkPermission(new FilePermission(fileName, "read"));
```

现在安全管理器询问所有适用的权限是否隐含了该权限。如果其中某个隐含了该权限，就通过了检查。

特别地，AllPermission 隐含了其他所有的权限。

如果你定义了自己的权限类，那么必须对权限对象定义一个合适的隐含法则。例如，假设你为采用 Java 技术的机顶盒定义一个 TVPermission，那么下面这个访问权限

```
new TVPermission("Tommy:2-12:1900-2200", "watch,record")
```

将允许 Tommy 在 19 点到 22 点之间对 2 至 12 频道的电视节目进行观看和录像。必须实现 implies 方法，以隐含像下面这样的更具体的权限。

```
new TVPermission("Tommy:4:2000-2100", "watch")
```

10.2.5 实现权限类

在下面这个示例程序中，我们实现了一个新的权限，用于监视将文本插入到文本域的操作。该程序会确保你不能输入"不良单词"，例如 sex, drugs 以及 C++ 等。我们使用了一个定制的权限类，以便在策略文件中提供这些不良单词。

下面这个 JTextArea 的子类询问安全管理器是否准备好了去添加新文本。

```
class WordCheckTextArea extends JTextArea
{
    public void append(String text)
    {
        var p = new WordCheckPermission(text, "insert");
        SecurityManager manager = System.getSecurityManager();
        if (manager != null) manager.checkPermission(p);
        super.append(text);
    }
}
```

如果安全管理器赋予了 WordCheckPermission 权限，那么该文本就可以追加。否则，checkPermission 方法就会抛出一个异常。

单词检查权限有两个可能的操作，一个是 insert（用于插入具体文本的权限），另一个是 avoid（添加不包含某些不良单词的任何文本的权限）。应该用下面的策略文件运行这个程序：

```
grant
{
    permission permissions.WordCheckPermission "sex,drugs,C++", "avoid";
};
```

这个策略文件赋予的权限是可以插入不包含不良单词 sex、drugs 和 C++ 的任何文本。

当设计 WordCheckPermission 类时，我们必须特别注意 implies 方法，下面是控制权限 p1 是否隐含 p2 的规则：

- 如果 p1 有 avoid 操作，p2 有 insert 操作，那么 p2 的目标必须避开 p1 中的所有单词。例如，下面这个权限：

  permissions.WordCheckPermission "sex,drugs,C++", "avoid"

  隐含了下面这个权限：

  permissions.WordCheckPermission "Mary had a little lamb", "insert"

- 如果 p1 和 p2 都有 avoid 操作，那么 p2 的单词集合必须包含 p1 单词集合中的所有单词。例如，下面这个权限：

  permissions.WordCheckPermission "sex,drugs", "avoid"

  隐含了下面这个权限：

  permissions.WordCheckPermission "sex,drugs,C++", "avoid"

- 如果 p1 和 p2 都有 insert 操作，那么 p1 的文本必须包含 p2 的文本。例如，下面这个权限：

  permissions.WordCheckPermission "Mary had a little lamb", "insert"

  包含了下面这个权限：

  permissions.WordCheckPermission "a little lamb", "insert"

可以在程序清单 10-4 中看到该类的具体实现。

请注意，可以用 Permission 类中名字容易混淆的 getName 方法来获取权限的目标。

由于在策略文件中权限是由一对字符串来表示的，因此，权限类需要准备好解析这些字符串。特别地，我们应该使用下面的方法，将用逗号分隔的 avoid 权限的不良单词表转换为一个真正的 Set。

```
public Set<String> badWordSet()
{
    var set = new HashSet<String>();
    set.addAll(List.of(getName().split(",")));
    return set;
}
```

该代码允许我们用 equals 和 containsAll 方法来比较这些集。正如我们在卷 I 第 9 章中所介绍的那样，如果两个集包含任意次序的相同元素，那么集类的 equals 方法可以判定它们相等。例如，由 "sex,drugs,C++" 和 "C++,drugs,sex" 产生的两个集是相等的集。

> ⚠️ **警告**：务必要把你的权限类设为 public。策略文件加载器不能加载具有包可视性的类，并且它会悄悄忽略其无法找到的所有类。

程序清单 10-5 中的程序展示了 WordCheck Permission 类是如何工作的。请在文本框内输入任意文本，然后按下 Insert 按钮。如果文本通过了安全检查，该文本就会被添加到文本区域中。如果没有通过检查，就会弹出一个消息（参见图 10-7）。

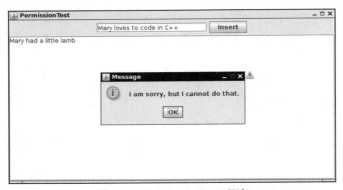

图 10-7　PermissionTest 程序

你现在已经看到应该如何配置 Java 平台的安全性了。更常见的情况是，你只需微调标准的权限集。对于其他额外的控制，你可以定义自制的权限，它们应该可以按照与标准权限相同的方式配置。

程序清单 10-4　permissions/WordCheckPermission.java

```
1  package permissions;
2
3  import java.security.*;
4  import java.util.*;
5
6  /**
7   * A permission that checks for bad words.
8   */
```

```java
 9  public class WordCheckPermission extends Permission
10  {
11     private String action;
12
13     /**
14      * Constructs a word check permission.
15      * @param target a comma separated word list
16      * @param anAction "insert" or "avoid"
17      */
18     public WordCheckPermission(String target, String anAction)
19     {
20        super(target);
21        action = anAction;
22     }
23
24     public String getActions()
25     {
26        return action;
27     }
28
29     public boolean equals(Object other)
30     {
31        if (other == null) return false;
32        if (!getClass().equals(other.getClass())) return false;
33        var b = (WordCheckPermission) other;
34        if (!Objects.equals(action, b.action)) return false;
35        if ("insert".equals(action)) return Objects.equals(getName(), b.getName());
36        else if ("avoid".equals(action)) return badWordSet().equals(b.badWordSet());
37        else return false;
38     }
39
40     public int hashCode()
41     {
42        return Objects.hash(getName(), action);
43     }
44
45     public boolean implies(Permission other)
46     {
47        if (!(other instanceof WordCheckPermission)) return false;
48        var b = (WordCheckPermission) other;
49        if (action.equals("insert"))
50        {
51           return b.action.equals("insert") && getName().indexOf(b.getName()) >= 0;
52        }
53        else if (action.equals("avoid"))
54        {
55           if (b.action.equals("avoid")) return b.badWordSet().containsAll(badWordSet());
56           else if (b.action.equals("insert"))
57           {
58              for (String badWord : badWordSet())
59                 if (b.getName().indexOf(badWord) >= 0) return false;
60              return true;
61           }
62           else return false;
```

```
 63        }
 64        else return false;
 65     }
 66
 67     /**
 68      * Gets the bad words that this permission rule describes.
 69      * @return a set of the bad words
 70      */
 71     public Set<String> badWordSet()
 72     {
 73        var set = new HashSet<String>();
 74        set.addAll(List.of(getName().split(",")));
 75        return set;
 76     }
 77  }
```

程序清单 10-5   permissions/PermissionTest.java

```
 1  package permissions;
 2
 3  import java.awt.*;
 4
 5  import javax.swing.*;
 6
 7  /**
 8   * This class demonstrates the custom WordCheckPermission.
 9   * @version 1.05 2018-05-01
10   * @author Cay Horstmann
11   */
12  public class PermissionTest
13  {
14     public static void main(String[] args)
15     {
16        System.setProperty("java.security.policy", "permissions/PermissionTest.policy");
17        System.setSecurityManager(new SecurityManager());
18        EventQueue.invokeLater(() ->
19           {
20              var frame = new PermissionTestFrame();
21              frame.setTitle("PermissionTest");
22              frame.setDefaultCloseOperation(JFrame.EXIT_ON_CLOSE);
23              frame.setVisible(true);
24           });
25     }
26  }
27
28  /**
29   * This frame contains a text field for inserting words into a text area that is protected
30   * from "bad words".
31   */
32  class PermissionTestFrame extends JFrame
33  {
34     private JTextField textField;
35     private WordCheckTextArea textArea;
36     private static final int TEXT_ROWS = 20;
```

```java
37      private static final int TEXT_COLUMNS = 60;
38
39      public PermissionTestFrame()
40      {
41         textField = new JTextField(20);
42         var panel = new JPanel();
43         panel.add(textField);
44         var openButton = new JButton("Insert");
45         panel.add(openButton);
46         openButton.addActionListener(event -> insertWords(textField.getText()));
47
48         add(panel, BorderLayout.NORTH);
49
50         textArea = new WordCheckTextArea();
51         textArea.setRows(TEXT_ROWS);
52         textArea.setColumns(TEXT_COLUMNS);
53         add(new JScrollPane(textArea), BorderLayout.CENTER);
54         pack();
55      }
56
57      /**
58       * Tries to insert words into the text area. Displays a dialog if the attempt fails.
59       * @param words the words to insert
60       */
61      public void insertWords(String words)
62      {
63         try
64         {
65            textArea.append(words + "\n");
66         }
67         catch (SecurityException ex)
68         {
69            JOptionPane.showMessageDialog(this, "I am sorry, but I cannot do that.");
70            ex.printStackTrace();
71         }
72      }
73   }
74
75   /**
76    * A text area whose append method makes a security check to see that no bad words are added.
77    */
78   class WordCheckTextArea extends JTextArea
79   {
80      public void append(String text)
81      {
82         var p = new WordCheckPermission(text, "insert");
83         SecurityManager manager = System.getSecurityManager();
84         if (manager != null) manager.checkPermission(p);
85         super.append(text);
86      }
87   }
```

API java.security.Permission 1.2

- Permission(String name)

用指定的目标名构建一个权限。

- String getName()

  返回该权限的对象名称。

- boolean implies(Permission other)

  检查该权限是否隐含了 other 权限。如果 other 权限描述了一个更加具体的条件，而这个具体条件是由该权限所描述的条件所产生的结果，那么该权限就隐含这个 other 权限。

10.3 用户认证

Java API 提供了一个名为 Java 认证和授权服务的框架，它将平台提供的认证与权限管理集成起来。我们将在以下各节中讨论 JAAS 框架。

10.3.1 JAAS 框架

正如其名字所表示的，Java 认证和授权服务（JAAS，Java Authentication and Authorization Service）包含两部分："认证"部分主要负责确定程序使用者的身份，而"授权"将各个用户映射到相应的权限。

JAAS 是一个可插拔的 API，可以将 Java 应用程序与实现认证的特定技术分离开来。除此之外，JAAS 还支持 UNIX 登录、NT 登录、Kerberos 认证和基于证书的认证。

一旦用户通过认证，就可以为其附加一组权限。例如，这里我们赋予 Harry 一个特定的权限集，而其他用户则没有，它的语法规则如下：

```
grant principal com.sun.security.auth.UnixPrincipal "harry"
{
    permission java.util.PropertyPermission "user.*", "read";
    ...
};
```

com.sun.security.auth.UnixPrincipal 类检查运行该程序的 UNIX 用户的名字，它的 getName 方法将返回 UNIX 登录名，然后我们就可以检查该名称是否等于 "harry"。

可以使用一个 LoginContext 以使得安全管理器能够检查这样的授权语句。下面是登录代码的基本轮廓：

```
try
{
    System.setSecurityManager(new SecurityManager());
    var context = new LoginContext("Login1"); // defined in JAAS configuration file
    context.login();
    // get the authenticated Subject
    Subject subject = context.getSubject();
    ...
    context.logout();
}
```

```
catch (LoginException exception) // thrown if login was not successful
{
    exception.printStackTrace();
}
```

这里，subject 是指已经被认证的个体。

LoginContext 构造器中的字符串参数 "Login1" 是指 JAAS 配置文件中具有相同名字的项。下面是一个简单的配置文件：

```
Login1
{
    com.sun.security.auth.module.UnixLoginModule required;
    com.whizzbang.auth.module.RetinaScanModule sufficient;
};

Login2
{
    ...
};
```

当然，JDK 中没有包含任何使用 biometric 的登录模块。JDK 在 com.sun.security.auth.module 包中包含以下模块：

```
UnixLoginModule
NTLoginModule
Krb5LoginModule
JndiLoginModule
KeyStoreLoginModule
```

一个登录策略由一个登录模块序列组成，每个模块被标记为 required、sufficient、requisite 或 optional。这些关键字的含义在下面的算法中进行了描述：

登录时要对登录的主体（subject）进行认证，该主体可以拥有多个特征（principal）。特征描述了主体的某些属性，比如用户名、组 ID 或角色等。我们在 grant 语句中可以看到，特征管制着各个权限。com.sun.security.auth.UnixPrincipal 类描述了 UNIX 登录名，UnixNumericGroupPrincipal 类可以用来检测用户是否归属于某个 UNIX 用户组。

使用下面的语法，grant 语句可以对一个特征进行测试：

grant *principalClass* "*principalName*"

例如：

grant com.sun.security.auth.UnixPrincipal "harry"

当用户登录后，就会在独立的访问控制上下文中，运行要求检查用户特征的代码。使用静态的 doAs 或 doAsPrivileged 方法，启动一个新的 PrivilegedAction，其 run 方法就会执行这段代码。

这两个方法都可以通过使用主体特征的权限来调用某个对象的 run 方法去执行特定操作，而该对象必须是实现了 PrivilegedAction 接口的对象。

```
PrivilegedAction<T> action = () ->
    {
```

```
          // run with permissions of subject principals
          ...
    };
T result
    = Subject.doAs(subject, action); // or Subject.doAsPrivileged(subject, action, null)
```

如果该操作会抛出受检查的异常，那么必须改为实现 PrivilegedExceptionAction 接口。

doAs 和 doAsPrivileged 方法之间的区别是微小的。doAs 方法开始于当前的访问控制上下文，而 doAsPrivileged 方法则开始于一个新的上下文。后者允许将登录代码和"业务逻辑"的权限相分离。在我们的示例应用程序中，登录代码有如下权限：

```
permission javax.security.auth.AuthPermission "createLoginContext.Login1";
permission javax.security.auth.AuthPermission "doAsPrivileged";
```

通过认证的用户有一个权限：

```
permission java.util.PropertyPermission "user.*", "read";
```

如果我们用 doAs 代替了 doAsPrivileged，那么登录代码也需要这个权限！

程序清单 10-6 和程序清单 10-7 的程序展示了如何限制某些用户的权限。AuthTest 程序对用户的身份进行了认证，然后运行了一个简单的操作，以获得一个系统属性。

程序清单 10-6　auth/AuthTest.java

```
 1  package auth;
 2
 3  import javax.security.auth.*;
 4  import javax.security.auth.login.*;
 5
 6  /**
 7   * This program authenticates a user via a custom login and then executes the SysPropAction
 8   * with the user's privileges.
 9   * @version 1.02 2018-05-01
10   * @author Cay Horstmann
11   */
12  public class AuthTest
13  {
14     public static void main(final String[] args)
15     {
16        System.setSecurityManager(new SecurityManager());
17        try
18        {
19           var context = new LoginContext("Login1");
20           context.login();
21           System.out.println("Authentication successful.");
22           Subject subject = context.getSubject();
23           System.out.println("subject=" + subject);
24           var action = new SysPropAction("user.home");
25           String result = Subject.doAsPrivileged(subject, action, null);
26           System.out.println(result);
27           context.logout();
28        }
29        catch (LoginException e)
```

```
30        {
31           e.printStackTrace();
32        }
33     }
34 }
```

程序清单 10-7 auth/SysPropAction.java

```
 1 package auth;
 2
 3 import java.security.*;
 4
 5 /**
 6  * This action looks up a system property.
 7  * @version 1.01 2007-10-06
 8  * @author Cay Horstmann
 9  */
10 public class SysPropAction implements PrivilegedAction<String>
11 {
12    private String propertyName;
13
14    /**
15       Constructs an action for looking up a given property.
16       @param propertyName the property name (such as "user.home")
17    */
18    public SysPropAction(String propertyName)
19    {
20       this.propertyName = propertyName;
21    }
22
23    public String run()
24    {
25       return System.getProperty(propertyName);
26    }
27 }
```

要使该例子能够运行，必须将登录类和操作类的代码封装到两个独立的 JAR 文件中：

```
javac auth/*.java
jar cvf login.jar auth/AuthTest.class
jar cvf action.jar auth/SysPropAction.class
```

如果查看程序清单 10-8 中的策略文件，将会看到名为 harry 的 UNIX 用户拥有读取所有文件的权限。将 harry 改为你自己的登录名，然后运行下面的命令：

```
java -classpath login.jar:action.jar \
   -Djava.security.policy=auth/AuthTest.policy \
   -Djava.security.auth.login.config=auth/jaas.config \
   auth.AuthTest
```

程序清单 10-8 auth/AuthTest.policy

```
1 grant codebase "file:login.jar"
2 {
```

```
 3      permission javax.security.auth.AuthPermission "createLoginContext.Login1";
 4      permission javax.security.auth.AuthPermission "doAsPrivileged";
 5  };
 6
 7  grant principal com.sun.security.auth.UnixPrincipal "harry"
 8  {
 9      permission java.util.PropertyPermission "user.*", "read";
10  };
```

程序清单 10-9 展示了登录的配置。

程序清单 10-9　auth/jaas.config

```
1  Login1
2  {
3      com.sun.security.auth.module.UnixLoginModule required;
4  };
```

在 Windows 下运行时，请将 AuthTest.policy 中的 UnixPrincipal 改为 NTUser-Principal，并将 jaas.config 中的 UnixLoginModule 改为 NTLoginModule。运行该程序时，请用分号来分隔各个 JAR 文件：

```
java -classpath login.jar;action.jar ...
```

AuthTest 程序现在将显示 user.home 属性的值。但是，如果用不同的名字登录，那么就应该抛出一个安全异常，因为你不再拥有必需的权限了。

> **警告**：必须严格按照这些指令来运行。如果对程序进行了一些看上去无关紧要的更改，那就很容易使你的设置出错。

API javax.security.auth.login.LoginContext 1.4

- LoginContext(String name)
  创建一个登录上下文。name 对应于 JAAS 配置文件中的登录描述符。
- void login()
  建立一个登录操作，如果登录失败，则抛出一个 LoginException 异常。它会调用 JAAS 配置文件中的管理器上的 login 方法。
- void logout()
  Subject 退出登录。它会调用 JAAS 配置文件中的管理器上的 logout 方法。
- Subject getSubject()
  返回认证过的 Subject。

API javax.security.auth.Subject 1.4

- Set<Principal> getPrincipals()
  获取该 Subject 的各个 Principal。

- static Object doAs(Subject subject, PrivilegedAction action)
- static Object doAs(Subject subject, PrivilegedExceptionAction action)
- static Object doAsPrivileged(Subject subject, PrivilegedAction action, AccessControlContext context)
- static Object doAsPrivileged(Subject subject, PrivilegedExceptionAction action, AccessControlContext context)

  以 subject 的身份执行特许操作。它将返回 run 方法的返回值。doAsPrivileged 方法在给定的访问控制上下文中执行该操作，你可以提供一个在前面调用静态方法 AccessController.getContext() 时所获得的"上下文快照"，或者指定为 null，以便使其在一个新的上下文中执行该代码。

API java.security.PrivilegedAction 1.4

- Object run()

  必须定义该方法，以执行你想要代表某个主体去执行的代码。

API java.security.PrivilegedExceptionAction 1.4

- Object run()

  必须定义该方法，以执行你想要代表某个主体去执行的代码。本方法可以抛出任何受检查的异常。

API java.security.Principal 1.1

- String getName()

  返回该特征的身份标识。

10.3.2 JAAS 登录模块

在本节中，我们将要用一个 JAAS 例子向读者介绍：
- 如何实现你自己的登录模块；
- 如何实现基于角色的认证。

如果登录信息存储在数据库中，那么使用自己的登录模块就非常有用。尽管你可能很喜欢默认的登录模块，但是学习如何定制自己的模块将有助于你理解 JAAS 配置文件的各个选项。

基于角色的认证对于大量用户的管理来说是十分必要的。将所有合法用户的名字都写入策略文件是不切实际的。而登录模块应该将用户映射到诸如"admin"或"HR"等角色，并且权限的赋予也要基于这些角色。

登录模块的工作之一是组装被认证的主体的特征集。如果一个登录模块支持某些角色，该模块就会添加 Principal 对象来描述这些角色。JDK 并没有提供相应的类，所以我们写了自己的类（见程序清单 10-10）。该类直接存储了一个描述/值对，例如 role=admin。该类的 getName 方法用于返回该描述/值对，因此我们就可以添加基于角色的权限到策略文件中：

```
grant principal SimplePrincipal "role=admin" { . . . }
```

程序清单 10-10 jaas/SimplePrincipal.java

```java
 1  package jaas;
 2
 3  import java.security.*;
 4  import java.util.*;
 5
 6  /**
 7   * A principal with a named value (such as "role=HR" or "username=harry").
 8   */
 9  public class SimplePrincipal implements Principal
10  {
11     private String descr;
12     private String value;
13
14     /**
15      * Constructs a SimplePrincipal to hold a description and a value.
16      * @param descr the description
17      * @param value the associated value
18      */
19     public SimplePrincipal(String descr, String value)
20     {
21        this.descr = descr;
22        this.value = value;
23     }
24
25     /**
26      * Returns the role name of this principal.
27      * @return the role name
28      */
29     public String getName()
30     {
31        return descr + "=" + value;
32     }
33
34     public boolean equals(Object otherObject)
35     {
36        if (this == otherObject) return true;
37        if (otherObject == null) return false;
38        if (getClass() != otherObject.getClass()) return false;
39        var other = (SimplePrincipal) otherObject;
40        return Objects.equals(getName(), other.getName());
41     }
42
43     public int hashCode()
44     {
45        return Objects.hashCode(getName());
46     }
47  }
```

我们的登录模块会在包含如下行的文本文件中查找用户、密码和角色：

```
harry|secret|admin
carl|guessme|HR
```

当然,在实际的登录模块中,你可能会将这些信息存储在数据库或者目录中。

在程序清单 10-11 中可以找到 SimpleLoginModule 的代码,其 checkLogin 方法用于检查输入的用户名和密码是否与密码文件中的用户记录相匹配。如果匹配成功,则会添加两个 SimplePrincipal 对象到主体的特征集中。

```
Set<Principal> principals = subject.getPrincipals();
principals.add(new SimplePrincipal("username", username));
principals.add(new SimplePrincipal("role", role));
```

程序清单 10-11　jaas/SimpleLoginModule.java

```java
 1  package jaas;
 2
 3  import java.io.*;
 4  import java.nio.charset.*;
 5  import java.nio.file.*;
 6  import java.security.*;
 7  import java.util.*;
 8  import javax.security.auth.*;
 9  import javax.security.auth.callback.*;
10  import javax.security.auth.login.*;
11  import javax.security.auth.spi.*;
12
13  /**
14   * This login module authenticates users by reading usernames, passwords, and roles from
15   * a text file.
16   */
17  public class SimpleLoginModule implements LoginModule
18  {
19     private Subject subject;
20     private CallbackHandler callbackHandler;
21     private Map<String, ?> options;
22
23     public void initialize(Subject subject, CallbackHandler callbackHandler,
24           Map<String, ?> sharedState, Map<String, ?> options)
25     {
26        this.subject = subject;
27        this.callbackHandler = callbackHandler;
28        this.options = options;
29     }
30
31     public boolean login() throws LoginException
32     {
33        if (callbackHandler == null) throw new LoginException("no handler");
34
35        var nameCall = new NameCallback("username: ");
36        var passCall = new PasswordCallback("password: ", false);
37        try
38        {
39           callbackHandler.handle(new Callback[] { nameCall, passCall });
```

```java
40        }
41        catch (UnsupportedCallbackException e)
42        {
43           var e2 = new LoginException("Unsupported callback");
44           e2.initCause(e);
45           throw e2;
46        }
47        catch (IOException e)
48        {
49           var e2 = new LoginException("I/O exception in callback");
50           e2.initCause(e);
51           throw e2;
52        }
53
54        try
55        {
56           return checkLogin(nameCall.getName(), passCall.getPassword());
57        }
58        catch (IOException ex)
59        {
60           var ex2 = new LoginException();
61           ex2.initCause(ex);
62           throw ex2;
63        }
64     }
65
66     /**
67      * Checks whether the authentication information is valid. If it is, the subject acquires
68      * principals for the user name and role.
69      * @param username the user name
70      * @param password a character array containing the password
71      * @return true if the authentication information is valid
72      */
73     private boolean checkLogin(String username, char[] password)
74        throws LoginException, IOException
75     {
76        try (var in = new Scanner(
77           Paths.get("" + options.get("pwfile")), StandardCharsets.UTF_8))
78        {
79           while (in.hasNextLine())
80           {
81              String[] inputs = in.nextLine().split("\\|");
82              if (inputs[0].equals(username)
83                 && Arrays.equals(inputs[1].toCharArray(), password))
84              {
85                 String role = inputs[2];
86                 Set<Principal> principals = subject.getPrincipals();
87                 principals.add(new SimplePrincipal("username", username));
88                 principals.add(new SimplePrincipal("role", role));
89                 return true;
90              }
91           }
92           return false;
93        }
```

```
 94        }
 95
 96        public boolean logout()
 97        {
 98           return true;
 99        }
100
101        public boolean abort()
102        {
103           return true;
104        }
105
106        public boolean commit()
107        {
108           return true;
109        }
110     }
```

SimpleLoginModule 剩余的部分就非常直截了当了。initialize 方法接收下面几个参数：
- 用于认证的 Subject。
- 一个获取登录信息的 handler。
- 一个 sharedState 映射表，它可以用于登录模块之间的通信。
- 一个 options 映射表，它包含了登录配置文件中设置的名/值对。

例如，我们将模块做如下配置：

```
SimpleLoginModule required pwfile="password.txt";
```

则登录模块可以从 options 映射表中获取 pwfile 设置。

该登录模块并没有收集用户名和密码，这是单独的 handler 需要做的工作。这种功能上的分离有助于在各种情况下使用相同的登录模块，而不用关心登录信息是来自 GUI 对话框、控制台提示符还是配置文件。

handler 是在创建 LoginContext 时指定的。例如，

```
var context = new LoginContext("Login1",
    new com.sun.security.auth.callback.DialogCallbackHandler());
```

DialogCallbackHandler 会弹出一个简单的 GUI 对话框，以获取用户名和密码。而 com.sun.security.auth.callback.TextCallbackHandler 则从控制台获取这些信息。

但是，在我们的应用程序中，是通过自己编写的 GUI 来获得用户名和密码的（参见图 10-8）。我们创建了一个简单的 handler，仅仅用于存储和返回这些信息（见程序清单 10-12）。

图 10-8　一个定制的登录模块

程序清单 10-12　jaas/SimpleCallbackHandler.java

```
1  package jaas;
2
```

```
 3   import javax.security.auth.callback.*;
 4
 5   /**
 6    * This simple callback handler presents the given user name and password.
 7    */
 8   public class SimpleCallbackHandler implements CallbackHandler
 9   {
10      private String username;
11      private char[] password;
12
13      /**
14       * Constructs the callback handler.
15       * @param username the user name
16       * @param password a character array containing the password
17       */
18      public SimpleCallbackHandler(String username, char[] password)
19      {
20         this.username = username;
21         this.password = password;
22      }
23
24      public void handle(Callback[] callbacks)
25      {
26         for (Callback callback : callbacks)
27         {
28            if (callback instanceof NameCallback)
29            {
30               ((NameCallback) callback).setName(username);
31            }
32            else if (callback instanceof PasswordCallback)
33            {
34               ((PasswordCallback) callback).setPassword(password);
35            }
36         }
37      }
38   }
```

该 handler 有一个简单的方法 handle，用于处理 Callback 对象数组。有很多预定义类，比如 NameCallback 和 PasswordCallback 等，都实现了 Callback 接口。也可以添加自己的类，比如 RetinaScanCallback 等。下面这段 handler 代码可能有些不雅致，因为它要分析 callback 对象的类型：

```
public void handle(Callback[] callbacks)
{
   for (Callback callback : callbacks)
   {
      if (callback instanceof NameCallback) . . .
      else if (callback instanceof PasswordCallback) . . .
      else . . .
   }
}
```

登录模块提供 callback 数组以满足认证的需要。

```
var nameCall = new NameCallback("username: ");
var passCall = new PasswordCallback("password: ", false);
callbackHandler.handle(new Callback[] { nameCall, passCall });
```

然后它从 callback 中获取所要的信息。

程序清单 10-13 中的程序将显示一个窗体，用于输入登录信息和系统属性名。如果用户通过了认证，属性值会在 PrivilegedAction 中被取出。从程序清单 10-14 的策略文件中可以看到，只有具有 admin 角色的用户才具有对属性的读取权限。

程序清单 10-13　jaas/JAASTest.java

```
1  package jaas;
2
3  import java.awt.*;
4  import javax.swing.*;
5
6  /**
7   * This program authenticates a user via a custom login and then looks up a system property
8   * with the user's privileges.
9   * @version 1.03 2018-05-01
10  * @author Cay Horstmann
11  */
12 public class JAASTest
13 {
14    public static void main(final String[] args)
15    {
16       System.setSecurityManager(new SecurityManager());
17       EventQueue.invokeLater(() ->
18          {
19             var frame = new JAASFrame();
20             frame.setDefaultCloseOperation(JFrame.EXIT_ON_CLOSE);
21             frame.setTitle("JAASTest");
22             frame.setVisible(true);
23          });
24    }
25 }
```

程序清单 10-14　jaas/JAASTest.policy

```
1  grant codebase "file:login.jar"
2  {
3     permission java.awt.AWTPermission "showWindowWithoutWarningBanner";
4     permission java.awt.AWTPermission "accessEventQueue";
5     permission javax.security.auth.AuthPermission "createLoginContext.Login1";
6     permission javax.security.auth.AuthPermission "doAsPrivileged";
7     permission javax.security.auth.AuthPermission "modifyPrincipals";
8     permission java.io.FilePermission "jaas/password.txt", "read";
9  };
10
11 grant principal jaas.SimplePrincipal "role=admin"
12 {
13    permission java.util.PropertyPermission "*", "read";
14 };
```

正如前一节中所讲到的,必须将登录和操作代码分开。因此,首先创建两个 JAR 文件:

```
javac *.java
jar cvf login.jar JAAS*.class Simple*.class
jar cvf action.jar SysPropAction.class
```

然后以如下方式运行程序:

```
java -classpath login.jar:action.jar \
    -Djava.security.policy=JAASTest.policy \
    -Djava.security.auth.login.config=jaas.config \
    JAASTest
```

程序清单 10-15 说明了登录的配置。

> **注释**:有些应用有可能需要支持更复杂的两阶段协议,即只有登录配置文件中的所有模块都认证成功,该登录才会被提交。更多详细信息,请参阅下面地址的登录模块开发指南: http://docs.oracle.com/javase/8/docs/technotes/guides/security/jaas/JAASLMDevGuide.html。

程序清单 10-15 jaas/jaas.config

```
1  Login1
2  {
3      jaas.SimpleLoginModule required pwfile="jaas/password.txt" debug=true;
4  };
```

API *javax.security.auth.callback.CallbackHandler* 1.4

- void handle(Callback[] callbacks)

  处理给定的 callback,如果愿意,可以与用户进行交互,并且将安全信息存储到 callback 对象中。

API *javax.security.auth.callback.NameCallback* 1.4

- NameCallback(String prompt)
- NameCallback(String prompt, String defaultName)

  用给定的提示符和默认的名字构建一个 NameCallback。

- String getName()
- void setName(String name)

  设置或者获取该 callback 所收集到的名字。

- String getPrompt()

  获取查询该名字时所使用的提示符。

- String getDefaultName()

  获取查询该名字时所使用的默认名字。

API *javax.security.auth.callback.PasswordCallback* 1.4

- PasswordCallback(String prompt, boolean echoOn)

用给定提示符和回显标记构建一个 PasswordCallback。
- char[] getPassword()
- void setPassword(char[] password)

  设置或者获取该 callback 所收集到的密码。
- String getPrompt()

  获取查询该密码时所使用的提示符。
- boolean isEchoOn()

  获取查询该密码时所使用的回显标记。

API *javax.security.auth.spi.LoginModule* 1.4

- void initialize(Subject subject, CallbackHandler handler, Map<String,?> sharedState, Map<String,?> options)

  为了认证给定的 subject，初始化该 LoginModule。在登录处理期间，用给定的 handler 来收集登录信息；使用 sharedState 映射表与其他登录模块进行通信；options 映射表包含该模块实例的登录配置中指定的名/值对。
- boolean login()

  执行认证过程，并组装主体的特征集。如果登录成功，则返回 true。
- boolean commit()

  对于需要两阶段提交的登录场景，当所有的登录模块都成功后，调用该方法。如果操作成功，则返回 true。
- boolean abort()

  如果某一登录模块失败导致登录过程中断，就调用该方法。如果操作成功，则返回 true。
- boolean logout()

  注销当前的主体。如果操作成功，则返回 true。

10.4 数字签名

正如我们前面所说，applet 是在 Java 平台上开始流行起来的。实际上，人们发现尽管他们可以编写出像著名的 "nervous text" 那样栩栩如生的 applet，但是在 JDK 1.0 安全模式下无法发挥其一整套非常有用的作用。例如，由于 JDK 1.0 下的 applet 要受到严密的监管，因此，即使 applet 在公司安全内部网上运行时风险相对较小，applet 也无法在企业内部网上发挥很大的作用。Sun 公司很快就认识到，要使 applet 真正变得非常有用，用户必须可以根据 applet 的来源为其分配不同的安全级别。如果 applet 来自值得信赖的提供商，并且没有被篡改过，那么 applet 的用户就可以决定是否给 applet 授予更多的运行特权。

如果要给予一个 applet 更多的信任，你必须知道下面两件事：

1. 这个 applet 来自哪里？
2. 在传输过程中代码是否被破坏？

在过去的 50 年里，数学家和计算机科学家已经开发出各种各样成熟的算法，用于确保数据和电子签名的完整性，在 java.security 包中包含了许多这类算法的实现，而且幸运的是，你无须掌握相应的数学基础知识，就可以使用 java.security 包中的算法。在下面几节中，我们将要介绍消息摘要是如何检测数据文件中的变化的，以及数字签名是如何证明签名者的身份的。

10.4.1 消息摘要

消息摘要（message digest）是数据块的数字指纹。例如，所谓的 SHA1（安全散列算法 #1）可将任何数据块，无论其数据有多长，都压缩为 160 位（20 字节）的序列。与真实的指纹一样，人们希望任何两条不同的消息都不会有相同的 SHA1 指纹。当然，这是不可能的—因为只存在 2^{160} 个 SHA1 指纹，所以肯定会有某些消息具有相同的指纹。因为 2^{160} 是一个很大的数字，所以存在重复指纹的可能性微乎其微，那么这种重复的可能性到底小到什么程度呢？根据 James Walsh 在他的 *True Odds: How Risks Affect Your Everyday Life*（Merritt Publishing 出版社 1996 年出版）一书中所叙述的，人死于雷击的概率为三万分之一。现在，假设有 9 个人，比如你不喜欢的 9 个经理或者教授，你和他们所有的人都死于雷击的概率，比伪造的消息与原有消息具有相同的 SHA1 指纹的概率还要高。（当然，可能有你不认识的其他 10 个以上的人会死于雷击，但这里我们讨论的是你选择的特定的人的死亡概率。）

消息摘要具有两个基本属性：

1. 如果数据的 1 位或者几位改变了，那么消息摘要也将改变。
2. 拥有给定消息的伪造者无法创建与原消息具有相同摘要的假消息。

当然，第二个属性又是一个概率问题。让我们来看看下面这位亿万富翁留下的遗嘱：

"我死了之后，我的财产将由我的孩子平分，但是，我的儿子 George 应该拿不到一个子。"

这份遗嘱的 SHA1 指纹为：

12 5F 09 03 E7 31 30 19 2E A6 E7 E4 90 43 84 B4 38 99 8F 67

这位有疑心病的父亲将这份遗嘱交给一位律师保存，而将指纹交给另一位律师保存。现在，假设 George 能够贿赂那位保存遗嘱的律师，他想修改这份遗嘱，使得 Bill 一无所得。当然，这需要将原指纹改为下面这样完全不同的位模式：

7D F6 AB 08 EB 40 EC CD AB 74 ED E9 86 F9 ED 99 D1 45 B1 57

那么 George 能够找到与该指纹相匹配的其他措辞吗？如果从地球形成之时，他就很自豪地拥有 10 亿台计算机，每台计算机每秒钟能处理一百万条信息，他依然无法找到一个能够替换的遗嘱。

人们已经设计出大量的算法，用于计算这些消息摘要，其中最著名的两种算法是 SHA1 和 MD5。SHA1 是由美国国家标准和技术学会开发的加密散列算法，MD5 是由麻省理工学院的 Ronald Rivest 发明的算法。这两种算法都使用了独特巧妙的方法对消息中的各个位进行扰乱。如果要了解这些方法的详细信息，请参阅 William Stallings 撰写的 *Cryptography and Network Security*（第 7 版）一书，该书由 Prentice Hall 出版社于 2017 年出版。但是，人们在这两种算法中发现了某些微妙的规律性，因此美国国家标准和技术学会建议切换到更强的加密算法上，Java 支持 SHA-2 和 SHA-3 算法集。

MessageDigest 类是用于创建封装了指纹算法的对象的"工厂"，它的静态方法 getInstance 返回继承了 MessageDigest 类的某个类的对象。这意味着 MessageDigest 类能够承担下面的双重职责：

- 作为一个工厂类。
- 作为所有消息摘要算法的超类。

例如，下面是如何获取一个能够计算 SHA 指纹的对象的方法：

```
MessageDigest alg = MessageDigest.getInstance("SHA-1");
```

在获取 MessageDigest 对象之后，可以通过反复调用 update 方法，将信息中的所有字节提供给该对象。例如，下面的代码将文件中的所有字节传给上面创建的 alg 对象，以执行指纹算法：

```
InputStream in = . . .;
int ch;
while ((ch = in.read()) != -1)
    alg.update((byte) ch);
```

另外，如果这些字节存放在一个数组中，那就可以一次完成整个数组的更新：

```
byte[] bytes = . . .;
alg.update(bytes);
```

当完成上述操作后，调用 digest 方法。该方法按照指纹算法的要求补齐输入，并且进行相应的计算，然后以字节数组的形式返回消息摘要。

```
byte[] hash = alg.digest();
```

程序清单 10-16 中的程序计算了一个消息摘要，可以在命令行中指定文件和算法：

```
java hash.Digest hash/input.txt SHA-1
```

如果没有提供命令行参数，那么就会提示你输入文件名和算法名。

程序清单 10-16　hash/Digest.java

```java
1  package hash;
2
3  import java.io.*;
4  import java.nio.file.*;
5  import java.security.*;
6  import java.util.*;
7
8  /**
9   * This program computes the message digest of a file.
10  * @version 1.21 2018-04-10
11  * @author Cay Horstmann
```

```java
12    */
13   public class Digest
14   {
15      /**
16       * @param args args[0] is the filename, args[1] is optionally the algorithm
17       * (SHA-1, SHA-256, or MD5)
18       */
19      public static void main(String[] args) throws IOException, GeneralSecurityException
20      {
21         var in = new Scanner(System.in);
22         String filename;
23         if (args.length >= 1)
24            filename = args[0];
25         else
26         {
27            System.out.print("File name: ");
28            filename = in.nextLine();
29         }
30         String algname;
31         if (args.length >= 2)
32            algname = args[1];
33         else
34         {
35            System.out.println("Select one of the following algorithms: ");
36            for (Provider p : Security.getProviders())
37               for (Provider.Service s : p.getServices())
38                  if (s.getType().equals("MessageDigest"))
39                     System.out.println(s.getAlgorithm());
40            System.out.print("Algorithm: ");
41            algname = in.nextLine();
42         }
43         MessageDigest alg = MessageDigest.getInstance(algname);
44         byte[] input = Files.readAllBytes(Paths.get(filename));
45         byte[] hash = alg.digest(input);
46         for (int i = 0; i < hash.length; i++)
47            System.out.printf("%02X ", hash[i] & 0xFF);
48         System.out.println();
49      }
50   }
```

API java.security.MessageDigest 1.1

- static MessageDigest getInstance(String algorithmName)

  返回实现指定算法的 MessageDigest 对象。如果没有提供该算法，则抛出一个 NoSuch-AlgorithmException 异常。

- void update(byte input)

- void update(byte[] input)

- void update(byte[] input, int offset, int len)

  使用指定的字节来更新摘要。

- byte[] digest()

完成散列计算，返回计算所得的摘要，并复位算法对象。
- void reset()

  重置摘要。

10.4.2 消息签名

在上一节中，我们介绍了如何计算消息摘要，即原始消息的指纹的方法。如果消息改变了，那么改变后的消息的指纹与原消息的指纹将不匹配。如果消息和它的指纹是分开传送的，那么接收者就可以检查消息是否被篡改过。但是，如果消息和指纹同时被截获了，对消息进行修改，再重新计算指纹，就是一件很容易的事情。毕竟，消息摘要算法是公开的，不需要使用任何密钥。在这种情况下，假消息和新指纹的接收者永远不会知道消息已经被篡改。数字签名解决了这个问题。

为了了解数字签名的工作原理，我们需要解释关于公共密钥加密技术领域中的几个概念。公共密钥加密技术是基于公共密钥和私有密钥这两个基本概念的。它的设计思想是你可以将公共密钥告诉世界上的任何人，但是，只有自己才持有私有密钥，重要的是你要保护你的私有密钥，不将它泄漏给其他任何人。这些密钥之间存在一定的数学关系，但是这种关系的具体性质对于实际的编程来说并不重要。（如果你有兴趣，可以参阅 http://www.cacr.math.uwaterloo.ca/hac/ 站点上的 *The Handbook of Applied Cryptography* 一书。）

密钥非常长，而且很复杂。例如，下面是一对匹配的数字签名算法（DSA）的公共密钥和私有密钥。

公共密钥：

p: fca682ce8e12caba26efccf7110e526db078b05edecbcd1eb4a208f3ae1617ae01f35b91a47e6df63413c5e12ed0899bcd132acd50d99151bdc43ee737592e17

q: 962eddcc369cba8ebb260ee6b6a126d9346e38c5

g: 678471b27a9cf44ee91a49c5147db1a9aaf244f05a434d6486931d2d14271b9e35030b71fd73da179069b32e2935630e1c2062354d0da20a6c416e50be794ca4

y: c0b6e67b4ac098eb1a32c5f8c4c1f0e7e6fb9d832532e27d0bdab9ca2d2a8123ce5a8018b8161a760480fadd040b927281ddb22cb9bc4df596d7de4d1b977d50

私有密钥：

p: fca682ce8e12caba26efccf7110e526db078b05edecbcd1eb4a208f3ae1617ae01f35b91a47e6df63413c5e12ed0899bcd132acd50d99151bdc43ee737592e17

q: 962eddcc369cba8ebb260ee6b6a126d9346e38c5

g: 678471b27a9cf44ee91a49c5147db1a9aaf244f05a434d6486931d2d14271b9e35030b71fd73da179069b32e2935630e1c2062354d0da20a6c416e50be794ca4

x: 146c09f881656cc6c51f27ea6c3a91b85ed1d70a

在现实中，几乎不可能用一个密钥去推算出另一个密钥。也就是说，即使每个人都知道你的公共密钥，不管他们拥有多少计算资源，他们一辈子也无法计算出你的私有密钥。

任何人都无法根据公共密钥来推算私有密钥，这似乎让人难以置信。但是时至今日，还没有人能够找到一种算法，来为现在常用的加密算法进行这种推算。如果密钥足够长，那么要是使用穷举法——也就是直接试验所有可能的密钥——所需要的计算机将比用太阳系中的所有原子来制造的计算机还要多，而且还得花费数千年的时间。当然，可能会有人提出比穷举更灵活的计算密钥的算法。例如，RSA 算法（该加密算法由 Rivest、Shamir 和 Adleman 发明）就利用了对数值巨大的数字进行因数分解的困难性。在最近 20 年里，许多优秀的数学家都在尝试提出好的因数分解算法，但是迄今为止都没有成功。据此，大多数密码学者认为，拥有 2000 位或者更多位"模数"的密钥目前是完全安全的，可以抵御任何攻击。DSA 被认为具有类似的安全性。

图 10-9 展示了实践中这种机制是如何工作的。

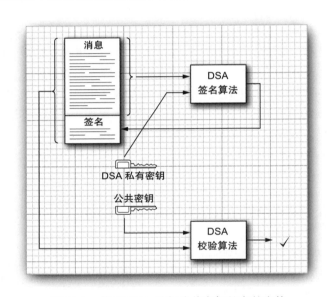

图 10-9 使用 DSA 进行公共密钥签名的交换

假设 Alice 想要给 Bob 发送一个消息，Bob 想知道该消息是否来自 Alice，而不是冒名顶替者。Alice 写好了消息，并且用她的私有密钥对该消息摘要签名。Bob 得到了她的公共密钥的拷贝，然后 Bob 用公共密钥对该签名进行校验。如果通过了校验，则 Bob 可以确认以下两个事实：

1. 原始消息没有被篡改过。

2. 该消息是由 Alice 签名的，她是私有密钥的持有者，该私有密钥就是与 Bob 用于校验的公共密钥相匹配的密钥。

你可以看到私有密钥的安全性为什么是最重要的。如果某个人偷了 Alice 的私有密钥，或者政府要求她交出私有密钥，那么她就麻烦了。小偷或者政府代表就可以假扮她的身份来发送消息，例如资金转账指令，而其他人则会相信这些消息确实来自于 Alice。

10.4.3 校验签名

JDK 配有一个 keytool 程序，该程序是一个命令行工具，用于生成和管理一组证书。我们期望该工具的功能最终能够被嵌入到其他更加用户友好的程序中去。但我们现在要做的是，使用 keytool 工具来展示 Alice 是如何对一个文档进行签名并且将它发送给 Bob 的，而 Bob 又是如何校验该文档确实是由 Alice 签名，而不是冒名顶替的。

keytool 程序负责管理密钥库、证书数据库和私有/公有密钥对。密钥库中的每一项都有一个"别名"。下面展示的是 Alice 如何创建一个密钥库 alice.certs 并且用别名生成一个密钥对。

```
keytool -genkeypair -keystore alice.certs -alias alice
```

当新建或者打开一个密钥库时，系统将提示输入密钥库口令，在下面的这个例子中，口令就使用 secret，如果要将 keytool 生成的密钥库用于重要的应用，那么需要选择一个好的口令来保护这个文件。

当生成一个密钥时，系统提示输入下面这些信息：

```
Enter keystore password: secret
Reenter new password: secret
What is your first and last name?
  [Unknown]: Alice Lee
What is the name of your organizational unit?
  [Unknown]: Engineering
What is the name of your organization?
  [Unknown]: ACME Software
What is the name of your City or Locality?
  [Unknown]: San Francisco
What is the name of your State or Province?
  [Unknown]: CA
What is the two-letter country code for this unit?
  [Unknown]: US
Is <CN=Alice Lee, OU=Engineering, O=ACME Software, L=San Francisco, ST=CA, C=US> correct?
  [no]: yes
```

keytool 工具使用 X.500 格式的名字，它包含常用名（CN）、机构单位（OU）、机构（O）、地点（L）、州（ST）和国别（C）等成分，以确定密钥持有者和证书发行者的身份。

最后，必须设定一个密钥口令，或者按回车键，将密钥库口令作为密钥口令来使用。

假设 Alice 想把她的公共密钥提供给 Bob，她必须导出一个证书文件：

```
keytool -exportcert -keystore alice.certs -alias alice -file alice.cer
```

这时，Alice 就可以把证书发送给 Bob。当 Bob 收到该证书时，他就可以将证书打印出来：

```
keytool -printcert -file alice.cer
```

打印的结果如下：

```
Owner: CN=Alice Lee, OU=Engineering, O=ACME Software, L=San Francisco, ST=CA, C=US
Issuer: CN=Alice Lee, OU=Engineering, O=ACME Software, L=San Francisco, ST=CA, C=US
Serial number: 470835ce
Valid from: Sat Oct 06 18:26:38 PDT 2007 until: Fri Jan 04 17:26:38 PST 2008
Certificate fingerprints:
```

```
MD5:  BC:18:15:27:85:69:48:B1:5A:C3:0B:1C:C6:11:B7:81
SHA1: 31:0A:A0:B8:C2:8B:3B:B6:85:7C:EF:C0:57:E5:94:95:61:47:6D:34
Signature algorithm name: SHA1withDSA
Version: 3
```

如果 Bob 想检查他是否得到了正确的证书，可以给 Alice 打电话，让她在电话里读出证书的指纹。

> **注释**：有些证书发放者将证书指纹公布在他们的网站上。例如，要检查 jre/lib/security/cacerts 目录中的密钥库里的 DigiCert 公司的证书，可以使用 -list 选项：
>
> ```
> keytool -list -v -keystore jre/lib/security/cacerts
> ```
>
> 该密钥库的口令是 changeit。在该密钥库中有一个证书是：
>
> ```
> Owner: CN=DigiCert Assured ID Root G3, OU=www.digicert.com, O=DigiCert Inc, C=US
> Issuer: CN=DigiCert Assured ID Root G3, OU=www.digicert.com, O=DigiCert Inc, C=US
> Serial number: ba15afa1ddfa0b54944afcd24a06cec
> Valid from: Thu Aug 01 14:00:00 CEST 2013 until: Fri Jan 15 13:00:00 CET 2038
> Certificate fingerprints:
>       SHA1: F5:17:A2:4F:9A:48:C6:C9:F8:A2:00:26:9F:DC:0F:48:2C:AB:30:89
>       SHA256: 7E:37:CB:8B:4C:47:09:0C:AB:36:55:1B:A6:F4:5D:B8:40:68:0F:BA:
>               16:6A:95:2D:B1:00:71:7F:43:05:3F:C2
> ```
>
> 通过访问网址 www.digicert.com/digicert-root-certificates.htm.，就可以核实该证书的有效性。

一旦 Bob 信任该证书，他就可以将它导入密钥库中。

```
keytool -importcert -keystore bob.certs -alias alice -file alice.cer
```

> **警告**：绝对不要将你并不完全信任的证书导入到密钥库中。一旦证书添加到密钥库中，使用密钥库的任何程序都会认为这些证书可以用来对签名进行校验。

现在 Alice 就可以给 Bob 发送签过名的文档了。jarsigner 工具负责对 JAR 文件进行签名和校验，Alice 只需要将文档添加到要签名的 JAR 文件中。

```
jar cvf document.jar document.txt
```

然后她使用 jarsigner 工具将签名添加到文件中，她必须指定要使用的密钥库、JAR 文件和密钥的别名。

```
jarsigner -keystore alice.certs document.jar alice
```

当 Bob 收到 JAR 文件时，他可以使用 jarsigner 程序的 -verify 选项，对文件进行校验。

```
jarsigner -verify -keystore bob.certs document.jar
```

Bob 不需要设定密钥别名。jarsigner 程序会在数字签名中找到密钥所有者的 X.500 名字，并在密钥库中搜寻匹配的证书。

如果 JAR 文件没有受到破坏而且签名匹配，那么 jarsigner 程序将打印：

```
jar verified.
```

否则，程序将显示一个出错消息。

10.4.4 认证问题

假设你从朋友 Alice 那接收到一个消息,该消息是 Alice 用她的私有密钥签名的,使用的签名方法就是我们刚刚介绍的方法。你可能已经有了她的公共密钥,或者你能够容易地获得她的公共密钥,比如问她要一个密钥拷贝,或者从她的 Web 页中获得密钥。这时,你就可以校验该消息是否是 Alice 签过名的,并且有没有被破坏过。现在,假设你从一个声称代表某著名软件公司的陌生人那里获得了一个消息,他要求你运行消息附带的程序。这个陌生人甚至将他的公共密钥的拷贝发送给你,以便让你校验他是否是该消息的作者。你检查后会发现该签名是有效的,这就证明该消息是用匹配的私有密钥签名的,并且没有遭到破坏。

此时你要小心:你仍然不清楚谁写的这条消息。任何人都可以生成一对公共密钥和私有密钥,再用私有密钥对消息进行签名,然后把签名好的消息和公共密钥发送给你。这种确定发送者身份的问题称为"认证问题"。

解决这个认证问题的通常做法是比较简单的。假设陌生人和你有一个你们俩都值得信赖的共同熟人。假设陌生人亲自约见了该熟人,将包含公共密钥的磁盘交给了他。后来,你的熟人与你见面,向你担保他与该陌生人见了面,并且该陌生人确实在那家著名的软件公司工作,然后将磁盘交给你(参见图 10-10)。这样一来,你的熟人就证明了陌生人身份的真实性。

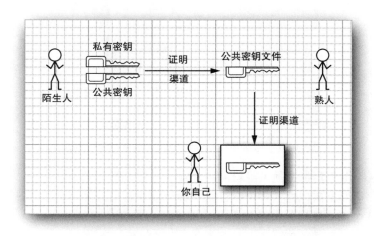

图 10-10 通过一个值得信赖的中间人进行认证

事实上,你的熟人并不需要与你见面。取而代之的是,他可以将他的私有签名应用于陌生人的公共密钥文件之上即可(参见图 10-11)。

当你拿到公共密钥文件之后,就可以检验你的熟人的签名是否真实,由于你信任他,因此你确信他在添加他的签名之前,确实核实了陌生人的身份。

然而,你们之间可能没有共同的熟人。有些信任模型假设你们之间总是存在一个"信任链"——即一个共同熟人的链路——这样你就可以信任该链中的每个成员。当然,实际情况并不总是这样。你可能信任你的熟人 Alice，而且你知道 Alice 信任 Bob，但是你不了解 Bob，因此你没有把握究竟是不是该信任他。其他的信任模型则假设有一个我们大家都信任的慈善大佬，即一家我们大家都信任的公司。在这样的公司中，如雷贯耳的有 DigiCert、GlobalSign 和 Entrust，它们都提供认证服务。

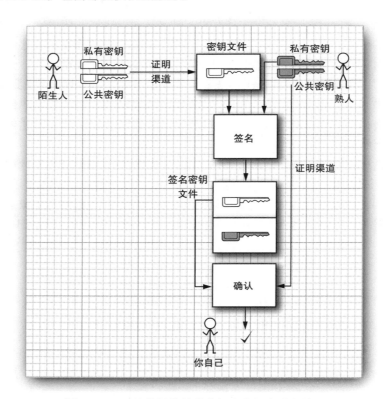

图 10-11　通过值得信赖的中间人的签名进行认证

你常常会遇到由负责担保他人身份的一个或多个实体签署的数字签名，你必须评估一下究竟能够在多大程度上信任这些身份认证人。你可能非常信赖某种特定的证书授权，因为也许你在许多网页中都看到过他们的标志，或者你曾经听说过，每当有新的万能密钥产生时，他们就会要求在一个非常保密的会议室中聚集众多揣着黑色公文包的人进行磋商。

然而，对于实际被认证的对象，你应该抱有一个符合实际的期望：直接在 Web 页面上填一份表格，并支付少量的费用，就可以获得一个 "第一类"（class 1）ID，包含在证书中的密钥将被发送到指定的邮件地址。因此，你有理由相信该电子邮件是真实的，但是密钥申请人也可能填入任意名字和机构。还有其他对身份信息的检验更加严格的 ID 类别。例如，如果是 "第三类"（class 3）ID，证书授权将要求密钥申请人必须进行身份公证，公证机构将要核实企业申请者的财务信用资质。其他认证机构将采用不同的认证程序。因此，当你收到一条经过认证的消息时，重要的是你应该明白它实际上认证了什么。

10.4.5 证书签名

在 10.4.3 节中，你已经看到了 Alice 如何使用自签名的证书向 Bob 分发公共密钥。但是，Bob 需要通过校验 Alice 的指纹以确保这个证书是有效的。

假设 Alice 想要给同事 Cindy 发送一条经过签名的消息，但是 Cindy 并不希望因为要校验许多签名指纹而受到困扰。因此，假设有一个 Cindy 信任的实体来校验这些签名。在这个例子中，Cindy 信任 ACME 软件公司的信息资源部。

这个部门负责证书授权（CA）的运作。ACME 的每个人在其密钥库中都有 CA 的公共密钥，这是由一个专门负责详细核查密钥指纹的系统管理员安装的。CA 对 ACME 雇员的密钥进行签名，当他们在安装彼此的密钥时，密钥库将隐含地信任这些密钥，因为它们是由一个可信任的密钥签名的。

下面展示了可以如何模仿这个过程。首先需要创建一个密钥库 acmesoft.Certs，生成一个密钥对并导出公共密钥。

```
keytool -genkeypair -keystore acmesoft.certs -alias acmeroot
keytool -exportcert -keystore acmesoft.certs -alias acmeroot -file acmeroot.cer
```

其中的公共密钥被导入到了一个自签名的证书中，然后将其添加到每个雇员的密钥库中：

```
keytool -importcert -keystore cindy.certs -alias acmeroot -file acmeroot.cer
```

如果 Alice 要发送消息给 Cindy 以及 ACME 软件公司的其他任何人，她需要将她自己的证书签名 - 并提交给信息资源部。但是，这个功能在 keytool 程序中是缺失的。在本书附带的代码中，我们提供了一个 CertificateSigner 类来弥补这个问题。ACME 软件公司的授权机构成员将负责核实 Alice 的身份，并且生成如下的签名证书：

```
java CertificateSigner -keystore acmesoft.certs -alias acmeroot \
    -infile alice.cer -outfile alice_signedby_acmeroot.cer
```

证书签名器程序必须拥有对 ACME 软件公司密钥库的访问权限，并且该公司成员必须知道密钥库的口令，显然这是一项敏感的操作。

现在 Alice 将文件 alice_signedby_acmeroot.cert 交给 Cindy 和 ACME 软件公司的其他任何人。或者，ACME 软件公司直接将该文件存储在公司的目录中。请记住，该文件包含了 Alice 的公共密钥和 ACME 软件公司的声明，证明该密钥确实属于 Alice。

现在，Cindy 将签名的证书导入到她的密钥库中：

```
keytool -importcert -keystore cindy.certs -alias alice -file alice_signedby_acmeroot.cer
```

密钥库要进行校验，以确定该密钥是由密钥库中已有的受信任的根密钥签过名的。Cindy 就不必对证书的指纹进行校验了。

一旦 Cindy 添加了根证书和经常给她发送文档的人的证书后，她就再也不用担心密钥库了。

10.4.6 证书请求

在前一节中，我们用密钥库和 CertificateSigner 工具模拟了一个 CA。但是，大多数 CA

都运行着更加复杂的软件来管理证书,并且使用的证书格式也略有不同。本节将展示与这些软件包进行交互时需要增加的处理步骤。

我们将用 OpenSSL 软件包作为实例。许多 Linux 系统和 Mac OS X 都预装了这个软件,并且用于 Windows 的 Cygwin 端口也可用这个软件,你也可以到 http://www.openssl.org 网站下载。

为了创建一个 CA,需要运行 CA 脚本,其确切位置依赖于你的操作系统。在 Ubuntu 上,运行

```
/usr/lib/ssl/misc/CA.pl -newca
```

这个脚本会在当前目录中创建一个 demoCA 子目录,这个目录包含了一个根密钥对,并存储了证书与证书撤销列表。

你希望将这个公共密钥导入到所有雇员的 Java 密钥库中,但是它的格式是隐私增强型邮件(PEM)格式,而不是密钥库更容易接受的 DER 格式。将文件 demoCA/cacert.pem 复制成文件 acmeroot.pem,然后在文本编辑器中打开这个文件。移除下面这行之前的所有内容:

```
-----BEGIN CERTIFICATE-----
```

以及下面这行之后的所有内容:

```
-----END CERTIFICATE-----
```

现在可以按照通常的方式将 acmeroot.pem 导入到每个密钥库中了:

```
keytool -importcert -keystore cindy.certs -alias alice -file acmeroot.pem
```

这看起来有点不可思议,keytool 竟然不能自己去执行这种编辑操作。

要对 Alice 的公共密钥签名,需要生成一个证书请求,它包含这个 PEM 格式的证书:

```
keytool -certreq -keystore alice.store -alias alice -file alice.pem
```

要签名这个证书,需要运行:

```
openssl ca -in alice.pem -out alice_signedby_acmeroot.pem
```

与前面一样,在 alice_signedby_acmeroot.pem 中切除 BEGIN CERTIFICATE/END CERTIFICATE 标记之外的所有内容。然后,将其导入到密钥库中:

```
keytool -importcert -keystore cindy.certs -alias alice -file alice_signedby_acmeroot.pem
```

你可以使用相同的步骤,使一个证书得到公共证书权威机构的签名。

10.4.7 代码签名

认证技术最重要的一个应用是对可执行程序进行签名。如果从网上下载一个程序,自然会关心该程序可能带来的危害,例如,该程序可能已经感染了病毒。如果知道代码从何而来,并且它从离开源头后就没有被篡改过,那么放心程度会比不清楚这些信息时要高得多。

本节将展示如何对 JAR 文件签名,以及如何配置 Java 以校验这种签名。这种能力是为

applet 和 Java Web Start 应用而设计的。这些技术已经不再被广泛使用了，但是你仍旧需要在遗留产品中支持它们。

当 Java 首次发布时，applet 在加载之后就运行于具有有限权限的"沙盒"之中。如果用户想要使用可以访问本地文件系统、创建网络连接等诸如此类功能的 applet，那么就必须明确同意允许其运行。为了确保 applet 代码不会在传输过程中被篡改，必须对其进行数字签名。

下面是一个具体例子。假设当你在因特网上冲浪时，遇到了一个 Web 站点，倘若你为它授予了需要的权限，它就会运行一个来自不明提供商的 applet（参见图 10-12）。这样的程序是用由证书权威机构发放的"软件开发者"证书进行签名的。弹出的对话框用于确定软件开发者和证书发放者的身份。现在，你需要决定是否对该程序授权。

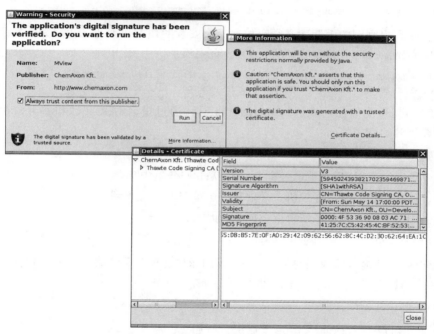

图 10-12　启动一个签过名的 applet

那么什么样的因素可能会影响你的决定呢？假设下面是你已经了解的情况：

1. Thawte 公司将一个证书卖给了软件开发人员。
2. 程序确实是用该证书签名的，并且在传输过程中没有被篡改过。
3. 该证书确实是由 Thawte 签名的，它是用本地 cacerts 文件中的公共密钥校验的。

当然，上面这些信息都不能告诉你代码是否可以安全运行。如果你只知道供应商的名字，以及 Thawte 公司卖给了他们一个软件开发者证书这个事实，那么你会信赖该供应商吗？这种方式显然没什么意义。

对内联网部署，证书更有用。管理员可以在本地机器上安装策略文件和证书，使得在启动从受信源而来的代码时可以无须任何用户交互。无论何时，只要 Java 插件工具加载了签名的文档，它就会向策略文件索要权限，并向密钥库索要签名。

在本节的剩余部分，我们将要介绍如何建立策略文件，来为已知来源的代码赋予特定的权限。

假设 ACME 软件公司想让它的用户运行某些需要具备本地文件访问权限的程序，并且想要通过浏览器将这些程序部署为 Web Start 应用。

正如在本章前面部分看到的那样，ACME 可以根据程序的代码基来确定它们的身份，但是那将意味着每当程序移动到不同的 Web 服务器时，ACME 都需要更新策略文件。为此，ACME 决定对含有程序代码的 JAR 文件进行签名。

首先，ACME 生成根证书：

```
keytool -genkeypair -keystore acmesoft.certs -alias acmeroot
```

当然，包含私有根密钥的密钥库必须存放在一个安全的地方。因此，我们为公共证书建立第二个密钥库 Client.certs，并将公共的 acmeroot 证书添加进去。

```
keytool -exportcert -keystore acmesoft.certs -alias acmeroot -file acmeroot.cer
keytool -importcert -keystore client.certs -alias acmeroot -file acmeroot.cer
```

ACME 中某个受信任的人运行 jarsigner 工具，通过指定 JAR 文件和私有密钥的别名来对任何想要签名的应用签名：

```
jarsigner -keystore acmesoft.certs ACMEApp.jar acmeroot
```

被签名的 Web Start 应用现在就已经准备好在 Web 服务器中部署了。

接着，让我们转而配置客户机，必须将一个策略文件发布到每一台客户机上。

为了引用密钥库，策略文件将以下面这行开头：

```
keystore "keystoreURL", "keystoreType";
```

其中，URL 可以是绝对的或相对的，其中相对 URL 是相对于策略文件的位置而言的。如果密钥库是由 keytool 工具生成的，则它的类型是 JKS。例如：

```
keystore "client.certs", "JKS";
```

grant 子句可以有 signedBy "*alias*" 后缀，例如：

```
grant signedBy "acmeroot"
{
    . . .
};
```

所有可以用与别名相关联的公共密钥进行校验的签名代码现在都已经在 grant 语句中被授予了权限。

10.5 加密

到现在为止，我们已经介绍了一种在 Java 安全 API 中实现的重要密码技术，即通过数字签名进行的认证。安全性的第二个重要方面是*加密*。当信息通过认证之后，该信息本身是直白可见的。数字签名只不过负责检验信息有没有被篡改过。相比之下，信息被加密后，是不可见的，只能用匹配的密钥进行解密。

认证对于代码签名已足够了——没必要将代码隐藏起来。但是，当 applet 或者应用程序传输机密信息时，比如信用卡号码和其他个人数据等，就有必要进行加密了。

过去，由于专利和出口控制的原因，许多公司被禁止提供高强度的加密技术。幸运的是，现在对加密技术的出口控制已经不是那么严格了，某些重要算法的专利也已到期。现在，Java 已经有了出色的加密支持，它已经成为标准类库的一部分。

10.5.1 对称密码

"Java 密码扩展"包含了一个 Cipher 类，该类是所有加密算法的超类。通过调用下面的 getInstance 方法可以获得一个密码对象：

```
Cipher cipher = Cipher.getInstance(algorithmName);
```

或者调用下面这个方法：

```
Cipher cipher = Cipher.getInstance(algorithmName, providerName);
```

JDK 中是由名为"SunJCE"的提供商提供密码的，如果没有指定其他提供商，则会默认为该提供商。如果要使用特定的算法，而对该算法 Oracle 公司没有提供支持，那么也可以指定其他的提供商。

算法名称是一个字符串，比如"AES"或者"DES/CBC/PKCS5Padding"。

DES，即数据加密标准，是一个密钥长度为 56 位的古老的分组密码。DES 加密算法在现在看来已经是过时了，因为可以用穷举法将它破译。更好的选择是采用它的后续版本，即高级加密标准（AES），更多详细信息，请访问网址 https://nvlpubs.nist.gov/nistpubs/FIPS/NIST.FIPS.197.pdf。我们在示例中使用了 AES。

一旦获得了一个密码对象，就可以通过设置模式和密钥来对它初始化。

```
int mode = ...;
Key key = ...;
cipher.init(mode, key);
```

模式有以下几种：

```
Cipher.ENCRYPT_MODE
Cipher.DECRYPT_MODE
Cipher.WRAP_MODE
Cipher.UNWRAP_MODE
```

wrap 和 unwrap 模式会用一个密钥对另一个密钥进行加密，具体例子请参见下一节。

现在可以反复调用 update 方法来对数据块进行加密。

```
int blockSize = cipher.getBlockSize();
var inBytes = new byte[blockSize];
... // read inBytes
int outputSize= cipher.getOutputSize(blockSize);
```

```
var outBytes = new byte[outputSize];
int outLength = cipher.update(inBytes, 0, outputSize, outBytes);
. . . // write outBytes
```

完成上述操作后,还必须调用一次 doFinal 方法。如果还有最后一个输入数据块(其字节数小于 blockSize),那么就要调用:

```
outBytes = cipher.doFinal(inBytes, 0, inLength);
```

如果所有的输入数据都已经加密,则用下面的方法调用来代替:

```
outBytes = cipher.doFinal();
```

对 doFinal 的调用是必需的,因为它会对最后的块进行"填充"。就拿 DES 密码来说,它的数据块的大小是 8 字节。假设输入数据的最后一个数据块少于 8 字节,当然我们可以将其余的字节全部用 0 填充,从而得到一个 8 字节的最终数据块,然后对它进行加密。但是,当对数据块进行解密时,数据块的结尾会附加若干个 0 字节,因此它与原始输入文件之间会略有不同。这肯定是个问题,我们需要一个填充方案来避免这个问题。常用的填充方案是 RSA Security 公司在公共密钥密码标准#5中(Public Key Cryptography Standard,PKCS)描述的方案(该方案的网址为 https://tools.ietf.org/html/rfc2898)。

在该方案中,最后一个数据块不是全部用填充值 0 进行填充,而是用等于填充字节数量的值作为填充值进行填充。换句话说,如果 L 是最后一个(不完整的)数据块,那么它将按如下方式进行填充:

```
L 01                         if length(L) = 7
L 02 02                      if length(L) = 6
L 03 03 03                   if length(L) = 5
. . .
L 07 07 07 07 07 07 07       if length(L) = 1
```

最后,如果输入的数据长度确实能被 8 整除,那么就会将下面这个数据块:

```
08 08 08 08 08 08 08 08
```

附加到数据块后,并进行加密。在解密时,明文的最后一个字节就是要丢弃的填充字符数。

10.5.2 密钥生成

为了加密,我们需要生成密钥。每个密码都有不同的用于密钥的格式,我们需要确保密钥的生成是随机的。这需要遵循下面的步骤:

1. 为加密算法获取 KeyGenerator。
2. 用随机源来初始化密钥发生器。如果密码块的长度是可变的,还需要指定期望的密码块长度。
3. 调用 generateKey 方法。

例如,下面是如何生成 AES 密钥的方法:

```
KeyGenerator keygen = KeyGenerator.getInstance("AES");
```

```
var random = new SecureRandom(); // see below
keygen.init(random);
Key key = keygen.generateKey();
```

或者，可以从一组固定的原生数据（也许是由口令或者随机击键产生的）中生成一个密钥，这时可以使用如下的 SecretKeyFactory：

```
byte[] keyData = . . .; // 16 bytes for AES
var key = new SecretKeySpec(keyData, "AES");
```

如果要生成密钥，必须使用"真正的随机"数。例如，在 Random 类中的常规的随机数发生器。是根据当前的日期和时间来产生随机数的，因此它不够随机。假设计算机时钟可以精确到 1/10 秒，那么，每天最多存在 864 000 个种子。如果攻击者知道发布密钥的日期（通常可以由消息日期或证书有效日期推算出来），那么就可以很容易地生成那一天所有可能的种子。

SecureRandom 类产生的随机数，远比由 Random 类产生的那些数字安全得多。你仍然需要提供一个种子，以便在一个随机点上开始生成数字序列。要这样做，最好的方法是从一个诸如白噪声发生器之类的硬件设备那里获取输入。另一个合理的随机输入源是请用户在键盘上进行随心所欲的盲打，但是每次敲击键盘只为随机种子提供 1 位或者 2 位。一旦你在字节数组中收集到这种随机位后，就可以将它传递给 setSeed 方法。

```
var secrand = new SecureRandom();
var b = new byte[20];
// fill with truly random bits
secrand.setSeed(b);
```

如果没有为随机数发生器提供种子，那么它将通过启动线程，使它们睡眠，然后测量它们被唤醒的准确时间，以此来计算自己的 20 个字节的种子。

> **注释**：这个算法仍然未被认为是安全的。而且，在过去，依靠对诸如硬盘访问时间之类的其他的计算机组件进行计时的算法，后来也被证明并不是完全随机的。

本节结尾处的示例程序将应用 AES 密码（参见程序清单 10-17）。程序清单 10-18 中的 Crpt 工具方法将会在其他示例中被复用。如果要使用该程序，首先要生成一个密钥，运行如下命令行：

密钥就被保存在 secret.key 文件中了。

```
java aes.AESTest -genkey secret.key
```

现在可以用如下命令进行加密：

```
java aes.AESTest -encrypt plaintextFile encryptedFile secret.key
```

用如下命令进行解密：

```
java aes.AESTest -decrypt encryptedFile decryptedFile secret.key
```

该程序非常直观。使用 -genkey 选项将产生一个新的密钥，并且将其序列化到给定的文件中。该操作需要花费较长的时间，因为密钥随机生成器的初始化非常耗费时间。-encrypt

和 -decrypt 选项都调用相同的 crypt 方法，而 crypt 方法会调用密码的 update 和 doFinal 方法。请注意 update 方法和 doFinal 方法是怎样被调用的，只要输入数据块具有全长度（长度能够被 8 整除），就要调用 update 方法，而如果输入数据块不具有全长度（长度不能被 8 整除，此时需要填充），或者没有更多额外的数据（以便生成一个填充字节），那么就要调用 doFinal 方法。

程序清单 10-17　aes/AESTest.java

```java
 1  package aes;
 2
 3  import java.io.*;
 4  import java.security.*;
 5  import javax.crypto.*;
 6
 7  /**
 8   * This program tests the AES cipher. Usage:<br>
 9   * java aes.AESTest -genkey keyfile<br>
10   * java aes.AESTest -encrypt plaintext encrypted keyfile<br>
11   * java aes.AESTest -decrypt encrypted decrypted keyfile<br>
12   * @author Cay Horstmann
13   * @version 1.02 2018-05-01
14   */
15  public class AESTest
16  {
17     public static void main(String[] args)
18        throws IOException, GeneralSecurityException, ClassNotFoundException
19     {
20        if (args[0].equals("-genkey"))
21        {
22           KeyGenerator keygen = KeyGenerator.getInstance("AES");
23           var random = new SecureRandom();
24           keygen.init(random);
25           SecretKey key = keygen.generateKey();
26           try (var out = new ObjectOutputStream(new FileOutputStream(args[1])))
27           {
28              out.writeObject(key);
29           }
30        }
31        else
32        {
33           int mode;
34           if (args[0].equals("-encrypt")) mode = Cipher.ENCRYPT_MODE;
35           else mode = Cipher.DECRYPT_MODE;
36
37           try (var keyIn = new ObjectInputStream(new FileInputStream(args[3]));
38                var in = new FileInputStream(args[1]);
39                var out = new FileOutputStream(args[2]))
40           {
41              var key = (Key) keyIn.readObject();
42              Cipher cipher = Cipher.getInstance("AES");
43              cipher.init(mode, key);
44              Util.crypt(in, out, cipher);
```

程序清单 10-18 aes/Util.java

```java
package aes;

import java.io.*;
import java.security.*;
import javax.crypto.*;

public class Util
{
   /**
    * Uses a cipher to transform the bytes in an input stream and sends the transformed bytes
    * to an output stream.
    * @param in the input stream
    * @param out the output stream
    * @param cipher the cipher that transforms the bytes
    */
   public static void crypt(InputStream in, OutputStream out, Cipher cipher)
         throws IOException, GeneralSecurityException
   {
      int blockSize = cipher.getBlockSize();
      int outputSize = cipher.getOutputSize(blockSize);
      var inBytes = new byte[blockSize];
      var outBytes = new byte[outputSize];

      int inLength = 0;
      var done = false;
      while (!done)
      {
         inLength = in.read(inBytes);
         if (inLength == blockSize)
         {
            int outLength = cipher.update(inBytes, 0, blockSize, outBytes);
            out.write(outBytes, 0, outLength);
         }
         else done = true;
      }
      if (inLength > 0) outBytes = cipher.doFinal(inBytes, 0, inLength);
      else outBytes = cipher.doFinal();
      out.write(outBytes);
   }
}
```

API javax.crypto.Cipher 1.4

- static Cipher getInstance(String algorithmName)
- static Cipher getInstance(String algorithmName, String providerName)

  返回实现了指定加密算法的 Cipher 对象。如果未提供该算法，则抛出一个 NoSuchAlgorithm-

Exception 异常。
- int getBlockSize()
  返回密码块的大小，如果该密码不是一个分组密码，则返回 0。
- int getOutputSize(int inputLength)
  如果下一个输入数据块拥有给定的字节数，则返回所需的输出缓冲区的大小。本方法的运行要考虑到密码对象中所有已缓冲的字节数量。
- void init(int mode, Key key)
  对加密算法对象进行初始化。Mode 是 ENCRYPT_MODE, DECRYPT_MODE, WRAP_MODE, 或者 UNWRAP_MODE 之一。
- byte[] update(byte[] in)
- byte[] update(byte[] in, int offset, int length)
- int update(byte[] in, int offset, int length, byte[] out)
  对输入数据块进行转换。前两个方法返回输出，第三个方法返回放入 out 的字节数。
- byte[] doFinal()
- byte[] doFinal(byte[] in)
- byte[] doFinal(byte[] in, int offset, int length)
- int doFinal(byte[] in, int offset, int length, byte[] out)
  转换输入的最后一个数据块，并刷新该加密算法对象的缓冲。前三个方法返回输出，第四个方法返回放入 out 的字节数。

API javax.crypto.KeyGenerator 1.4

- static KeyGenerator getInstance(String algorithmName)
  返回实现指定加密算法的 KeyGenerator 对象。如果未提供该加密算法，则抛出一个 NoSuchAlgorithmException 异常。
- void init(SecureRandom random)
- void init(int keySize, SecureRandom random)
  对密钥生成器进行初始化。
- SecretKey generateKey()
  生成一个新的密钥。

API javax.crypto.spec.SecretKeySpec 1.4

- SecretKeySpec(byte[] key, String algorithmName)
  创建一个密钥描述规格说明。

10.5.3 密码流

JCE 库提供了一组使用便捷的流类，用于对流数据进行自动加密或解密。例如，下面是对文件数据进行加密的方法：

```
Cipher cipher = . . .;
cipher.init(Cipher.ENCRYPT_MODE, key);
var out = new CipherOutputStream(new FileOutputStream(outputFileName), cipher);
var bytes = new byte[BLOCKSIZE];
int inLength = getData(bytes); // get data from data source
while (inLength != -1)
{
   out.write(bytes, 0, inLength);
   inLength = getData(bytes); // get more data from data source
}
out.flush();
```

同样地，可以使用 CipherInputStream，对文件的数据进行读取和解密：

```
Cipher cipher = . . .;
cipher.init(Cipher.DECRYPT_MODE, key);
var in = new CipherInputStream(new FileInputStream(inputFileName), cipher);
var bytes = new byte[BLOCKSIZE];
int inLength = in.read(bytes);
while (inLength != -1)
{
   putData(bytes, inLength); // put data to destination
   inLength = in.read(bytes);
}
```

密码流类能够透明地调用 update 和 doFinal 方法，所以非常方便。

API javax.crypto.CipherInputStream 1.4

- CipherInputStream(InputStream in, Cipher cipher)
  构建一个输入流，以读取 in 中的数据，并且使用指定的密码对数据进行解密和加密。
- int read()
- int read(byte[] b, int off, int len)
  读取输入流中的数据，该数据会被自动解密和加密。

API javax.crypto.CipherOutputStream 1.4

- CipherOutputStream(OutputStream out, Cipher cipher)
  构建一个输出流，以便将数据写入 out，并且使用指定的密码对数据进行加密和解密。
- void write(int ch)
- void write(byte[] b, int off, int len)
  将数据写入输出流，该数据会被自动加密和解密。
- void flush()
  刷新密码缓冲区，如果需要的话，执行填充操作。

10.5.4 公共密钥密码

在前面的小节中看到的 AES 密码是一种对称密码，加密和解密都使用相同的密钥。对称密码的致命缺点在于密码的分发。如果 Alice 给 Bob 发送了一个加密的方法，那么 Bob 需要使用与 Alice 相同的密钥。如果 Alice 修改了密钥，那么她必须在给 Bob 发送信息的同时，还要通过安全信道发送新的密钥，但是也许她并没有到达 Bob 的安全信道，这也正是她必须对她发送给 Bob 的信息进行加密的原因。

公共密钥密码技术解决了这个问题。在公共密钥密码中，Bob 拥有一个密钥对，包括一个公共密钥和一个相匹配的私有密钥。Bob 可以在任何地方发布公共密钥，但是他必须严格保守他的私有密钥。Alice 只需要使用公共密钥对她发送给 Bob 的信息进行加密即可。

实际上，加密过程并没有那么简单。所有已知的公共密钥算法的操作速度都比对称密钥算法（比如 DES 或 AES 等）慢得多，使用公共密钥算法对大量的信息进行加密是不切实际的。但是，如果像下面这样，将公共密钥密码与快速的对称密码结合起来，这个问题就可以得到解决：

1. Alice 生成一个随机对称加密密钥，她用该密钥对明文进行加密。
2. Alice 用 Bob 的公共密钥给对称密钥进行加密。
3. Alice 将加密后的对称密钥和加密后的明文同时发送给 Bob。
4. Bob 用他的私有密钥给对称密钥解密。
5. Bob 用解密后的对称密钥给信息解密。

除了 Bob 之外，其他人无法给对称密钥进行解密，因为只有 Bob 拥有解密的私有密钥。这样，昂贵的公共密钥加密技术就可以只应用于少量的关键数据的加密。

最常见的公共密钥算法是 Rivest、Shamir 和 Adleman 发明的 RSA 算法。直到 2000 年 10 月，该算法一直受 RSA Security 公司授予的专利保护。该专利的转让许可证价格昂贵，通常要支付 3% 的专利权使用费，每年至少付款 50 000 美元。现在该加密算法已经公开。

如果要使用 RSA 算法，就需要一对公共/私有密钥。你可以按如下方法使用 KeyPairGenerator 来获得：

```
KeyPairGenerator pairgen = KeyPairGenerator.getInstance("RSA");
var random = new SecureRandom();
pairgen.initialize(KEYSIZE, random);
KeyPair keyPair = pairgen.generateKeyPair();
Key publicKey = keyPair.getPublic();
Key privateKey = keyPair.getPrivate();
```

程序清单 10-19 中的程序有三个选项。-genkey 选项用于产生一个密钥对，-encrypt 选项用于生成 AES 密钥，并且用公共密钥对其进行包装。

```
Key key = . . .; // an AES key
Key publicKey = . . .; // a public RSA key
Cipher cipher = Cipher.getInstance("RSA");
cipher.init(Cipher.WRAP_MODE, publicKey);
byte[] wrappedKey = cipher.wrap(key);
```

然后它会生成一个包含下列内容的文件：

- 包装过的密钥的长度。
- 包装过的密钥字节。
- 用 AES 密钥加密的明文。

-decrypt 选项用于对这样的文件进行解密。请试运行该程序，首先生成 RSA 密钥：

java rsa.RSATest -genkey public.key private.key

然后对一个文件进行加密：

java rsa.RSATest -encrypt plaintextFile encryptedFile public.key

最后，对该文件进行解密，并且检验解密后的文件是否与明文相匹配：

java rsa.RSATest -decrypt encryptedFile decryptedFile private.key

程序清单 10-19　rsa/RSATest.java

```java
package rsa;

import java.io.*;
import java.security.*;
import javax.crypto.*;

/**
 * This program tests the RSA cipher. Usage:<br>
 * java rsa.RSATest -genkey public private<br>
 * java rsa.RSATest -encrypt plaintext encrypted public<br>
 * java rsa.RSATest -decrypt encrypted decrypted private<br>
 * @author Cay Horstmann
 * @version 1.02 2018-05-01
 */
public class RSATest
{
    private static final int KEYSIZE = 512;

    public static void main(String[] args)
        throws IOException, GeneralSecurityException, ClassNotFoundException
    {
        if (args[0].equals("-genkey"))
        {
            KeyPairGenerator pairgen = KeyPairGenerator.getInstance("RSA");
            var random = new SecureRandom();
            pairgen.initialize(KEYSIZE, random);
            KeyPair keyPair = pairgen.generateKeyPair();
            try (var out = new ObjectOutputStream(new FileOutputStream(args[1])))
            {
                out.writeObject(keyPair.getPublic());
            }
            try (var out = new ObjectOutputStream(new FileOutputStream(args[2])))
            {
                out.writeObject(keyPair.getPrivate());
            }
        }
        else if (args[0].equals("-encrypt"))
        {
            KeyGenerator keygen = KeyGenerator.getInstance("AES");
            var random = new SecureRandom();
            keygen.init(random);
            SecretKey key = keygen.generateKey();
```

```
43
44          // wrap with RSA public key
45          try (var keyIn = new ObjectInputStream(new FileInputStream(args[3]));
46              var out = new DataOutputStream(new FileOutputStream(args[2]));
47              var in = new FileInputStream(args[1]) )
48          {
49              var publicKey = (Key) keyIn.readObject();
50              Cipher cipher = Cipher.getInstance("RSA");
51              cipher.init(Cipher.WRAP_MODE, publicKey);
52              byte[] wrappedKey = cipher.wrap(key);
53              out.writeInt(wrappedKey.length);
54              out.write(wrappedKey);
55
56              cipher = Cipher.getInstance("AES");
57              cipher.init(Cipher.ENCRYPT_MODE, key);
58              Util.crypt(in, out, cipher);
59          }
60      }
61      else
62      {
63          try (var in = new DataInputStream(new FileInputStream(args[1]));
64              var keyIn = new ObjectInputStream(new FileInputStream(args[3]));
65              var out = new FileOutputStream(args[2]))
66          {
67              int length = in.readInt();
68              var wrappedKey = new byte[length];
69              in.read(wrappedKey, 0, length);
70
71              // unwrap with RSA private key
72              var privateKey = (Key) keyIn.readObject();
73
74              Cipher cipher = Cipher.getInstance("RSA");
75              cipher.init(Cipher.UNWRAP_MODE, privateKey);
76              Key key = cipher.unwrap(wrappedKey, "AES", Cipher.SECRET_KEY);
77
78              cipher = Cipher.getInstance("AES");
79              cipher.init(Cipher.DECRYPT_MODE, key);
80
81              Util.crypt(in, out, cipher);
82          }
83      }
84  }
85 }
```

你现在已经看到了 Java 安全模型是如何允许我们去控制代码的执行的，这是 Java 平台的一个独一无二且越来越重要的方面。你也已经看到了 Java 类库提供的认证和加密服务。

下一章我们将深入讨论高级 Swing 编程和图形化。

第 11 章 高级 Swing 和图形化编程

▲ 表格
▲ 树
▲ 高级 AWT

▲ 像素图
▲ 打印

在本章中，我们继续对卷 I 的 Swing 用户界面工具包和 AWT 图形化编程进行讨论。我们聚焦于可以同时应用于客户端用户界面和服务器端图形图像生成的技术。Swing 有很多复杂的构件来绘制表格和树。通过使用 2D 图形化 API，我们可以产生具有任意复杂度的向量艺术品。ImageIO API 使我们可以操作光栅图像。最终，可以使用打印 API 来生成打印资料和 PostScript 文件。

11.1 表格

JTable 构件用于显示二维对象表格。当然，表格在用户界面中很常见。Swing 开发小组将大量的精力投入到了表格控件上。表格本身比较复杂，但是它可能比其他 Swing 类更为成功，因为 JTable 构件隐藏了更多的复杂性。只需编写几行代码就能够产生具有完整功能的、行为丰富的表格。当然，还可以编写更多的代码，为具体应用定制显示外观和运行特性。

在本节中，我们将着重讲解怎样产生简单表格，用户怎样与它们交互，以及怎样进行一些最常见的调整操作。与其他一些复杂的 Swing 构件一样，我们不可能覆盖所有的细节。如果想获得详细信息，请查阅 David M. Geary 撰写的 *Graphic Java*（第 3 版，Prentice Hall，1999）或 Kim Topley 撰写的 *Core Swing*（Prentice Hall，1999）。

11.1.1 一个简单表格

JTable 并不存储它自己的数据，而是从一个表格模型中获取数据。JTable 类有一个构造器，能够将二维对象数组包装进一个默认的模型。这也正是我们第一个示例程序要用到的策略。在本章的后续部分，我们将转向介绍表格模型。

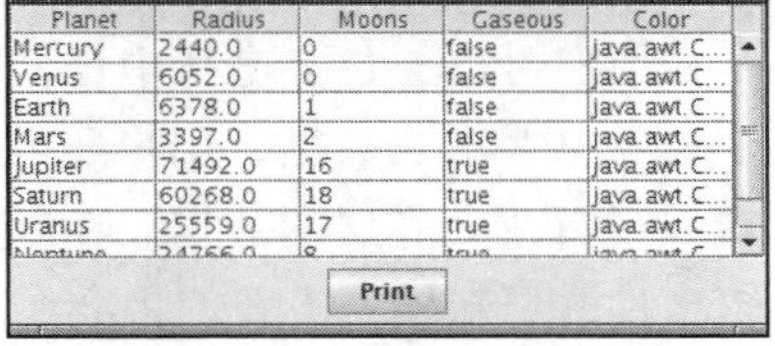

图 11-1 展示了一个典型的表格，用于描述太阳系各个行星的属性。（如果一个行星主要由氢气和氦气组成，那么它就是气态行星。对于 "Color" 项，你不必太当真，我们之所以将它添加为一列是因为在后面的

图 11-1 简单表格

示例代码中它会很有用。）

正如在程序清单 11-1 中看到的那样，表格中的数据是以 Object 值的二维数组的形式存储的：

```
Object[][] cells =
{
   { "Mercury", 2440.0, 0, false, Color.YELLOW },
   { "Venus", 6052.0, 0, false, Color.YELLOW },
   ...
}
```

> **注释**：这里，我们充分利用了自动装箱机制。第二列、第三列、第四列中的项会自动转换成类型为 Double、Integer 和 Boolean 的对象。

该表格直接调用每个对象上的 toString 方法来显示它们，这也正是颜色显示为 java.awt.Color[r = ..., g = ..., b = ...] 的原因所在。

可以用一个单独的字符串数组来提供列名：

```
String[] columnNames = { "Planet", "Radius", "Moons", "Gaseous", "Color" };
```

接着，就可以从单元格和列名数组中构建一个表格：

```
var table = new JTable(cells, columnNames);
```

最后，通过将表格包装到一个 JScrollPane 中这种常用方法来添加滚动条：

```
var pane = new JScrollPane(table);
```

在滚动表格时，列表头并不会滑到视图的外面。

接着，单击列表头的某一列，并且向左或向右拖拉。看看整个列是怎样移开的（参见图 11-2），可以将它放到别的位置上。这种列的重新排列只是视图上的重新排列，对数据模型没有任何影响。

如果要调整列的尺寸大小，只需将鼠标移到两列之间，直到鼠标的形状变成箭头为止，然后将列的边界拖移到你期望的位置上（参见图 11-3）。

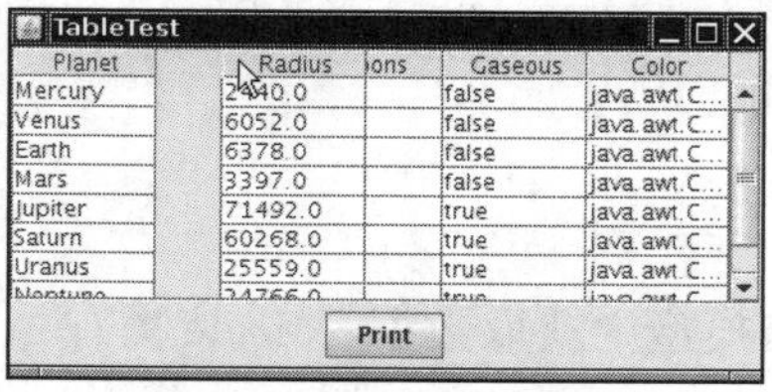

图 11-2　移动表格中的一列 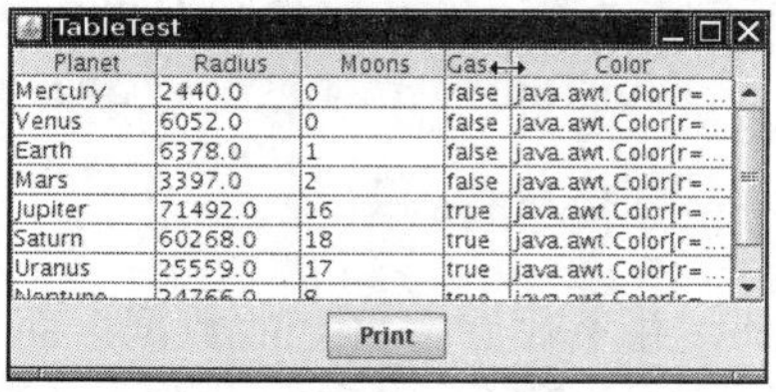　　　　　　　图 11-3　调整列的尺寸大小

用户可以通过点击行中任何一个地方来选中一行，而选中的行会高亮显示。通过单击一个单元格并键入数据，用户还可以编辑表格中的各个项。不过，在这个代码示例中，这些编辑并没有改变底层的数据。在程序中，应该要么使这些单元格不可编辑，要么处理单元格编辑事件并更新模型。我们将会在本节的后面对这些问题进行讨论。

最后，点击列的头，行就会自动排序。如果再次点击，排序顺序就会反过来。这个行为是通过下面的调用激活的：

```
table.setAutoCreateRowSorter(true);
```

可以使用下面的调用对表格进行打印：

```
table.print();
```

> **警告**：如果没有将表格包装在滚动面板中，那么就需要显式地添加表头：
> ```
> add(table.getTableHeader(), BorderLayout.NORTH);
> ```

程序清单 11-1　table/TableTest.java

```java
package table;

import java.awt.*;
import java.awt.print.*;

import javax.swing.*;

/**
 * This program demonstrates how to show a simple table.
 * @version 1.14 2018-05-01
 * @author Cay Horstmann
 */
public class TableTest
{
   public static void main(String[] args)
   {
      EventQueue.invokeLater(() ->
         {
            var frame = new PlanetTableFrame();
            frame.setTitle("TableTest");
            frame.setDefaultCloseOperation(JFrame.EXIT_ON_CLOSE);
            frame.setVisible(true);
         });
   }
}

/**
 * This frame contains a table of planet data.
 */
class PlanetTableFrame extends JFrame
{
   private String[] columnNames = { "Planet", "Radius", "Moons", "Gaseous", "Color" };
   private Object[][] cells =
   {
      { "Mercury", 2440.0, 0, false, Color.YELLOW },
      { "Venus", 6052.0, 0, false, Color.YELLOW },
      { "Earth", 6378.0, 1, false, Color.BLUE },
      { "Mars", 3397.0, 2, false, Color.RED },
      { "Jupiter", 71492.0, 16, true, Color.ORANGE },
      { "Saturn", 60268.0, 18, true, Color.ORANGE },
      { "Uranus", 25559.0, 17, true, Color.BLUE },
```

```
42          { "Neptune", 24766.0, 8, true, Color.BLUE },
43          { "Pluto", 1137.0, 1, false, Color.BLACK }
44       };
45
46       public PlanetTableFrame()
47       {
48          var table = new JTable(cells, columnNames);
49          table.setAutoCreateRowSorter(true);
50          add(new JScrollPane(table), BorderLayout.CENTER);
51          var printButton = new JButton("Print");
52          printButton.addActionListener(event ->
53             {
54                try { table.print(); }
55                catch (SecurityException | PrinterException ex) { ex.printStackTrace(); }
56             });
57          var buttonPanel = new JPanel();
58          buttonPanel.add(printButton);
59          add(buttonPanel, BorderLayout.SOUTH);
60          pack();
61       }
62    }
```

API javax.swing.JTable 1.2

- JTable(Object[][] entries, Object[] columnNames)
  用默认的表格模型构建一个表格。

- void print() 5.0
  显示打印对话框，并打印该表格。

- boolean getAutoCreateRowSorter() 6

- void setAutoCreateRowSorter(boolean newValue) 6
  获取或设置 autoCreateRowSorter 属性，默认值为 false。如果进行了设置，只要模型发生变化，就会自动设置一个默认的行排序器。

- boolean getFillsViewportHeight() 6

- void setFillsViewportHeight(boolean newValue) 6
  获取或设置 fillsViewportHeight 属性，默认值为 false。如果进行了设置，该表格就总是会填充其外围的视图。

11.1.2 表格模型

在上一个示例中，表格数据是存储在一个二维数组中的。不过，通常不应该在自己的代码中使用这种策略。如果你发现自己在将数据装入一个数组中，然后作为一个表格显示出来，那么就应该考虑实现自己的表格模型了。

表格模型实现起来特别简单，因为可以充分利用 AbstractTableModel 类，它实现了大部分必需的方法。你仅仅需要提供下面三个方法便可：

```
public int getRowCount();
```

```
public int getColumnCount();
public Object getValueAt(int row, int column);
```

实现 getValueAt 方法有多种途径。例如，如果想显示包含数据库查询结果的 RowSet 的内容，只需提供下面的方法：

```
public Object getValueAt(int r, int c)
{
   try
   {
      rowSet.absolute(r + 1);
      return rowSet.getObject(c + 1);
   }
   catch (SQLException e)
   {
      e.printStackTrace();
      return null;
   }
}
```

我们的示例程序相当简单。我们构建了一个只是用来显示某些计算结果的表格，这些计算结果也就是在不同利率条件下的投资增长额（参见图 11-4）。

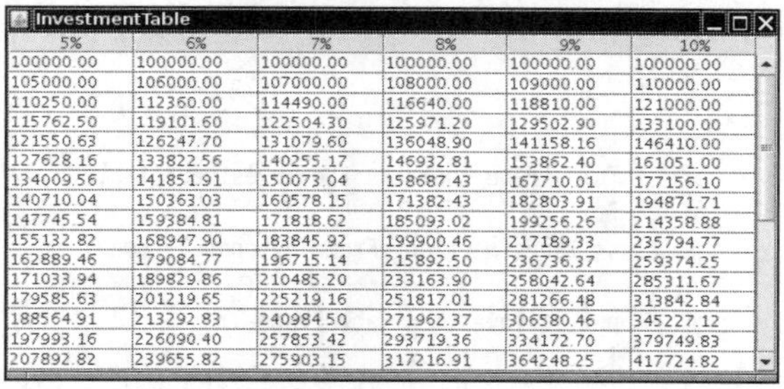

getValueAt 方法计算出正确值，并将其格式化：

```
public Object getValueAt(int r, int c)
{
   double rate = (c + minRate) / 100.0;
   int nperiods = r;
   double futureBalance = INITIAL_BALANCE * Math.pow(1 + rate, nperiods);
   return String.format("%.2f", futureBalance);
}
```

图 11-4　投资增长额

getrowCount 和 getColumnCount 方法只是返回行数和列数。

```
public int getRowCount() { return years; }
public int getColumnCount() {  return maxRate - minRate + 1; }
```

如果不提供列名，那么 AbstractTableModel 的 getColumnName 方法会将列命名为 A、B、C 等。如果要改变列名，请覆盖 getColumnName 方法。通常需要覆盖默认的行为。在这个示例中，我们只是将每列用利率标识了出来。

```
public String getColumnName(int c) { return (c + minRate) + "%"; }
```

程序清单 11-2 中显示了完整的源代码。

程序清单 11-2　tableModel/InvestmentTable.java

```
1 package tableModel;
2
3 import java.awt.*;
```

```java
4
5   import javax.swing.*;
6   import javax.swing.table.*;
7
8   /**
9    * This program shows how to build a table from a table model.
10   * @version 1.04 2018-05-01
11   * @author Cay Horstmann
12   */
13  public class InvestmentTable
14  {
15     public static void main(String[] args)
16     {
17        EventQueue.invokeLater(() ->
18           {
19              var frame = new InvestmentTableFrame();
20              frame.setTitle("InvestmentTable");
21              frame.setDefaultCloseOperation(JFrame.EXIT_ON_CLOSE);
22              frame.setVisible(true);
23           });
24     }
25  }
26
27  /**
28   * This frame contains the investment table.
29   */
30  class InvestmentTableFrame extends JFrame
31  {
32     public InvestmentTableFrame()
33     {
34        var model = new InvestmentTableModel(30, 5, 10);
35        var table = new JTable(model);
36        add(new JScrollPane(table));
37        pack();
38     }
39  }
40
41  /**
42   * This table model computes the cell entries each time they are requested. The table contents
43   * shows the growth of an investment for a number of years under different interest rates.
44   */
45  class InvestmentTableModel extends AbstractTableModel
46  {
47     private static double INITIAL_BALANCE = 100000.0;
48
49     private int years;
50     private int minRate;
51     private int maxRate;
52
53     /**
54      * Constructs an investment table model.
55      * @param y the number of years
56      * @param r1 the lowest interest rate to tabulate
57      * @param r2 the highest interest rate to tabulate
58      */
```

```
59      public InvestmentTableModel(int y, int r1, int r2)
60      {
61         years = y;
62         minRate = r1;
63         maxRate = r2;
64      }
65
66      public int getRowCount()
67      {
68         return years;
69      }
70
71      public int getColumnCount()
72      {
73         return maxRate - minRate + 1;
74      }
75
76      public Object getValueAt(int r, int c)
77      {
78         double rate = (c + minRate) / 100.0;
79         int nperiods = r;
80         double futureBalance = INITIAL_BALANCE * Math.pow(1 + rate, nperiods);
81         return String.format("%.2f", futureBalance);
82      }
83
84      public String getColumnName(int c)
85      {
86         return (c + minRate) + "%";
87      }
88   }
```

API *javax.swing.table.TableModel* 1.2

- int getRowCount()
- int getColumnCount()

  获取表模型中的行和列的数量。

- Object getValueAt(int row, int column)

  获取在给定的行和列所确定的位置处的值。

- void setValueAt(Object newValue, int row, int column)

  设置在给定的行和列所确定的位置处的值。

- boolean isCellEditable(int row, int column)

  如果在给定的行和列所确定的位置处的值是可编辑的，则返回 true。

- String getColumnName(int column)

  获取列的名字。

11.1.3 对行和列的操作

在本小节中，你会看到怎样操作一个表格中的行和列。在你阅读本材料的整个过程中，要牢记 Swing 中的表格是相当不对称的,也就是你可以实施的行操作和列操作会有所不同。表格构件已经被优化过,以便能够显示具有相同结构的行信息,例如一次数据库查询的结果,而不是任意的二维对象表格。你将会看到,这种不对称性贯穿于本小节。

11.1.3.1 各种列类

在下一个示例中,我们将再次展示行星数据,不过这次我们会给出更多的有关表格列类型的信息。这是通过在表格模型中定义下面这个方法来实现的:

```
Class<?> getColumnClass(int columnIndex)
```

这个方法可以返回一个描述列类型的类。

JTable 类会为该类选取合适的绘制器,表 11-1 显示了默认的绘制动作。

表 11-1 默认的绘制动作

类型	绘制结果
Boolean	复选框
Icon	图像
Object	字符串

可以在图 11-5 中看到复选框和图像。(感谢 Jim Evins 提供了这些行星图像。)

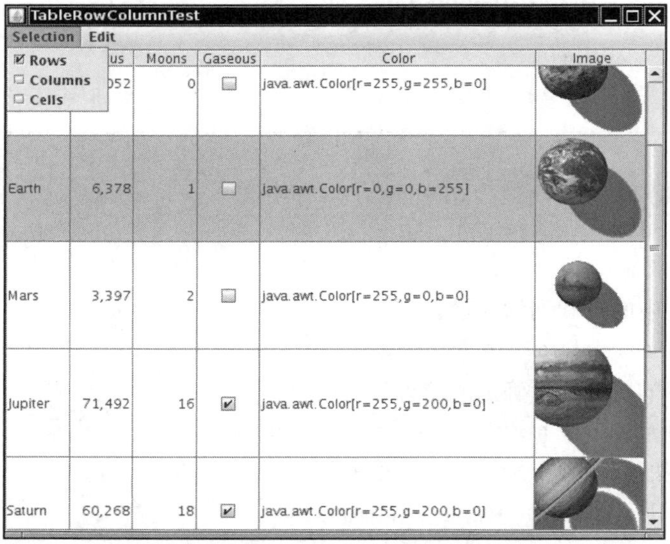

图 11-5 具有单元格绘制器的表格

要绘制其他类型,需要安装定制的绘制器,请参见 11.1.4 节。

11.1.3.2 访问表格列

JTable 类将有关表格列的信息存放在类型为 TableColumn 的对象中,由一个 TableColumn-Model 对象负责管理这些列。(图 11-6 展示了最重要的表格类之间的关系。)如果不想动态地插入或删除，那么最好不要过多地使用表格列模型。列模型最常见的用法是直接获取一个 TableColumn 对象：

```
int columnIndex = . . .;
TableColumn column = table.getColumnModel().getColumn(columnIndex);
```

11.1.3.3 改变列的大小

TableColumn 类可以控制更改列的大小的行为。使用下面这些方法，可以设置首选的、最小的以及最大的宽度：

```
void setPreferredWidth(int width)
void setMinWidth(int width)
void setMaxWidth(int width)
```

这些信息将提供给表格构件，以便对列进行布局。

使用方法

```
void setResizable(boolean resizable)
```

可以控制是否允许用户改变列的大小。

可以使用下面这个方法在程序中改变列的大小：

```
void setWidth(int width)
```

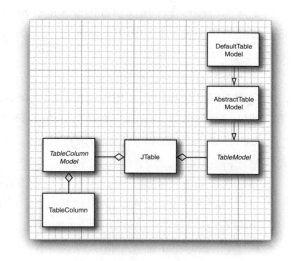

图 11-6　表格类之间的关系图

调整一个列的大小时，默认情况下表格的总体大小会保持不变。当然，更改过大小的列的宽度的增加值或减小值会分摊到其他列上。默认方式是更改那些在被改变了大小的列右边的所有列的大小。这是一种很好的默认方式，因为这样使得用户可以通过将所有列从左到右移动，将它们调整为自己所期望的宽度。

使用下面这个方法，可以设置表 11-2 中列出的 JTable 类的其他行为：

```
void setAutoResizeMode(int mode)
```

表 11-2　变更列大小的模式

模式	行为
AUTO_RESIZE_OFF	不更改其他列的大小，而是更改整个表格的宽度
AUTO_RESIZE_NEXT_COLUMN	只更改下一列的大小
AUTO_RESIZE_SUBSEQUENT_COLUMNS	均匀地更改后续列的大小，这是默认的行为
AUTO_RESIZE_LAST_COLUMN	只更改最后一列的大小
AUTO_RESIZE_ALL_COLUMNS	更改表格中的所有列的大小，这并不是一种很明智的选择，因为这阻碍了用户只对数列而不是整个表进行调整以达到自己期望大小的行为

11.1.3.4 改变行的大小

行的高度是直接由 JTable 类管理的。如果单元格比默认值高，那么可以像下面这样设置行的高度：

```
table.setRowHeight(height);
```

默认情况下，表格中的所有行都具有相同的高度，可以用下面的调用来为每一行单独设置高度：

```
table.setRowHeight(row, height);
```

实际的行高度等于用这些方法设置的行高度减去行边距，其中行边距的默认值是 1 个像素，但是可以通过下面的调用来修改它：

```
table.setRowMargin(margin);
```

11.1.3.5 选择行、列和单元格

利用不同的选择模式，用户可以分别选择表格中的行、列或者单个的单元格。默认情况下，使能的是行选择，点击一个单元格的内部就可以选择整行（参见图 11-5）。调用

```
table.setRowSelectionAllowed(false);
```

可以禁用行选择。

当行选择功能可用时，可以控制用户是否可以选择单一行、连续几行或者任意几行。此时，需要获取选择模式，然后调用它的 setSelectionMode 方法：

```
table.getSelectionModel().setSelectionMode(mode);
```

在这里，mode 是下面三个值的其中一个：

```
ListSelectionModel.SINGLE_SELECTION
ListSelectionModel.SINGLE_INTERVAL_SELECTION
ListSelectionModel.MULTIPLE_INTERVAL_SELECTION
```

默认情况下，列选择是禁用的。不过可以调用下面这个方法启用列选择：

```
table.setColumnSelectionAllowed(true);
```

同时启用行选择和列选择等价于启用单元格选择，这样用户就可以选择一定范围内的单元格（参见图 11-7）。也可以使用下面的调用完成这项设置：

```
table.setCellSelectionEnabled(true);
```

可以运行程序清单 11-3 中的程序，观察一下单元格选择的运行情况。启用 Selection 菜单中的行、列或单元格选项，然后观察选择行为是如何改变的。

可以通过调用 getSelectedRows 方法和 getSelectedColumns 方法来查看选中了哪些行及哪些列。这两个方法都返回一个由被选定项的索引构成的 int[] 数组。注意，这些索引值是表格视图中的索引值，而不是底层表格模型中的索引值。尝试着选择一些行和列，然后将列拖拽到不同的位置，并通过点击列头来对这些行进行排序。使用 Print Selection 菜单项来查看它会报告哪些行和列被选中。

如果要将表格索引值转译为表格模型索引值，可以使用 JTable 的 ConvertRowIndexToModel 和 convertColumnIndexToModel 方法。

11.1.3.6 对行排序

正如在第一个表格示例中看到的那样，向 JTable 中添加行排序机制是很容易的，只需调

用 setAutoCreateRowSorter 方法。但是，要对排序行为进行细粒度的控制，就必须向 JTable 中安装一个 TableRowSorter<M> 对象，并对其进行定制化。类型参数 M 表示表格模型，它必须是 TableModel 接口的子类型。

```
var sorter = new TableRowSorter<TableModel>(model);
table.setRowSorter(sorter);
```

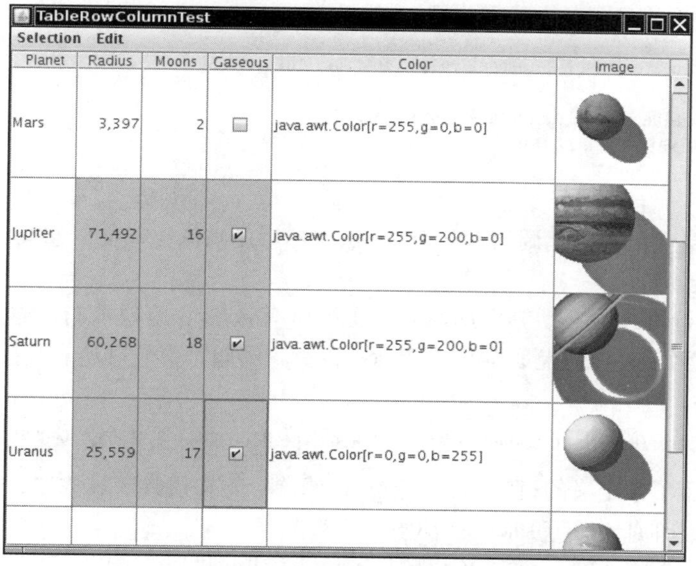

图 11-7　选择一个单元格范围

某些列是不可排序的，例如，在我们的行星数据中的图像列，可以通过下面的调用来关闭排序机制：

```
sorter.setSortable(IMAGE_COLUMN, false);
```

可以对每个列都安装一个定制的比较器。在我们的示例中，将对 Color 列中的颜色进行排序，因为我们相对于红色来说，更喜欢蓝色和绿色。当点击 Color 列时，将会看到蓝色行星出现在表格底部，这是通过下面的调用完成的：

```
sorter.setComparator(COLOR_COLUMN, new Comparator<Color>()
    {
        public int compare(Color c1, Color c2)
        {
            int d = c1.getBlue() - c2.getBlue();
            if (d != 0) return d;
            d = c1.getGreen() - c2.getGreen();
            if (d != 0) return d;
            return c1.getRed() - c2.getRed();
        }
    });
```

如果不指定列的比较器，那么排列顺序就是按照下面的原则确定的：

1. 如果列所属的类是 String，就使用 Collator.getInstance() 方法返回的默认比较器。它按照适用于当前 locale 的方式对字符串排序。（参见第 7 章以了解 locale 和比较器的更多信息）。

2. 如果列所属的类型实现了 Comparable，则使用它的 compareTo 方法。

3. 如果已经为排序器设置过 TableStringConverter，就用默认比较器对转换器的 toString 方法返回的字符串进行排序。如果要使用该方法，可以像下面这样定义转换器：

```
sorter.setStringConverter(new TableStringConverter()
  {
     public String toString(TableModel model, int row, int column)
     {
        Object value = model.getValueAt(row, column);
        convert value to a string and return it
     }
  });
```

4. 否则，在单元格的值上调用 toString 方法，然后用默认比较器对它们进行比较。

11.1.3.7 过滤行

除了可以对行排序之外，TableRowSorter 还可以有选择性地隐藏行，这种处理称为过滤（filtering）。要想激活过滤机制，需要设置 RowFilter。例如，要包含所有至少有一个卫星的行星行，可以调用：

```
sorter.setRowFilter(RowFilter.numberFilter(ComparisonType.NOT_EQUAL, 0, MOONS_COLUMN));
```

这里我们使用了预定义的过滤器，即数字过滤器。要构建数字过滤器，需要提供：

- 比较类型（EQUAL、NOT_EQUAL、AFTER 和 BEFORE 之一）。
- Number 的某个子类的一个对象（例如 Integer 和 Double），只有与给定的 Number 对象属于相同的类的对象才在考虑的范围内。
- 0 或多列的索引值，如果不提供任何索引值，那么所有的列都被搜索。

静态的 RowFilter.dateFilter 方法以相同的方式构建了日期过滤器，这里需要提供 Date 对象而不是 Number 对象。

最后，静态的 RowFilter.regexFilter 方法构建的过滤器可以查找匹配某个正则表达式的字符串。例如：

```
sorter.setRowFilter(RowFilter.regexFilter(".*[^s]$", PLANET_COLUMN));
```

将只显示那些名字以"s"结尾的行星（参见第 2 章以了解有关正则表达式的更多信息）。

还可以用 andFilter、orFilter 和 notFilter 方法来组合过滤器，例如，要过滤掉名字不是以"s"结尾，并且至少有一颗卫星的行星，可以使用下面的过滤器组合：

```
sorter.setRowFilter(RowFilter.andFilter(List.of(
   RowFilter.regexFilter(".*[^s]$", PLANET_COLUMN),
   RowFilter.numberFilter(ComparisonType.NOT_EQUAL, 0, MOONS_COLUMN))));
```

要实现自己的过滤器，需要提供 RowFilter 的一个子类，并实现 include 方法来表示哪些行应该显示。这很容易实现，但是 RowFilter 类卓越的普适性令它有点可怕。

RowFilter<M, I> 类有两个类型参数：模型的类型和行标识符的类型。在处理表格时，模型总是 TableModel 的某个子类型，而标识符类型总是 Integer。(在将来的某个时刻，其他构件可能也会支持行过滤机制。例如，要过滤 JTree 中的行，就可能可以使用 RowFilter <TreeModel, TreePath> 了。)

行过滤器必须实现下面的方法：

```
public boolean include(RowFilter.Entry<? extends M, ? extends I> entry)
```

RowFilter.Entry 类提供了获取模型、行标识符和给定索引处的值等内容的方法，因此，按照行标识符和行的内容都可以进行过滤。

例如，下面的过滤器将隔行显示：

```
var filter = new RowFilter<TableModel, Integer>()
{
    public boolean include(Entry<? extends TableModel, ? extends Integer> entry)
    {
        return entry.getIdentifier() % 2 == 0;
    }
};
```

如果想要只包含那些具有偶数个卫星的行星，可以将上面的测试条件替换为下面的内容：

```
((Integer) entry.getValue(MOONS_COLUMN)) % 2 == 0
```

在我们的示例程序中，允许用户隐藏任意多行，我们在一个 set 中存储了所有隐藏的行的索引。而其中的行过滤器将包含那些索引不在这个 set 中的所有行。

过滤机制并不是为那些过滤标准在不时地发生变化的过滤器而设计的。因此，在我们的示例程序中，只要隐藏行的 set 发生了变化，我们就会调用下面的语句：

```
sorter.setRowFilter(filter);
```

过滤器一旦被设置，就会立即得到应用。

11.1.3.8 隐藏和显示列

正如在前一节中看到的，可以根据内容或标识符来过滤表格行，而隐藏表格列使用的是完全不同的机制。

JTable 类的 removeColumn 方法可以将一列从表格视图中移除。该列的数据实际上并没有从模型中移除，它们只是在视图中被隐藏了起来。removeColumn 方法接受一个 TableColumn 参数，如果你有的是一个列号（比如来自 getSelectedColumns 的调用结果），那就需要向表格模型请求实际的列对象：

```
TableColumnModel columnModel = table.getColumnModel();
TableColumn column = columnModel.getColumn(i);
table.removeColumn(column);
```

如果你记得住该列，那么将来就可以再把它添加回去：

```
table.addColumn(column);
```

该方法将该列添加到表格的最后面。如果想让它出现在表格中的其他任何地方，那么可以调用 moveColumn 方法。

通过添加一个新的 TableColumn 对象,还可以添加一个对应于表格模型中的一个列索引的新列:

```
table.addColumn(new TableColumn(modelColumnIndex));
```

可以让多个表格列展示模型中的同一列。

程序清单 11-3 展示了如何选择和过滤行与列。

程序清单 11-3 tableRowColumn/planetTableFrame.java

```java
 1  package tableRowColumn;
 2
 3  import java.awt.*;
 4  import java.util.*;
 5
 6  import javax.swing.*;
 7  import javax.swing.table.*;
 8
 9  /**
10   * This frame contains a table of planet data.
11   */
12  public class PlanetTableFrame extends JFrame
13  {
14     private static final int DEFAULT_WIDTH = 600;
15     private static final int DEFAULT_HEIGHT = 500;
16
17     public static final int COLOR_COLUMN = 4;
18     public static final int IMAGE_COLUMN = 5;
19
20     private JTable table;
21     private HashSet<Integer> removedRowIndices;
22     private ArrayList<TableColumn> removedColumns;
23     private JCheckBoxMenuItem rowsItem;
24     private JCheckBoxMenuItem columnsItem;
25     private JCheckBoxMenuItem cellsItem;
26
27     private String[] columnNames = { "Planet", "Radius", "Moons", "Gaseous", "Color", "Image" };
28
29     private Object[][] cells =
30     {
31        { "Mercury", 2440.0, 0, false, Color.YELLOW,
32           new ImageIcon(getClass().getResource("Mercury.gif")) },
33        { "Venus", 6052.0, 0, false, Color.YELLOW,
34           new ImageIcon(getClass().getResource("Venus.gif")) },
35        { "Earth", 6378.0, 1, false, Color.BLUE,
36           new ImageIcon(getClass().getResource("Earth.gif")) },
37        { "Mars", 3397.0, 2, false, Color.RED,
38           new ImageIcon(getClass().getResource("Mars.gif")) },
39        { "Jupiter", 71492.0, 16, true, Color.ORANGE,
40           new ImageIcon(getClass().getResource("Jupiter.gif")) },
41        { "Saturn", 60268.0, 18, true, Color.ORANGE,
42           new ImageIcon(getClass().getResource("Saturn.gif")) },
43        { "Uranus", 25559.0, 17, true, Color.BLUE,
44           new ImageIcon(getClass().getResource("Uranus.gif")) },
```

```java
            { "Neptune", 24766.0, 8, true, Color.BLUE,
                new ImageIcon(getClass().getResource("Neptune.gif")) },
            { "Pluto", 1137.0, 1, false, Color.BLACK,
                new ImageIcon(getClass().getResource("Pluto.gif")) }
      };

   public PlanetTableFrame()
   {
      setSize(DEFAULT_WIDTH, DEFAULT_HEIGHT);

      var model = new DefaultTableModel(cells, columnNames)
         {
            public Class<?> getColumnClass(int c)
            {
               return cells[0][c].getClass();
            }
         };

      table = new JTable(model);

      table.setRowHeight(100);
      table.getColumnModel().getColumn(COLOR_COLUMN).setMinWidth(250);
      table.getColumnModel().getColumn(IMAGE_COLUMN).setMinWidth(100);

      var sorter = new TableRowSorter<TableModel>(model);
      table.setRowSorter(sorter);
      sorter.setComparator(COLOR_COLUMN, Comparator.comparing(Color::getBlue)
         .thenComparing(Color::getGreen).thenComparing(Color::getRed));
      sorter.setSortable(IMAGE_COLUMN, false);
      add(new JScrollPane(table), BorderLayout.CENTER);

      removedRowIndices = new HashSet<>();
      removedColumns = new ArrayList<>();

      var filter = new RowFilter<TableModel, Integer>()
         {
            public boolean include(Entry<? extends TableModel, ? extends Integer> entry)
            {
               return !removedRowIndices.contains(entry.getIdentifier());
            }
         };

      // create menu

      var menuBar = new JMenuBar();
      setJMenuBar(menuBar);

      var selectionMenu = new JMenu("Selection");
      menuBar.add(selectionMenu);

      rowsItem = new JCheckBoxMenuItem("Rows");
      columnsItem = new JCheckBoxMenuItem("Columns");
      cellsItem = new JCheckBoxMenuItem("Cells");

      rowsItem.setSelected(table.getRowSelectionAllowed());
```

```java
            columnsItem.setSelected(table.getColumnSelectionAllowed());
            cellsItem.setSelected(table.getCellSelectionEnabled());

            rowsItem.addActionListener(event ->
                {
                    table.clearSelection();
                    table.setRowSelectionAllowed(rowsItem.isSelected());
                    updateCheckboxMenuItems();
                });
            selectionMenu.add(rowsItem);

            columnsItem.addActionListener(event ->
                {
                    table.clearSelection();
                    table.setColumnSelectionAllowed(columnsItem.isSelected());
                    updateCheckboxMenuItems();
                });
            selectionMenu.add(columnsItem);

            cellsItem.addActionListener(event ->
                {
                    table.clearSelection();
                    table.setCellSelectionEnabled(cellsItem.isSelected());
                    updateCheckboxMenuItems();
                });
            selectionMenu.add(cellsItem);

            var tableMenu = new JMenu("Edit");
            menuBar.add(tableMenu);

            var hideColumnsItem = new JMenuItem("Hide Columns");
            hideColumnsItem.addActionListener(event ->
                {
                    int[] selected = table.getSelectedColumns();
                    TableColumnModel columnModel = table.getColumnModel();

                    // remove columns from view, starting at the last
                    // index so that column numbers aren't affected

                    for (int i = selected.length - 1; i >= 0; i--)
                    {
                        TableColumn column = columnModel.getColumn(selected[i]);
                        table.removeColumn(column);

                        // store removed columns for "show columns" command

                        removedColumns.add(column);
                    }
                });
            tableMenu.add(hideColumnsItem);

            var showColumnsItem = new JMenuItem("Show Columns");
            showColumnsItem.addActionListener(event ->
                {
```

```
154            // restore all removed columns
155            for (TableColumn tc : removedColumns)
156               table.addColumn(tc);
157            removedColumns.clear();
158         });
159      tableMenu.add(showColumnsItem);
160
161      var hideRowsItem = new JMenuItem("Hide Rows");
162      hideRowsItem.addActionListener(event ->
163         {
164            int[] selected = table.getSelectedRows();
165            for (int i : selected)
166               removedRowIndices.add(table.convertRowIndexToModel(i));
167            sorter.setRowFilter(filter);
168         });
169      tableMenu.add(hideRowsItem);
170
171      var showRowsItem = new JMenuItem("Show Rows");
172      showRowsItem.addActionListener(event ->
173         {
174            removedRowIndices.clear();
175            sorter.setRowFilter(filter);
176         });
177      tableMenu.add(showRowsItem);
178
179      var printSelectionItem = new JMenuItem("Print Selection");
180      printSelectionItem.addActionListener(event ->
181         {
182            int[] selected = table.getSelectedRows();
183            System.out.println("Selected rows: " + Arrays.toString(selected));
184            selected = table.getSelectedColumns();
185            System.out.println("Selected columns: " + Arrays.toString(selected));
186         });
187      tableMenu.add(printSelectionItem);
188   }
189
190   private void updateCheckboxMenuItems()
191   {
192      rowsItem.setSelected(table.getRowSelectionAllowed());
193      columnsItem.setSelected(table.getColumnSelectionAllowed());
194      cellsItem.setSelected(table.getCellSelectionEnabled());
195   }
196 }
```

API *javax.swing.table.TableModel* 1.2

- Class getColumnClass(int columnIndex)

    获取该列中的值的类。该信息用于排序或绘制。

API *javax.swing.JTable* 1.2

- TableColumnModel getColumnModel()

    获取描述表格列布局安排的"列模式"。

- void setAutoResizeMode(int mode)

  设置自动更改表格列大小的模式。

  参数：mode　　AUTO_RESIZE_OFF、AUTO_RESIZE_NEXT_COLUMN、AUTO_RESIZE_SUBSEQUENT_COLUMNS、AUTO_RESIZE_LAST_COLUMN 以及 AUTO_RESIZE_ALL_COLUMNS 其中之一。

- int getRowHeight()
- void setRowMargin(int margin)

  获取和设置相邻行中单元格之间的间隔大小。

- int getRowHeight()
- void setRowHeight(int height)

  获取和设置表格中所有行的默认高度。

- int getRowHeight(int row)
- void setRowHeight(int row, int height)

  获取和设置表格中给定行的高度。

- ListSelectionModel getSelectionModel()

  返回列表的选择模式。你需要该模式以便在行、列以及单元格之间进行选择。

- boolean getRowSelectionAllowed()
- void setRowSelectionAllowed(boolean b)

  获取和设置 rowSelectionAllowed 属性。如果为 true，那么当用户点击单元格的时候，可以选定行。

- boolean getColumnSelectionAllowed()
- void setColumnSelectionAllowed(boolean b)

  获取和设置 columnSelectionAllowed 属性。如果为 true，那么当用户点击单元格的时候，可以选定列。

- boolean getCellSelectionEnabled()

  如果既允许选定行又允许选定列，则返回 true。

- void setCellSelectionEnabled(boolean b)

  同时将 rowSelectionAllowed 和 columnSelectionAllowed 设置为 b。

- void addColumn(TableColumn column)

  向表格视图中添加一列作为最后一列。

- void moveColumn(int from, int to)

  移动表格 from 索引位置中的列，使它的索引变成 to。该操作仅仅影响到视图。

- void removeColumn(TableColumn column)

  将给定的列从视图中移除。

- int convertRowIndexToModel(int index)　6
- int convertColumnIndexToModel(int index)

  返回具有给定索引的行或列的模型索引，这个值与行被排序和过滤，以及列被移动和移除时的索引不同。
- void setRowSorter(RowSorter<? extends TableModel> sorter)
  设置行排序器。

API *javax.swing.table.TableColumnModel* 1.2

- TableColumn getColumn(int index)
  获取表格的列对象，用于描述给定索引的列。

API *javax.swing.table.TableColumn* 1.2

- TableColumn(int modelColumnIndex)
  构建一个表格列，用以显示给定索引位置上的模型列。
- void setPreferredWidth(int width)
- void setMinWidth(int width)
- void setMaxWidth(int width)
  将表格的首选宽度、最小宽度以及最大宽度设置为 width。
- void setWidth(int width)
  设置该列的实际宽度为 width。
- void setResizable(boolean b)
  如果 b 为 true，那么该列可以更改大小。

API *javax.swing.ListSelectionModel* 1.2

- void setSelectionMode(int mode)
  参数：mode SINGLE_SELECTION、SINGLE_INTERVAL_SELECTION 与 MULTIPLE_INTERVAL_SELECTION 之一。

API *javax.swing.DefaultRowSorter<M, I>* 6

- void setComparator(int column, Comparator<?> comparator)
  设置用于给定列的比较器。
- void setSortable(int column, boolean enabled)
  使对给定列的排序可用或禁用。
- void setRowFilter(RowFilter<? super M,? super I> filter)
  设置行过滤器。

API *javax.swing.table.TableRowSorter<M extends TableModel>* 6

- void setStringConverter(TableStringConverter stringConverter)
  设置用于排序和过滤的字符串转换器。

API *javax.swing.table.TableStringConverter* 6

- abstract String toString(TableModel model, int row, int column)

将给定位置的模型值转换为字符串，你可以覆盖这个方法。

API javax.swing.RowFilter<M, I> 6

- boolean include(RowFilter.Entry<? extends M,? extends I> entry)
  指定要保留的行，你可以覆盖这个方法。
- static <M,I> RowFilter<M,I> numberFilter(RowFilter.ComparisonType type, Number number, int... indices)
- static <M,I> RowFilter<M,I> dateFilter(RowFilter.ComparisonType type, Date date, int... indices)
  返回一个过滤器，它包含的行是那些与给定的数字或日期进行给定比较后匹配的行。比较类型是 EQUAL、NOT_EQUAL、AFTER 或 BEFORE 之一。如果给定了列模型索引，则只搜索这些列。否则，将搜索所有列。对于数字过滤器，单元格的值所属的类必须与给定数字的类匹配。
- static <M,I> RowFilter<M,I> regexFilter(String regex, int... indices)
  返回一个过滤器，它包含的行含有与给定的正则表达式匹配的字符串。如果给定了列模型索引，则只搜索这些列。否则，将搜索所有列。注意，RowFilter.Entry 的 getStringValue 方法返回的字符串是匹配的。
- static <M,I> RowFilter<M,I> andFilter(Iterable<? extends RowFilter<? super M,? super I>> filters)
- static <M,I> RowFilter<M,I> orFilter(Iterable<? extends RowFilter<? super M,? super I>> filters)
  返回一个过滤器，它包含的项是那些包含在所有的过滤器或至少包含在一个过滤器中的项。
- static <M,I> RowFilter<M,I> notFilter(RowFilter<M,I> filter)
  返回一个过滤器，它包含的项是那些不包含在给定过滤器中的项。

API javax.swing.RowFilter.Entry<M, I> 6

- I getIdentifier()
  返回这个行的标识符。
- M getModel()
  返回这个行的模型。
- Object getValue(int index)
  返回在这个行的给定索引处存储的值。
- int getValueCount()
  返回在这个行中存储的值的数量。
- String getStringValue()
  返回在这个行的给定索引处存储的值转换成的字符串。由 TableRowSorter 产生的项的 getStringValue 方法会调用排序器的字符串转换器。

11.1.4 单元格的绘制和编辑

正如在 11.1.3.2 节中看到的，列的类型确定了单元格应该如何绘制。Boolean 和 Icon 类型有默认的绘制器，它们将绘制复选框或图标，而对于其他所有类型，都需要安装定制的绘制器。

11.1.4.1 绘制单元格

表格的单元格绘制器与你在前面看到的列表单元格绘制器类似。它们都实现了 TableCellRenderer 接口，并只有一个方法：

```
Component getTableCellRendererComponent(JTable table, Object value,
    boolean isSelected, boolean hasFocus, int row, int column)
```

该方法在表格需要绘制单元格的时候被调用。它会返回一个构件，接着该构件的 paint 方法会被调用，以填充单元格区域。

在图 11-8 中的表格包含类型为 Color 的单元格，绘制器直接返回一个面板，其背景颜色设置为存储在该单元格中的颜色对象，该颜色是作为 value 参数传递的。

```
class ColorTableCellRenderer extends JPanel implements TableCellRenderer
{
    public Component getTableCellRendererComponent(JTable table, Object value,
        boolean isSelected, boolean hasFocus, int row, int column)
    {
        setBackground((Color) value);
        if (hasFocus)
            setBorder(UIManager.getBorder("Table.focusCellHighlightBorder"));
        else
            setBorder(null);
        return this;
    }
}
```

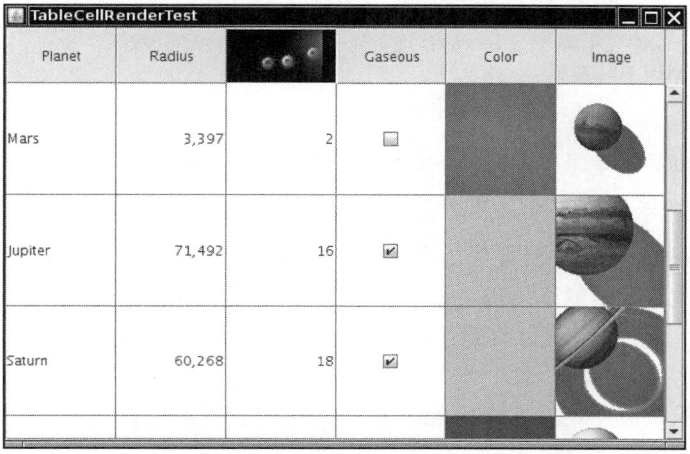

图 11-8 具有单元格绘制器的表格

正如你看到的那样，当该单元格获得焦点的时候，绘制器会画出一个边框。(我们可以向

UIManager 寻求合适的边框。为了发现查找的关键所在，我们可以深入 DefaultTableCellRenderer 类的源码内部看个究竟。)

> 提示：如果你的绘制器只是绘制一个文本字符串或者一个图标，那么可以继承 DefaultTableCellRenderer 这个类。该类会负责绘制焦点和选择状态。

你必须告诉表格要使用这个绘制器去绘制所有类型为 Color 的对象。JTable 类的 setDefaultRenderer 方法可以让你建立它们之间的这种联系。你需要提供一个 Class 对象和绘制器。

```
table.setDefaultRenderer(Color.class, new ColorTableCellRenderer());
```

现在这个绘制器就可以用于表格中具有给定类型的所有对象了。

如果想要基于其他标准选择绘制器，则需要从 JTable 类中扩展子类，并覆盖 getCellRender 方法。

11.1.4.2 绘制表头

为了在表头中显示图标，需要设置表头值。

```
moonColumn.setHeaderValue(new ImageIcon("Moons.gif"));
```

然而，表头还未智能到可以为表头值选择一个合适的绘制器，因此，绘制器需要手工安装。例如，要在列头显示图像图标，可以调用：

```
moonColumn.setHeaderRenderer(table.getDefaultRenderer(ImageIcon.class));
```

11.1.4.3 单元格编辑

为了使单元格可编辑，表格模型必须通过定义 isCellEditable 方法来指明哪些单元格是可编辑的。最常见的情况是，你可能想使某几列可编辑。在这个示例程序中，我们允许对表格中的四列进行编辑。

```
public boolean isCellEditable(int r, int c)
{
   return c == PLANET_COLUMN || c == MOONS_COLUMN || c == GASEOUS_COLUMN
      || c == COLOR_COLUMN;
}
```

> 注释：AbstractTableModel 定义的 isCellEditable 方法总是返回 false。DefaultTableModel 覆盖了该方法以便总是返回 true。

运行一下程序清单 11-4 到程序清单 11-7 的程序就会注意到，可以点击 Gaseous 列中的复选框，并能选中或取消复选标记。如果点击 Moons 列中的某个单元格，就会出现一个组合框（参见图 11-9）。你很快就会看到怎样将这样一个组合框作为一个单元格编辑器安装到表格上。

最后，点击第一列中的某个单元格，该单元格就会获取焦点。你就可以开始键入数据，而该单元格的内容也会随之更改。

你刚刚看到的是 DefaultCellEditor 类的三种变型。DefaultCellEditor 可以用 JTextField、JCheckBox 或者 JComboBox 来构造。JTable 类会自动为 Boolean 类型的单元格安装一个复选框编辑

器，并为所有可编辑但未提供它们自己的绘制器的单元格安装一个文本编辑器。文本框可以让用户去编辑那些对表格模型 getValueAt 方法的返回值执行 toString 操作而产生的字符串。

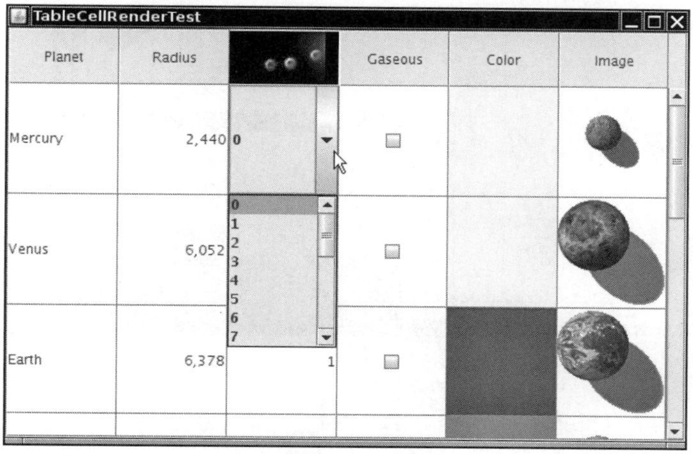

图 11-9 单元格编辑器

一旦编辑完成，通过调用编辑器的 getCellEditorValue 方法就可以读取编辑过的值。该方法应该返回一个正确类型的值（也就是模型的 getColumnType 方法返回的类型）。

为了获得一个组合框编辑器，你需要手动设置单元格编辑器，因为 JTable 构件并不知道什么样的值对某一特殊类型来说是适合的。对于 Moons 列来说，我们希望可以让用户选择 0 ～ 20 之间的任何值。下面是对组合框进行初始化的代码。

```
var moonCombo = new JComboBox();
for (int i = 0; i <= 20; i++)
    moonCombo.addItem(i);
```

为了构造一个 DefaultCellEditor，需要在该构造器中提供一个组合框。

```
var moonEditor = new DefaultCellEditor(moonCombo);
```

接下来，我们需要安装这个编辑器。与颜色单元格绘制器不同，这个编辑器不依赖于对象类型，我们未必想要把它作用于类型为 Integer 的所有对象上。相反，我们需要把它安装到一个特定列中：

```
moonColumn.setCellEditor(moonEditor);
```

11.1.4.4 定制编辑器

再次运行一下示例程序并点击一种颜色。这时会弹出一个颜色选择器让你为行星选择一种新颜色。选中一种颜色，然后点击 OK，单元格颜色就会随之更新（参见图 11-10）。

颜色单元格编辑器并不是一种标准的表格单元格编辑器，而是一种定制实现的编辑器。为了创建一个定制的单元格编辑器，需要实现 TableCellEditor 接口。这个接口有点拖沓冗长，从 Java SE 1.3 开始，提供了 AbstractCellEditor 类，用于负责事件处理的细节。

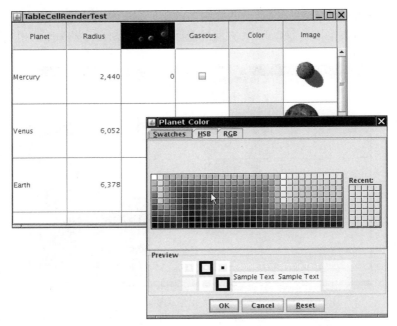

图 11-10 使用颜色选择器对单元格的颜色进行编辑

TableCellEditor 接口的 getTableCellEditorComponent 方法请求某个构件去绘制单元格。除了没有 focus 参数之外，它和 TableCellRenderer 接口的 getTableCellRendererComponent 方法极为相似。因为我们要编辑单元格，所以假设它获得了焦点。在编辑过程中，编辑器构件会暂时取代绘制器。在我们的示例中，返回的是一个没有颜色的空面板。这只是告诉用户该单元格正在被编辑。

接下来，当用户点击单元格时，你希望能弹出自己的编辑器。

JTable 类用一个事件（例如鼠标点击）去调用你的编辑器，以便确定该事件是否可以被接受去启动编辑过程。AbstractCellEditor 将该方法定义为能够接收所有的事件类型。

```
public boolean isCellEditable(EventObject anEvent)
{
    return true;
}
```

然而，如果你将该方法覆盖成 false，那么表格模型就不会遇到插入编辑器构件这样的麻烦了。

一旦安装了编辑器构件，假设我们使用的是相同的事件，那么 shouldSelectCell 方法就会被调用。应该在这个方法中启动编辑过程，例如，弹出一个外部的编辑对话框。

```
public boolean shouldSelectCell(EventObject anEvent)
{
    colorDialog.setVisible(true);
    return true;
}
```

如果用户取消编辑，表格会调用 cancelCellEditing 方法。如果用户已经点击了另一个表格单元，那么表格会调用 stopCellEditing 方法。在这两种情况中，都应该将对话框隐藏起来。当 stopCellEditing 方法被调用时，表格可能会使用被部分编辑的值。如果当前值有效，那么应该返回 true。在颜色选择器中，任何值都是有效的。但是如果编辑的是其他数据，那么应该保证只有有效的数据才能从编辑器中读取出来。

另外，应该调用超类的方法，以便进行事件的触发，否则，编辑事件就无法正确地取消。

```
public void cancelCellEditing()
{
   colorDialog.setVisible(false);
   super.cancelCellEditing();
}
```

最后，必须提供一个方法，以便产生用户在编辑过程中所提供的值。

```
public Object getCellEditorValue()
{
   return colorChooser.getColor();
}
```

总结一下，定制编辑器应该遵循下面几点：

1. 继承 AbstractCellEditor 类，并实现 TableCellEditor 接口。
2. 定义 getTableCellEditorComponent 方法以提供一个构件。它可以是一个哑构件（如果弹出一个对话框）或者是用于就地编辑的构件，例如复选框或文本框。
3. 定义 shouldSelectCell、stopCellEditing 及 cancelCellEditing 方法，来处理编辑过程的启动、完成以及取消。stopCellEditing 和 cancelCellEditing 方法应该调用超类方法以保证监听器能够接收到通知。
4. 定义 getCellEditorValue 方法返回编辑结果的值。

最后，通过调用 stopCellEditing 和 cancelCellEditing 方法，以表明用户什么时间完成了编辑操作。在构建颜色对话框的时候，我们安装了接受和取消的回调，用于触发这些事件。

```
colorDialog = JColorChooser.createDialog(null, "Planet Color", false, colorChooser,
   EventHandler.create(ActionListener.class, this, "stopCellEditing"),
   EventHandler.create(ActionListener.class, this, "cancelCellEditing"));
```

这样就完成了定制编辑器的实现过程。

你现在已经知道了怎样使一个单元格可编辑，以及怎样安装一个编辑器。还剩下一个问题，即怎样使用用户编辑过的值来更新表格模型。当编辑完成的时候，JTable 类会调用表格模型的下面这个方法：

```
void setValueAt(Object value, int r, int c)
```

需要将这个方法覆盖掉以便存储新值。value 参数是单元格编辑器返回的对象。如果实现了单元格编辑器，那么你就知道从 getCellEditorValue 方法返回的是什么类型的对象。在 DefaultCellEditor 这种情况中，这个值有三种可能：如果单元格编辑器是复选框，那么它就是

Boolean 值；如果是一个文本框，那么它就是一个字符串；如果这个值来源于组合框，那么就是用户选定的对象。

如果 value 对象不具有合适的类型，那么需要对它进行转换。例如，在一个文本框中编辑一个数字，这种情况最常发生。在我们的示例中，我们是将组合框组装成了 Integer 对象，所以不需要任何转换。

程序清单 11-4　tableCellRender/TableCellRenderFrame.java

```java
 1  package tableCellRender;
 2
 3  import java.awt.*;
 4  import javax.swing.*;
 5  import javax.swing.table.*;
 6
 7  /**
 8   * This frame contains a table of planet data.
 9   */
10  public class TableCellRenderFrame extends JFrame
11  {
12     private static final int DEFAULT_WIDTH = 600;
13     private static final int DEFAULT_HEIGHT = 400;
14
15     public TableCellRenderFrame()
16     {
17        setSize(DEFAULT_WIDTH, DEFAULT_HEIGHT);
18
19        var model = new PlanetTableModel();
20        var table = new JTable(model);
21        table.setRowSelectionAllowed(false);
22
23        // set up renderers and editors
24
25        table.setDefaultRenderer(Color.class, new ColorTableCellRenderer());
26        table.setDefaultEditor(Color.class, new ColorTableCellEditor());
27
28        var moonCombo = new JComboBox<Integer>();
29        for (int i = 0; i <= 20; i++)
30           moonCombo.addItem(i);
31
32        TableColumnModel columnModel = table.getColumnModel();
33        TableColumn moonColumn = columnModel.getColumn(PlanetTableModel.MOONS_COLUMN);
34        moonColumn.setCellEditor(new DefaultCellEditor(moonCombo));
35        moonColumn.setHeaderRenderer(table.getDefaultRenderer(ImageIcon.class));
36        moonColumn.setHeaderValue(new ImageIcon(getClass().getResource("Moons.gif")));
37
38        // show table
39
40        table.setRowHeight(100);
41        add(new JScrollPane(table), BorderLayout.CENTER);
42     }
43  }
```

程序清单 11-5 tableCellRender/PlanetTableModel.java

```java
package tableCellRender;

import java.awt.*;
import javax.swing.*;
import javax.swing.table.*;

/**
 * The planet table model specifies the values, rendering and editing properties for the
 * planet data.
 */
public class PlanetTableModel extends AbstractTableModel
{
   public static final int PLANET_COLUMN = 0;
   public static final int MOONS_COLUMN = 2;
   public static final int GASEOUS_COLUMN = 3;
   public static final int COLOR_COLUMN = 4;

   private Object[][] cells =
   {
      { "Mercury", 2440.0, 0, false, Color.YELLOW,
         new ImageIcon(getClass().getResource("Mercury.gif")) },
      { "Venus", 6052.0, 0, false, Color.YELLOW,
         new ImageIcon(getClass().getResource("Venus.gif")) },
      { "Earth", 6378.0, 1, false, Color.BLUE,
         new ImageIcon(getClass().getResource("Earth.gif")) },
      { "Mars", 3397.0, 2, false, Color.RED,
         new ImageIcon(getClass().getResource("Mars.gif")) },
      { "Jupiter", 71492.0, 16, true, Color.ORANGE,
         new ImageIcon(getClass().getResource("Jupiter.gif")) },
      { "Saturn", 60268.0, 18, true, Color.ORANGE,
         new ImageIcon(getClass().getResource("Saturn.gif")) },
      { "Uranus", 25559.0, 17, true, Color.BLUE,
         new ImageIcon(getClass().getResource("Uranus.gif")) },
      { "Neptune", 24766.0, 8, true, Color.BLUE,
         new ImageIcon(getClass().getResource("Neptune.gif")) },
      { "Pluto", 1137.0, 1, false, Color.BLACK,
         new ImageIcon(getClass().getResource("Pluto.gif")) }
   };

   private String[] columnNames = { "Planet", "Radius", "Moons", "Gaseous",
      "Color", "Image" };

   public String getColumnName(int c)
   {
      return columnNames[c];
   }

   public Class<?> getColumnClass(int c)
   {
      return cells[0][c].getClass();
   }

```

```java
53    public int getColumnCount()
54    {
55       return cells[0].length;
56    }
57
58    public int getRowCount()
59    {
60       return cells.length;
61    }
62
63    public Object getValueAt(int r, int c)
64    {
65       return cells[r][c];
66    }
67
68    public void setValueAt(Object obj, int r, int c)
69    {
70       cells[r][c] = obj;
71    }
72
73    public boolean isCellEditable(int r, int c)
74    {
75       return c == PLANET_COLUMN || c == MOONS_COLUMN || c == GASEOUS_COLUMN
76          || c == COLOR_COLUMN;
77    }
78 }
```

程序清单 11-6 tableCellRender/ColorTableCellRenderer.java

```java
1  package tableCellRender;
2
3  import java.awt.*;
4  import javax.swing.*;
5  import javax.swing.table.*;
6
7  /**
8   * This renderer renders a color value as a panel with the given color.
9   */
10 public class ColorTableCellRenderer extends JPanel implements TableCellRenderer
11 {
12    public Component getTableCellRendererComponent(JTable table, Object value,
13          boolean isSelected, boolean hasFocus, int row, int column)
14    {
15       setBackground((Color) value);
16       if (hasFocus) setBorder(UIManager.getBorder("Table.focusCellHighlightBorder"));
17       else setBorder(null);
18       return this;
19    }
20 }
```

程序清单 11-7 tableCellRender/ColorTableCellEditor.java

```java
1  package tableCellRender;
2
```

```java
3   import java.awt.*;
4   import java.awt.event.*;
5   import java.beans.*;
6   import java.util.*;
7   import javax.swing.*;
8   import javax.swing.table.*;
9
10  /**
11   * This editor pops up a color dialog to edit a cell value.
12   */
13  public class ColorTableCellEditor extends AbstractCellEditor implements TableCellEditor
14  {
15     private JColorChooser colorChooser;
16     private JDialog colorDialog;
17     private JPanel panel;
18
19     public ColorTableCellEditor()
20     {
21        panel = new JPanel();
22        // prepare color dialog
23
24        colorChooser = new JColorChooser();
25        colorDialog = JColorChooser.createDialog(null, "Planet Color", false, colorChooser,
26              EventHandler.create(ActionListener.class, this, "stopCellEditing"),
27              EventHandler.create(ActionListener.class, this, "cancelCellEditing"));
28     }
29
30     public Component getTableCellEditorComponent(JTable table, Object value,
31           boolean isSelected, int row, int column)
32     {
33        // this is where we get the current Color value. We store it in the dialog in case the
34        // user starts editing
35        colorChooser.setColor((Color) value);
36        return panel;
37     }
38
39     public boolean shouldSelectCell(EventObject anEvent)
40     {
41        // start editing
42        colorDialog.setVisible(true);
43
44        // tell caller it is ok to select this cell
45        return true;
46     }
47
48     public void cancelCellEditing()
49     {
50        // editing is canceled--hide dialog
51        colorDialog.setVisible(false);
52        super.cancelCellEditing();
53     }
54
55     public boolean stopCellEditing()
56     {
```

```
57          // editing is complete--hide dialog
58          colorDialog.setVisible(false);
59          super.stopCellEditing();
60
61          // tell caller is is ok to use color value
62          return true;
63       }
64
65       public Object getCellEditorValue()
66       {
67          return colorChooser.getColor();
68       }
69    }
```

API `javax.swing.JTable` 1.2

- TableCellRenderer getDefaultRenderer(Class<?> type)
  获取给定类型的默认绘制器。

- TableCellEditor getDefaultEditor(Class<?> type)
  获取给定类型的默认编辑器。

API `javax.swing.table.TableCellRenderer` 1.2

- Component getTableCellRendererComponent(JTable table, Object value, boolean selected, boolean hasFocus, int row, int column)
  返回一个构件，它的 paint 方法将被调用以便绘制一个表格单元格。

  参数： table   该表格包含要绘制的单元格
  　　　 value   要绘制的单元格
  　　　 selected   如果该单元格当前已被选中，则为 true
  　　　 hasFocus   如果该单元格当前具有焦点，则为 true
  　　　 row, column 单元格的行及列

API `javax.swing.table.TableColumn` 1.2

- void setCellEditor(TableCellEditor editor)
- void setCellRenderer(TableCellRenderer renderer)
  为该列中的所有单元格设置单元格编辑器或绘制器。

- void setHeaderRenderer(TableCellRenderer renderer)
  为该列中的所有表头单元格设置单元格绘制器。

- void setHeaderValue(Object value)
  为该列中的表头设置用于显示的值。

API `javax.swing.DefaultCellEditor` 1.2

- DefaultCellEditor(JComboBox comboBox)
  构建一个单元格编辑器，并以一个组合框的形式显示出来，用于选择单元格的值。

API *javax.swing.table.TableCellEditor* 1.2

- Component getTableCellEditorComponent(JTable table, Object value, boolean selected, int row, int column)
  返回一个构件，它的 paint 方法用于绘制表格的单元格。
  参数：table　　　　包含要绘制的单元格的表格
  　　　value　　　　要绘制的单元格
  　　　selected　　 如果该单元格已被当前选中，则为 true
  　　　row,colum   单元格的行及列

API *javax.swing.CellEditor* 1.2

- boolean isCellEditable(EventObject event)
  如果该事件能够启动对该单元格的编辑过程，那么返回 true。
- boolean shouldSelectCell(EventObject anEvent)
  启动编辑过程。如果被编辑的单元格应该被选中，则返回 true。通常情况下，你希望返回的是 true，不过，如果你不希望在编辑过程中改变单元格被选中的情况，那么你可以返回 false。
- void cancelCellEditing()
  取消编辑过程。你可以放弃已进行了部分编辑的操作。
- boolean stopCellEditing()
  出于使用编辑结果的目的，停止编辑过程。如果被编辑的值对读取来说处于适合的状态，则返回 true。
- Object getCellEditorValue()
  返回编辑结果。
- void addCellEditorListener(CellEditorListener l)
- void removeCellEditorListener(CellEditorListener l)
  添加或移除必需的单元格编辑器的监听器。

11.2 树

每个使用过分层结构的文件系统的计算机用户都见过树状显示。当然，目录和文件形式仅仅是树状组织结构中的一种。日常生活中还有很多这样的树结构，例如国家、州以及城市之间的层次结构，如图 11-11 所示。

作为一名编程人员，我们经常需要显示这些树形结构。幸运的是，Swing 类库中有一个正是用于此目的的 JTree 类。JTree 类（以及它的辅助类）负责布局树状结构，按照用户请求展开或折叠树的节点。在本节中，我们将介绍怎样使用 JTree 类。

与其他复杂的 Swing 构件一样，我们必须集中介绍一些常用方法，无法涉及所有的细节。如果读者想获得与众不同的效果，我们推荐你参考 David M. Geary 撰写的 *Graphic Java*

(第 3 版)，Kim Topley 编写的 *Core Swing*。

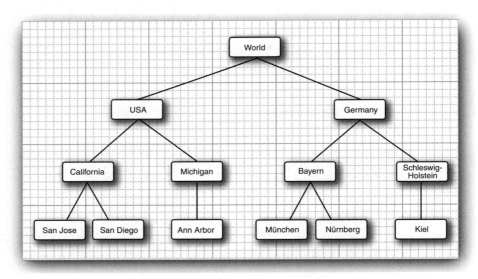

图 11-11 国家、州及城市的层次结构

在我们深入展开之前，先介绍一些术语（参见图 11-12）。一棵树由一些节点（node）组成。每个节点要么是叶节点（leaf）要么是有孩子节点（child node）的节点。除了根节点（root node），每一个节点都有一个唯一的父节点（parent node）。一棵树只有一个根节点。有时，你可能有一个树的集合，其中每棵树都有自己的根节点。这样的集合称作森林（forest）。

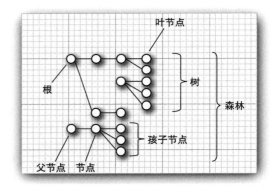

图 11-12 树中的术语

11.2.1 简单的树

在第一个示例程序中，我们仅仅展示了一个具有几个节点的树（参见图 11-14）。如同大多数 Swing 构件一样，只要提供一个数据模型，构件就可以将它显示出来。为了构建 JTree，需要在构造器中提供这样一个树模型：

```
TreeModel model = . . .;
var tree = new JTree(model);
```

> 注释：还有一些构造器可以用一些元素的集合来构建树。
> ```
> JTree(Object[] nodes)
> JTree(Vector<?> nodes)
> JTree(Hashtable<?, ?> nodes) // the values become the nodes
> ```

> 这些构造器不是特别有用。它们仅仅是创建出一个包含了若干棵树的森林，其中每棵树只有一个节点。第三个构造器显得特别没用，因为这些节点实际的显示次序是由键的散列码确定的。

怎样才能获得一个树模型呢？可以通过创建一个实现了 TreeModel 接口的类来构建自己的树模型。在本章的后面部分，将会介绍应该如何实现。现在，我们仍坚持使用 Swing 类库提供的 DefaultTreeModel 模型。

为了构建一个默认的树模型，必须提供一个根节点。

```
TreeNode root = ...;
var model = new DefaultTreeModel(root);
```

TreeNode 是另外一个接口。可以将任何实现了这个接口的类的对象组装到默认的树模型中。这里，我们使用的是 Swing 提供的具体节点类，叫做 DefaultMutableTreeNode。这个类实现了 MutableTreeNode 接口，该接口是 TreeNode 的一个子接口（参见图 11-13）。

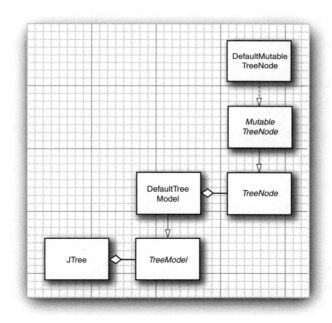

图 11-13 有关树的类

任何一个默认的可变树节点都存放着一个对象，即用户对象（user object）。树会为所有的节点绘制这些用户对象。除非指定一个绘制器，否则树将直接显示执行完 toString 方法之后的结果字符串。

在第一个示例程序中，我们使用了字符串作为用户对象。实际应用中，通常会在树中组装更具表现力的用户对象。例如，当显示一个目录树时，将 File 对象用于节点将具有实际意义。

可以在构造器中设定用户对象，也可以稍后在 setUserObject 方法中设定用户对象：

```
var node = new DefaultMutableTreeNode("Texas");
...
node.setUserObject("California");
```

接下来，可以建立节点之间的父/子关系。从根节点开始，使用 add 方法来添加子节点：

```
var root = new DefaultMutableTreeNode("World");
var country = new DefaultMutableTreeNode("USA");
root.add(country);
var state = new DefaultMutableTreeNode("California");
country.add(state);
```

图 11-14 显示了这棵树的外观。

按照这种方式将所有的节点链接起来。然后用根节点构建一个 DefaultTreeModel。最后，用这个树模型构建一个 JTree。

```
var treeModel = new DefaultTreeModel(root);
var tree = new JTree(treeModel);
```

图 11-14 一棵简单的树

或者，使用快捷方式，直接将根节点传递给 JTree 构造器。那么这棵树就会自动构建一个默认的树模型：

```
var tree = new JTree(root);
```

程序清单 11-8 给出了完整的代码。

程序清单 11-8　tree/SimpleTreeFrame.java

```java
1  package tree;
2
3  import javax.swing.*;
4  import javax.swing.tree.*;
5
6  /**
7   * This frame contains a simple tree that displays a manually constructed tree model.
8   */
9  public class SimpleTreeFrame extends JFrame
10 {
11     private static final int DEFAULT_WIDTH = 300;
12     private static final int DEFAULT_HEIGHT = 200;
13
14     public SimpleTreeFrame()
15     {
16         setSize(DEFAULT_WIDTH, DEFAULT_HEIGHT);
17
18         // set up tree model data
19
20         var root = new DefaultMutableTreeNode("World");
21         var country = new DefaultMutableTreeNode("USA");
22         root.add(country);
23         var state = new DefaultMutableTreeNode("California");
24         country.add(state);
```

```
25      var city = new DefaultMutableTreeNode("San Jose");
26      state.add(city);
27      city = new DefaultMutableTreeNode("Cupertino");
28      state.add(city);
29      state = new DefaultMutableTreeNode("Michigan");
30      country.add(state);
31      city = new DefaultMutableTreeNode("Ann Arbor");
32      state.add(city);
33      country = new DefaultMutableTreeNode("Germany");
34      root.add(country);
35      state = new DefaultMutableTreeNode("Schleswig-Holstein");
36      country.add(state);
37      city = new DefaultMutableTreeNode("Kiel");
38      state.add(city);
39
40      // construct tree and put it in a scroll pane
41
42      var tree = new JTree(root);
43      add(new JScrollPane(tree));
44   }
45 }
```

运行这段程序代码时，最初的树外观如图 11-15 所示。只有根节点和它的子节点可见。单击圆圈图标（把手）展开子树。当子树折叠起来时，把手图标的线伸出指向右边，当子树展开时，把手图标的线伸出指向下方（参见图 11-16）。虽然我们无法得知 Metal 外观的设计者当时是如何构想的，但是我们可以将这个图标看作一个门把手，按下把手就可以打开子树。

图 11-15　最初的树的显示

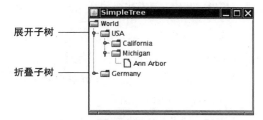

图 11-16　折叠和展开后的子树

> **注释**：当然，树的显示还依赖于所选择的外观模式。我们这里只讨论 Metal 这种外观模式。在 Windows 或 Motif 外观模式中，把手则具有我们更熟悉的外观，即带有 "－" 或 "＋" 的框结构（参见图 11-17）。

可以使用下面这句神奇的代码取消父子节点之间的连接线（参见图 11-18）：

```
tree.putClientProperty("JTree.lineStyle", "None");
```

相反地，如果要确保显示这些线条，则可以使用：

```
tree.putClientProperty("JTree.lineStyle", "Angled");
```

图 11-17　一棵具有 Windows 外观的树　　　　图 11-18　不带连接线的树

另一种线条样式，"水平线"，如图 11-19 所示。这棵树显示有水平线，而这些水平线只是用来将根节点的孩子节点分离开来。我们很难说清这样做的好处。

默认情况下，这种树中的根节点没有用于折叠的把手。如果需要的话，可以通过下面的调用来添加一个把手：

tree.setShowsRootHandles(true);

图 11-20 显示了调用后的结果。现在就可以将整棵树折叠到根节点中了。

 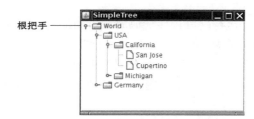

图 11-19　具有水平线样式的树　　　　图 11-20　具有一个根把手的树

相反地，也可以将根节点完全隐藏起来。这样做只是为了显示一个森林，即一个树集，每棵树都有它自己的根节点。但是仍然必须将森林中的所有树都放到一个公共节点下。因此，可以使用下面这条指令将根节点隐藏起来。

tree.setRootVisible(false);

请观察图 11-21。它看起来似乎有两个根节点，分别用"USA"和"Germany"标识了出来，而实际上将二者合并起来的根节点是不可见的。

让我们将注意力从树的根节点转移到叶节点。注意，这些叶节点的图标和其他节点的图标是不同的（参见图 11-22）。

在显示这棵树的时候，每个节点都绘有一个图标。实际上一共有三种图标：叶节点图标、展开的非叶节点图标以及闭合的非叶节点图标。为了简化起见，我们将后面两种图标称为文件夹图标。

节点绘制器必须知道每个节点要使用什么样的图标。默认情况下，这个决策过程是这样的：如果某个节点的 isLeaf 方法返回的是 true，那么就使用叶节点图标，否则，使用文件夹图标。

图 11-21　一个森林

图 11-22　叶节点和折叠节点的图标

如果某个节点没有任何儿子节点，那么 DefaultMutableTreeNode 类的 isLeaf 方法将返回 true。因此，具有儿子节点的节点使用文件夹图标，没有儿子节点的节点使用叶节点图标。

有时，这种做法并不合适。假设我们要向我们那棵简单的树中添加一个 "Montana" 节点，但是我们还不知道要添加什么城市。此时，我们并不希望一个州节点使用叶节点图标，因为从概念上来讲，只有城市才使用叶节点。

JTree 类无法知道哪些节点是叶节点，它要询问树模型。如果一个没有任何子节点的节点不应该自动地被设置为概念上的叶节点，那么可以让树模型对这些叶节点使用一个不同的标准，即可以查询其 "允许有子节点" 的节点属性。

对于那些不应该有子节点的节点，调用

```
node.setAllowsChildren(false);
```

然后，告诉树模型去查询 "允许有子节点" 的属性值以确定一个节点是否应该显示成叶子图标。可以使用 DefaultTreeModel 类中的方法 setAsksAllowsChildren 设定此动作：

```
model.setAsksAllowsChildren(true);
```

有了这个判定规则，允许有子节点的节点就可以获得文件夹图标，而不允许有子节点的节点将获得叶子图标。

另外，如果是通过提供根节点来构建一棵树的，那么请在构造器中直接提供 "询问允许有子节点" 属性值的设置。

```
var tree = new JTree(root, true); // nodes that don't allow children get leaf icons
```

API　javax.swing.JTree　1.2

- JTree(TreeModel model)

  根据一个树模型构造一棵树。

- JTree(TreeNode root)

- JTree(TreeNode root, boolean asksAllowChildren)

  使用默认的树模型构造一棵树，显示根节点和它的子节点。

  参数：root　　　　　　　　根节点

  asksAllowChildren　如果设置为 true，则使用 "允许有子节点" 的节点属性来确定一个节点是否是叶节点

- void setShowsRootHandles(boolean b)

如果 b 为 true，则根节点具有折叠或展开它的子节点的把手图标。
- void setRootVisible(boolean b)
  如果 b 为 true，则显示根节点，否则隐藏根节点。

API *javax.swing.tree.TreeNode* 1.2

- boolean isLeaf()
  如果该节点是一个概念上的叶节点，则返回 true。
- boolean getAllowsChildren()
  如果该节点可以拥有子节点，则返回 true。

API *javax.swing.tree.MutableTreeNode* 1.2

- void setUserObject(Object userObject)
  设置树节点用于绘制的"用户对象"。

API *javax.swing.tree.TreeModel* 1.2

- boolean isLeaf(Object node)
  如果该节点应该以叶节点的形式显示，则返回 true。

API *javax.swing.tree.DefaultTreeModel* 1.2

- void setAsksAllowsChildren(boolean b)
  如果 b 为 true，那么当节点的 getAllowsChildren 方法返回 false 时，这些节点显示为叶节点。否则，当节点的 isLeaf 方法返回 true 时，它们显示为叶节点。

API *javax.swing.tree.DefaultMutableTreeNode* 1.2

- DefaultMutableTreeNode(Object userObject)
  用给定的用户对象构建一个可变树节点。
- void add(MutableTreeNode child)
  将一个节点添加为该节点最后一个子节点。
- void setAllowsChildren(boolean b)
  如果 b 为 true，则可以向该节点添加子节点。

API *javax.swing.JComponent* 1.2

- void putClientProperty(Object key, Object value)
  将一个键/值对添加到一个小表格中，每一个构件都管理着这样的一个小表格。这是一种"紧急逃生"机制，很多 Swing 构件用它来存放与外观相关的属性。

11.2.1.1 编辑树和树的路径

在下面的一个示例程序中，将会看到怎样编辑一棵树。图 11-23 显示了用户界面。如果点击"Add Sibling"（添加兄弟节点）或"Add Child"（添加子节点）按钮，该程序将向树中添加一个新节点（带有"New"标题）。如果你点击"Delete"（删除）按钮，该程序将删除当前选中的节点。

为了实现这种行为，需要弄清楚当前选定的是哪个节点。JTree 类用的是一种令人惊讶的方式来标识树中的节点。它并不处理树的节点，而是处理对象路径（称为树路径）。一个树路径从根节点开始，由一个子节点序列构成，参见图 11-24。

图 11-23　编辑一棵树

图 11-24　一个树路径

你可能要怀疑 JTree 类为什么需要整个路径。它不能只获得一个 TreeNode，然后不断调用 getParent 方法吗？实际上，JTree 类一点都不清楚 TreeNode 接口的情况。该接口从来没有被 TreeModel 接口用到过，它只被 DefaultTreeModel 的实现用到了。你完全可以拥有其他的树模型，这些树模型中的节点可能根本就没有实现 TreeNode 接口。如果你使用的是一个管理其他类型对象的树模型，那么这些对象有可能根本就没有 getParent 和 getChild 方法。它们彼此之间当然会有其他某种连接。将其他节点连接起来这是树模型的职责，JTree 类本身并没有节点之间连接属性的任何线索。因此，JTree 类总是需要用完整的路径来工作。

TreePath 类管理着一个 Object（不是 TreeNode！）引用序列。有很多 JTree 的方法都可以返回 TreePath 对象。当拥有一个树路径时，通常只需要知道其终端节点，该节点可以通过 getLastPathComponent 方法得到。例如，如果要查找一棵树中当前选定的节点，可以使用 JTree 类中的 getSelectionPath 方法。它将返回一个 TreePath 对象，根据这个对象就可以检索实际节点。

```
TreePath selectionPath = tree.getSelectionPath();
var selectedNode = (DefaultMutableTreeNode) selectionPath.getLastPathComponent();
```

实际上，由于这种特定查询经常被使用到，因此还提供了一个更方便的方法，它能够立即给出选定的节点。

```
var selectedNode = (DefaultMutableTreeNode) tree.getLastSelectedPathComponent();
```

该方法之所以没有被称为 getSelectedNode，是因为这棵树并不了解它包含的节点，它的树模型只处理对象的路径。

> 📝 **注释**：树路径是 JTree 类描述节点的两种方式之一。JTree 有许多方法可以接收或返回一个整数索引——行的位置。行的位置仅仅是节点在树中显示的一个行号（从 0 开始）。只有那些可视节点才有行号，并且如果一个节点之前的其他节点展开、折叠或者被修改过，这个节点的行号也会随之改变。因此，你应该避免使用行的位置。相反地，所有使用行的 JTree 方法都有一个与之等价的使用树路径的方法。

一旦选定了某个节点，那么就可以对它进行编辑了。不过，不能直接向树节点添加子节点：

```
selectedNode.add(newNode); // No!
```

如果改变了节点的结构，那么改变的只是树模型，而相关的视图却没有被通知到。可以自己发送一个通知消息，但是如果使用 DefaultTreeModel 类的 insertNodeInto 方法，那么该模型类会全权负责这件事情。例如，下面的调用可以将一个新节点作为选定节点的最后子节点添加到树中，并通知树的视图。

```
model.insertNodeInto(newNode, selectedNode, selectedNode.getChildCount());
```

类似的调用 removeNodeFromParent 可以移除一个节点并通知树的视图：

```
model.removeNodeFromParent(selectedNode);
```

如果想保持节点结构，但是要改变用户对象，那么可以调用下面这个方法：

```
model.nodeChanged(changedNode);
```

自动通知是使用 DefaultTreeModel 的主要优势。如果你提供自己的树模型，那么必须自己动手实现这种自动通知。（详见 Kim Topley 撰写的 *Core Swing*。）

> ⚠️ **警告**：DefaultTreeModel 有一个 reload 方法能够将整个模型重新载入。但是，不要在进行了少数几个修改之后，只是为了更新树而调用 reload 方法。在重建一棵树的时候，根节点的子节点之后的所有节点将全部再次折叠起来。如果你的用户在每次修改之后都要不断地展开整棵树，这确实是一件令人烦心的事。

当视图接收到节点结构被改变的通知时，它会更新显示树的视图，但是不会自动展开某个节点以展现新添加的子节点。特别是在我们上面那个示例程序中，如果用户将一个新节点添加到其子节点正处于折叠状态的节点上，那么这个新添加的节点就被悄无声息地添加到了一个处于折叠状态的子树中，这就没有给用户提供任何反馈信息以告诉用户已经执行了该命令。在这种情况下，你可能需要特别费劲地展开所有的父节点，以便让新添加的节点成为可视节点。可以使用类 JTree 中的方法 makeVisible 实现这个目的。makeVisible 方法将接受一个树路径作为参数，该树路径指向应该变为可视的节点。

因此，需要构建一个从根节点到新添加节点的树路径。为了获得一个这样的树路径，首先要调用 DefaultTreeModel 类中的 getPathToRoot 方法，它返回一个包含了某一节点到根节点之间所有节点的数组 TreeNode[]。可以将这个数组传递给一个 TreePath 构造器。

例如，下面展示了怎样将一个新节点变成可见的：

```
TreeNode[] nodes = model.getPathToRoot(newNode);
var path = new TreePath(nodes);
tree.makeVisible(path);
```

> 📄 **注释**：令人惊奇的是，DefaultTreeModel 类好像完全忽视了 TreePath 类，尽管它的职责是与一个 JTree 通信。JTree 类大量地使用到了树路径，而它从不使用节点对象数组。

但是，现在假设你的树是放在一个滚动面板里面，在展开树节点之后，新节点仍是不可见的，因为它落在视图之外。为了克服这个问题，请调用

```
tree.scrollPathToVisible(path);
```

而不是调用 makeVisible。这个调用将展开路径中的所有节点，并告诉外围的滚动面板将路径末端的节点滚动到视图中（参见图 11-25）。

默认情况下，这些树节点是不可编辑的。不过，如果调用

```
tree.setEditable(true);
```

那么，用户就可以编辑某一节点了。可以先双击该节点，然后编辑字符串，最后按下回车键。双击操作会调用默认单元格编辑器，它实现了 DefaultCellEditor 类（参见图 11-26）。也可以安装其他一些单元格编辑器，其过程与表格单元格编辑器中讨论的过程一样。

图 11-25 滚动以显示新节点的滚动面板

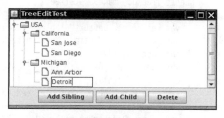

图 11-26 默认的单元格编辑器

程序清单 11-9 展示了树编辑程序的完整源代码。运行该程序，添加几个新节点，然后通过双击它们进行编辑操作。请观察折叠的节点是怎样展开以显现添加的子节点的，以及滚动面板是怎样让添加的节点保持在视图中的。

程序清单 11-9 treeEdit/TreeEditFrame.java

```java
 1  package treeEdit;
 2
 3  import java.awt.*;
 4  import javax.swing.*;
 5  import javax.swing.tree.*;
 6
 7  /**
 8   * A frame with a tree and buttons to edit the tree.
 9   */
10  public class TreeEditFrame extends JFrame
11  {
12     private static final int DEFAULT_WIDTH = 400;
13     private static final int DEFAULT_HEIGHT = 200;
14
15     private DefaultTreeModel model;
16     private JTree tree;
17
18     public TreeEditFrame()
19     {
20        setSize(DEFAULT_WIDTH, DEFAULT_HEIGHT);
```

```java
21
22     // construct tree
23
24     TreeNode root = makeSampleTree();
25     model = new DefaultTreeModel(root);
26     tree = new JTree(model);
27     tree.setEditable(true);
28
29     // add scroll pane with tree
30
31     var scrollPane = new JScrollPane(tree);
32     add(scrollPane, BorderLayout.CENTER);
33
34     makeButtons();
35  }
36
37  public TreeNode makeSampleTree()
38  {
39     var root = new DefaultMutableTreeNode("World");
40     var country = new DefaultMutableTreeNode("USA");
41     root.add(country);
42     var state = new DefaultMutableTreeNode("California");
43     country.add(state);
44     var city = new DefaultMutableTreeNode("San Jose");
45     state.add(city);
46     city = new DefaultMutableTreeNode("San Diego");
47     state.add(city);
48     state = new DefaultMutableTreeNode("Michigan");
49     country.add(state);
50     city = new DefaultMutableTreeNode("Ann Arbor");
51     state.add(city);
52     country = new DefaultMutableTreeNode("Germany");
53     root.add(country);
54     state = new DefaultMutableTreeNode("Schleswig-Holstein");
55     country.add(state);
56     city = new DefaultMutableTreeNode("Kiel");
57     state.add(city);
58     return root;
59  }
60
61  /**
62   * Makes the buttons to add a sibling, add a child, and delete a node.
63   */
64  public void makeButtons()
65  {
66     var panel = new JPanel();
67     var addSiblingButton = new JButton("Add Sibling");
68     addSiblingButton.addActionListener(event ->
69        {
70           var selectedNode = (DefaultMutableTreeNode) tree.getLastSelectedPathComponent();
71
72           if (selectedNode == null) return;
73
74           var parent = (DefaultMutableTreeNode) selectedNode.getParent();
```

```java
75
76                    if (parent == null) return;
77
78                    var newNode = new DefaultMutableTreeNode("New");
79
80                    int selectedIndex = parent.getIndex(selectedNode);
81                    model.insertNodeInto(newNode, parent, selectedIndex + 1);
82
83                    // now display new node
84
85                    TreeNode[] nodes = model.getPathToRoot(newNode);
86                    var path = new TreePath(nodes);
87                    tree.scrollPathToVisible(path);
88                });
89          panel.add(addSiblingButton);
90
91          var addChildButton = new JButton("Add Child");
92          addChildButton.addActionListener(event ->
93                {
94                    var selectedNode = (DefaultMutableTreeNode) tree.getLastSelectedPathComponent();
95
96                    if (selectedNode == null) return;
97
98                    var newNode = new DefaultMutableTreeNode("New");
99                    model.insertNodeInto(newNode, selectedNode, selectedNode.getChildCount());
100
101                   // now display new node
102
103                   TreeNode[] nodes = model.getPathToRoot(newNode);
104                   var path = new TreePath(nodes);
105                   tree.scrollPathToVisible(path);
106               });
107         panel.add(addChildButton);
108
109         var deleteButton = new JButton("Delete");
110         deleteButton.addActionListener(event ->
111               {
112                   var selectedNode = (DefaultMutableTreeNode) tree.getLastSelectedPathComponent();
113
114                   if (selectedNode != null && selectedNode.getParent() != null) model
115                       .removeNodeFromParent(selectedNode);
116              });
117         panel.add(deleteButton);
118         add(panel, BorderLayout.SOUTH);
119    }
120 }
```

API javax.swing.JTree 1.2

- TreePath getSelectionPath()

  获取到当前选定节点的路径,如果选定多个节点,则获取到第一个选定节点的路径。如果没有选定任何节点,则返回 null。

- Object getLastSelectedPathComponent()

获取表示当前选定节点的节点对象，如果选定多个节点，则获取第一个选定的节点。如果没有选定任何节点，则返回 null。
- void makeVisible(TreePath path)
展开该路径中的所有节点。
- void scrollPathToVisible(TreePath path)
展开该路径中的所有节点，如果这棵树是置于滚动面板中的，则滚动以确保该路径中的最后一个节点是可见的。

API *javax.swing.tree.TreePath* 1.2

- Object getLastPathComponent()
获取该路径中最后一个节点，也就该路径代表的节点对象。

API *javax.swing.tree.TreeNode* 1.2

- TreeNode getParent()
返回该节点的父节点。
- TreeNode getChildAt(int index)
查找给定索引号上的子节点。该索引号必须在 0 和 getChildCount()-1 之间。
- int getChildCount()
返回该节点的子节点个数。
- Enumeration children()
返回一个枚举对象，可以迭代遍历该节点的所有子节点。

API *javax.swing.tree.DefaultTreeModel* 1.2

- void insertNodeInto(MutableTreeNode newChild, MutableTreeNode parent, int index)
将 newChild 作为 parent 的新子节点添加到给定的索引位置上，并通知树模型的监听器。
- void removeNodeFromParent(MutableTreeNode node)
将节点 node 从该模型中删除，并通知树模型的监听器。
- void nodeChanged(TreeNode node)
通知树模型的监听器：节点 node 发生了改变。
- void nodesChanged(TreeNode parent, int[] changedChildIndexes)
通知树模型的监听器：节点 parent 所有在给定索引位置上的子节点发生了改变。
- void reload()
将所有节点重新载入到树模型中。这是一项动作剧烈的操作，只有当由于一些外部作用，导致树的节点完全改变时，才应该使用该方法。

11.2.2 节点枚举

有时为了查找树中一个节点，必须从根节点开始，遍历所有子节点直到找到相匹配的节点。DefaultMutableTreeNode 类有几个很方便的方法用于迭代遍历所有节点。

breadthFirstEnumeration 方法和 depthFirstEnumeration 方法分别使用广度优先或深度优先的遍历方式，返回枚举对象，它们的 nextElement 方法能够访问当前节点的所有子节点。图 11-27 显示了对示例树进行遍历的情况，节点标签则指示遍历节点时的先后次序。

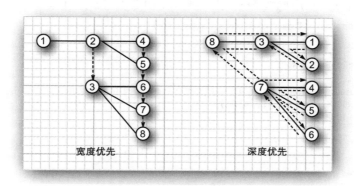

图 11-27　树的遍历顺序

按照广度优先的方式进行枚举是最容易可视化的。树是以层的形式遍历的，首先访问根节点，然后是它的所有子节点，接着是它的孙子节点，依此类推。

为了可视化深度优先的枚举，让我们想象一只老鼠陷入一个树状陷阱的情形。它沿着第一条路径迅速爬行，直到到达一个叶节点位置。然后，原路返回并转入下一条路径，依此类推。

计算机科学家也将其称为后序遍历（postorder traversal），因为整个查找过程是先访问到子节点，然后才访问到父节点。postOrderTraversal 方法是 depthFirstTraversal 的同义语。为了完整性，还存在一个 preOrderTraversal 方法，它也是一种深度优先搜索方法，但是它首先枚举父节点，然后是子节点。

下面是一种典型的使用模式：

```
Enumeration breadthFirst = node.breadthFirstEnumeration();
while (breadthFirst.hasMoreElements())
   do something with breadthFirst.nextElement();
```

最后，还有一个相关方法 pathFromAncestorEnumeration，用于查找一条从祖先节点到给定节点之间的路径，然后枚举出该路径中的所有节点。整个过程并不需要大量的处理操作，只需要不断调用 getParent 直到发现祖先节点，然后将该路径倒置过来存放即可。

在我们的下个示例程序中，将运用到节点枚举。该程序显示了类之间的继承树。向窗体最下面的文本框中输入一个类名，该类以及它的所有父类就会添加到树中（参见图 11-28）。

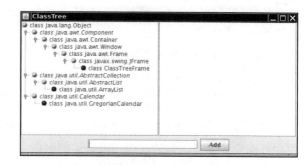

图 11-28　一棵继承树

在这个示例中,我们充分利用了这个事实,即树节点的用户对象可以是任何类型的对象。因为我们这里的节点是用来描述类的,因此我们在这些节点中存储的是 Class 对象。

当然,我们不想对同一个类对象添加两次,因此我们必须检查一个类是否已经存在于树中。如果在树中存在给定用户对象的节点,那么下面这个方法就可以用来查找该节点。

```
public DefaultMutableTreeNode findUserObject(Object obj)
{
   Enumeration e = root.breadthFirstEnumeration();
   while (e.hasMoreElements())
   {
      DefaultMutableTreeNode node = (DefaultMutableTreeNode) e.nextElement();
      if (node.getUserObject().equals(obj))
         return node;
   }
   return null;
}
```

11.2.3 绘制节点

在应用中可能会经常需要改变树构件绘制节点的方式,最常见的改变当然是为节点和叶节点选取不同的图标,其他一些改变可能涉及节点标签的字体或节点上的图像绘制等方面。所有这些改变都可以通过向树中安装一个新的树单元格绘制器来实现。在默认情况下,JTree 类使用 DefaultTreeCellRenderer 对象来绘制每个节点。DefaultTreeCellRenderer 类继承自 JLabel 类,该标签包含节点图标和节点标签。

> **注释**:单元格绘制器并不能绘制用于展开或折叠子树的"把手"图标。这些把手是外观模式的一部分,建议最好不要试图改变它们。

可以通过以下三种方式定制显示外观:
- 可以使用 DefaultTreeCellRenderer 改变图标、字体以及背景颜色。这些设置适用于树中所有节点。
- 可以安装一个继承了 DefaultTreeCellRenderer 类的绘制器,用于改变每个节点的图标、字体以及背景颜色。
- 可以安装一个实现了 TreeCellRenderer 接口的绘制器,为每个节点绘制自定义的图像。

让我们逐个研究这几种可能。最简单的定制方法是构建一个 DefaultTreeCellRenderer 对象,改变图标,然后将它安装到树中:

```
var renderer = new DefaultTreeCellRenderer();
renderer.setLeafIcon(new ImageIcon("blue-ball.gif")); // used for leaf nodes
renderer.setClosedIcon(new ImageIcon("red-ball.gif")); // used for collapsed nodes
renderer.setOpenIcon(new ImageIcon("yellow-ball.gif")); // used for expanded nodes
tree.setCellRenderer(renderer);
```

可以在图 11-28 中看到运行效果。我们只是使用"球"图标作为占位符,这里假设你的用户界面设计者会为你的应用提供合适的图标。

我们不建议改变整棵树中的字体或背景颜色,因为这实际上是外观设置的职责所在。

不过，改变树中个别节点的字体，以突显某些节点还是很有用的。如果仔细观察图11-28，你会看到抽象类是设成斜体字的。

为了改变单个节点的外观，需要安装一个树单元格绘制器。树单元格绘制器与我们在本章前一节讨论的列表单元格绘制器很相似。TreeCellRenderer 接口只有下面这个单一方法：

```
Component getTreeCellRendererComponent(JTree tree, Object value, boolean selected,
    boolean expanded, boolean leaf, int row, boolean hasFocus)
```

DefaultTreeCellRenderer 类的 getTreeCellRendererComponent 方法返回的是 this，换句话说，就是一个标签（DefaultTreeCellRenderer 类继承了 JLabel 类）。如果要定制一个构件，需要继承 DefaultTreeCellRenderer 类。按照以下方式覆盖 getTreeCell RendererComponent 方法：调用超类中的方法，以便准备标签的数据，然后定制标签属性，最后返回 this。

```
class MyTreeCellRenderer extends DefaultTreeCellRenderer
{
    public Component getTreeCellRendererComponent(JTree tree, Object value, boolean selected,
        boolean expanded, boolean leaf, int row, boolean hasFocus)
    {
        Component comp = super.getTreeCellRendererComponent(tree, value, selected,
            expanded, leaf, row, hasFocus);
        DefaultMutableTreeNode node = (DefaultMutableTreeNode) value;
        look at node.getUserObject();
        Font font = appropriate font;
        comp.setFont(font);
        return comp;
    }
};
```

> ⚠️ **警告**：getTreeCellRendererComponent 方法的 value 参数是节点对象，而不是用户对象！请记住，用户对象是 DefaultMutableTreeNode 的一个特性，而 JTree 可以包含任意类型的节点。如果树使用的是 DefaultMutableTreeNode 节点，那么必须在第二个步骤中获取这个用户对象，正如我们在上一个代码示例中所做的那样。

> ⚠️ **警告**：DefaultTreeCellRenderer 为所有节点使用的是相同的标签对象，仅仅是为每个节点改变标签文本而已。如果想为某个特定节点更改字体，那么必须在该方法再次调用的时候将它设置回默认值。否则，随后的所有节点都会以更改过的字体进行绘制！见程序清单11-10中的程序代码，看看它是怎样将字体恢复到其默认值的。

根据 Class 对象有无 ABSTRACT 修饰符，程序清单11-10 中的 ClassNameTreeCellRenderer 会将类名设置为标准字体或斜体字体。我们不想设置成特殊的字体，因为我们不想改变任何通常用于显示标签的字体外观。因此，我们使用来自标签本身的字体以及从它衍生而来的一个斜体字体。请回忆一下，全部的调用只返回一个共享的单一的 JLabel 对象。因此，我们需要保存初始字体，并在下一次调用 gettreeCellRendererComponent 方法时将其恢复为初始值。

同时，注意一下我们是如何改变 ClassTreeFrame 构造器中的节点图标的。

API javax.swing.tree.DefaultMutableTreeNode 1.2

- Enumeration breadthFirstEnumeration()
- Enumeration depthFirstEnumeration()
- Enumeration preOrderEnumeration()
- Enumeration postOrderEnumeration()

返回枚举对象，用于按照某种特定顺序访问树模型中的所有节点。在广度优先遍历中，先访问离根节点更近的子节点，再访问那些离根节点远的节点。在深度优先遍历中，先访问一个节点的所有子节点，然后再访问它的兄弟节点。postOrderEnumeration 方法与 depthFirstEnumeration 基本上相似。除了先访问父节点，后访问子节点之外，先序遍历和后序遍历基本上一样。

API javax.swing.tree.TreeCellRenderer 1.2

- Component getTreeCellRendererComponent(JTree tree, Object value, boolean selected, boolean expanded, boolean leaf, int row, boolean hasFocus)

返回一个 paint 方法被调用的构件，以便绘制树的一个单元格。

参数：tree   包含要绘制节点的树
     value   要绘制的节点
     selected   如果该节点是当前选定的节点，则为 true
     expanded   如果该节点的子节点可见，则为 true
     leaf   如果该节点应该显示为叶节点，则为 true
     row   显示包含该节点的那行
     hasFocus   如果当前选定的节点拥有输入焦点，则为 true

API javax.swing.tree.DefaultTreeCellRenderer 1.2

- void setLeafIcon(Icon icon)
- void setOpenIcon(Icon icon)
- void setClosedIcon(Icon icon)

设置叶节点、展开节点以及折叠节点的显示图标。

11.2.4 监听树事件

通常情况下，一个树构件会成对地伴随着其他某个构件。当用户选定了一些树节点时，某些信息就会在其他窗口中显示出来。参见图 11-29 的示例。当用户选定一个类时，这个类的实例及静态变量信息就会在右边的文本区显示出来。

为了获得这项功能，可以安装一个

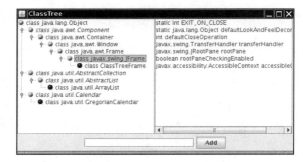

图 11-29　一个类浏览器

树选择监听器。该监听器必须实现 TreeSelection-Listener 接口，这是一个只有下面这个单一方法的接口：

```
void valueChanged(TreeSelectionEvent event)
```

每当用户选定或者撤销选定树节点的时候，这个方法就会被调用。

可以按照下面这种通常方式向树中添加监听器：

```
tree.addTreeSelectionListener(listener);
```

可以设定是否允许用户选定一个单一的节点、连续区间内的节点或者一个任意的、可能不连续的节点集。JTree 类使用 TreeSelectionModel 来管理节点的选择。必须检索整个模型，以便将选择状态设置为 SINGLE_TREE_SELECTION、CONTIGUOUS_TREE_SELECTION 或 DISCONTIGUOUS_TREE_SELECTION 三种状态之一。（在默认情况下是非连续的选择模式。）例如，在我们的类浏览器中，我们希望只允许选择单个类：

```
int mode = TreeSelectionModel.SINGLE_TREE_SELECTION;
tree.getSelectionModel().setSelectionMode(mode);
```

除了设置选择模式之外，并不需要担心树的选择模型。

> 📖 **注释**：用户怎样选定多个选项则依赖于外观。在 Metal 外观中，按下 CTRL 键，同时点击一个选项将它添加到选项集中，如果当前已经选定了该选项，则将其从选项集中删除。按下 SHIFT 键，同时点击一个选项，可以选定一个选项范围，它从先前已选定的选项延伸到新选定的选项。

要找出当前的选项集，可以用 getSelectionPaths 方法来查询树：

```
TreePath[] selectedPaths = tree.getSelectionPaths();
```

如果想限制用户只能做单项选择，那么可以使用便捷的 getSelectionPath 方法，它将返回第一个被选择的路径，或者是 null（如果没有任何路径被选）。

> ⚠ **警告**：TreeSelectionEvent 类具有一个 getPaths 方法，它将返回一个 TreePath 对象数组，但是该数组描述的是选项集的变化，而不是当前的选项集。

程序清单 11-10 显示了类树这个程序的窗体类。该程序可以显示继承的层次结构，并且将抽象类定制显示为斜体字（参见程序清单 11-11 的单元格绘制器）。可以在窗体底部的文本框中输入任何类名，按下 Enter 键或者点击"Add"按钮，将该类及其超类添加到树中。必须输入完整的包名，例如 java.util.ArrayList。

程序清单 11-10　treeRender/ClassTreeFrame.java

```
1  package treeRender;
2
3  import java.awt.*;
4  import java.awt.event.*;
5  import java.lang.reflect.*;
6  import java.util.*;
```

```java
import javax.swing.*;
import javax.swing.tree.*;

/**
 * This frame displays the class tree, a text field, and an "Add" button to add more classes
 * into the tree.
 */
public class ClassTreeFrame extends JFrame
{
   private static final int DEFAULT_WIDTH = 400;
   private static final int DEFAULT_HEIGHT = 300;

   private DefaultMutableTreeNode root;
   private DefaultTreeModel model;
   private JTree tree;
   private JTextField textField;
   private JTextArea textArea;

   public ClassTreeFrame()
   {
      setSize(DEFAULT_WIDTH, DEFAULT_HEIGHT);

      // the root of the class tree is Object
      root = new DefaultMutableTreeNode(java.lang.Object.class);
      model = new DefaultTreeModel(root);
      tree = new JTree(model);

      // add this class to populate the tree with some data
      addClass(getClass());

      // set up node icons
      var renderer = new ClassNameTreeCellRenderer();
      renderer.setClosedIcon(new ImageIcon(getClass().getResource("red-ball.gif")));
      renderer.setOpenIcon(new ImageIcon(getClass().getResource("yellow-ball.gif")));
      renderer.setLeafIcon(new ImageIcon(getClass().getResource("blue-ball.gif")));
      tree.setCellRenderer(renderer);

      // set up selection mode
      tree.addTreeSelectionListener(event ->
         {
            // the user selected a different node--update description
            TreePath path = tree.getSelectionPath();
            if (path == null) return;
            var selectedNode = (DefaultMutableTreeNode) path.getLastPathComponent();
            Class<?> c = (Class<?>) selectedNode.getUserObject();
            String description = getFieldDescription(c);
            textArea.setText(description);
         });
      int mode = TreeSelectionModel.SINGLE_TREE_SELECTION;
      tree.getSelectionModel().setSelectionMode(mode);

      // this text area holds the class description
      textArea = new JTextArea();
```

```java
          // add tree and text area
          var panel = new JPanel();
          panel.setLayout(new GridLayout(1, 2));
          panel.add(new JScrollPane(tree));
          panel.add(new JScrollPane(textArea));

          add(panel, BorderLayout.CENTER);

          addTextField();
       }

       /**
        * Add the text field and "Add" button to add a new class.
        */
       public void addTextField()
       {
          var panel = new JPanel();

          ActionListener addListener = event ->
             {
                // add the class whose name is in the text field
                try
                {
                   String text = textField.getText();
                   addClass(Class.forName(text)); // clear text field to indicate success
                   textField.setText("");
                }
                catch (ClassNotFoundException e)
                {
                   JOptionPane.showMessageDialog(null, "Class not found");
                }
             };

          // new class names are typed into this text field
          textField = new JTextField(20);
          textField.addActionListener(addListener);
          panel.add(textField);

          var addButton = new JButton("Add");
          addButton.addActionListener(addListener);
          panel.add(addButton);

          add(panel, BorderLayout.SOUTH);
       }

       /**
        * Finds an object in the tree.
        * @param obj the object to find
        * @return the node containing the object or null if the object is not present in the tree
        */
       public DefaultMutableTreeNode findUserObject(Object obj)
       {
          // find the node containing a user object
```

```java
115         var e = (Enumeration<TreeNode>) root.breadthFirstEnumeration();
116         while (e.hasMoreElements())
117         {
118            var node = (DefaultMutableTreeNode) e.nextElement();
119            if (node.getUserObject().equals(obj)) return node;
120         }
121         return null;
122      }
123
124      /**
125       * Adds a new class and any parent classes that aren't yet part of the tree.
126       * @param c the class to add
127       * @return the newly added node
128       */
129      public DefaultMutableTreeNode addClass(Class<?> c)
130      {
131         // add a new class to the tree
132
133         // skip non-class types
134         if (c.isInterface() || c.isPrimitive()) return null;
135
136         // if the class is already in the tree, return its node
137         DefaultMutableTreeNode node = findUserObject(c);
138         if (node != null) return node;
139
140         // class isn't present--first add class parent recursively
141
142         Class<?> s = c.getSuperclass();
143
144         DefaultMutableTreeNode parent;
145         if (s == null) parent = root;
146         else parent = addClass(s);
147
148         // add the class as a child to the parent
149         var newNode = new DefaultMutableTreeNode(c);
150         model.insertNodeInto(newNode, parent, parent.getChildCount());
151
152         // make node visible
153         var path = new TreePath(model.getPathToRoot(newNode));
154         tree.makeVisible(path);
155
156         return newNode;
157      }
158
159      /**
160       * Returns a description of the fields of a class.
161       * @param the class to be described
162       * @return a string containing all field types and names
163       */
164      public static String getFieldDescription(Class<?> c)
165      {
166         // use reflection to find types and names of fields
167         var r = new StringBuilder();
168         Field[] fields = c.getDeclaredFields();
```

```java
169      for (int i = 0; i < fields.length; i++)
170      {
171         Field f = fields[i];
172         if ((f.getModifiers() & Modifier.STATIC) != 0) r.append("static ");
173         r.append(f.getType().getName());
174         r.append(" ");
175         r.append(f.getName());
176         r.append("\n");
177      }
178      return r.toString();
179   }
180 }
```

程序清单 11-11   treeRender/ClassNameTreeCellRenderer.java

```java
1  package treeRender;
2
3  import java.awt.*;
4  import java.lang.reflect.*;
5  import javax.swing.*;
6  import javax.swing.tree.*;
7
8  /**
9   * This class renders a class name either in plain or italic. Abstract classes are italic.
10  */
11 public class ClassNameTreeCellRenderer extends DefaultTreeCellRenderer
12 {
13    private Font plainFont = null;
14    private Font italicFont = null;
15
16    public Component getTreeCellRendererComponent(JTree tree, Object value, boolean selected,
17          boolean expanded, boolean leaf, int row, boolean hasFocus)
18    {
19       super.getTreeCellRendererComponent(tree, value, selected, expanded, leaf,
20             row, hasFocus);
21       // get the user object
22       var node = (DefaultMutableTreeNode) value;
23       Class<?> c = (Class<?>) node.getUserObject();
24
25       // the first time, derive italic font from plain font
26       if (plainFont == null)
27       {
28          plainFont = getFont();
29          // the tree cell renderer is sometimes called with a label that has a null font
30          if (plainFont != null) italicFont = plainFont.deriveFont(Font.ITALIC);
31       }
32
33       // set font to italic if the class is abstract, plain otherwise
34       if ((c.getModifiers() & Modifier.ABSTRACT) == 0) setFont(plainFont);
35       else setFont(italicFont);
36       return this;
37    }
38 }
```

这个程序用到了一点小小的技巧，它是通过反射机制来构建这棵类树的。这项操作包含在 addClass 方法内。(细节倒不那么重要，在这个例子中，我们之所以使用类树，是因为继承树不需要怎么费劲地编码就能生成一棵丰满的树。如果想在自己的应用中显示树，那么你需要准备自己的层次结构数据的来源。) 该方法使用广度优先的搜索算法，通过调用我们在前一节实现的 findUserObject 方法，来确定当前的类是否已经存在于树中。如果这个类还不存在于树中，那么我们将其超类添加到这棵树中，然后将新节点作为它的子节点，并使该节点成为可见的。

在选择树的一个节点时，右侧的文本域将填充为选中的类的属性。在窗体构造器中，限制用户只能进行单个选项的选择，并添加了一个树选择监听器。当调用 valueChanged 方法时，我们忽略它的事件参数，只向该树询问当前的选定路径。正如通常情况那样，我们必须获得路径中的最后一个节点，并且查看它的用户对象。然后调用 getFieldDescription 方法，该方法使用反射机制将所选类的所有属性组装成一个字符串。

API javax.swing.JTree 1.2

- TreePath getSelectionPath()
- TreePath[] getSelectionPaths()

  返回第一个选定的路径，或者一个包含所有选定节点的数组。如果没有选定任何路径，这两个方法都返回为 null。

API javax.swing.event.TreeSelectionListener 1.2

- void valueChanged(TreeSelectionEvent event)

  每当选定节点或撤销选定的时候，该方法就被调用。

API javax.swing.event.TreeSelectionEvent 1.2

- TreePath getPath()
- TreePath[] getPaths()

  获取在该选择事件中已经发生更改的第一个路径或所有路径。如果你想知道当前的选择路径，而不是选择路径的更改情况，那么应该调用 JTree.getSelectionPaths。

11.2.5 定制树模型

在最后一个示例中，我们实现了一个能够查看变量内容的程序，正如调试器所做的那样（参见图 11-30）。

在继续深入之前，请先编译运行这个示例程序。其中每个节点对应于一个实例域。如果该域是一个对象，那么可以展开该节点以便查看它自己的实例域。该程序会审视窗体中的内容。如果你浏览了好几个实例域，那么你将会发现一些熟悉的类，还会对复杂的 Swing 用户界面构件有所了解。

图 11-30 对象查看树

该程序的不同之处在于它的树并没有使用 DefaultTreeModel。如果你已经拥有按照层次结构组织的数据，那么你可能并不想花精力去再创建一棵副本树，而且创建副本树还要担心怎样保持两棵树的一致性。这正是我们要讨论的情形：通过对象的引用，被审视的对象已经彼此连接起来了，因此在这里就不需要复制这种连接结构了。

TreeModel 接口只有几个方法。第一组方法使得 JTree 能够按照先是根节点，然后是子节点的顺序找到树中的节点。JTree 类只在用户真正展开一个节点的时候才会调用这些方法。

```
Object getRoot()
int getChildCount(Object parent)
Object getChild(Object parent, int index)
```

这个示例显示了为什么 TreeModel 接口像 JTree 类那样，不需要明确的用于描述节点的概念。根节点和子节点可以是任何对象，TreeModel 负责告知 JTree 它们是怎样联系起来的。

TreeModel 接口的下一个方法与 getChild 相反：

```
int getIndexOfChild(Object parent, Object child)
```

实际上，这个方法可以用前面的三个方法实现，参见程序清单 11-12 中的代码。

程序清单 11-12 treeModel/ObjectInspectorFrame.java

```java
 1  package treeModel;
 2
 3  import java.awt.*;
 4  import javax.swing.*;
 5
 6  /**
 7   * This frame holds the object tree.
 8   */
 9  public class ObjectInspectorFrame extends JFrame
10  {
11     private JTree tree;
12     private static final int DEFAULT_WIDTH = 400;
13     private static final int DEFAULT_HEIGHT = 300;
14
15     public ObjectInspectorFrame()
16     {
17        setSize(DEFAULT_WIDTH, DEFAULT_HEIGHT);
18
19        // we inspect this frame object
20
21        var v = new Variable(getClass(), "this", this);
22        var model = new ObjectTreeModel();
23        model.setRoot(v);
24
25        // construct and show tree
26
27        tree = new JTree(model);
28        add(new JScrollPane(tree), BorderLayout.CENTER);
29     }
30  }
```

树模型会告诉 JTree 哪些节点应该显示成叶节点：

```
boolean isLeaf(Object node)
```

如果你的代码更改了树模型，那么必须告知这棵树以便它能够对自己进行重新绘制。树是将它自己作为一个 TreeModelListener 添加到模型中的，因此，模型必须支持通常的监听器管理方法：

```
void addTreeModelListener(TreeModelListener l)
void removeTreeModelListener(TreeModelListener l)
```

可以在程序清单 11-13 中看到这些方法的具体实现。

程序清单 11-13　treeModel/ObjectTreeModel.java

```java
 1  package treeModel;
 2
 3  import java.lang.reflect.*;
 4  import java.util.*;
 5  import javax.swing.event.*;
 6  import javax.swing.tree.*;
 7
 8  /**
 9   * This tree model describes the tree structure of a Java object. Children are the objects
10   * that are stored in instance variables.
11   */
12  public class ObjectTreeModel implements TreeModel
13  {
14     private Variable root;
15     private EventListenerList listenerList = new EventListenerList();
16
17     /**
18      * Constructs an empty tree.
19      */
20     public ObjectTreeModel()
21     {
22        root = null;
23     }
24
25     /**
26      * Sets the root to a given variable.
27      * @param v the variable that is being described by this tree
28      */
29     public void setRoot(Variable v)
30     {
31        Variable oldRoot = v;
32        root = v;
33        fireTreeStructureChanged(oldRoot);
34     }
35
36     public Object getRoot()
37     {
38        return root;
39     }
40
```

```java
41    public int getChildCount(Object parent)
42    {
43       return ((Variable) parent).getFields().size();
44    }
45
46    public Object getChild(Object parent, int index)
47    {
48       ArrayList<Field> fields = ((Variable) parent).getFields();
49       var f = (Field) fields.get(index);
50       Object parentValue = ((Variable) parent).getValue();
51       try
52       {
53          return new Variable(f.getType(), f.getName(), f.get(parentValue));
54       }
55       catch (IllegalAccessException e)
56       {
57          return null;
58       }
59    }
60
61    public int getIndexOfChild(Object parent, Object child)
62    {
63       int n = getChildCount(parent);
64       for (int i = 0; i < n; i++)
65          if (getChild(parent, i).equals(child)) return i;
66       return -1;
67    }
68
69    public boolean isLeaf(Object node)
70    {
71       return getChildCount(node) == 0;
72    }
73
74    public void valueForPathChanged(TreePath path, Object newValue)
75    {
76    }
77
78    public void addTreeModelListener(TreeModelListener l)
79    {
80       listenerList.add(TreeModelListener.class, l);
81    }
82
83    public void removeTreeModelListener(TreeModelListener l)
84    {
85       listenerList.remove(TreeModelListener.class, l);
86    }
87
88    protected void fireTreeStructureChanged(Object oldRoot)
89    {
90       var event = new TreeModelEvent(this, new Object[] { oldRoot });
91       for (TreeModelListener l : listenerList.getListeners(TreeModelListener.class))
92          l.treeStructureChanged(event);
93    }
94 }
```

当模型修改了树的内容时,它会调用 TreeModelListener 接口中下面 4 个方法中的某一个:

```
void treeNodesChanged(TreeModelEvent e)
void treeNodesInserted(TreeModelEvent e)
void treeNodesRemoved(TreeModelEvent e)
void treeStructureChanged(TreeModelEvent e)
```

TreeModelEvent 对象用于描述修改的位置。对描述插入或移除事件的树模型事件进行组装的细节是相当技术性的。如果树中确实有要添加或移除的节点,只需要考虑如何触发这些事件。在程序清单 11-12 中,我们展示了怎么触发一个事件:将根节点替换为一个新的对象。

> ✓ **提示**:为了简化事件触发的代码,我们使用了 javax.swing.EventListenerList 这个使用方便、能够收集监听器的类。程序清单 11-13 中最后 3 个方法展示了如何使用这个类。

最后,如果用户要编辑树节点,那么模型会随着这种修改而被调用:

```
void valueForPathChanged(TreePath path, Object newValue)
```

如果不允许编辑,则永远不会调用到该方法。

如果不支持编辑功能,那么构建一个树模型就变得相当容易了。我们要实现下面 3 个方法:

```
Object getRoot()
int getChildCount(Object parent)
Object getChild(Object parent, int index)
```

这 3 个方法用于描述树的结构。还要提供另外 5 个方法的常规实现,如程序清单 10-16 那样,然后就可以准备显示你的树了。

现在让我们转向示例程序的具体实现,我们的树将包含类型为 Variable 的对象。

> 📖 **注释**:一旦使用了 DefaultTreeModel,我们的节点就可以具有类型为 DefaultMutable-TreeNode、用户对象类型为 Variable 的对象。

例如,假设我们查看下面这个变量

```
Employee joe;
```

该变量的类型为 Employee.class,名字为 joe,值为对象引用 joe 的值。在程序清单 11-14 中,我们定义了 Variable 这个类,用来描述程序中的变量:

```
var v = new Variable(Employee.class, "joe", joe);
```

程序清单 11-14　treeModel/Variable.java

```
1  package treeModel;
2
3  import java.lang.reflect.*;
4  import java.util.*;
5
6  /**
7   * A variable with a type, name, and value.
8   */
```

```java
 9  public class Variable
10  {
11     private Class<?> type;
12     private String name;
13     private Object value;
14     private ArrayList<Field> fields;
15
16     /**
17      * Construct a variable.
18      * @param aType the type
19      * @param aName the name
20      * @param aValue the value
21      */
22     public Variable(Class<?> aType, String aName, Object aValue)
23     {
24        type = aType;
25        name = aName;
26        value = aValue;
27        fields = new ArrayList<>();
28
29        // find all fields if we have a class type except we don't expand strings and
30        // null values
31
32        if (!type.isPrimitive() && !type.isArray() && !type.equals(String.class)
33              && value != null)
34        {
35           // get fields from the class and all superclasses
36           for (Class<?> c = value.getClass(); c != null; c = c.getSuperclass())
37           {
38              Field[] fs = c.getDeclaredFields();
39              AccessibleObject.setAccessible(fs, true);
40
41              // get all nonstatic fields
42              for (Field f : fs)
43                 if ((f.getModifiers() & Modifier.STATIC) == 0) fields.add(f);
44           }
45        }
46     }
47
48     /**
49      * Gets the value of this variable.
50      * @return the value
51      */
52     public Object getValue()
53     {
54        return value;
55     }
56
57     /**
58      * Gets all nonstatic fields of this variable.
59      * @return an array list of variables describing the fields
60      */
61     public ArrayList<Field> getFields()
62     {
```

```
63        return fields;
64     }
65
66     public String toString()
67     {
68        String r = type + " " + name;
69        if (type.isPrimitive()) r += "=" + value;
70        else if (type.equals(String.class)) r += "=" + value;
71        else if (value == null) r += "=null";
72        return r;
73     }
74  }
```

如果该变量的类型为基本类型，必须为这个值使用对象包装器。

```
new Variable(double.class, "salary", new Double(salary));
```

如果变量的类型是一个类，那么该变量就会拥有一些域。使用反射机制可以将所有域枚举出来，并将它们收集存放到一个 ArrayList 中。因为 Class 类的 getFields 方法不返回超类的任何域，因此还必须调用超类中的 getFields 方法，可以在 Variable 构造器中找到这些代码。Variable 类的 getFields 方法将返回一个包含了各类域的一个数组。最后，Variable 类的 toString 方法将节点格式化为标签，这个标签通常包含变量的类型和名称。如果变量不是一个类，那么该标签还将包含变量的值。

> **注释**：如果类型是一个数组，那么我们不会显示数组中的元素。这并不难实现，因此我们就把它留作众所周知的"读者练习"了。

让我们继续介绍树模型，头两个方法很简单。

```
public Object getRoot()
{
   return root;
}

public int getChildCount(Object parent)
{
   return ((Variable) parent).getFields().size();
}
```

getChild 方法返回一个新的 Variable 对象，用于描述给定索引位置上的域。Field 类的 getType 方法和 getName 方法用于产生域的类型和名称。通过使用反射机制，就可以按照 f.get(parentValue) 这种方式读取域的值。该方法可以抛出一个异常 IllegalAccessException，不过，我们可以让所有域在 Variable 构造器中都是可访问的，这样在实际应用中就不会发生这种抛出异常的情况。

下面是 getChild 方法的完整代码。

```
public Object getChild(Object parent, int index)
{
   ArrayList fields = ((Variable) parent).getFields();
   var f = (Field) fields.get(index);
```

```java
         Object parentValue = ((Variable) parent).getValue();
         try
         {
            return new Variable(f.getType(), f.getName(), f.get(parentValue));
         }
         catch (IllegalAccessException e)
         {
            return null;
         }
      }
```

这 3 个方法展示了对象树到 JTree 构件之间的结构，其余的方法是一些常规方法，源代码请见程序清单 11-13。

关于该树模型，有一个不同寻常之处：它实际上描述的是一棵无限树。可以通过追踪 WeakReference 对象来证实这一点。当你点击名字为 referent 的变量时，它会引导你回到初始的对象。你将获得一棵相同的子树，并且可以再次展开它的 WeakReference 对象，周而复始，无穷无尽。当然，你无法存储一个无限的节点集合。树模型只是在用户展开父节点时，按照需要来产生这些节点。

程序清单 11-12 展示了样例程序的框体类。

API javax.swing.tree.TreeModel 1.2

- Object getRoot()
  返回根节点。
- int getChildCount(Object parent)
  获取 parent 节点的子节点个数。
- Object getChild(Object parent, int index)
  获取给定索引位置上 parent 节点的子节点。
- int getIndexOfChild(Object parent, Object child)
  获取 parent 节点的子节点 child 的索引位置。如果在树模型中 child 节点不是 parent 的一个子节点，则返回 -1。
- boolean isLeaf(Object node)
  如果节点 node 从概念上讲是一个叶节点，则返回 true。
- void addTreeModelListener(TreeModelListener l)
- void removeTreeModelListener(TreeModelListener l)
  当模型中的信息发生变化时，告知添加和移除监听器。
- void valueForPathChanged(TreePath path, Object newValue)
  当一个单元格编辑器修改了节点值的时候，该方法被调用。
  参数：path   到被编辑节点的树路径
       newValue 编辑器返回的修改值

API javax.swing.event.TreeModelListener 1.2

- void treeNodesChanged(TreeModelEvent e)

- void treeNodesInserted(TreeModelEvent e)
- void treeNodesRemoved(TreeModelEvent e)
- void treeStructureChanged(TreeModelEvent e)

  如果树被修改过，树模型将调用该方法。

API javax.swing.event.TreeModelEvent 1.2

- TreeModelEvent(Object eventSource, TreePath node)

  构建一个树模型事件。

  参数： eventSource 产生该事件的树模型

  　　　node 到达要修改节点的树路径

11.3 高级 AWT

Graphics 类有多种方法可以用来创建简单的图形。这些方法对于简单的 applet 和应用来说已经绰绰有余了，但是当你创建复杂的图形或者需要全面控制图形的外观时，它们就显得力不从心了。Java 2D API 是一个更加成熟的类库，可以用它产生高质量的图形。下面我们将概要地介绍一下该 API。

11.3.1 绘图操作流程

在最初的 JDK 1.0 中，用来绘制形状的是一种非常简单的机制，即选择颜色和画图的模式，并调用 Graphics 类的各种方法，比如 drawRect 或者 fillOval。而 Java 2D API 支持更多的功能：

- 可以很容易地绘制各式各样的形状。
- 可以控制绘制形状的笔画，即控制跟踪形状边界的绘图笔。
- 可以用单色、变化的色调和重复的模式来填充各种形状。
- 可以使用变换法，对各种形状进行移动、缩放、旋转和拉伸。
- 可以对形状进行剪切，将其限制在任意的区域内。
- 可以选择各种组合规则，来描述如何将新形状的像素与现有的像素组合起来。
- 可以提供绘制图形提示，以便在速度与绘图质量之间实现平衡。

如果要绘制一个形状，可以按照如下步骤操作：

1. 获得一个 Graphics2D 类的对象，该类是 Graphics 类的子类。自 Java SE 1.2 以来，paint 和 paintComponent 等方法就能够自动地接收一个 Graphics2D 类的对象，这时可以直接使用如下的转型：

```
public void paintComponent(Graphics g)
{
   var g2 = (Graphics2D) g;
   . . .
}
```

2. 使用 setRenderingHints 方法来设置绘图提示，它提供了速度与绘图质量之间的一种平衡。

```
RenderingHints hints = ...;
g2.setRenderingHints(hints);
```

3. 使用 setStroke 方法来设置笔画，笔画用于绘制形状的边框。可以选择边框的粗细和线段的虚实。

```
Stroke stroke = ...;
g2.setStroke(stroke);
```

4. 使用 setPaint 方法来设置着色法，着色法用于填充诸如笔画路径或者形状内部等区域的颜色。可以创建单色、渐变色或者平铺的填充模式。

```
Paint paint = ...;
g2.setPaint(paint);
```

5. 使用 clip 方法来设置剪切区域。

```
Shape clip = ...;
g2.clip(clip);
```

6. 使用 transform 方法设置一个从用户空间到设备空间的变换方式。如果使用变换方式比使用像素坐标更容易定义在定制坐标系统中的形状，那么就可以使用变换方式。

```
AffineTransform transform = ...;
g2.transform(transform);
```

7. 使用 setComposite 方法设置一个组合规则，用来描述如何将新像素与现有的像素组合起来。

```
Composite composite = ...;
g2.setComposite(composite);
```

8. 建立一个形状，Java 2D API 提供了用来组合各种形状的许多形状对象和方法。

```
Shape shape = ...;
```

9. 绘制或者填充该形状。如果要绘制该形状，那么它的边框就会用笔画画出来。如果要填充该形状，那么它的内部就会被着色。

```
g2.draw(shape);
g2.fill(shape);
```

当然，在许多实际的环境中，并不需要采用所有这些操作步骤。Java 2D 图形上下文中有合理的默认设置。只有当你确实想要改变设置时，再去修改这些默认设置。

在下面的几节中，我们将要介绍如何描绘形状、笔画、着色、变换及组合的规则。

各种不同的 set 方法只是用于设置 2D 图形上下文的状态，它们并不进行任何实际的绘图操作。同样，在构建 shape 对象时，也不进行任何绘图操作。只有在调用 draw 或者 fill 方法时，才会绘制出图形的形状，而就在此刻，这个新的图形由绘图操作流程计算出来（参见图 11-31）。

在绘图流程中，需要以下这些操作步骤来绘制一个形状：

1. 用笔画画出形状的线条；
2. 对形状进行变换操作；

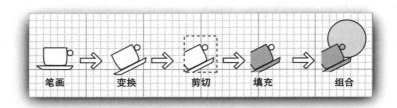

图 11-31 绘图操作流程

3. 对形状进行剪切。如果形状与剪切区域之间没有任何相交的地方，那么就不用执行该操作；

4. 对剪切后的形状进行填充；

5. 把填充后的形状与已有的形状进行组合（在图 11-31 中，圆形是已有像素部分，杯子的形状加在了它的上面）。

在下一节中，将会讲述如何对形状进行定义。然后，我们将转而对 2D 图形上下文设置进行介绍。

API java.awt.Graphics2D 1.2

- void draw(Shape s)

用当前的笔画来绘制给定形状的边框。

- void fill(Shape s)

用当前的着色方案来填充给定形状的内部。

11.3.2 形状

下面是 Graphics 类中绘制形状的若干方法：

```
drawLine
drawRectangle
drawRoundRect
draw3DRect
drawPolygon
drawPolyline
drawOval
drawArc
```

它们还有对应的 fill 方法，这些方法从 JDK 1.0 起就被纳入到 Graphics 类中了。Java 2D API 使用了一套完全不同的面向对象的处理方法，即不再使用方法，而是使用下面的这些类：

```
Line2D
Rectangle2D
RoundRectangle2D
Ellipse2D
Arc2D
QuadCurve2D
```

```
CubicCurve2D
GeneralPath
```

这些类全部都实现了 Shape 接口，我们将在下面各小节中一一审视它们。

11.3.2.1 形状类层次结构

Line2D、Rectangle2D、RoundRectangle2D、Ellipse2D 和 Arc2D 等这些类对应于 drawLine、drawRectangle、drawRoundRect、drawOval 和 drawArc 等方法。（"3D 矩形"的概念已经理所当然地过时了，因而没有与 draw3DRect 方法相对应的类。）Java 2D API 提供了两个补充类，即二次曲线类和三次曲线类。我们将在本节的后面部分阐释这些形状。Java 2D API 中没有任何 Polygon2D 类。相反，它用 GeneralPath 类来描述由线条、二次曲线、三次曲线构成的线条路径。可以使用 GeneralPath 来描述一个多边形；我们将在本节的后面部分对它进行介绍。

如果要绘制一个形状，首先要创建一个实现了 Shape 接口的类的对象，然后调用 Graphics2D 类的 draw 方法。

下面这些类：

```
Rectangle2D
RoundRectangle2D
Ellipse2D
Arc2D
```

都是从一个公共超类 RectangularShape 继承而来的。诚然，椭圆形和弧形都不是矩形，但是它们都有一个矩形的边界框（参见图 11-32）。

名字以"2D"结尾的每个类都有两个子类，用于指定坐标是 float 类型的还是 double 类型的。在本书的卷 I 中，我们已经介绍了 Rectangle2D.Float 和 Rectangle2D.Double。

其他类也使用了相同的模式，比如 Arc2D.Float 和 Arc2D.Double。

从内部来讲，所有的图形类使用的都是 float 类型的坐标，因为 float 类型的数

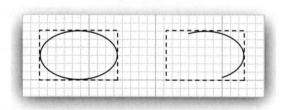

图 11-32 椭圆形和弧形的矩形边界框

占用较少的存储空间，而且它们有足够高的几何计算精度。然而，Java 编程语言使得对 float 类型的数的操作要稍微复杂些。由于这个原因，图形类的大多数方法使用的都是 double 类型的参数和返回值。只有在创建一个 2D 对象的时候，才需要选择究竟是使用带有 float 类型坐标的构造器，还是使用带有 double 类型坐标的构造器。例如：

```
var floatRect = new Rectangle2D.Float(5F, 10F, 7.5F, 15F);
var doubleRect = new Rectangle2D.Double(5, 10, 7.5, 15);
```

*Xxx*2D.Float 和 *Xxx*2D.Double 两个类都是 *Xxx*2D 类的子类，在对象被构建之后，再记住其确切的子类型实质上已经没有任何额外的好处了，因此可以将刚被构建的对象存储为一个超类变量，正如上面代码示例中所阐释的那样。

从这些类古怪的名字中就可以判断出，*Xxx*2D.Float 和 *Xxx*2D.Double 两个类同时也是 *Xxx*2D

类的内部类。这只是为了在语法上比较方便,以避免外部类的名字变得太长。

最后,还有一个 Point2D 类,它用 x 和 y 坐标来描述一个点。点对于定义形状非常有用,不过它们本身并不是形状。

图 11-33 显示了各个形状类之间的关系。不过图中省略了 Double 和 Float 子类,并且来自以前的 2D 类库的遗留类用灰色的填充色标识。

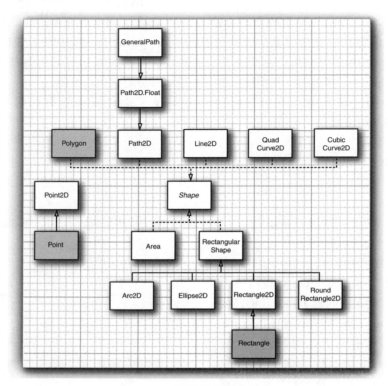

图 11-33 形状类之间的关系

11.3.2.2 使用形状类

我们在本书的卷 I 第 10 章中介绍了如何使用 Rectangle2D、Ellipse2D 和 Line2D 类的方法。本节将介绍如何建立其他的 2D 形状。

如果要建立一个 RoundRectangle2D 形状,应该设定左上角、宽度、高度及应该变成圆角的边角区的 x 和 y 的坐标尺寸(参见图 11-34)。例如,调用下面的方法:

```
var r = new RoundRectangle2D.Double(150, 200, 100, 50, 20, 20);
```

便产生了一个带圆角的矩形,每个角的圆半径为 20。

如果要建立一个弧形,首先应该设定边界框,接着设定它的起始角度和弧形跨越的角度(见图 11-35),并且设定弧形闭合的类型,即 Arc2D.OPEN、Arc2D.PIE 或者 Arc2D.CHORD 这几种类型中的一个。

```
var a = new Arc2D(x, y, width, height, startAngle, arcAngle, closureType);
```

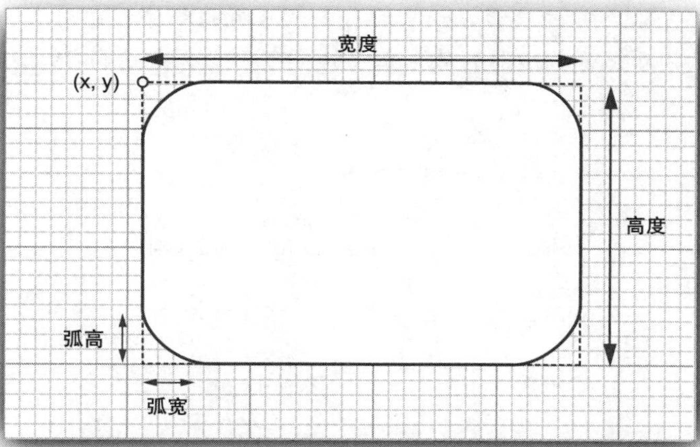

图 11-34　构建一个 RoundRectangle2D

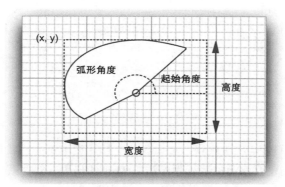

图 11-35　构建一个椭圆弧形

图 11-36 显示了几种弧形的类型。

> **警告**：如果弧形是椭圆的，那么弧形角的计算就不是很直接了。API 文档中描述到："角是相对于非正方形的矩形边框指定的，以使得 45 度总是落到了从椭圆中心指向矩形边框右上角的方向上。因此，如果矩形边框的一条轴比另一条轴明显长许多，那么弧形段的起始点和终止点就会与边框中的长轴斜交。"但是，文档中并没有说明如何计算这种"斜交"。下面是其细节：
>
> 假设弧形的中心是原点，而且点(x, y)在弧形上。那么我们可以用下面的公式来获得这个斜交角：
>
> ```
> skewedAngle = Math.toDegrees(Math.atan2(-y * height, x * width));
> ```
>
> 这个值介于 −180 到 180 之间。按照这种方式计算斜交的起始角和终止角，然后，计

算两个斜交角之间的差,如果起始角或角的差是负数,则加上360。之后,将起始角和角的差提供给弧形的构造器。如果运行本节末尾的程序,你用肉眼就能观察到这种计算所产生的用于弧形构造器的值是正确的。可参见本章图11-39。

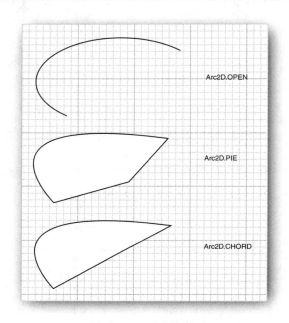

图 11-36 弧形的类型

Java 2D API 提供了对二次曲线和三次曲线的支持。在本章中,我们并不会深入介绍这些曲线的数学特征。我们建议你通过运行程序清单 11-1 的代码,对曲线的形状有一个感性的认识。正如在图 11-37 和图 11-38 中看到的那样,二次曲线和三次曲线是由两个端点和一个或两个控制点来设定的。移动控制点,曲线的形状就会改变。

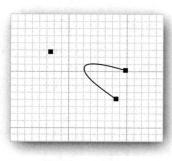

图 11-37 二次曲线

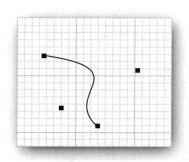

图 11-38 三次曲线

如果要构建二次曲线和三次曲线,需要给出两个端点和控制点的坐标。例如,

```
var q = new QuadCurve2D.Double(startX, startY, controlX, controlY, endX, endY);
var c = new CubicCurve2D.Double(startX, startY, control1X, control1Y,
   control2X, control2Y, endX, endY);
```

二次曲线不是非常灵活，所以实际上它并不常用。三次曲线（比如用 CubicCurve2D 类绘制的贝塞尔（Bézier）曲线）却是非常常用的。通过将三次曲线组合起来，使得连接点的各个斜率互相匹配，就能够创建复杂的、外观平滑的曲线形状。如果要了解这方面的详细信息，请参阅 James D. Foley、Andries van Dam 和 Steven K. Feiner 等人合作撰写的 *Computer Graphics: Principles and Practice*（第 3 版），Addison Wesley 出版社 2013 年出版。

可以建立线段、二次曲线和三次曲线的任意序列，并把它们存放到一个 GeneralPath 对象中去。可以用 moveTo 方法来指定路径的第一个坐标，例如，

```
var path = new GeneralPath();
path.moveTo(10, 20);
```

然后，可以通过调用 lineTo、quadTo 或 curveTo 三种方法之一来扩展路径，这些方法分别用线条、二次曲线或者三次曲线来扩展路径。如果要调用 lineTo 方法，需要提供它的端点。而对两个曲线方法的调用，应该先提供控制点，然后提供端点。例如，

```
path.lineTo(20, 30);
path.curveTo(control1X, control1Y, control2X, control2Y, endX, endY);
```

可以调用 closePath 方法来闭合路径，它能够绘制一条回到路径起始点的线条。

如果要绘制一个多边形，只需调用 moveTo 方法，以到达第一个拐角点，然后反复调用 lineTo 方法，以便到达其他的拐角点。最后调用 closePath 方法来闭合多边形。程序清单 11-15 更加详细地展示了构建多边形的方法。

程序清单 11-15　shape/ShapeTest.java

```java
 1  package shape;
 2
 3  import java.awt.*;
 4  import java.awt.event.*;
 5  import java.awt.geom.*;
 6  import java.util.*;
 7  import javax.swing.*;
 8
 9  /**
10   * This program demonstrates the various 2D shapes.
11   * @version 1.04 2018-05-01
12   * @author Cay Horstmann
13   */
14  public class ShapeTest
15  {
16     public static void main(String[] args)
17     {
18        EventQueue.invokeLater(() ->
19           {
20              var frame = new ShapeTestFrame();
21              frame.setTitle("ShapeTest");
```

```java
22              frame.setDefaultCloseOperation(JFrame.EXIT_ON_CLOSE);
23              frame.setVisible(true);
24           });
25    }
26 }
27
28 /**
29  * This frame contains a combo box to select a shape and a component to draw it.
30  */
31 class ShapeTestFrame extends JFrame
32 {
33    public ShapeTestFrame()
34    {
35       var comp = new ShapeComponent();
36       add(comp, BorderLayout.CENTER);
37       var comboBox = new JComboBox<ShapeMaker>();
38       comboBox.addItem(new LineMaker());
39       comboBox.addItem(new RectangleMaker());
40       comboBox.addItem(new RoundRectangleMaker());
41       comboBox.addItem(new EllipseMaker());
42       comboBox.addItem(new ArcMaker());
43       comboBox.addItem(new PolygonMaker());
44       comboBox.addItem(new QuadCurveMaker());
45       comboBox.addItem(new CubicCurveMaker());
46       comboBox.addActionListener(event ->
47          {
48             ShapeMaker shapeMaker = comboBox.getItemAt(comboBox.getSelectedIndex());
49             comp.setShapeMaker(shapeMaker);
50          });
51       add(comboBox, BorderLayout.NORTH);
52       comp.setShapeMaker((ShapeMaker) comboBox.getItemAt(0));
53       pack();
54    }
55 }
56
57 /**
58  * This component draws a shape and allows the user to move the points that define it.
59  */
60 class ShapeComponent extends JComponent
61 {
62    private static final Dimension PREFERRED_SIZE = new Dimension(300, 200);
63    private Point2D[] points;
64    private static Random generator = new Random();
65    private static int SIZE = 10;
66    private int current;
67    private ShapeMaker shapeMaker;
68
69    public ShapeComponent()
70    {
71       addMouseListener(new MouseAdapter()
72          {
73             public void mousePressed(MouseEvent event)
74             {
75                Point p = event.getPoint();
```

```java
76              for (int i = 0; i < points.length; i++)
77              {
78                 double x = points[i].getX() - SIZE / 2;
79                 double y = points[i].getY() - SIZE / 2;
80                 var r = new Rectangle2D.Double(x, y, SIZE, SIZE);
81                 if (r.contains(p))
82                 {
83                    current = i;
84                    return;
85                 }
86              }
87           }
88
89           public void mouseReleased(MouseEvent event)
90           {
91              current = -1;
92           }
93        });
94     addMouseMotionListener(new MouseMotionAdapter()
95        {
96           public void mouseDragged(MouseEvent event)
97           {
98              if (current == -1) return;
99              points[current] = event.getPoint();
100             repaint();
101          }
102       });
103    current = -1;
104 }
105
106 /**
107  * Set a shape maker and initialize it with a random point set.
108  * @param aShapeMaker a shape maker that defines a shape from a point set
109  */
110 public void setShapeMaker(ShapeMaker aShapeMaker)
111 {
112    shapeMaker = aShapeMaker;
113    int n = shapeMaker.getPointCount();
114    points = new Point2D[n];
115    for (int i = 0; i < n; i++)
116    {
117       double x = generator.nextDouble() * getWidth();
118       double y = generator.nextDouble() * getHeight();
119       points[i] = new Point2D.Double(x, y);
120    }
121    repaint();
122 }
123
124 public void paintComponent(Graphics g)
125 {
126    if (points == null) return;
127    var g2 = (Graphics2D) g;
128    for (int i = 0; i < points.length; i++)
129    {
```

```java
130            double x = points[i].getX() - SIZE / 2;
131            double y = points[i].getY() - SIZE / 2;
132            g2.fill(new Rectangle2D.Double(x, y, SIZE, SIZE));
133         }
134
135         g2.draw(shapeMaker.makeShape(points));
136      }
137
138      public Dimension getPreferredSize() { return PREFERRED_SIZE; }
139 }
140
141 /**
142  * A shape maker can make a shape from a point set. Concrete subclasses must return a shape in
143  * the makeShape method.
144  */
145 abstract class ShapeMaker
146 {
147    private int pointCount;
148
149    /**
150     * Constructs a shape maker.
151     * @param ointCount the number of points needed to define this shape
152     */
153    public ShapeMaker(int pointCount)
154    {
155       this.pointCount = pointCount;
156    }
157
158    /**
159     * Gets the number of points needed to define this shape.
160     * @return the point count
161     */
162    public int getPointCount()
163    {
164       return pointCount;
165    }
166
167    /**
168     * Makes a shape out of the given point set.
169     * @param p the points that define the shape
170     * @return the shape defined by the points
171     */
172    public abstract Shape makeShape(Point2D[] p);
173
174    public String toString()
175    {
176       return getClass().getName();
177    }
178 }
179
180 /**
181  * Makes a line that joins two given points.
182  */
183 class LineMaker extends ShapeMaker
```

```java
184 {
185     public LineMaker()
186     {
187         super(2);
188     }
189
190     public Shape makeShape(Point2D[] p)
191     {
192         return new Line2D.Double(p[0], p[1]);
193     }
194 }
195
196 /**
197  * Makes a rectangle that joins two given corner points.
198  */
199 class RectangleMaker extends ShapeMaker
200 {
201     public RectangleMaker()
202     {
203         super(2);
204     }
205
206     public Shape makeShape(Point2D[] p)
207     {
208         var s = new Rectangle2D.Double();
209         s.setFrameFromDiagonal(p[0], p[1]);
210         return s;
211     }
212 }
213
214 /**
215  * Makes a round rectangle that joins two given corner points.
216  */
217 class RoundRectangleMaker extends ShapeMaker
218 {
219     public RoundRectangleMaker()
220     {
221         super(2);
222     }
223
224     public Shape makeShape(Point2D[] p)
225     {
226         var s = new RoundRectangle2D.Double(0, 0, 0, 0, 20, 20);
227         s.setFrameFromDiagonal(p[0], p[1]);
228         return s;
229     }
230 }
231
232 /**
233  * Makes an ellipse contained in a bounding box with two given corner points.
234  */
235 class EllipseMaker extends ShapeMaker
236 {
237     public EllipseMaker()
```

```java
238    {
239       super(2);
240    }
241
242    public Shape makeShape(Point2D[] p)
243    {
244       var s = new Ellipse2D.Double();
245       s.setFrameFromDiagonal(p[0], p[1]);
246       return s;
247    }
248 }
249
250 /**
251  * Makes an arc contained in a bounding box with two given corner points, and with starting
252  * and ending angles given by lines emanating from the center of the bounding box and ending
253  * in two given points. To show the correctness of the angle computation, the returned shape
254  * contains the arc, the bounding box, and the lines.
255  */
256 class ArcMaker extends ShapeMaker
257 {
258    public ArcMaker()
259    {
260       super(4);
261    }
262
263    public Shape makeShape(Point2D[] p)
264    {
265       double centerX = (p[0].getX() + p[1].getX()) / 2;
266       double centerY = (p[0].getY() + p[1].getY()) / 2;
267       double width = Math.abs(p[1].getX() - p[0].getX());
268       double height = Math.abs(p[1].getY() - p[0].getY());
269
270       double skewedStartAngle = Math.toDegrees(Math.atan2(-(p[2].getY() - centerY) * width,
271          (p[2].getX() - centerX) * height));
272       double skewedEndAngle = Math.toDegrees(Math.atan2(-(p[3].getY() - centerY) * width,
273          (p[3].getX() - centerX) * height));
274       double skewedAngleDifference = skewedEndAngle - skewedStartAngle;
275       if (skewedStartAngle < 0) skewedStartAngle += 360;
276       if (skewedAngleDifference < 0) skewedAngleDifference += 360;
277
278       var s = new Arc2D.Double(0, 0, 0, 0,
279          skewedStartAngle, skewedAngleDifference, Arc2D.OPEN);
280       s.setFrameFromDiagonal(p[0], p[1]);
281
282       var g = new GeneralPath();
283       g.append(s, false);
284       var r = new Rectangle2D.Double();
285       r.setFrameFromDiagonal(p[0], p[1]);
286       g.append(r, false);
287       var center = new Point2D.Double(centerX, centerY);
288       g.append(new Line2D.Double(center, p[2]), false);
289       g.append(new Line2D.Double(center, p[3]), false);
290       return g;
291    }
```

```java
292    }
293
294    /**
295     * Makes a polygon defined by six corner points.
296     */
297    class PolygonMaker extends ShapeMaker
298    {
299       public PolygonMaker()
300       {
301          super(6);
302       }
303
304       public Shape makeShape(Point2D[] p)
305       {
306          var s = new GeneralPath();
307          s.moveTo((float) p[0].getX(), (float) p[0].getY());
308          for (int i = 1; i < p.length; i++)
309             s.lineTo((float) p[i].getX(), (float) p[i].getY());
310          s.closePath();
311          return s;
312       }
313    }
314
315    /**
316     * Makes a quad curve defined by two end points and a control point.
317     */
318    class QuadCurveMaker extends ShapeMaker
319    {
320       public QuadCurveMaker()
321       {
322          super(3);
323       }
324
325       public Shape makeShape(Point2D[] p)
326       {
327          return new QuadCurve2D.Double(p[0].getX(), p[0].getY(), p[1].getX(), p[1].getY(),
328             p[2].getX(), p[2].getY());
329       }
330    }
331
332    /**
333     * Makes a cubic curve defined by two end points and two control points.
334     */
335    class CubicCurveMaker extends ShapeMaker
336    {
337       public CubicCurveMaker()
338       {
339          super(4);
340       }
341
342       public Shape makeShape(Point2D[] p)
343       {
344          return new CubicCurve2D.Double(p[0].getX(), p[0].getY(), p[1].getX(), p[1].getY(),
345             p[2].getX(), p[2].getY(), p[3].getX(), p[3].getY());
```

```
346     }
347 }
```

普通路径没有必要一定要连接在一起,我们随时可以调用 moveTo 方法来建立一个新的路径段。

最后,可以使用 append 方法,向普通路径添加任意个 Shape 对象。如果新建的形状应该连接到路径的最后一个端点,那么 append 方法的第二个参数值就是 true,如果不应该连接,那么该参数值就是 false。例如,调用下面的方法:

```
Rectangle2D r = . . .;
path.append(r, false);
```

可以把矩形的边框添加到该路径中,但并不与现有的路径连接在一起。而下面的方法调用:

```
path.append(r, true);
```

则是在路径的终点和矩形的起点之间添加了一条直线,然后将矩形的边框添加到该路径中。

程序清单 11-15 中的程序使你能够构建许多示例路径。图 11-37 和图 11-38 显示了运行该程序的示例结果。你可以从组合框中选择一个形状绘制器,该程序包含的形状绘制器可以用来绘制:

- 直线;
- 矩形、圆角矩形和椭圆形;
- 弧形(除了显示弧形本身外,还可以显示矩形边框的线条和起始角度及结束角度);
- 多边形(使用 GeneralPath 方法);
- 二次曲线和三次曲线。

可以用鼠标来调整控制点。当移动控制点时,形状会连续地重绘。

该程序有些复杂,因为它可以用来处理多种不同的形状,并且支持对控制点的拖拽操作。

抽象超类 ShapeMaker 封装了形状绘制器类的共性特征。每个形状都拥有固定数量的控制点,用户可以在控制点周围随意移动,而 getPointCount 方法用于返回控制点的数量。下面这个抽象方法:

```
Shape makeShape(Point2D[] points)
```

将在给定控制点的当前位置的情况下,计算实际的形状。toString 方法用于返回类的名字,这样,ShapeMaker 对象就能够放置到一个 JComboBox 中。

为了激活控制点的拖拽特征,ShapePanel 类要同时处理鼠标事件和鼠标移动事件。当鼠标在一个矩形上面被按下时,那么拖拽鼠标就可以移动该矩形了。

大部分形状绘制器类都很简单,它们的 makeShape 方法只是用于构建和返回需要的形状。然而,当使用 ArcMaker 类的时候,需要计算弧形的变形起始角度和结束角度。此外,为了说明这些计算确实是正确的,返回的形状应该是包含该弧本身、矩形边框和从弧形中心到角度控制点之间的线条等的 GeneralPath(参见图 11-39)。

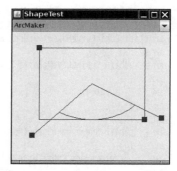

图 11-39 ShapeTest 程序的运行结果

API java.awt.geom.RoundRectangle2D.Double 1.2

- RoundRectangle2D.Double(double x, double y, double width, double height, double arcWidth, double arcHeight)

  用给定的矩形边框和弧形尺寸构建一个圆角矩形。参见图 11-34 有关 arcWidth 和 arcHeight 参数的解释。

API java.awt.geom.Arc2D.Double 1.2

- Arc2D.Double(double x, double y, double w, double h, double startAngle, double arcAngle, int type)

  用给定的矩形边框、起始角度、弧形角度和弧形类型构建一个弧形。startAngle 和 arcAngle 在图 11-35 中已做介绍，type 是 Arc2D.OPEN、Arc2D.PIE 和 Arc2D.CHORD 之一。

API java.awt.geom.QuadCurve2D.Double 1.2

- QuadCurve2D.Double(double x1, double y1, double ctrlx, double ctrly, double x2, double y2)

  用起始点、控制点和结束点构建一条二次曲线。

API java.awt.geom.CubicCurve2D.Double 1.2

- CubicCurve2D.Double(double x1, double y1, double ctrlx1, double ctrly1, double ctrlx2, double ctrly2, double x2, double y2)

  用起始点、两个控制点和结束点构建一条三次曲线。

API java.awt.geom.GeneralPath 1.2

- GeneralPath()

  构建一条空的普通路径。

API java.awt.geom.Path2D.Float 6

- void moveTo(float x, float y)

  使（x，y）成为当前点，也就是下一个线段的起始点。

- void lineTo(float x, float y)

- void quadTo(float ctrlx, float ctrly, float x, float y)

- void curveTo(float ctrl1x, float ctrl1y, float ctrl2x, float ctrl2y, float x, float y)

  从当前点绘制一个线条、二次曲线或者三次曲线到达结束点（x，y），并且使该结束点成为当前点。

API java.awt.geom.Path2D 6

- void append(Shape s, boolean connect)

  将给定形状的边框添加到普通路径中去。如果布尔型变量 connect 的值是 true，那么该普通路径的当前点与添加进来的形状的起始点之间用一条直线连接起来。

- void closePath()

  从当前点到路径的第一点之间绘制一条直线，从而使路径闭合。

11.3.3 区域

在上一节中，我们介绍了如何通过建立由线条和曲线构成的普通路径来绘制复杂的形状。通过使用足够数量的线条和曲线可以绘制出任何一种形状，例如，在屏幕上和打印文件上看到的字符的各种字体形状，都是由线条和三次曲线构成的。

有时候，使用各种不同形状的区域，比如矩形、多边形和椭圆形来建立形状，可能会更加容易描述。Java 2D API 支持四种区域几何作图（constructive area geometry）操作，用于将两个区域组合成一个区域。

- add：组合区域包含了所有位于第一个区域或第二个区域内的点。
- subtract：组合区域包含了所有位于第一个区域内的点，但是不包括任何位于第二个区域内的点。
- intersect：组合区域包含了所有既位于第一个区域内，又位于第二个区域内的点。
- exclusiveOr：组合区域包含了所有位于第一个区域内，或者是位于第二个区域内的所有点，但是这些点不能同时位于两个区域内。

图 11-40 显示了这些操作的结果。

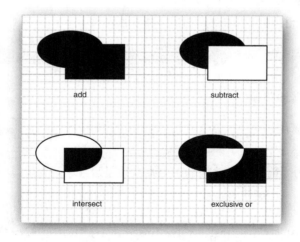

图 11-40 区域几何作图操作

如果要构建一个复杂的区域，可以使用下面的方法先创建一个默认的区域对象。

```
var a = new Area();
```

然后，将该区域和其他的形状组合起来：

```
a.add(new Rectangle2D.Double(. . .));
a.subtract(path);
. . .
```

Area 类实现了 Shape 接口。可以用 draw 方法勾勒出该区域的边界，或者使用 Graphics2D 类的 fill 方法给区域的内部着色。

API java.awt.geom.Area

- void add(Area other)
- void subtract(Area other)
- void intersect(Area other)
- void exclusiveOr(Area other)

对该区域和 other 所代表的另一个区域执行区域几何作图操作，并且将该区域设置为执行后的结果。

11.3.4 笔画

Graphics2D 类的 draw 操作通过使用当前选定的笔画来绘制一个形状的边界。在默认的情况下，笔画是一条宽度为一个像素的实线。可以通过调用 setStroke 方法来选定不同的笔画，此时要提供一个实现了 Stroke 接口的类的对象。Java 2D API 只定义了一个这样的类，即 BasicStroke 类。在本节中，我们将介绍 BasicStroke 类的功能。

你可以构建任意粗细的笔画。例如，下面的方法就绘制了一条粗细为 10 个像素的线条。

```
g2.setStroke(new BasicStroke(10.0F));
g2.draw(new Line2D.Double(. . .));
```

当一个笔画的粗细大于一个像素的宽度时，笔画的末端可采用不同的样式。图 11-41 显示了这些所谓的端头样式。端头样式有下面三种：

- 平头样式（butt cap）在笔画的末端处就结束了；
- 圆头样式（round cap）在笔画的末端处加了一个半圆；
- 方头样式（square cap）在笔画的末端处加了半个方块。

当两个较粗的笔画相遇时，有三种笔画的连接样式可供选择（参见图 11-42）：

图 11-41 笔画的端头样式

图 11-42 笔画的连接样式

- 斜连接（bevel join），用一条直线将两个笔画连接起来，该直线与两个笔画之间的夹角的平分线相垂直。
- 圆连接（round join），延长了每个笔画，并使其带有一个圆头。
- 斜尖连接（miter join），通过增加一个尖峰，从而同时延长了两个笔画。

斜尖连接不适合小角度连接的线条。如果两条线连接后的角度小于斜尖连接的最小角度，那么应该使用斜连接。斜连接的使用，可以防止出现太长的尖峰。默认情况下，斜尖连接的最小角度是 10 度。

可以在 BasicStroke 构造器中设定这些选择，例如：

```
g2.setStroke(new BasicStroke(10.0F, BasicStroke.CAP_ROUND, BasicStroke.JOIN_ROUND));
g2.setStroke(new BasicStroke(10.0F, BasicStroke.CAP_BUTT, BasicStroke.JOIN_MITER,
    15.0F /* miter limit */));
```

最后，通过设置一个虚线模式，可以指定需要使用的虚线。在程序清单 11-16 的程序中，可以选择一个虚线模式，拼出摩斯电码中的 SOS 代码。虚线模式是一个 float[] 类型的数组，它包含了笔画中"连接"（on）和"断开"（off）的长度（见图 11-43）。

图 11-43　一种虚线图案

程序清单 11-16　stroke/StrokeTest.java

```
 1  package stroke;
 2
 3  import java.awt.*;
 4  import java.awt.event.*;
 5  import java.awt.geom.*;
 6  import javax.swing.*;
 7
 8  /**
 9   * This program demonstrates different stroke types.
10   * @version 1.05 2018-05-01
11   * @author Cay Horstmann
12   */
13  public class StrokeTest
14  {
15     public static void main(String[] args)
16     {
17        EventQueue.invokeLater(() ->
18           {
19              var frame = new StrokeTestFrame();
20              frame.setTitle("StrokeTest");
21              frame.setDefaultCloseOperation(JFrame.EXIT_ON_CLOSE);
22              frame.setVisible(true);
```

```java
23              });
24      }
25  }
26
27  /**
28   * This frame lets the user choose the cap, join, and line style, and shows the resulting
29   * stroke.
30   */
31  class StrokeTestFrame extends JFrame
32  {
33      private StrokeComponent canvas;
34      private JPanel buttonPanel;
35
36      public StrokeTestFrame()
37      {
38          canvas = new StrokeComponent();
39          add(canvas, BorderLayout.CENTER);
40
41          buttonPanel = new JPanel();
42          buttonPanel.setLayout(new GridLayout(3, 3));
43          add(buttonPanel, BorderLayout.NORTH);
44
45          var group1 = new ButtonGroup();
46          makeCapButton("Butt Cap", BasicStroke.CAP_BUTT, group1);
47          makeCapButton("Round Cap", BasicStroke.CAP_ROUND, group1);
48          makeCapButton("Square Cap", BasicStroke.CAP_SQUARE, group1);
49
50          var group2 = new ButtonGroup();
51          makeJoinButton("Miter Join", BasicStroke.JOIN_MITER, group2);
52          makeJoinButton("Bevel Join", BasicStroke.JOIN_BEVEL, group2);
53          makeJoinButton("Round Join", BasicStroke.JOIN_ROUND, group2);
54
55          var group3 = new ButtonGroup();
56          makeDashButton("Solid Line", false, group3);
57          makeDashButton("Dashed Line", true, group3);
58      }
59
60      /**
61       * Makes a radio button to change the cap style.
62       * @param label the button label
63       * @param style the cap style
64       * @param group the radio button group
65       */
66      private void makeCapButton(String label, final int style, ButtonGroup group)
67      {
68          // select first button in group
69          boolean selected = group.getButtonCount() == 0;
70          var button = new JRadioButton(label, selected);
71          buttonPanel.add(button);
72          group.add(button);
73          button.addActionListener(event -> canvas.setCap(style));
74          pack();
75      }
76
```

```java
 77     /**
 78      * Makes a radio button to change the join style.
 79      * @param label the button label
 80      * @param style the join style
 81      * @param group the radio button group
 82      */
 83     private void makeJoinButton(String label, final int style, ButtonGroup group)
 84     {
 85         // select first button in group
 86         boolean selected = group.getButtonCount() == 0;
 87         var button = new JRadioButton(label, selected);
 88         buttonPanel.add(button);
 89         group.add(button);
 90         button.addActionListener(event -> canvas.setJoin(style));
 91     }
 92
 93     /**
 94      * Makes a radio button to set solid or dashed lines.
 95      * @param label the button label
 96      * @param style false for solid, true for dashed lines
 97      * @param group the radio button group
 98      */
 99     private void makeDashButton(String label, final boolean style, ButtonGroup group)
100     {
101         // select first button in group
102         boolean selected = group.getButtonCount() == 0;
103         var button = new JRadioButton(label, selected);
104         buttonPanel.add(button);
105         group.add(button);
106         button.addActionListener(event -> canvas.setDash(style));
107     }
108 }
109
110 /**
111  * This component draws two joined lines, using different stroke objects, and allows the user
112  * to drag the three points defining the lines.
113  */
114 class StrokeComponent extends JComponent
115 {
116     private static final Dimension PREFERRED_SIZE = new Dimension(400, 400);
117     private static int SIZE = 10;
118
119     private Point2D[] points;
120     private int current;
121     private float width;
122     private int cap;
123     private int join;
124     private boolean dash;
125
126     public StrokeComponent()
127     {
128         addMouseListener(new MouseAdapter()
129             {
130                 public void mousePressed(MouseEvent event)
```

```java
            {
               Point p = event.getPoint();
               for (int i = 0; i < points.length; i++)
               {
                  double x = points[i].getX() - SIZE / 2;
                  double y = points[i].getY() - SIZE / 2;
                  var r = new Rectangle2D.Double(x, y, SIZE, SIZE);
                  if (r.contains(p))
                  {
                     current = i;
                     return;
                  }
               }
            }

            public void mouseReleased(MouseEvent event)
            {
               current = -1;
            }
         });

      addMouseMotionListener(new MouseMotionAdapter()
         {
            public void mouseDragged(MouseEvent event)
            {
               if (current == -1) return;
               points[current] = event.getPoint();
               repaint();
            }
         });

      points = new Point2D[3];
      points[0] = new Point2D.Double(200, 100);
      points[1] = new Point2D.Double(100, 200);
      points[2] = new Point2D.Double(200, 200);
      current = -1;
      width = 8.0F;
   }

   public void paintComponent(Graphics g)
   {
      var g2 = (Graphics2D) g;
      var path = new GeneralPath();
      path.moveTo((float) points[0].getX(), (float) points[0].getY());
      for (int i = 1; i < points.length; i++)
         path.lineTo((float) points[i].getX(), (float) points[i].getY());
      BasicStroke stroke;
      if (dash)
      {
         float miterLimit = 10.0F;
         float[] dashPattern = { 10F, 10F, 10F, 10F, 10F, 10F, 30F, 10F, 30F, 10F, 30F, 10F,
            10F, 10F, 10F, 10F, 10F, 30F };
         float dashPhase = 0;
         stroke = new BasicStroke(width, cap, join, miterLimit, dashPattern, dashPhase);
```

```
185         }
186         else stroke = new BasicStroke(width, cap, join);
187         g2.setStroke(stroke);
188         g2.draw(path);
189      }
190
191      /**
192       * Sets the join style.
193       * @param j the join style
194       */
195      public void setJoin(int j)
196      {
197         join = j;
198         repaint();
199      }
200
201      /**
202       * Sets the cap style.
203       * @param c the cap style
204       */
205      public void setCap(int c)
206      {
207         cap = c;
208         repaint();
209      }
210
211      /**
212       * Sets solid or dashed lines.
213       * @param d false for solid, true for dashed lines
214       */
215      public void setDash(boolean d)
216      {
217         dash = d;
218         repaint();
219      }
220
221      public Dimension getPreferredSize() { return PREFERRED_SIZE; }
222   }
```

当构建 BasicStroke 时，可以指定虚线模式和虚线相位（dash phase）。虚线相位用来表示每条线应该从虚线模式的何处开始。通常情况下，应该把它的值设置为 0。

```
float[] dashPattern = { 10, 10, 10, 10, 10, 10, 30, 10, 30, . . . };
g2.setStroke(new BasicStroke(10.0F, BasicStroke.CAP_BUTT, BasicStroke.JOIN_MITER,
   10.0F /* miter limit */, dashPattern, 0 /* dash phase */));
```

> **注释：** 在虚线模式中，每一条虚线的末端都可以应用端头样式。

程序清单 11-16 中的程序可以设定端头样式、连接样式和虚线（见图 11-44）。可以移动线段的端头，用以测试斜尖连接的最小角度：首先选定斜尖连接；然后，移动线段末端形成一个非常尖的锐角。可以看到斜尖连接变成了一个斜连接。

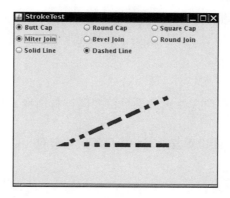

图 11-44 StrokeTest 程序

这个程序类似于程序清单 11-15 的程序。当点击一个线段的末端时，鼠标监听器就会记下操作，而鼠标动作监听器则监听对端点的拖曳操作。一组单选按钮用以表示用户选择的端头样式、连接样式以及实线或虚线。StrokePanel 类的 paintComponent 方法构建了一个 GeneralPath，它由连接着用户可以用鼠标移动的三个点的两条线段构成。然后，它根据用户的选择构建一个 BasicStroke，最后绘制出这个路径。

API java.awt.Graphics2D 1.2

- void setStroke(Stroke s)
  将该图形上下文的笔画设置为实现了 Stroke 接口的给定对象。

API java.awt.BasicStroke 1.2

- BasicStroke(float width)
- BasicStroke(float width, int cap, int join)
- BasicStroke(float width, int cap, int join, float miterlimit)
- BasicStroke(float width, int cap, int join, float miterlimit, float[] dash, float dashPhase)
  用给定的属性构建一个笔画对象。

  参数：width   画笔的宽度
  　　　cap   端头样式，它是 CAP_BUTT、CAP_ROUND 和 CAP_SQUARE 三种样式中的一个
  　　　join   连接样式，它是 JOIN_BEVEL、JOIN_MITER 和 JOIN_ROUND 三种样式中的一个
  　　　miterlimit 用度数表示的角度，如果小于这个角度，斜尖连接将呈现为斜连接
  　　　dash   虚线笔画的填充部分和空白部分交替出现的一组长度
  　　　dashPhase   虚线模式的"相位"；位于笔画起始点前面的这段长度被假设为已经应用了该虚线模式

11.3.5 着色

当填充一个形状时，该形状的内部就上了颜色。使用 setPaint 方法，可以把颜色的样式

设定为一个实现了 Paint 接口的类的对象。Java 2D API 提供了三个这样的类：
- Color 类实现了 Paint 接口。如果要用单色填充形状，只需要用 Color 对象调用 setPaint 方法即可，例如：

    g2.setPaint(Color.red);

- GradientPaint 类通过在两个给定的颜色值之间进行渐变，从而改变使用的颜色（参见图 11-45）。
- TexturePaint 类用一个图像重复地对一个区域进行着色（见图 11-46）。

图 11-45　渐变着色

图 11-46　纹理着色

可以通过指定两个点以及在这两个点上想使用的颜色来构建一个 GradientPaint 对象，即：

g2.setPaint(new GradientPaint(p1, Color.RED, p2, Color.YELLOW));

上面语句将沿着连接两个点之间的直线的方向对颜色进行渐变，而沿着与该连接线垂直方向上的线条颜色则是不变的。超过线条端点的各个点被赋予端点上的颜色。

另外，如果调用 GradientPaint 构造器时 cyclic 参数的值为 true，即：

g2.setPaint(new GradientPaint(p1, Color.RED, p2, Color.YELLOW, true));

那么颜色将循环变换，并且在端点之外仍然保持这种变换。

如果要构建一个 TexturePaint 对象，需要指定一个 BufferedImage 和一个锚位矩形。

g2.setPaint(new TexturePaint(bufferedImage, anchorRectangle));

在本章后面部分详细讨论图像时，我们再介绍 BufferedImage 类。获取缓冲图像最简单的方式就是读入图像文件：

bufferedImage = ImageIO.read(new File("blue-ball.gif"));

锚位矩形在 x 和 y 方向上将不断地重复延伸，使之平铺到整个坐标平面。图像可以伸缩，以便纳入该锚位，然后复制到每一个平铺显示区中。

API　java.awt.Graphics2D　1.2

- void setPaint(Paint s)

    将图形上下文的着色设置为实现了 Paint 接口的给定对象。

API　java.awt.GradientPaint　1.2

- GradientPaint(float x1, float y1, Color color1, float x2, float y2, Color color2)

- GradientPaint(float x1, float y1, Color color1, float x2, float y2, Color color2, boolean cyclic)
- GradientPaint(Point2D p1, Color color1, Point2D p2, Color color2)
- GradientPaint(Point2D p1, Color color1, Point2D p2, Color color2, boolean cyclic)

构建一个渐变着色的对象，以便用颜色来填充各个形状，其中，起始点的颜色为color1，结束点的颜色为color2，而两个点之间的颜色则是以线性的方式渐变。沿着连接起始点和结束点之间的线条相垂直的方向上的线条颜色是恒定不变的。在默认的情况下，渐变着色不是循环变换的。也就是说，起始点和结束点之外的各个点的颜色是分别与起始点和结束点的颜色相同的。如果渐变着色是循环的，那么颜色是连续变换的，首先返回到起始点的颜色，然后在两个方向上无限地重复。

API java.awt.TexturePaint 1.2

- TexturePaint(BufferedImage texture, Rectangle2D anchor)
  建立纹理着色对象。锚位矩形定义了色的平铺空间，该矩形在 x 和 y 方向上不断地重复延伸，纹理图像则被缩放，以便填充每个平铺空间。

11.3.6 坐标变换

假设我们要绘制一个对象，比如汽车。从制造商的规格说明书中可以了解到汽车的高度、轴距和整个车身的长度。如果设定了每米的像素个数，当然就可以计算出所有像素的位置。但是，可以使用更加容易的方法：让图形上下文来执行这种转换。

```
g2.scale(pixelsPerMeter, pixelsPerMeter);
g2.draw(new Line2D.Double(coordinates in meters)); // converts to pixels and
                                                   // draws scaled line
```

Graphics2D 类的 scale 方法可以将图形上下文中的坐标变换设置为一个比例变换。这种变换能够将用户坐标（用户设定的单元）转换成设备坐标（pixel，即像素）。图 11-47 显示了如何进行这种变换的方法。

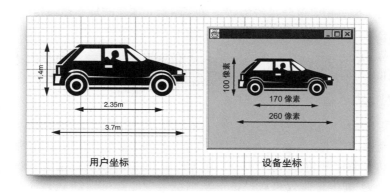

图 11-47　用户坐标与设备坐标

坐标变换在实际应用中非常有用,程序员可以使用方便的坐标值进行各种操作,图形上下文则负责执行将坐标值变换成像素的复杂工作。

这里有四种基本的变换:

- 比例缩放:放大和缩小从一个固定点出发的所有距离。
- 旋转:环绕着一个固定中心旋转所有点。
- 平移:将所有的点移动一个固定量。
- 切变:使一个线条固定不变,再按照与该固定线条之间的距离,成比例地将与该线条平行的各个线条"滑动"一个距离量。

图 11-48 显示了对一个单位的正方形进行这四种基本变换操作的效果。

Graphics2D 类的 scale、rotate、translate 和 shear 等方法用以将图形上下文中的坐标变换设置成为以上这些基本变换中的一种。

可以组合不同的变换操作。例如,你可能想对图形进行旋转和两倍尺寸放大的操作,这时,可以同时提供旋转和比例缩放的变换:

```
g2.rotate(angle);
g2.scale(2, 2);
g2.draw(...);
```

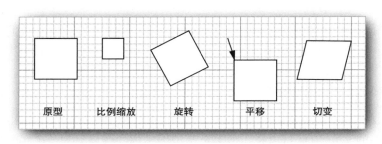

图 11-48 基本的变换

在这种情况下,变换方法的顺序是无关紧要的。然而,在大多数变换操作中,顺序却是很重要的。例如,如果想对形状进行旋转和切变操作,那么两种变换操作的不同执行序列,将会产生不同的图形。你必须明确想要得到的是什么样的图形,图形上下文将按照你所提供的相反顺序来应用这些变换操作。也就是说,你最后提供的方法会被最先应用。

可以根据你的需要提供任意多的变换操作。例如,假设你提供了下面这个变换操作序列:

```
g2.translate(x, y);
g2.rotate(a);
g2.translate(-x, -y);
```

最后一个变换操作(它是第一个被应用的)将把某个形状从点(x, y)移动到原点,第二个变换将使该形状围绕着原点旋转一个角度 a,最后一个变换方法又重新把该形状从原点移动到点(x, y)处。总体效果就是该形状围绕着中心点(x, y)进行了一次旋转(参见图 11-49)。围绕着原点之外的任意点进行旋转是一个很常见的操作,所以我们采用下面的快捷方法：

```
g2.rotate(a, x, y);
```

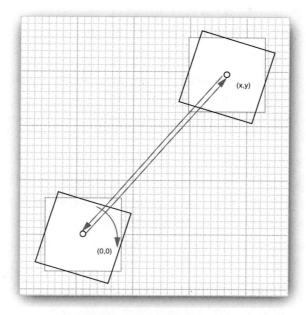

图 11-49　组合变换操作的应用

如果对矩阵论有所了解，那么就会知道所有操作（诸如旋转、平移、缩放、切变）和由这些操作组合起来的操作都能够以如下矩阵变换的形式表示出来：

$$\begin{bmatrix} x_{new} \\ y_{new} \\ 1 \end{bmatrix} = \begin{bmatrix} a & c & e \\ b & d & f \\ 0 & 0 & 1 \end{bmatrix} \cdot \begin{bmatrix} x \\ y \\ 1 \end{bmatrix}$$

这种变换称为仿射变换（affine transformation）。Java 2D API 中的 AffineTransform 类就是用于描述这种变换的。如果你知道某个特定变换矩阵的组成元素，就可以用下面的方法直接构造它：

```
var t = new AffineTransform(a, b, c, d, e, f);
```

另外，工厂方法 getRotateInstance、getScaleInstance、getTranslateInstance 和 getShearInstance 能够构建出表示相应变换类型的矩阵。例如，调用下面的方法：

```
t = AffineTransform.getScaleInstance(2.0F, 0.5F);
```

将返回一个与下面这个矩阵相一致的变换。

$$\begin{bmatrix} 2 & 0 & 0 \\ 0 & 0.5 & 0 \\ 0 & 0 & 1 \end{bmatrix}$$

最后，实例方法 setToRotation、setToScale、setToTranslation 和 setToShear 用于将变换对象设置为一个新的类型。下面是一个例子：

```
t.setToRotation(angle); // sets t to a rotation
```

可以把图形上下文的坐标变换设置为一个 AffineTransform 对象：

```
g2.setTransform(t); // replaces current transformation
```

不过，在实际运用中，不要调用 setTransform 操作，因为它会取代图形上下文中可能存在的任何现有的变换。例如，一个用以横向打印的图形上下文已经有了一个 90° 的旋转变换，如果调用方法 setTransfrom，就会删除这样的旋转操作。可以调用 transform 方法作为替代方案：

```
g2.transform(t); // composes current transformation with t
```

它会把现有的变换操作和新的 AffineTransform 对象组合起来。

如果只想临时应用某个变换操作，那么应该首先获得旧的变换操作，然后和新的变换操作组合起来，最后当你完成操作时，再还原旧的变换操作：

```
AffineTransform oldTransform = g2.getTransform(); // save old transform
g2.transform(t); // apply temporary transform
draw on g2
g2.setTransform(oldTransform); // restore old transform
```

API java.awt.geom.AffineTransform 1.2

- AffineTransform(double a, double b, double c, double d, double e, double f)
- AffineTransform(float a, float b, float c, float d, float e, float f)
  用下面的矩阵构建该仿射变换。

$$\begin{bmatrix} a & c & e \\ b & d & f \\ 0 & 0 & 1 \end{bmatrix}$$

- AffineTransform(double[] m)
- AffineTransform(float[] m)
  用下面的矩阵构建该仿射变换。

$$\begin{bmatrix} m[0] & m[2] & m[4] \\ m[1] & m[3] & m[5] \\ 0 & 0 & 1 \end{bmatrix}$$

- static AffineTransform getRotateInstance(double a)
  创建一个围绕原点、旋转角度为 a（弧度）的旋转变换。其变换矩阵是：

$$\begin{bmatrix} \cos(a) & -\sin(a) & 0 \\ \sin(a) & \cos(a) & 0 \\ 0 & 0 & 1 \end{bmatrix}$$

如果 a 在 0 到 π/2 之间,那么图形将沿着 x 轴正半轴向 y 轴正半轴的方向旋转。
- static AffineTransform getRotateInstance(double a, double x, double y)

  创建一个围绕点 (x, y)、旋转角度为 a (弧度) 的旋转变换。
- static AffineTransform getScaleInstance(double sx, double sy)

  创建一个比例缩放变换。x 轴缩放幅度为 sx;y 轴缩放幅度为 sy。其变换矩阵是:

$$\begin{bmatrix} sx & 0 & 0 \\ 0 & sy & 0 \\ 0 & 0 & 1 \end{bmatrix}$$

- static AffineTransform getShearInstance(double shx, double shy)

  创建一个切变变换。x 轴切变 shx;y 轴切变 shy。其变换矩阵是:

$$\begin{bmatrix} 1 & shx & 0 \\ shy & 1 & 0 \\ 0 & 0 & 1 \end{bmatrix}$$

- static AffineTransform getTranslateInstance(double tx, double ty)

  创建一个平移变换。x 轴平移 tx;y 轴平移 ty。其变换矩阵是:

$$\begin{bmatrix} 1 & 0 & tx \\ 0 & 1 & ty \\ 0 & 0 & 1 \end{bmatrix}$$

- void setToRotation(double a)
- void setToRotation(double a, double x, double y)
- void setToScale(double sx, double sy)
- void setToShear(double sx, double sy)
- void setToTranslation(double tx, double ty)

  用给定的参数将该变换设置为一个的基本变换。如果要了解基本变换和它们的参数说明,请参见 getXxxInstance 方法。

API java.awt.Graphics2D 1.2

- void setTransform(AffineTransform t)

  以 t 来取代该图形上下文中现有的坐标变换。
- void transform(AffineTransform t)

  将该图形上下文的现有坐标变换和 t 组合起来。
- void rotate(double a)
- void rotate(double a, double x, double y)
- void scale(double sx, double sy)
- void shear(double sx, double sy)
- void translate(double tx, double ty)

将该图形上下文中现有的坐标变换和一个带有给定参数的基本变换组合起来。如果要了解基本变换和它们的参数说明，请参见 AffineTransform.getXxxInstance 方法。

11.3.7 剪切

通过在图形上下文中设置一个剪切形状，就可以将所有的绘图操作限制在该剪切形状内部来进行。

```
g2.setClip(clipShape); // but see below
g2.draw(shape); // draws only the part that falls inside the clipping shape
```

但是，在实际应用中，不应该调用这个 setClip 操作，因为它会取代图形上下文中可能存在的任何剪切形状。例如，正如在本章的后面部分所看到的那样，用于打印操作的图形上下文就具有一个剪切矩形，以确保你不会在页边距上绘图。相反，你应该调用 clip 方法。

```
g2.clip(clipShape); // better
```

clip 方法将你所提供的新的剪切形状同现有的剪切形状相交。

如果只想临时地使用一个剪切区域的话，那么应该首先获得旧的剪切形状，然后添加新的剪切形状，最后，在完成操作时，再还原旧的剪切形状：

```
Shape oldClip = g2.getClip(); // save old clip
g2.clip(clipShape); // apply temporary clip
draw on g2
g2.setClip(oldClip); // restore old clip
```

在图 11-50 的例子中，我们炫耀了一下剪切的功能，它绘制了一个按照复杂形状进行剪切的相当出色的线条图案，即一组字符的轮廓。

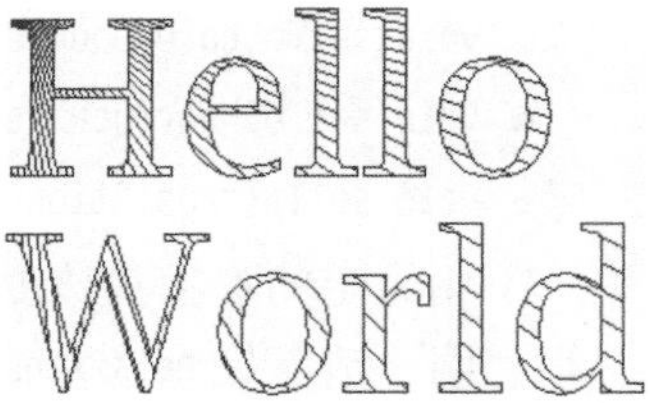

如果要获得字符的外形，需要一个字体渲染上下文（font render context）。请使用 Graphics2D 类的 getFontRenderContext 方法：

```
FontRenderContext context = g2.getFontRenderContext();
```

接着，使用某个字符串、某种字体和字体渲染上下文来创建一个 TextLayout 对象：

```
var layout = new TextLayout("Hello", font, context);
```

图 11-50 按照字母形状剪切出的线条图案

这个文本布局对象用于描述由特定字体渲染上下文所渲染的一个字符序列的布局。这种布局依赖于字体渲染上下文，相同的字符在屏幕上或者打印机上看起来会有不同的显示。

对我们当前的应用来说，更重要的是，getOutline 方法将会返回一个 Shape 对象，这个 Shape 对象用以描述在文本布局中的各个字符轮廓的形状。字符轮廓的形状从原点（0，0）开始，这并不适合大多数的绘图操作。因此，必须为 getOutline 操作提供一个仿射变换操作，以便设定想要的字体轮廓所显示的位置：

```
AffineTransform transform = AffineTransform.getTranslateInstance(0, 100);
Shape outline = layout.getOutline(transform);
```

接着，我们把字体的轮廓附加给剪切的形状：

```
var clipShape = new GeneralPath();
clipShape.append(outline, false);
```

最后，我们设置剪切形状，并且绘制一组线条。线条仅仅在字符边界的内部显示：

```
g2.setClip(clipShape);
var p = new Point2D.Double(0, 0);
for (int i = 0; i < NLINES; i++)
{
   double x = . . .;
   double y = . . .;
   var q = new Point2D.Double(x, y);
   g2.draw(new Line2D.Double(p, q)); // lines are clipped
}
```

API java.awt.Graphics 1.0

- void setClip(Shape s) 1.2

  将当前的剪切形状设置为形状 s。

- Shape getClip() 1.2

  返回当前的剪切形状。

API java.awt.Graphics2D 1.2

- void clip(Shape s)

  将当前的剪切形状和形状 s 相交。

- FontRenderContext getFontRenderContext()

  返回一个构建 TextLayout 对象所必需的字体渲染上下文。

API java.awt.font.TextLayout 1.2

- TextLayout(String s, Font f, FontRenderContext context)

  根据给定的字符串和字体来构建文本布局对象。方法中使用字体渲染上下文来获取特定设备的字体属性。

- float getAdvance()

  返回该文本布局的宽度。

- float getAscent()
- float getDescent()

  返回基准线上方和下方该文本布局的高度。

- float getLeading()

  返回该文本布局使用的字体中相邻两行之间的距离。

11.3.8　透明与组合

在标准的 RGB 颜色模型中，每种颜色都是由它的红、绿和蓝这三种成分来描述的。但是，用它来描述透明或者部分透明的图像区域也是非常方便的。当你将一个图像置于现有图像的上面时，透明的像素完全不会遮挡它们下面的像素，而部分透明的像素则与它们下面的像素相混合。图 11-51 显示了一个部分透明的矩形和一个图像相重叠时所产生的效果，我们仍然可以透过矩形看到该图像的细节。

图 11-51　一个部分透明的矩形和一个图像相重叠时所显示的效果

在 Java 2D API 中，透明是由一个透明度通道（alpha channel）来描述的。每个像素，除了它的红、绿和蓝色部分外，还有一个介于 0（完全透明）和 1（部分透明）之间的透明度（alpha）值。例如，图 11-51 中的矩形填充了一种淡黄色，透明度为 50%：

```
new Color(0.7F, 0.7F, 0.0F, 0.5F);
```

现在让我们看一看如果将两个形状重叠在一起时将会出现什么情况。必须把源像素和目标像素的颜色和透明度值混合或者组合起来。从事计算机图形学研究的 Porter 和 Duff 已经阐明了在这个混合过程中的 12 种可能的组合原则，Java 2D API 实现了所有的这些原则。在继续介绍这个问题之前，需要指出的是，这些原则中只有两个原则有实际的意义。如果你发现这些原则晦涩难懂或者难以搞清楚，那么只使用 SRC_OVER 原则就可以了。它是 Graphics2D 对象的默认原则，并且它产生的结果最直接。

下面是这些规则的原理。假设你有了一个透明度值为 a_S 的源像素，在该图像中，已经存在了一个透明度值为 a_D 的目标像素，你想把两个像素组合起来。图 11-52 的示意图显示了如何设计一个像素的组合原则。

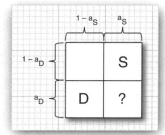

Porter 和 Duff 将透明度值作为像素颜色将被使用的概率。从源像素的角度来看，存在一个概率 a_S，它是源像素颜色被使用的概率；还存在一个概率 $1 - a_S$，它是不在乎是否使用该像素颜色的概率。同样的原则也适用于目标像素。当组合颜色时，我们假设源像素的概率和目标像素的概率是不相关的。那么正如图 11-52 所示，有四种组合情况。如果源像素想要使用它的颜色，而目标像素也不在乎，那么很自然的，我们就只使用源像素的颜色。这也是为什么右上角的矩形框用"S"来标志的原因了，这种情况的概率为 $a_S \cdot (1 - a_D)$。同理，左下角的矩形框用"D"来标志。如果源像素和目标像素都想选择自己的颜色，那该怎么办才好呢？这里就要应用 Porter-Duff 原则了。如果我们认为源像素比较重要，那么我们在右下角的矩形框内也标志上一个"S"。这个规则被称为 SRC_OVER。在这个规则中，我们赋予源像素颜色的权值 a_S，目标像素颜色的权值为 $(1 - a_S) \cdot a_D$，然后将它们组合起来。

图 11-52　设计一个像素组合的原则

这样产生的视觉效果是源像素与目标像素相混合的结果，并且优先选择给定的源像素的颜色。特别是，如果 a_S 为 1，那么根本就不用考虑目标像素的颜色。如果 a_S 为 0，那么源像素将是完全透明的，而目标像素颜色则是不变的。

还有其他的规则，可以根据置于概率示意图各个框中的字母来理解这些规则的概念。表 11-3 和图 11-53 显示了 Java 2D API 支持的所有这些规则。图 11-53 中的各个图像显示了当你使用透明度值为 0.75 的矩形源区域和透明度值为 1.0 的椭圆目标区域组合时，所显示的各种组合效果。

表 11-3　Porter-Duff 组合规则

规则	解释
CLEAR	源像素清除目标像素
SRC	源像素覆盖目标像素和空像素
DST	源像素不影响目标像素
SRC_OVER	源像素和目标像素混合，并且覆盖空像素
DST_OVER	源像素不影响目标像素，并且不覆盖空像素
SRC_IN	源像素覆盖目标像素
SRC_OUT	源像素清除目标像素，并且覆盖空像素
DST_IN	源像素的透明度值修改目标像素的透明度值
DST_OUT	源像素的透明度值取反修改目标像素的透明度值
SRC_ATOP	源像素和目标像素相混合
DST_ATOP	源像素的透明度值修改目标像素的透明度值。源像素覆盖空像素
XOR	源像素的透明度值取反修改目标像素的透明度值。源像素覆盖空像素

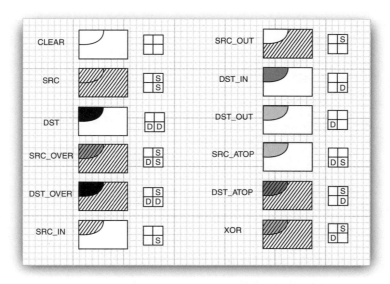

图 11-53　Porter-Duff 组合规则

如你所见，大多数规则并不是非常有用。例如，DST_IN 规则就是一个极端的例子。它根本不考虑源像素颜色,但是却使用了源像素的透明度值来影响目标像素。SRC 规则可能是有用的,它强制使用源像素颜色,而且关闭了与目标像素相混合的特性。

如果要了解更多的关于 Porter-Duff 规则的信息,请参阅 Foley、Dam 和 Feiner 等撰写的 *Computer Graphics: Principles and Practice, Second Edition*。

你可以使用 Graphics2D 类的 setComposite 方法安装一个实现了 Composite 接口的类的对象。Java 2D API 提供了这样的一个类,即 AlphaComposite 它实现了图 11-53 中的所有的 Porter-Duff 规则。

AlphaComposite 类的工厂方法 getInstance 用来产生 AlphaComposite 对象,此时需要提供用于源像素的规则和透明度值。例如,可以考虑使用下面的代码:

```
int rule = AlphaComposite.SRC_OVER;
float alpha = 0.5f;
g2.setComposite(AlphaComposite.getInstance(rule, alpha));
g2.setPaint(Color.blue);
g2.fill(rectangle);
```

这时,矩形将使用蓝色和值为 0.5 的透明度进行着色。因为该组合规则是 SRC_OVER,所以它透明地置于现有图像的上面。

程序清单 11-17 中的程序深入地研究了这些组合规则。可以从组合框中选择一个规则,调节滑动条来设置 AlphaComposite 对象的透明度值。

程序清单 11-17　composite/CompositeTestFrame.java

```
 1  package composite;
 2
 3  import java.awt.*;
 4  import javax.swing.*;
 5
 6  /**
 7   * This frame contains a combo box to choose a composition rule, a slider to change the
 8   * source alpha channel, and a component that shows the composition.
 9   */
10  class CompositeTestFrame extends JFrame
11  {
12     private static final int DEFAULT_WIDTH = 400;
13     private static final int DEFAULT_HEIGHT = 400;
14
15     private CompositeComponent canvas;
16     private JComboBox<Rule> ruleCombo;
17     private JSlider alphaSlider;
18     private JTextField explanation;
19
20     public CompositeTestFrame()
21     {
22        setSize(DEFAULT_WIDTH, DEFAULT_HEIGHT);
23
24        canvas = new CompositeComponent();
25        add(canvas, BorderLayout.CENTER);
26
27        ruleCombo = new JComboBox<>(new Rule[] { new Rule("CLEAR", " ", " "),
```

```
28            new Rule("SRC", " S", " S"), new Rule("DST", "  ", "DD"),
29            new Rule("SRC_OVER", " S", "DS"), new Rule("DST_OVER", " S", "DD"),
30            new Rule("SRC_IN", "  ", " S"), new Rule("SRC_OUT", " S", "  "),
31            new Rule("DST_IN", "  ", " D"), new Rule("DST_OUT", "  ", "D "),
32            new Rule("SRC_ATOP", "  ", "DS"), new Rule("DST_ATOP", " S", " D"),
33            new Rule("XOR", " S", "D "), });
34         ruleCombo.addActionListener(event ->
35            {
36               var r = (Rule) ruleCombo.getSelectedItem();
37               canvas.setRule(r.getValue());
38               explanation.setText(r.getExplanation());
39            });
40
41         alphaSlider = new JSlider(0, 100, 75);
42         alphaSlider.addChangeListener(event -> canvas.setAlpha(alphaSlider.getValue()));
43         var panel = new JPanel();
44         panel.add(ruleCombo);
45         panel.add(new JLabel("Alpha"));
46         panel.add(alphaSlider);
47         add(panel, BorderLayout.NORTH);
48
49         explanation = new JTextField();
50         add(explanation, BorderLayout.SOUTH);
51
52         canvas.setAlpha(alphaSlider.getValue());
53         Rule r = ruleCombo.getItemAt(ruleCombo.getSelectedIndex());
54         canvas.setRule(r.getValue());
55         explanation.setText(r.getExplanation());
56      }
57   }
```

此外，对每一条规则该程序都显示了一条文字描述。请注意，描述是根据组合规则表计算而来的。例如，第二行中的"DS"表示的就是"与目标像素相混合"。

该程序有一个重要的缺陷：它不能保证和屏幕相对应的图形上下文一定具有透明通道。（实际上，它通常没有这个透明通道）。当像素被放到没有透明通道的目标像素之上的时候，这些像素的颜色会与目标像素的透明度值相乘，而其透明度值却被弃用了。因为许多 Porter-Duff 规则都使用目标像素的透明度值，因此目标像素的透明通道是很重要的。由于这个原因，我们使用了一个采用 ARGB 颜色模型的缓存图像来组合各种形状。在图像被组合后，我们就将产生的图像在屏幕上绘制出来：

```
var image = new BufferedImage(getWidth(), getHeight(), BufferedImage.TYPE_INT_ARGB);
Graphics2D gImage = image.createGraphics();
// now draw to gImage
g2.drawImage(image, null, 0, 0);
```

程序清单 11-17 和程序清单 11-18 展示了框体和构件类，程序清单 11-19 中的 Rule 类提供了对每条规则的简要解释，如图 11-54 所示。在运行这个程序的时候，从左到右地移动 Alpha 滑动条，就可以观察到所产生的组合形状的效果。特别是，请注意 DST_IN 与 DST_OUT 规则之间唯一的差别，那就是，当你改变源像素的透明度值时，目标（！）颜色将会发生什么样的变化。

程序清单 11-18　composite/CompositeComponent.java

```java
 1  package composite;
 2
 3  import java.awt.*;
 4  import java.awt.geom.*;
 5  import java.awt.image.*;
 6  import javax.swing.*;
 7
 8  /**
 9   * This component draws two shapes, composed with a composition rule.
10   */
11  class CompositeComponent extends JComponent
12  {
13     private int rule;
14     private Shape shape1;
15     private Shape shape2;
16     private float alpha;
17
18     public CompositeComponent()
19     {
20        shape1 = new Ellipse2D.Double(100, 100, 150, 100);
21        shape2 = new Rectangle2D.Double(150, 150, 150, 100);
22     }
23
24     public void paintComponent(Graphics g)
25     {
26        var g2 = (Graphics2D) g;
27
28        var image = new BufferedImage(getWidth(), getHeight(), BufferedImage.TYPE_INT_ARGB);
29        Graphics2D gImage = image.createGraphics();
30        gImage.setPaint(Color.red);
31        gImage.fill(shape1);
32        AlphaComposite composite = AlphaComposite.getInstance(rule, alpha);
33        gImage.setComposite(composite);
34        gImage.setPaint(Color.blue);
35        gImage.fill(shape2);
36        g2.drawImage(image, null, 0, 0);
37     }
38
39     /**
40      * Sets the composition rule.
41      * @param r the rule (as an AlphaComposite constant)
42      */
43     public void setRule(int r)
44     {
45        rule = r;
46        repaint();
47     }
48
49     /**
50      * Sets the alpha of the source.
```

```java
51      * @param a the alpha value between 0 and 100
52      */
53     public void setAlpha(int a)
54     {
55        alpha = (float) a / 100.0F;
56        repaint();
57     }
58  }
```

程序清单 11-19   composite/Rule.java

```java
1   package composite;
2
3   import java.awt.*;
4
5   /**
6    * This class describes a Porter-Duff rule.
7    */
8   class Rule
9   {
10     private String name;
11     private String porterDuff1;
12     private String porterDuff2;
13
14     /**
15      * Constructs a Porter-Duff rule.
16      * @param n the rule name
17      * @param pd1 the first row of the Porter-Duff square
18      * @param pd2 the second row of the Porter-Duff square
19      */
20     public Rule(String n, String pd1, String pd2)
21     {
22        name = n;
23        porterDuff1 = pd1;
24        porterDuff2 = pd2;
25     }
26
27     /**
28      * Gets an explanation of the behavior of this rule.
29      * @return the explanation
30      */
31     public String getExplanation()
32     {
33        var r = new StringBuilder("Source ");
34        if (porterDuff2.equals("  ")) r.append("clears");
35        if (porterDuff2.equals(" S")) r.append("overwrites");
36        if (porterDuff2.equals("DS")) r.append("blends with");
37        if (porterDuff2.equals(" D")) r.append("alpha modifies");
38        if (porterDuff2.equals("D ")) r.append("alpha complement modifies");
39        if (porterDuff2.equals("DD")) r.append("does not affect");
40        r.append(" destination");
41        if (porterDuff1.equals(" S")) r.append(" and overwrites empty pixels");
42        r.append(".");
43        return r.toString();
```

```
 44     }
 45
 46     public String toString()
 47     {
 48        return name;
 49     }
 50
 51     /**
 52      * Gets the value of this rule in the AlphaComposite class.
 53      * @return the AlphaComposite constant value, or -1 if there is no matching constant
 54      */
 55     public int getValue()
 56     {
 57        try
 58        {
 59           return (Integer) AlphaComposite.class.getField(name).get(null);
 60        }
 61        catch (Exception e)
 62        {
 63           return -1;
 64        }
 65     }
 66 }
```

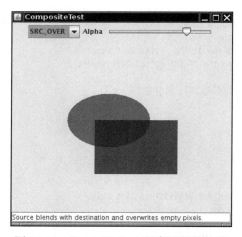

图 11-54 CompositeTest 程序运行的结果

API java.awt.Graphics2D 1.2

- void setComposite(Composite s)

  把图形上下文的组合方式设置为实现了 Composite 接口的给定对象。

API java.awt.AlphaComposite 1.2

- static AlphaComposite getInstance(int rule)
- static AlphaComposite getInstance(int rule, float sourceAlpha)

构建一个透明度（alpha）值的组合对象。规则是 CLEAR, SRC, SRC_OVER, DST_OVER, SRC_IN, SRC_OUT, DST_IN, DST_OUT, DST, DST_ATOP, SRC_ATOP, XOR 等值之一。

11.4 像素图

Java2D API 使得我们可以创建由直线、曲线和区域构成的图。它是一个"向量"API，因为我们需要指定各种形状的数学属性。但是，对于处理由像素构成的图像，我们希望能够操作由颜色数据构成的"栅格"。下面将展示 Java 中的像素图。

11.4.1 图像的读取器和写入器

javax.imageio 包包含了对读取和写入数种常用文件格式进行支持的"非常方便的"特性。同时还包含了一个框架，使得第三方能够为其他图像格式的文件添加读取器和写入器。GIF、JPEG、PNG、BMP（Windows 位图）和 WBMP（无线位图）等文件格式都得到了支持。

该类库的基本应用是极其直接的。要想装载一个图像，可以使用 ImageIO 类的静态 read 方法。

```
File f = . . .;
BufferedImage image = ImageIO.read(f);
```

ImageIO 类会根据文件的类型，选择一个合适的读取器。它可以参考文件的扩展名和文件开头的专用于此目的的"幻数"（magic number）来选择读取器。如果没有找到合适的读取器或者读取器不能解码文件的内容，那么 read 方法将返回 null。

把图像写入到文件中也是一样地简单。

```
File f = . . .;
String format = . . .;
ImageIO.write(image, format, f);
```

这里，format 字符串用来标识图像的格式，比如"JPEG"或者"PNG"。ImageIO 类将选择一个合适的写入器以存储文件。

11.4.1.1 获得适合图像文件类型的读取器和写入器

对于那些超出 ImageIO 类的静态 read 和 write 方法能力范围的高级图像读取和写入操作来说，首先需要获得合适的 ImageReader 和 ImageWriter 对象。ImageIO 类枚举了匹配下列条件之一的读取器和写入器。

- 图像格式（比如"JPEG"）
- 文件后缀（比如"jpg"）
- MIME 类型（比如"image/jpeg"）

> 注释：MIME（Multipurpose Internet Mail Extensions standard）是"多用途因特网邮件扩展标准"的英文缩写。MIME 标准定义了常用的数据格式，比如"image/jpeg"和"application/pdf"等。

例如，可以用下面的代码来获取一个 JPEG 格式文件的读取器。

```
ImageReader reader = null;
Iterator<ImageReader> iter = ImageIO.getImageReadersByFormatName("JPEG");
if (iter.hasNext()) reader = iter.next();
```

getImageReadersBySuffix 和 getImageReadersByMIMEType 这两个方法用于枚举与文件扩展名或 MIME 类型相匹配的读取器。

ImageIO 类可能会找到多个读取器，而它们都能够读取某一特殊类型的图像文件。在这种情况下，必须从中选择一个，但是也许你不清楚怎样才能选择一个最好的。如果要了解更多的关于读取器的信息，就要获取它的服务提供者接口：

```
ImageReaderSpi spi = reader.getOriginatingProvider();
```

然后，可以获得供应商的名字和版本号：

```
String vendor = spi.getVendor();
String version = spi.getVersion();
```

也许该信息能够帮助你决定选择哪一种读取器，或者你可以为你的程序用户提供一个读取器的列表，让他们做出选择。然而，目前来说，我们假定第一个列出来的读取器就能够满足用户的需求。

在程序清单 11-20 中，我们想查找所有可获得的读取器能够处理的文件的所有后缀，这样我们就可以在文件过滤器中使用它们。我们可以使用静态的 ImageIO.getReader-FileSuffixes 方法来达到此目的：

```
String[] extensions = ImageIO.getWriterFileSuffixes();
chooser.setFileFilter(new FileNameExtensionFilter("Image files", extensions));
```

对于保存文件，相对来说更麻烦一些：我们希望为用户展示一个支持所有图像类型的菜单。可惜，IOImage 类的 getWriterFormateNames 方法返回了一个相当奇怪的列表，里边包含了许多冗余的名字，比如：

jpg, BMP, bmp, JPG, jpeg, wbmp, png, JPEG, PNG, WBMP, GIF, gif

这些并不是人们想要在菜单中显示的东西，我们所需要的是"首选"格式名列表。我们提供了一个用于此目的的助手方法 getWriterFormats（参见程序清单 11-7）。我们查找与每一种格式名相关的第一个写入器，然后，询问该写入器它支持的格式名是什么，从而希望它能够将最流行的一个格式名列在首位。实际上，对 JPEG 写入器来说，这种方法确实很有效：它将"JPEG"列在其他选项的前面。（另一方面，PNG 写入器把小写字母的"png"列在"PNG"的前面。我们希望这种行为能够在将来的某个时候得以解决。同时，我们强制将全小写名字转换为大写）。一旦挑选了首选名，我们就会将所有其他的候选名从最初的名字集中移除。之后，我们会继续执行直至所有的格式名都得到处理。

11.4.1.2 读取和写入带有多个图像的文件

有些文件，特别是 GIF 动画文件，都包含了多个图像。ImageIO 类的 read 方法只能够读取单个图像。为了读取多个图像，应该将输入源（例如，输入流或者输入文件）转换成一个

ImageInputStream。

```
InputStream in = . . .;
ImageInputStream imageIn = ImageIO.createImageInputStream(in);
```

接着把图像输入流作为参数传递给读取器的 setInput 方法：

```
reader.setInput(imageIn, true);
```

方法中的第二个参数值表示输入的方式是"只向前搜索"，否则，就采用随机访问的方式，要么是在读取时缓冲输入流，要么是使用随机文件访问。对于某些操作来说，必须使用随机访问的方法。例如，为了在一个 GIF 文件中查寻图像的个数，就需要读入整个文件。这时，如果想获取某一图像的话，必须再次读入该输入文件。

只有当从一个流中读取图像，并且输入流中包含多个图像，而且在文件头中的图像格式部分没有所需要的信息（比如图像的个数）时，考虑使用上面的方法才是合适的。如果要从一个文件中读取图像信息的话，可直接使用下面的方法：

```
File f = . . .;
ImageInputStream imageIn = ImageIO.createImageInputStream(f);
reader.setInput(imageIn);
```

一旦拥有了一个读取器后，就可以通过调用下面的方法来读取输入流中的图像。

```
BufferedImage image = reader.read(index);
```

其中 index 是图像的索引，其值从 0 开始。

如果输入流采用"只向前搜索"的方式，那么应该持续不断地读取图像，直到 read 方法抛出一个 IndexOutOfBoundsException 为止。否则，可以调用 getNumImages 方法：

```
int n = reader.getNumImages(true);
```

在该方法中，它的参数表示允许搜索输入流以确定图像的数目。如果输入流采用"只向前搜索"的方式，那么该方法将抛出一个 IllegalStateException 异常。要不然，可以把是否"允许搜索"参数设置为 false。如果 getNumImages 方法在不搜索输入流的情况下无法确定图像的数目，那么它将返回 -1。在这种情况下，必须转换到 B 方案，那就是持续不断地读取图像，直到获得一个 IndexOutOfBoundsException 异常为止。

有些文件包含一些缩略图，也就是图像用来预览的小版本。可以通过调用下面的方法来获得某个图像的缩略图数量。

```
int count = reader.getNumThumbnails(index);
```

然后可以按如下方式得到一个特定索引：

```
BufferedImage thumbnail = reader.getThumbnail(index, thumbnailIndex);
```

另一个问题是，有时你想在实际获得图像之前，了解该图像的大小。特别是，当图像很大，或者是从一个较慢的网络连接中获取的时候，你更加希望能够事先了解到该图像的大小。那么请使用下面的方法：

```
int width = reader.getWidth(index);
int height = reader.getHeight(index);
```

通过上面两个方法可以获得具有给定索引的图像的大小。

如果要将多个图像写入到一个文件中，首先需要一个 ImageWriter。ImageIO 类能够枚举可以写入某种特定图像格式的所有写入器。

```
String format = . . .;
ImageWriter writer = null;
Iterator<ImageWriter> iter = ImageIO.getImageWritersByFormatName(format);
if (iter.hasNext()) writer = iter.next();
```

接着，将一个输出流或者输出文件转换成 ImageOutputStream，并且将其作为参数传给写入器。例如，

```
File f = . . .;
ImageOutputStream imageOut = ImageIO.createImageOutputStream(f);
writer.setOutput(imageOut);
```

必须将每一个图像都包装到 IIOImage 对象中。可以根据情况提供一个缩略图和图像元数据（比如，图像的压缩算法和颜色信息）的列表。在本例中，我们把两者都设置为 null；如果要了解详细信息，请参阅 API 文档。

```
var iioImage = new IIOImage(images[i], null, null);
```

使用 write 方法，可以写出第一个图像：

```
writer.write(new IIOImage(images[0], null, null));
```

对于后续的图像，使用下面的方法：

```
if (writer.canInsertImage(i))
   writer.writeInsert(i, iioImage, null);
```

上面方法中的第三个参数可以包含一个 ImageWriteParam 对象，用以设置图像写入的详细信息，比如是平铺还是压缩；可以用 null 作为其默认值。

并不是所有的图像格式都能够处理多个图像。在这种情况下，如果 i>0，canInsertImage 方法将返回 false 值，而且只保存单一图像。

程序清单 11-20 中的程序使用 Java 类库所提供的读取器和写入器支持的格式来加载和保持文件。该程序显示了多个图像（见图 11-55），但是没有缩略图。

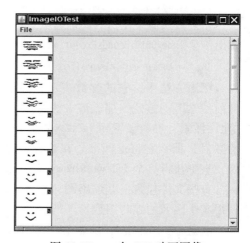

图 11-55　一个 GIF 动画图像

程序清单 11-20　imageIO/ImageIOFrame.java

```
1  package imageIO;
2
3  import java.awt.image.*;
4  import java.io.*;
5  import java.util.*;
6
```

```java
   7  import javax.imageio.*;
   8  import javax.imageio.stream.*;
   9  import javax.swing.*;
  10  import javax.swing.filechooser.*;
  11
  12  /**
  13   * This frame displays the loaded images. The menu has items for loading and saving files.
  14   */
  15  public class ImageIOFrame extends JFrame
  16  {
  17     private static final int DEFAULT_WIDTH = 400;
  18     private static final int DEFAULT_HEIGHT = 400;
  19
  20     private static Set<String> writerFormats = getWriterFormats();
  21
  22     private BufferedImage[] images;
  23
  24     public ImageIOFrame()
  25     {
  26        setSize(DEFAULT_WIDTH, DEFAULT_HEIGHT);
  27
  28        var fileMenu = new JMenu("File");
  29        var openItem = new JMenuItem("Open");
  30        openItem.addActionListener(event -> openFile());
  31        fileMenu.add(openItem);
  32
  33        var saveMenu = new JMenu("Save");
  34        fileMenu.add(saveMenu);
  35        Iterator<String> iter = writerFormats.iterator();
  36        while (iter.hasNext())
  37        {
  38           final String formatName = iter.next();
  39           var formatItem = new JMenuItem(formatName);
  40           saveMenu.add(formatItem);
  41           formatItem.addActionListener(event -> saveFile(formatName));
  42        }
  43
  44        var exitItem = new JMenuItem("Exit");
  45        exitItem.addActionListener(event -> System.exit(0));
  46        fileMenu.add(exitItem);
  47
  48        var menuBar = new JMenuBar();
  49        menuBar.add(fileMenu);
  50        setJMenuBar(menuBar);
  51     }
  52
  53     /**
  54      * Open a file and load the images.
  55      */
  56     public void openFile()
  57     {
  58        var chooser = new JFileChooser();
  59        chooser.setCurrentDirectory(new File("."));
  60        String[] extensions = ImageIO.getReaderFileSuffixes();
  61        chooser.setFileFilter(new FileNameExtensionFilter("Image files", extensions));
```

```java
62      int r = chooser.showOpenDialog(this);
63      if (r != JFileChooser.APPROVE_OPTION) return;
64      File f = chooser.getSelectedFile();
65      Box box = Box.createVerticalBox();
66      try
67      {
68         String name = f.getName();
69         String suffix = name.substring(name.lastIndexOf('.') + 1);
70         Iterator<ImageReader> iter = ImageIO.getImageReadersBySuffix(suffix);
71         ImageReader reader = iter.next();
72         ImageInputStream imageIn = ImageIO.createImageInputStream(f);
73         reader.setInput(imageIn);
74         int count = reader.getNumImages(true);
75         images = new BufferedImage[count];
76         for (int i = 0; i < count; i++)
77         {
78            images[i] = reader.read(i);
79            box.add(new JLabel(new ImageIcon(images[i])));
80         }
81      }
82      catch (IOException e)
83      {
84         JOptionPane.showMessageDialog(this, e);
85      }
86      setContentPane(new JScrollPane(box));
87      validate();
88   }
89
90   /**
91    * Save the current image in a file.
92    * @param formatName the file format
93    */
94   public void saveFile(final String formatName)
95   {
96      if (images == null) return;
97      Iterator<ImageWriter> iter = ImageIO.getImageWritersByFormatName(formatName);
98      ImageWriter writer = iter.next();
99      var chooser = new JFileChooser();
100     chooser.setCurrentDirectory(new File("."));
101     String[] extensions = writer.getOriginatingProvider().getFileSuffixes();
102     chooser.setFileFilter(new FileNameExtensionFilter("Image files", extensions));
103
104     int r = chooser.showSaveDialog(this);
105     if (r != JFileChooser.APPROVE_OPTION) return;
106     File f = chooser.getSelectedFile();
107     try
108     {
109        ImageOutputStream imageOut = ImageIO.createImageOutputStream(f);
110        writer.setOutput(imageOut);
111
112        writer.write(new IIOImage(images[0], null, null));
113        for (int i = 1; i < images.length; i++)
114        {
115           var iioImage = new IIOImage(images[i], null, null);
116           if (writer.canInsertImage(i)) writer.writeInsert(i, iioImage, null);
```

```
117                }
118            }
119            catch (IOException e)
120            {
121                JOptionPane.showMessageDialog(this, e);
122            }
123        }
124
125        /**
126         * Gets a set of "preferred" format names of all image writers. The preferred format name
127         * is the first format name that a writer specifies.
128         * @return the format name set
129         */
130        public static Set<String> getWriterFormats()
131        {
132            var writerFormats = new TreeSet<String>();
133            var formatNames = List.of(ImageIO.getWriterFormatNames());
134            while (formatNames.size() > 0)
135            {
136                String name = formatNames.iterator().next();
137                Iterator<ImageWriter> iter = ImageIO.getImageWritersByFormatName(name);
138                ImageWriter writer = iter.next();
139                String[] names = writer.getOriginatingProvider().getFormatNames();
140                String format = names[0];
141                if (format.equals(format.toLowerCase())) format = format.toUpperCase();
142                writerFormats.add(format);
143                formatNames.removeAll(List.of(names));
144            }
145            return writerFormats;
146        }
147 }
```

API javax.imageio.ImageIO 1.4

- static BufferedImage read(File input)
- static BufferedImage read(InputStream input)
- static BufferedImage read(URL input)

  从 input 中读取一个图像。

- static boolean write(RenderedImage image, String formatName, File output)
- static boolean write(RenderedImage image, String formatName, OutputStream output)

  将给定格式的图像写入 output 中。如果没有找到合适的写入器，则返回 false。

- static Iterator<ImageReader> getImageReadersByFormatName(String formatName)
- static Iterator<ImageReader> getImageReadersBySuffix(String fileSuffix)
- static Iterator<ImageReader> getImageReadersByMIMEType(String mimeType)
- static Iterator<ImageWriter> getImageWritersByFormatName(String formatName)
- static Iterator<ImageWriter> getImageWritersBySuffix(String fileSuffix)
- static Iterator<ImageWriter> getImageWritersByMIMEType(String mimeType)

获得能够处理给定格式（例如 "JPEG"）、文件后缀（例如 "jpg"）或者 MIME 类型（例如 "image/jpeg"）的所有读取器和写入器。

- static String[] getReaderFormatNames()
- static String[] getReaderMIMETypes()
- static String[] getWriterFormatNames()
- static String[] getWriterMIMETypes()
- static String[] getReaderFileSuffixes() 6
- static String[] getWriterFileSuffixes() 6

获取读取器和写入器所支持的所有格式名、MIME 类型名和文件后缀。

- ImageInputStream createImageInputStream(Object input)
- ImageOutputStream createImageOutputStream(Object output)

根据给定的对象来创建一个图像输入流或者图像输出流。该对象可能是一个文件、一个流、一个 RandomAccessFile 或者某个服务提供商能够处理的其他类型的对象。如果没有任何注册过的服务提供器能够处理这个对象，那么返回 null 值。

API javax.imageio.ImageReader 1.4

- void setInput(Object input)
- void setInput(Object input, boolean seekForwardOnly)

设置读取器的输入源。

参数：input　　　　　　一个 ImageInputStream 对象或者是这个读取器能够接受的其他对象
　　　seekForwardOnly　如果读取器只应该向前读取，则返回 true。默认地，读取器会采用随机访问的方式，如果有必要，将会缓存图像数据

- BufferedImage read(int index)

读取给定索引的图像（索引从 0 开始）。如果没有这个图像，则抛出一个 IndexOutOfBounds-Exception 异常。

- int getNumImages(boolean allowSearch)

获取读取器中图像的数目。如果 allowSearch 值为 false，并且不向前阅读就无法确定图像的数目，那么它将返回 — 1。如果 allowSearch 值是 true，并且读取器采用了"只向前搜索"方式，那么就会抛出 IllegalStateException 异常。

- int getNumThumbnails(int index)

获取给定索引的图像的缩略图的数量。

- BufferedImage readThumbnail(int index, int thumbnailIndex)

获取给定索引的图像的索引号为 thumbnailIndex 的缩略图。

- int getWidth(int index)
- int getHeight(int index)

获取图像的宽度和高度。如果没有这样的图像，就抛出一个 IndexOutOfBounds-Exception

异常。
- ImageReaderSpi getOriginatingProvider()
获取构建该读取器的服务提供者。

API javax.imageio.spi.IIOServiceProvider 1.4
- String getVendorName()
- String getVersion()

获取该服务提供者的提供商的名字和版本。

API javax.imageio.spi.ImageReaderWriterSpi 1.4
- String[] getFormatNames()
- String[] getFileSuffixes()
- String[] getMIMETypes()

获取由该服务提供者创建的读取器或者写入器所支持的图像格式名、文件的后缀和 MIME 类型。

API javax.imageio.ImageWriter 1.4
- void setOutput(Object output)
  设置该写入器的输出目标。
  参数：output   一个 ImageOutputSteam 对象或者这个写入器能够接受的其他对象。
- void write(IIOImage image)
- void write(RenderedImage image)
  把单一的图像写入到输出流中。
- void writeInsert(int index, IIOImage image, ImageWriteParam param)
  把一个图像写入到一个包含多个图像的文件中。
- boolean canInsertImage(int index)
  如果在给定的索引处可以插入一个图像的话，则返回 true 值。
- ImageWriterSpi getOriginatingProvider()
  获取构建该写入器的服务提供者。

API javax.imageio.IIOImage 1.4
- IIOImage(RenderedImage image, List thumbnails, IIOMetadata metadata)
  根据一个图像、可选的缩略图和可选的元数据来构建一个 IIOImage 对象。

11.4.2 图像处理

假设你有一个图像，并且希望改善图像的外观。这时需要访问该图像的每一个像素，并用其他的像素来取代这些像素。或者，你也许想要从头计算某个图像的像素，例如，你想显示一下物理测量或者数学计算的结果。BufferedImage 类提供了对图像中像素的控制能力，而实现了 BufferedImageOP 接口的类都可以对图像进行变换操作。

> **注释**：JDK1.0 有一个完全不同且复杂得多的图像框架，它得到了优化，以支持对从 Web 下载的图像进行增量渲染（incremental rendering），即一次绘制一个扫描行。但是，操作这些图像很困难。我们在本书中不讨论这个框架。

11.4.2.1 构建像素图

你处理的大多数图像都是直接从图像文件中读入的。这些图像有的可能是数码相机产生的，有的是扫描仪扫描而产生的，还有的一些图像是绘图程序产生的。在本节中，我们将介绍一种不同的构建图像技术，也就是每次为图像增加一个像素。

为了创建一个图像，需要以通常的方法构建一个 BufferedImage 对象：

```
image = new BufferedImage(width, height, BufferedImage.TYPE_INT_ARGB);
```

现在，调用 getRaster 方法来获得一个类型为 WritableRaster 的对象，后面将使用这个对象来访问和修改该图像的各个像素：

```
WritableRaster raster = image.getRaster();
```

使用 setPixel 方法可以设置一个单独的像素。这项操作的复杂性在于不能只是为该像素设置一个 Color 值，还必须知道存放在缓冲中的图像是如何设定颜色的，这依赖于图像的类型。如果图像有一个 TYPE_INT_ARGB 类型，那么每一个像素都用四个值来描述，即：红、绿、蓝和透明度（alpha），每个值的取值范围都介于 0 和 255 之间，这需要以包含四个整数值的一个数组的形式给出：

```
int[] black = { 0, 0, 0, 255 };
raster.setPixel(i, j, black);
```

用 Java 2D API 的行话来说，这些值被称为像素的样本值。

> **警告**：还有一些参数值是 float[] 和 double[] 类型的 setPixel 方法。然而，需要在这些数组中放置的值并不是介于 0.0 和 1.0 之间的规格化的颜色值：
>
> ```
> float[] red = { 1.0F, 0.0F, 0.0F, 1.0F };
> raster.setPixel(i, j, red); // ERROR
> ```
>
> 无论数组属于什么类型，都必须提供介于 0 和 255 之间的某个值。

可以使用 setPixels 方法提供批量的像素。需要设置矩形的起始像素的位置和矩形的宽度和高度。接着，提供一个包含所有像素的样本值的一个数组。例如，如果你缓冲的图像有一个 TYPE_INT_ARGB 类型，那么就应该提供第一个像素的红、绿、蓝和透明度的值（alpha），然后，提供第二个像素的红、绿、蓝和透明度的值，以此类推：

```
var pixels = new int[4 * width * height];
pixels[0] = ...; // red value for first pixel
pixels[1] = ...; // green value for first pixel
pixels[2] = ...; // blue value for first pixel
pixels[3] = ...; // alpha value for first pixel
...
raster.setPixels(x, y, width, height, pixels);
```

反过来，如果要读入一个像素，可以使用 getPixel 方法。这需要提供一个含有四个整数的数组，用以存放各个样本值：

```
var sample = new int[4];
raster.getPixel(x, y, sample);
var color = new Color(sample[0], sample[1], sample[2], sample[3]);
```

可以使用 getPixels 方法来读取多个像素：

```
raster.getPixels(x, y, width, height, samples);
```

如果使用的图像类型不是 TYPE_INT_ARGB，并且已知该类型是如何表示像素值的，那么仍旧可以使用 getPixel/setPixel 方法。不过，必须要知道该特定图像类型的样本值是如何进行编码的。

如果需要对任意未知类型的图像进行处理，那么你就要费神了。每一个图像类型都有一个颜色模型，它能够在样本值数组和标准的 RGB 颜色模型之间进行转换。

> **注释**：RGB 颜色模型并不像你想象中的那么标准。颜色值的确切样子依赖于成像设备的特性。数码相机、扫描仪、控制器和 LCD 显示器等都有它们独有的特性。结果是，同样的 RGB 值在不同的设备上看上去就存在很大的差别。国际配色联盟（http://www.color.org）推荐，所有的颜色数据都应该配有一个 ICC 配置特性，它用以设定各种颜色是如何映射到标准格式的，比如 1931 CIE XYZ 颜色技术规范。该规范是由国际照明委员会即 CIE（Commission Internationale de l'Eclairage，其网址为：http://www.cie.co.at）制定的。该委员会是负责提供涉及照明和颜色等相关领域事务的技术指导的国际性机构。该规范是显示肉眼能够察觉到的所有颜色的一个标准化方法。它采用称为 X、Y、Z 三元组坐标的方式来显示颜色。（关于 1931 CIE XYZ 规范的详尽信息，可以参阅 Foley、van Dam 和 Feiner 等人所撰写的 *Computer Graphics: Principles and Practice* 一书的第 13 章。）
>
> ICC 配置特性非常复杂。然而，我们建议使用一个相对简单的标准，称为 sRGB（请访问其网址 http://www.w3.org/Color/sRGB.html）。它设定了 RGB 值与 1931 CIE XYZ 值之间的具体转换方法，它可以非常出色地在通用的彩色监视器上应用。当需要在 RGB 与其他颜色空间之间进行转换的时候，Java 2D API 就使用这种转换方式。getColorModel 方法返回一个颜色模型：
>
> ```
> ColorModel model = image.getColorModel();
> ```

为了了解一个像素的颜色值，可以调用 Raster 类的 getDataElements 方法。这个方法返回了一个 Object，它包含了有关该颜色值的与特定颜色模型相关的描述：

```
Object data = raster.getDataElements(x, y, null);
```

> **注释**：getDataElements 方法返回的对象实际上是一个样本值的数组。在处理这个对象时，不必要了解到这些。但是，它却解释了为什么这个方法名叫做 getDataElements 的原因。

颜色模型能够将该对象转换成标准的 ARGB 的值。getRGB 方法返回一个 int 类型的值，它把透明度（alpha）、红、绿和蓝的值打包成四个块，每块包含 8 位。也可以使用 Color(int argb, boolean hasAlpha) 构造器来构建一个颜色的值：

```
int argb = model.getRGB(data);
var color = new Color(argb, true);
```

如果要把一个像素设置为某个特定的颜色值，需要按与上述相反的步骤进行操作。Color 类的 getRGB 方法会产生一个包含透明度、红、绿和蓝值的 int 型值。把这个值提供给 ColorModel 类的 getDataElements 方法，其返回值是一个包含了该颜色值的特定颜色模型描述的 Object。再将这个对象传递给 WritableRaster 类的 setDataElements 方法：

```
int argb = color.getRGB();
Object data = model.getDataElements(argb, null);
raster.setDataElements(x, y, data);
```

为了阐明如何使用这些方法来用各个像素构建图像，我们按照传统，绘制了一个 Mandelbrot 集，如图 11-56 所示。

Mandelbrot 集的思想就是把平面上的每一点和一个数字序列关联在一起。如果数字序列是收敛的，该点就被着色。如果数字序列是发散的，该点就处于透明状态。

下面就是构建简单 Manderbrot 集的方法。对于每一个点（a, b），你都能按照如下的公式得到一个点集序列，其开始于点 (x, y) = (0, 0)，反复进行迭代：

$$x_{new} = x^2 - y^2 + a$$
$$y_{new} = 2 \cdot x \cdot y + b$$

结果证明，如果 x 或者 y 的值大于 2，那么序列就是发散的。仅有那些与导致数字序列收敛的点（a, b）相对应的像素才会被着色。（该数字序列的计算公式基本上是从复杂的数学概念中推导出来的。我们只使用现成的公式。）

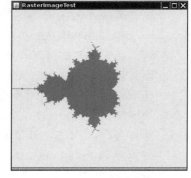

图 11-56　Mandelbrot 集

程序清单 11-21 显示了该代码。在此程序中，我们展示了如何使用 ColorModel 类将 Color 值转换成像素数据。这个过程和图像的类型是不相关的。为了增加些趣味，你可以把缓冲图像的颜色类型改变为 TYPE_BYTE_GRAY。不必改变程序中的任何代码，该图像的颜色模型会自动地负责把颜色转换为样本值。

程序清单 11-21　rasterImage/RasterImageFrame.java

```
1  package rasterImage;
2
3  import java.awt.*;
4  import java.awt.image.*;
5  import javax.swing.*;
6
7  /**
```

```java
 8    * This frame shows an image with a Mandelbrot set.
 9    */
10   public class RasterImageFrame extends JFrame
11   {
12      private static final double XMIN = -2;
13      private static final double XMAX = 2;
14      private static final double YMIN = -2;
15      private static final double YMAX = 2;
16      private static final int MAX_ITERATIONS = 16;
17      private static final int IMAGE_WIDTH = 400;
18      private static final int IMAGE_HEIGHT = 400;
19
20      public RasterImageFrame()
21      {
22         BufferedImage image = makeMandelbrot(IMAGE_WIDTH, IMAGE_HEIGHT);
23         add(new JLabel(new ImageIcon(image)));
24         pack();
25      }
26
27      /**
28       * Makes the Mandelbrot image.
29       * @param width the width
30       * @parah height the height
31       * @return the image
32       */
33      public BufferedImage makeMandelbrot(int width, int height)
34      {
35         var image = new BufferedImage(width, height, BufferedImage.TYPE_INT_ARGB);
36         WritableRaster raster = image.getRaster();
37         ColorModel model = image.getColorModel();
38
39         Color fractalColor = Color.RED;
40         int argb = fractalColor.getRGB();
41         Object colorData = model.getDataElements(argb, null);
42
43         for (int i = 0; i < width; i++)
44            for (int j = 0; j < height; j++)
45            {
46               double a = XMIN + i * (XMAX - XMIN) / width;
47               double b = YMIN + j * (YMAX - YMIN) / height;
48               if (!escapesToInfinity(a, b)) raster.setDataElements(i, j, colorData);
49            }
50         return image;
51      }
52
53      private boolean escapesToInfinity(double a, double b)
54      {
55         double x = 0.0;
56         double y = 0.0;
57         int iterations = 0;
58         while (x <= 2 && y <= 2 && iterations < MAX_ITERATIONS)
59         {
60            double xnew = x * x - y * y + a;
61            double ynew = 2 * x * y + b;
```

```
62              x = xnew;
63              y = ynew;
64              iterations++;
65           }
66           return x > 2 || y > 2;
67       }
68   }
```

API java.awt.image.BufferedImage 1.2

- BufferedImage(int width, int height, int imageType)
  构建一个被缓存的图像对象。
  参数：width, height 图像的尺寸
       imageType   图像的类型，最常用的类型是 TYPE_INT_RGB、TYPE_INT_ARGB、TYPE_BYTE_GRAY 和 TYPE_BYTE_INDEXED
- ColorModel getColorModel()
  返回被缓存图像的颜色模型。
- WritableRaster getRaster()
  获得访问和修改该缓存图像的像素栅格。

API java.awt.image.Raster 1.2

- Object getDataElements(int x, int y, Object data)
  返回某个栅格点的样本数据，该数据位于一个数组中，而该数组的长度和类型依赖于颜色模型。如果 data 不为 null，那么它将被视为是适合于存放样本数据的数组，从而被充填。如果 data 为 null，那么将分配一个新的数组，其元素的类型和长度依赖于颜色模型。
- int[] getPixel(int x, int y, int[] sampleValues)
- float[] getPixel(int x, int y, float[] sampleValues)
- double[] getPixel(int x, int y, double[] sampleValues)
- int[] getPixels(int x, int y, int width, int height, int[] sampleValues)
- float[] getPixels(int x, int y, int width, int height, float[] sampleValues)
- double[] getPixels(int x, int y, int width, int height, double[] sampleValues)
  返回某个栅格点或者是由栅格点组成的某个矩形的样本值，该数据位于一个数组中，数组的长度依赖于颜色模型。如果 sampleValues 不为 null，那么该数组被视为长度足够存放样本值，从而该数组被填充。如果 sampleValues 为 null，就要分配一个新数组。仅当你知道某一颜色模型的样本值的具体含义的时候，这些方法才会有用。

API java.awt.image.WritableRaster 1.2

- void setDataElements(int x, int y, Object data)
  设置栅格点的样本数据。data 是一个已经填入了某一像素样本值的数组。数组元素的

类型和长度依赖于颜色模型。
- void setPixel(int x, int y, int[] sampleValues)
- void setPixel(int x, int y, float[] sampleValues)
- void setPixel(int x, int y, double[] sampleValues)
- void setPixels(int x, int y, int width, int height, int[] sampleValues)
- void setPixels(int x, int y, int width, int height, float[] sampleValues)
- void setPixels(int x, int y, int width, int height, double[] sampleValues)

设置某个栅格点或由多个栅格点组成的矩形的样本值。只有当你知道颜色模型样本值的编码规则时，这些方法才会有用。

API java.awt.image.ColorModel 1.2

- int getRGB(Object data)

  返回对应于 data 数组中传递的样本数据的 ARGB 值。其元素的类型和长度依赖于颜色模型。

- Object getDataElements(int argb, Object data);

  返回某个颜色值的样本数据。如果 data 不为 null，那么该数组被视为非常适合于存放样本值，进而该数组被填充。如果 data 为 null，那么将分配一个新的数组。data 是一个填充了用于某个像素的样本数据的数组，其元素的类型和长度依赖于该颜色模型。

API java.awt.Color 1.0

- Color(int argb, boolean hasAlpha) 1.2

  如果 hasAlpha 的值是 true，则用指定的 ARGB 组合值创建一种颜色。如果 hasAlpha 的值是 false，则用指定的 RGB 值创建一种颜色。

- int getRGB()

  返回和该颜色相对应的 ARGB 颜色值。

11.4.2.2 图像过滤

在前面的章节中，我们介绍了从头开始构建图像的方法。然而，你常常是因为另一个原因去访问图像数据的：你已经拥有了一个图像，并且想从某些方面对图像进行改进。

当然，可以使用前一节中的 getPixel/getDataElements 方法来读取和处理图像数据，然后把图像数据写回到文件中。不过，幸运的是，Java 2D API 已经提供了许多过滤器，它们能够执行常用的图像处理操作。

图像处理都实现了 BufferedImageOp 接口。构建了图像处理的操作之后，只需调用 filter 方法，就可以把该图像转换成另一个图像。

```
BufferedImageOp op = . . .;
BufferedImage filteredImage
    = new BufferedImage(image.getWidth(), image.getHeight(), image.getType());
op.filter(image, filteredImage);
```

有些图像操作可以恰当地（通过 op.filter(image,image) 方法）转换一个图像，但是大多数的图像操作都做不到这一点。

以下五个类实现了 BufferedImageOp 接口。

AffineTransformOp
RescaleOp
LookupOp
ColorConvertOp
ConvolveOp

AffineTransformOp 类用于对各个像素执行仿射变换。例如，下面的代码就说明了如何使一个图像围绕着它的中心旋转。

```
AffineTransform transform = AffineTransform.getRotateInstance(Math.toRadians(angle),
    image.getWidth() / 2, image.getHeight() / 2);
var op = new AffineTransformOp(transform, interpolation);
op.filter(image, filteredImage);
```

AffineTransformOp 构造器需要一个仿射变换和一个渐变变换策略。如果源像素在目标像素之间的某处会发生变换的话，那么就必须使用渐变变换策略来确定目标图像的像素。例如，如果旋转源像素，那么通常它们不会精确地落在目标像素上。有两种渐变变换策略：AffineTransformOp.TYPE_BILINEAR 和 AffineTransformOp.TYPE_NEAREST_NEIGHBOR。双线性（Bilinear）渐变变换需要的时间较长，但是变换的效果却更好。

使用程序清单 11-22 的程序，可以把一个图像旋转 5°（参见图 11-57）。

RescaleOp 用于为图像中的所有的颜色构件执行一个调整其大小的变换操作（透明度构件不受影响）：

$$x_{\text{new}} = a \cdot x + b$$

用 a>1 进行调整，那么调整后的效果是使图像变亮。可以通过设定调整大小的参数和可选的绘图提示来构建 RescaleOp。在程序清单 11-22 中，我们使用下面的设置：

```
float a = 1.1f;
float b = 20.0f;
var op = new RescaleOp(a, b, null);
```

也可以为每个颜色构件提供单独的缩放值，参见 API 说明。

图 11-57　一个旋转的图像

使用 LookupOp 操作，可以为样本值设定任意的映射操作。你提供一张表格，用于设定每一个样本值应该如何进行映射操作。在示例程序中，我们计算了所有颜色的反，即将颜色 c 变成 255 — c。

LookupOp 构造器需要一个类型是 LookupTable 的对象和一个选项提示映射表。LookupTable 是抽象类，其有两个实体子类：ByteLookupTable 和 ShortLookupTable。因为 RGB 颜色值是由字节组成的，所以 ByteLookupTable 类应该就够用了。但是，考虑到在 http://bugs.sun.com/bugdatabase/view_bug.do?bug_id=6183251 中描述的缺陷，我们将使用 ShortLookupTable。下面的代码说明了我们在程序清单中是如何构建一个 LookupOp 类的：

```
var negative = new short[256];
for (int i = 0; i < 256; i++) negative[i] = (short) (255 - i);
```

```
var table = new ShortLookupTable(0, negative);
var op = new LookupOp(table, null);
```

此项操作可以分别应用于每个颜色构件,但是不能应用于透明度值。也可以为每个颜色构件提供单独的查找表,参见 API 说明。

> **注释**:不能将 LookupOp 用于带有索引颜色模型的图像。(在这些图像中,每个样本值都是调色板中的一个偏移量。)

ColorConvertOp 对于颜色空间的转换非常有用。我们不准备在这里讨论这个问题了。

ConvolveOp 是功能最强大的转换操作,它用于执行卷积变换。我们不想过分深入地介绍卷积变换的详尽细节。不过,其基本概念还是比较简单的。我们不妨看一下模糊过滤器的例子(见图 11-58)。

这种模糊的效果是通过用像素和该像素临近的 8 个像素的平均值来取代每一个像素值而达到的。凭借直观感觉,就可以知道为什么这种变换操作能使得图像变模糊了。从数学理论上来说,这种平均法可以表示为一个以下面这个矩阵为内核的卷积变换操作:

$$\begin{bmatrix} 1/9 & 1/9 & 1/9 \\ 1/9 & 1/9 & 1/9 \\ 1/9 & 1/9 & 1/9 \end{bmatrix}$$

卷积变换操作的内核是一个矩阵,用以说明在临近的像素点上应用的加权值。应用上面的内核进行卷积变换,就会产生一个模糊图像。下面这个不同的内核用以进行图像的边缘检测,查找图像颜色变化的区域:

$$\begin{bmatrix} 0 & -1 & 0 \\ -1 & 4 & -1 \\ 0 & -1 & 0 \end{bmatrix}$$

边缘检测是在分析摄影图片时使用的一项非常重要的技术(参见图 11-59)。

图 11-58 对图像进行模糊处理

图 11-59 边缘检测

如果要构建一个卷积变换操作,首先应为矩阵内核建立一个含有内核值的数组,并且构建一个 Kernel 对象。接着，根据内核对象建立一个 ConvolveOp 对象，进而执行过滤操作。

```
float[] elements =
   {
      0.0f, -1.0f, 0.0f,
      -1.0f,  4.f, -1.0f,
      0.0f, -1.0f, 0.0f
   };
var kernel = new Kernel(3, 3, elements);
var op = new ConvolveOp(kernel);
op.filter(image, filteredImage);
```

使用程序清单 11-22 的程序，用户可以装载一个 GIF 或者 JPEG 图像，并且执行我们已经介绍过的各种图像处理的操作。由于 Java 2D API 的图像处理的功能很强大，下面的程序非常简单。

程序清单 11-22　imageProcessing/ImageProcessingFrame.java

```java
 1  package imageProcessing;
 2
 3  import java.awt.*;
 4  import java.awt.geom.*;
 5  import java.awt.image.*;
 6  import java.io.*;
 7
 8  import javax.imageio.*;
 9  import javax.swing.*;
10  import javax.swing.filechooser.*;
11
12  /**
13   * This frame has a menu to load an image and to specify various transformations, and a
14   * component to show the resulting image.
15   */
16  public class ImageProcessingFrame extends JFrame
17  {
18     private static final int DEFAULT_WIDTH = 400;
19     private static final int DEFAULT_HEIGHT = 400;
20
21     private BufferedImage image;
22
23     public ImageProcessingFrame()
24     {
25        setTitle("ImageProcessingTest");
26        setSize(DEFAULT_WIDTH, DEFAULT_HEIGHT);
27
28        add(new JComponent()
29           {
30              public void paintComponent(Graphics g)
31              {
32                 if (image != null) g.drawImage(image, 0, 0, null);
33              }
34           });
35
36        var fileMenu = new JMenu("File");
```

```java
37         var openItem = new JMenuItem("Open");
38         openItem.addActionListener(event -> openFile());
39         fileMenu.add(openItem);
40
41         var exitItem = new JMenuItem("Exit");
42         exitItem.addActionListener(event -> System.exit(0));
43         fileMenu.add(exitItem);
44
45         var editMenu = new JMenu("Edit");
46         var blurItem = new JMenuItem("Blur");
47         blurItem.addActionListener(event ->
48            {
49               float weight = 1.0f / 9.0f;
50               float[] elements = new float[9];
51               for (int i = 0; i < 9; i++)
52                  elements[i] = weight;
53               convolve(elements);
54            });
55         editMenu.add(blurItem);
56
57         var sharpenItem = new JMenuItem("Sharpen");
58         sharpenItem.addActionListener(event ->
59            {
60               float[] elements = { 0.0f, -1.0f, 0.0f, -1.0f, 5.f, -1.0f, 0.0f, -1.0f, 0.0f };
61               convolve(elements);
62            });
63         editMenu.add(sharpenItem);
64
65         var brightenItem = new JMenuItem("Brighten");
66         brightenItem.addActionListener(event ->
67            {
68               float a = 1.1f;
69               float b = 20.0f;
70               var op = new RescaleOp(a, b, null);
71               filter(op);
72            });
73         editMenu.add(brightenItem);
74
75         var edgeDetectItem = new JMenuItem("Edge detect");
76         edgeDetectItem.addActionListener(event ->
77            {
78               float[] elements = { 0.0f, -1.0f, 0.0f, -1.0f, 4.f, -1.0f, 0.0f, -1.0f, 0.0f };
79               convolve(elements);
80            });
81         editMenu.add(edgeDetectItem);
82
83         var negativeItem = new JMenuItem("Negative");
84         negativeItem.addActionListener(event ->
85            {
86               short[] negative = new short[256 * 1];
87               for (int i = 0; i < 256; i++)
88                  negative[i] = (short) (255 - i);
89               var table = new ShortLookupTable(0, negative);
90               var op = new LookupOp(table, null);
```

```java
               filter(op);
            });
         editMenu.add(negativeItem);

         var rotateItem = new JMenuItem("Rotate");
         rotateItem.addActionListener(event ->
            {
               if (image == null) return;
               var transform = AffineTransform.getRotateInstance(Math.toRadians(5),
                  image.getWidth() / 2, image.getHeight() / 2);
               var op = new AffineTransformOp(transform,
                  AffineTransformOp.TYPE_BICUBIC);
               filter(op);
            });
         editMenu.add(rotateItem);

         var menuBar = new JMenuBar();
         menuBar.add(fileMenu);
         menuBar.add(editMenu);
         setJMenuBar(menuBar);
      }

      /**
       * Open a file and load the image.
       */
      public void openFile()
      {
         var chooser = new JFileChooser(".");
         chooser.setCurrentDirectory(new File(getClass().getPackage().getName()));
         String[] extensions = ImageIO.getReaderFileSuffixes();
         chooser.setFileFilter(new FileNameExtensionFilter("Image files", extensions));
         int r = chooser.showOpenDialog(this);
         if (r != JFileChooser.APPROVE_OPTION) return;

         try
         {
            Image img = ImageIO.read(chooser.getSelectedFile());
            image = new BufferedImage(img.getWidth(null), img.getHeight(null),
               BufferedImage.TYPE_INT_RGB);
            image.getGraphics().drawImage(img, 0, 0, null);
         }
         catch (IOException e)
         {
            JOptionPane.showMessageDialog(this, e);
         }
         repaint();
      }

      /**
       * Apply a filter and repaint.
       * @param op the image operation to apply
       */
      private void filter(BufferedImageOp op)
      {
```

```
145        if (image == null) return;
146        image = op.filter(image, null);
147        repaint();
148     }
149
150     /**
151      * Apply a convolution and repaint.
152      * @param elements the convolution kernel (an array of 9 matrix elements)
153      */
154     private void convolve(float[] elements)
155     {
156        var kernel = new Kernel(3, 3, elements);
157        var op = new ConvolveOp(kernel);
158        filter(op);
159     }
160  }
```

API *java.awt.image.BufferedImageOp* 1.2

- BufferedImage filter(BufferedImage source, BufferedImage dest)

  将图像操作应用于源图像,并且将操作的结果存放在目标图像中。如果 dest 为 null,一个新的目标图像将被创建。该目标图像将被返回。

API *java.awt.image.AffineTransformOp* 1.2

- AffineTransformOp(AffineTransform t, int interpolationType)

  构建一个仿射变换操作符。渐变变换的类型是 TYPE_BILINEAR、TYPE_BICUBIC 或者 TYPE_NEAREST_NEIGHBOR 中的一个。

API *java.awt.image.RescaleOp* 1.2

- RescaleOp(float a, float b, RenderingHints hints)
- RescaleOp(float[] as, float[] bs, RenderingHints hints)

  构建一个进行尺寸调整的操作符,它会执行缩放操作 $x_{new} = a \cdot x + b$。当使用第一个构造器时,所有的颜色构件(但不包括透明度构件)都将按照相同的系数进行缩放。当使用第二个构造器时,可以为每个颜色构件提供单独的值,在这种情况下,透明度构件不受影响,或者为每个颜色构件和透明度构件都提供单独的值。

API *java.awt.image.LookupOp* 1.2

- LookupOp(LookupTable table, RenderingHints hints)

  为给定的查找表构建一个查找操作符。

API *java.awt.image.ByteLookupTable* 1.2

- ByteLookupTable(int offset, byte[] data)
- ByteLookupTable(int offset, byte[][] data)

  为转化 byte 值构建一个字节查找表。在查找之前,从输入中减去偏移量。在第一个构造器中的值将提供给所有的颜色构件，但不包括透明度构件。当使用第二个构造器时，可以为每个颜色构件提供单独的值，在这种情况下，透明度构件不受影响，或者为每个颜色构件和透明度构件都提供单独的值。

API java.awt.image.ShortLookupTable 1.2

- ShortLookupTable(int offset, short[] data)
- ShortLookupTable(int offset, short[][] data)

为转化 short 值构建一个字节查找表。在查找之前，从输入中减去偏移量。在第一个构造器中的值将提供给所有的颜色构件，但不包括透明度构件。当使用第二个构造器时，可以为每个颜色构件提供单独的值，在这种情况下，透明度构件不受影响，或者为每个颜色构件和透明度构件都提供单独的值。

API java.awt.image.ConvolveOp 1.2

- ConvolveOp(Kernel kernel)
- ConvolveOp(Kernel kernel, int edgeCondition, RenderingHints hints)

构建一个卷积变换操作符。边界条件是 EDGE_NO_OP 和 EDGE_ZERO_FILL 两种方式之一。由于边界值没有足够的临近值来进行卷积变换的计算，所以边界值必须被特殊处理，其默认值是 EDGE_ZERO_FILL。

API java.awt.image.Kernel 1.2

- Kernel(int width, int height, float[] matrixElements)

为指定的矩阵构建一个内核。

11.5 打印

在本节中，我们将介绍如何在单页纸上轻松地打印出一幅图画，如何来管理多页打印输出，以及如何将打印内容存储为 PostScript 文件。

11.5.1 图形打印

在本节中，我们将处理最常用的打印情景，即打印一个 2D 图形，当然该图形可以含有不同字体组成的文本，甚至可能完全由文本构成。

如果要生成打印输出，必须完成下面这两个任务：

- 提供一个实现了 Printable 接口的对象。
- 启动一个打印作业。

Printable 接口只有下面一个方法：

```
int print(Graphics g, PageFormat format, int page)
```

每当打印引擎需要对某一页面进行排版以便打印时，都要调用这个方法。你的代码绘制了准备在图形上下文上打印的文本和图像,页面排版显示了纸张的大小和页边距,页号显示了将要打印的页。

如果要启动一个打印作业,需要使用 PrinterJob 类。首先,应该调用静态方法 getPrinterJob 来获取一个打印作业对象。然后,设置要打印的 Printable 对象。

```
Printable canvas = . . .;
PrinterJob job = PrinterJob.getPrinterJob();
job.setPrintable(canvas);
```

> ◆ **警告**:PrintJob 这个类处理的是 JDK1.1 风格的打印操作,这个类已经被弃用了。请不要把 PrinterJob 类同其混淆在一起。

在开始打印作业之前,应该调用 printDialog 方法来显示一个打印对话框(见图 11-60)。这个对话框为用户提供了机会去选择要使用的打印机(在有多个打印机可用的情况下),选择将要打印的页的范围,以及选择打印机的各种设置。

可以在一个实现了 PrintRequestAttributeSet 接口的类的对象中收集到各种打印机的设置,例如 HashPrintRequestAttributeSet 类:

```
var attributes = new HashPrintRequestAttributeSet();
```

你可以添加属性设置,并且把 attributes 对象传递给 printDialog 方法。

如果用户点击 OK,那么 printDialog 方法将返回 true;如果用户关掉对话框,那么该方法

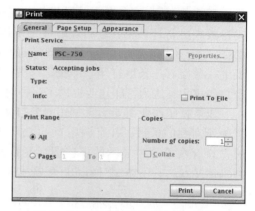

图 11-60 一个跨平台的打印对话框

将返回 false。如果用户接受了设置,那么就可以调用 PrinterJob 类的 print 方法来启动打印进程。print 方法可能会抛出一个 PrinterException 异常。下面是打印代码的基本框架:

```
if (job.printDialog(attributes))
{
   try
   {
      job.print(attributes);
   }
   catch (PrinterException exception)
   {
      . . .
   }
}
```

> 📖 **注释**:在 JDK1.4 之前,打印系统使用的都是宿主平台本地的打印和页面设置对话框。要展示本地打印对话框,可以调用没有任何参数的 printDialog 方法。(不存在任何方式可以用来将用户的设置收集到一个属性集中。)

在执行打印操作时，PrinterJob 类的 print 方法不断地调用和此项打印作业相关的 Printable 对象的 print 方法。

由于打印作业不知道用户想要打印的页数，所以它只是不断地调用 print 方法。只要该 print 方法的返回值是 Printable.PAGE_EXISTS，打印作业就不断地产生输出页。当 print 方法返回 Pringtable.NO_SUCH_PAGE 时，打印作业就停止。

> ⚠️ **警告**：打印作业传递到 print 方法的打印页号是从 0 开始的。

因此，在打印操作完成之前，打印作业并不知道准确的打印页数。为此，打印对话框无法显示正确的页码范围，而只能显示"Pages 1 to 1"（从第一页到第一页）。在下一节中，我们将介绍如何通过为打印作业提供一个 Book 对象来避免这个缺陷。

在打印的过程中，打印作业反复地调用 Printable 对象的 print 方法。打印作业可以对同一页面多次调用 print 方法，因此不应该在 print 方法内对页进行计数，而是应始终依赖于页码参数来进行计数操作。打印作业之所以能够对某一页反复地调用 print 方法是有一定道理的：一些打印机，尤其是点阵式打印机和喷墨式打印机，都使用条带打印技术，它们在打印纸上一条接着一条地打印。即使是每次打印一整页的激光打印机，打印作业都有可能使用条带打印技术。这为打印作业提供了一种对假脱机文件的大小进行管理的方法。

如果打印作业需要 printable 对象打印一个条带，那么它可以将图形上下文的剪切区域设置为所需要的条带，并且调用 print 方法。它的绘图操作将按照条带矩形区域进行剪切，同时，只有在条带中显示的那些图形元素才会被绘制出来。你的 print 方法不必晓得该过程，但是请注意：它不应该对剪切区域产生任何干扰。

> ⚠️ **警告**：你的 print 方法获得的 Graphics 对象也是按照页边距进行剪切的。如果替换了剪切区域，那么就可以在边距外面进行绘图操作。尤其是在打印机的绘图上下文中，剪切区域是被严格遵守的。如果想进一步地限制剪切区域，可以调用 clip 方法，而不是 setClip 方法。如果必须要移除一个剪切区域，那么请务必在你的 print 方法开始处调用 getClip 方法，并还原该剪切区域。

print 方法的 PageFormat 参数包含有关被打印页的信息。getWidth 方法和 getHeight 方法返回该纸张的大小，它以磅为计量单位。1 磅等于 1/72 英寸[⊖]。例如，A4 纸的大小大约是 595 × 842 磅，美国人使用的信纸大小为 612 × 792 磅。

磅是美国印刷业中通用的计量单位，让世界上其他地方的人感到苦恼的是，打印软件包使用的是磅这种计量单位。使用磅有两个原因，即纸张的大小和纸张的页边距都是用磅来计量的。对所有的图形上下文来说，默认的计量单位就是 1 磅。你可以在本节后面的示例程序中证明这一点。该程序打印了两行文本，这两行文本之间的距离为 72 磅。运行一下示例程序，并且测量一下基准线之间的距离。它们之间的距离恰好是 1 英寸或是 25.4 毫米。

PageFormat 类的 getWidth 和 getHeight 方法给你的信息是完整的页面大小，但并不是所有的

⊖ 1 英寸 = 0.0254 米。——编辑注

纸张区域都会被用来打印。通常的情况是，用户会选择页边距，即使他们没有选择页边距，打印机也需要用某种方法来夹住纸张，因此在纸张的周围就出现了一个不能打印的区域。

getImageableWidth 和 getImageableHeight 方法可以告诉你能够真正用来打印的区域的大小。然而，页边距没有必要是对称的，所以还必须知道可打印区域的左上角，见图 11-61，它们可以通过调用 getImageableX 和 getImageableY 方法来获得。

> **提示**：在 print 方法中接收到的图形上下文是经过剪切后的图形上下文，它不包括页边距。但是，坐标系统的原点仍然是纸张的左上角。应该将该坐标系统转换成可打印区域的左上角，并以其为起点。这只需让 print 方法以下面的代码开始即可：
>
>   g.translate(pageFormat.getImageableX(), pageFormat.getImageableY());

如果想让用户来设定页边距，或者让用户在纵向和横向打印方式之间切换，同时并不涉及设置其他打印属性，那么就应该调用 PrinterJob 类的 pageDialog 方法。

```
PageFormat format = job.pageDialog(attributes);
```

> **注释**：打印对话框中有一个选项卡包含了页面设置对话框（参见图 11-62）。在打印前，你仍然可以为用户提供选项来设置页面格式。特别是，如果你的程序给出了一个待打印页面的"所见即所得"的显示屏幕，那么就更应该提供这样的选项。pageDialog 方法返回了一个含有用户设置的 PageFormat 对象。

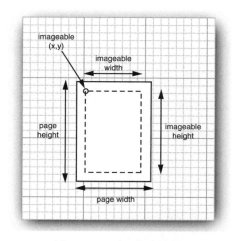

图 11-61　页面格式计量

图 11-62　一个跨平台的页面设置对话框

程序清单 11-23 和程序清单 11-24 显示了如何在屏幕和打印页面上绘制相同的一组形状的方法。Jpanel 类的一个子类实现了 Printable 接口，该类中的 paintComponent 和 print 方法都调用了相同的方法来执行实际的绘图操作。

```
class PrintPanel extends JPanel implements Printable
{
```

```java
   public void paintComponent(Graphics g)
   {
      super.paintComponent(g);
      var g2 = (Graphics2D) g;
      drawPage(g2);
   }

   public int print(Graphics g, PageFormat pf, int page) throws PrinterException
   {
      if (page >= 1) return Printable.NO_SUCH_PAGE;
      var g2 = (Graphics2D) g;
      g2.translate(pf.getImageableX(), pf.getImageableY());
      drawPage(g2);
      return Printable.PAGE_EXISTS;
   }

   public void drawPage(Graphics2D g2)
   {
      // shared drawing code goes here
      . . .
   }
   . . .
}
```

程序清单 11-23 print/PrintTestFrame.java

```
 1  package print;
 2
 3  import java.awt.*;
 4  import java.awt.print.*;
 5
 6  import javax.print.attribute.*;
 7  import javax.swing.*;
 8
 9  /**
10   * This frame shows a panel with 2D graphics and buttons to print the graphics and to set up
11   * the page format.
12   */
13  public class PrintTestFrame extends JFrame
14  {
15     private PrintComponent canvas;
16     private PrintRequestAttributeSet attributes;
17
18     public PrintTestFrame()
19     {
20        canvas = new PrintComponent();
21        add(canvas, BorderLayout.CENTER);
22
23        attributes = new HashPrintRequestAttributeSet();
24
25        var buttonPanel = new JPanel();
26        var printButton = new JButton("Print");
27        buttonPanel.add(printButton);
28        printButton.addActionListener(event ->
```

```java
29      {
30         try
31         {
32            PrinterJob job = PrinterJob.getPrinterJob();
33            job.setPrintable(canvas);
34            if (job.printDialog(attributes)) job.print(attributes);
35         }
36         catch (PrinterException ex)
37         {
38            JOptionPane.showMessageDialog(PrintTestFrame.this, ex);
39         }
40      });
41
42      var pageSetupButton = new JButton("Page setup");
43      buttonPanel.add(pageSetupButton);
44      pageSetupButton.addActionListener(event ->
45         {
46            PrinterJob job = PrinterJob.getPrinterJob();
47            job.pageDialog(attributes);
48         });
49
50      add(buttonPanel, BorderLayout.NORTH);
51      pack();
52   }
53 }
```

程序清单 11-24　print/PrintComponent.java

```java
 1 package print;
 2
 3 import java.awt.*;
 4 import java.awt.font.*;
 5 import java.awt.geom.*;
 6 import java.awt.print.*;
 7 import javax.swing.*;
 8
 9 /**
10  * This component generates a 2D graphics image for screen display and printing.
11  */
12 public class PrintComponent extends JComponent implements Printable
13 {
14    private static final Dimension PREFERRED_SIZE = new Dimension(300, 300);
15
16    public void paintComponent(Graphics g)
17    {
18       var g2 = (Graphics2D) g;
19       drawPage(g2);
20    }
21
22    public int print(Graphics g, PageFormat pf, int page) throws PrinterException
23    {
24       if (page >= 1) return Printable.NO_SUCH_PAGE;
25       var g2 = (Graphics2D) g;
26       g2.translate(pf.getImageableX(), pf.getImageableY());
```

```java
27          g2.draw(new Rectangle2D.Double(0, 0, pf.getImageableWidth(), pf.getImageableHeight()));
28
29          drawPage(g2);
30          return Printable.PAGE_EXISTS;
31       }
32
33       /**
34        * This method draws the page both on the screen and the printer graphics context.
35        * @param g2 the graphics context
36        */
37       public void drawPage(Graphics2D g2)
38       {
39          FontRenderContext context = g2.getFontRenderContext();
40          var f = new Font("Serif", Font.PLAIN, 72);
41          var clipShape = new GeneralPath();
42
43          var layout = new TextLayout("Hello", f, context);
44          AffineTransform transform = AffineTransform.getTranslateInstance(0, 72);
45          Shape outline = layout.getOutline(transform);
46          clipShape.append(outline, false);
47
48          layout = new TextLayout("World", f, context);
49          transform = AffineTransform.getTranslateInstance(0, 144);
50          outline = layout.getOutline(transform);
51          clipShape.append(outline, false);
52
53          g2.draw(clipShape);
54          g2.clip(clipShape);
55
56          final int NLINES = 50;
57          var p = new Point2D.Double(0, 0);
58          for (int i = 0; i < NLINES; i++)
59          {
60             double x = (2 * getWidth() * i) / NLINES;
61             double y = (2 * getHeight() * (NLINES - 1 - i)) / NLINES;
62             var q = new Point2D.Double(x, y);
63             g2.draw(new Line2D.Double(p, q));
64          }
65       }
66
67       public Dimension getPreferredSize() { return PREFERRED_SIZE; }
68    }
```

该示例代码显示并且打印了图 11-50，即被用作线条模式的剪切区域的消息 "Hello, World" 的边框。

可以点击 Print 按钮来启动打印，或者点击页面设置按钮来打开页面设置对话框。程序清单 11-23 显示了它的代码。

> **注释：**为了显示本地页面设置对话框，需要将默认的 PageFormat 对象传递给 pageDialog 方法。该方法会克隆这个对象，并根据用户在对话框中的选择来修改它，然后返回这个克隆的对象。

```
PageFormat defaultFormat = printJob.defaultPage();
PageFormat selectedFormat = printJob.pageDialog(defaultFormat);
```

API *java.awt.print.Printable* 1.2

- int print(Graphics g, PageFormat format, int pageNumber)

    绘制一个页面，并且返回 PAGE_EXISTS，或者返回 NO_SUCH_PAGE。

    参数：g   在上面绘制页面的图形上下文
       format   要绘制的页面的格式
       pageNumber 所请求页面的页码

API *java.awt.print.PrinterJob* 1.2

- static PrinterJob getPrinterJob()

    返回一个打印机作业对象。

- PageFormat defaultPage()

    为该打印机返回默认的页面格式。

- boolean printDialog(PrintRequestAttributeSet attributes)
- boolean printDialog()

    打开打印对话框，允许用户选择将要打印的页面，并且改变打印设置。第一个方法将显示一个跨平台的打印对话框，第二个方法将显示一个本地的打印对话框。第一个方法修改了 attributes 对象来反映用户的设置。如果用户接受默认的设置，两种方法都返回 true。

- PageFormat pageDialog(PrintRequestAttributeSet attributes)
- PageFormat pageDialog(PageFormat defaults)

    显示页面设置对话框。第一个方法将显示一个跨平台的对话框，第二个方法将显示一个本地的页面设置对话框。两种方法都返回了一个 PageFormat 对象，对象的格式是用户在对话框中所请求的格式。第一个方法修改了 attributes 对象以反映用户的设置。第二个对象不修改 defaults 对象。

- void setPrintable(Printable p)
- void setPrintable(Printable p, PageFormat format)

    设置该打印作业的 Printable 和可选的页面格式。

- void print()
- void print(PrintRequestAttributeSet attributes)

    反复地调用 print 方法，以打印当前的 Printable，并将绘制的页面发送给打印机，直到没有更多的页面需要打印为止。

API *java.awt.print.PageFormat* 1.2

- double getWidth()
- double getHeight()

返回页面的宽度和高度。
- double getImageableWidth()
- double getImageableHeight()

  返回可打印区域的页面宽度和高度。
- double getImageableX()
- double getImageableY()

  返回可打印区域的左上角的位置。
- int getOrientation()

  返回 PORTARIT、LANDSCAPE 和 REVERSE_LANDSCAPE 三者之一。页面打印的方向对程序员来说是透明的，因为打印格式和图形上下文自动地反映了页面的打印方向。

11.5.2 打印多页文件

在实际的打印操作中，通常不应该将原生的 Printable 对象传递给打印作业。相反，应该获取一个实现了 Pageable 接口的类的对象。Java 平台提供了这样的一个被称为 Book 的类。一本书是由很多章节组成的，而每个章节都是一个 Printable 对象。可以通过添加 Printable 对象和相应的页数来构建一个 Book 对象。

```
var book = new Book();
Printable coverPage = . . .;
Printable bodyPages = . . .;
book.append(coverPage, pageFormat); // append 1 page
book.append(bodyPages, pageFormat, pageCount);
```

然后，可以使用 setPageable 方法把 Book 对象传递给打印作业。

```
printJob.setPageable(book);
```

现在，打印作业就知道将要打印的确切页数了。然后，打印对话框显示一个准确的页面范围，用户可以选择整个页面范围或可选择它的一个子范围。

> ⚠️ **警告**：当打印作业调用 Printable 章节的 print 方法时，它传递的是该书的当前页码，而不是每个章节的页码。这让人非常痛苦，因为每个章节必须知道它之前所有章节的页数，这样才能使得页码参数有意义。

从程序员的视角来看，使用 Book 类最大的挑战就是，当你打印它时，必须知道每一个章节究竟有多少页。你的 Printable 类需要一个布局算法，以便用来计算在打印页面上的素材的布局。在打印开始前，要调用这个算法来计算出分页符的位置和页数。可以保留此布局信息，从而可以在打印的过程中方便地使用它。

必须警惕"用户已经修改过页面格式"这种情况的发生。如果用户修改了页面格式，即使是所打印的信息没有发生任何改变，也必须要重新计算布局。

程序清单 11-26 中显示了如何产生一个多页打印输出。该程序用很大的字符在多个页面上打印了一条消息（见图 11-63）。然后，可以剪裁掉页边缘，并将这些页面粘连起来，形成一个标语。

Banner 类的 layoutPages 方法用以计算页面的布局。我们首先展示了一个字体为 72 磅的消息字符串。然后，我们计算产生的字符串的高度，并且将其与该页面的可打印高度进行比较。我们根据这两个高度值得出一个比例因子，当打印该字符串时，我们按照比例因子来放大此字符串。

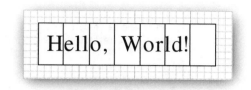

图 11-63　一幅标语

> **警告**：如果要准确地布局打印信息，通常需要访问打印机的图形上下文。遗憾的是，只有当打印真正开始时，才能获得打印机的图形上下文。在我们的示例程序中使用的是屏幕的图形上下文，并且希望屏幕的字体度量单位与打印机的相匹配。

Banner 类的 getPageCount 方法首先调用布局方法。然后，扩展字符串的宽度，并且将该宽度除以每一页的可打印宽度。得到的商向上取整，就是要打印的页数。

由于字符可以断开分布到多个页面上，所以上面打印标语的操作好像会有困难。然而，感谢 Java 2D API 提供的强大功能，这个问题现在不过是小菜一碟。当需要打印某一页时，我们只需要调用 Graphics2D 类的 translate 方法，将字符串的左上角向左平移。接着，设置一个大小是当前页面的剪切矩形（参见图 11-64）。最后，我们用布局方法计算出的比例因子来扩展该图形上下文。

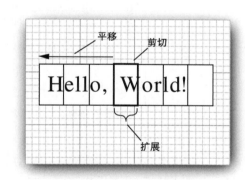

图 11-64　打印一个标语页面

这个例子显示了图形变换操作的强大功能。绘图代码很简单，而图形变换操作负责执行将图形放到恰当位置上的所有操作。最后，剪切操作负责将落在页面外面的图像剪切掉。这个程序展示了另一种必须使用变换操作的情况，即显示页面的打印预览。

程序清单 11-25　book/BookTestFrame.java

```
 1  package book;
 2
 3  import java.awt.*;
 4  import java.awt.print.*;
 5
 6  import javax.print.attribute.*;
 7  import javax.swing.*;
 8
 9  /**
10   * This frame has a text field for the banner text and buttons for printing, page setup, and
11   * print preview.
```

```java
   */
public class BookTestFrame extends JFrame
{
   private JTextField text;
   private PageFormat pageFormat;
   private PrintRequestAttributeSet attributes;

   public BookTestFrame()
   {
      text = new JTextField();
      add(text, BorderLayout.NORTH);

      attributes = new HashPrintRequestAttributeSet();

      var buttonPanel = new JPanel();

      var printButton = new JButton("Print");
      buttonPanel.add(printButton);
      printButton.addActionListener(event ->
         {
            try
            {
               PrinterJob job = PrinterJob.getPrinterJob();
               job.setPageable(makeBook());
               if (job.printDialog(attributes))
               {
                  job.print(attributes);
               }
            }
            catch (PrinterException e)
            {
               JOptionPane.showMessageDialog(BookTestFrame.this, e);
            }
         });

      var pageSetupButton = new JButton("Page setup");
      buttonPanel.add(pageSetupButton);
      pageSetupButton.addActionListener(event ->
         {
            PrinterJob job = PrinterJob.getPrinterJob();
            pageFormat = job.pageDialog(attributes);
         });

      var printPreviewButton = new JButton("Print preview");
      buttonPanel.add(printPreviewButton);
      printPreviewButton.addActionListener(event ->
         {
            var dialog = new PrintPreviewDialog(makeBook());
            dialog.setVisible(true);
         });

      add(buttonPanel, BorderLayout.SOUTH);
      pack();
   }
```

```java
 66
 67    /**
 68     * Makes a book that contains a cover page and the pages for the banner.
 69     */
 70    public Book makeBook()
 71    {
 72       if (pageFormat == null)
 73       {
 74          PrinterJob job = PrinterJob.getPrinterJob();
 75          pageFormat = job.defaultPage();
 76       }
 77       var book = new Book();
 78       String message = text.getText();
 79       var banner = new Banner(message);
 80       int pageCount = banner.getPageCount((Graphics2D) getGraphics(), pageFormat);
 81       book.append(new CoverPage(message + " (" + pageCount + " pages)"), pageFormat);
 82       book.append(banner, pageFormat, pageCount);
 83       return book;
 84    }
 85 }
```

程序清单 11-26　book/Banner.java

```java
 1  package book;
 2
 3  import java.awt.*;
 4  import java.awt.font.*;
 5  import java.awt.geom.*;
 6  import java.awt.print.*;
 7
 8  /**
 9   * A banner that prints a text string on multiple pages.
10   */
11  public class Banner implements Printable
12  {
13     private String message;
14     private double scale;
15
16     /**
17      * Constructs a banner.
18      * @param m the message string
19      */
20     public Banner(String m)
21     {
22        message = m;
23     }
24
25     /**
26      * Gets the page count of this section.
27      * @param g2 the graphics context
28      * @param pf the page format
29      * @return the number of pages needed
30      */
31     public int getPageCount(Graphics2D g2, PageFormat pf)
```

```java
32      {
33         if (message.equals("")) return 0;
34         FontRenderContext context = g2.getFontRenderContext();
35         var f = new Font("Serif", Font.PLAIN, 72);
36         Rectangle2D bounds = f.getStringBounds(message, context);
37         scale = pf.getImageableHeight() / bounds.getHeight();
38         double width = scale * bounds.getWidth();
39         int pages = (int) Math.ceil(width / pf.getImageableWidth());
40         return pages;
41      }
42
43      public int print(Graphics g, PageFormat pf, int page) throws PrinterException
44      {
45         var g2 = (Graphics2D) g;
46         if (page > getPageCount(g2, pf)) return Printable.NO_SUCH_PAGE;
47         g2.translate(pf.getImageableX(), pf.getImageableY());
48
49         drawPage(g2, pf, page);
50         return Printable.PAGE_EXISTS;
51      }
52
53      public void drawPage(Graphics2D g2, PageFormat pf, int page)
54      {
55         if (message.equals("")) return;
56         page--; // account for cover page
57
58         drawCropMarks(g2, pf);
59         g2.clip(new Rectangle2D.Double(0, 0, pf.getImageableWidth(), pf.getImageableHeight()));
60         g2.translate(-page * pf.getImageableWidth(), 0);
61         g2.scale(scale, scale);
62         FontRenderContext context = g2.getFontRenderContext();
63         var f = new Font("Serif", Font.PLAIN, 72);
64         var layout = new TextLayout(message, f, context);
65         AffineTransform transform = AffineTransform.getTranslateInstance(0, layout.getAscent());
66         Shape outline = layout.getOutline(transform);
67         g2.draw(outline);
68      }
69
70      /**
71       * Draws 1/2" crop marks in the corners of the page.
72       * @param g2 the graphics context
73       * @param pf the page format
74       */
75      public void drawCropMarks(Graphics2D g2, PageFormat pf)
76      {
77         final double C = 36; // crop mark length = 1/2 inch
78         double w = pf.getImageableWidth();
79         double h = pf.getImageableHeight();
80         g2.draw(new Line2D.Double(0, 0, 0, C));
81         g2.draw(new Line2D.Double(0, 0, C, 0));
82         g2.draw(new Line2D.Double(w, 0, w, C));
83         g2.draw(new Line2D.Double(w, 0, w - C, 0));
84         g2.draw(new Line2D.Double(0, h, 0, h - C));
85         g2.draw(new Line2D.Double(0, h, C, h));
```

```java
86              g2.draw(new Line2D.Double(w, h, w, h - C));
87              g2.draw(new Line2D.Double(w, h, w - C, h));
88          }
89      }
90
91      /**
92       * This class prints a cover page with a title.
93       */
94      class CoverPage implements Printable
95      {
96          private String title;
97
98          /**
99           * Constructs a cover page.
100          * @param t the title
101          */
102         public CoverPage(String t)
103         {
104             title = t;
105         }
106
107         public int print(Graphics g, PageFormat pf, int page) throws PrinterException
108         {
109             if (page >= 1) return Printable.NO_SUCH_PAGE;
110             var g2 = (Graphics2D) g;
111             g2.setPaint(Color.black);
112             g2.translate(pf.getImageableX(), pf.getImageableY());
113             FontRenderContext context = g2.getFontRenderContext();
114             Font f = g2.getFont();
115             var layout = new TextLayout(title, f, context);
116             float ascent = layout.getAscent();
117             g2.drawString(title, 0, ascent);
118             return Printable.PAGE_EXISTS;
119         }
120     }
```

程序清单 11-27　book/PrintPreviewDialog.java

```java
1   package book;
2
3   import java.awt.*;
4   import java.awt.print.*;
5
6   import javax.swing.*;
7
8   /**
9    * This class implements a generic print preview dialog.
10   */
11  public class PrintPreviewDialog extends JDialog
12  {
13      private static final int DEFAULT_WIDTH = 300;
14      private static final int DEFAULT_HEIGHT = 300;
15
16      private PrintPreviewCanvas canvas;
```

```java
17
18      /**
19       * Constructs a print preview dialog.
20       * @param p a Printable
21       * @param pf the page format
22       * @param pages the number of pages in p
23       */
24      public PrintPreviewDialog(Printable p, PageFormat pf, int pages)
25      {
26         var book = new Book();
27         book.append(p, pf, pages);
28         layoutUI(book);
29      }
30
31      /**
32       * Constructs a print preview dialog.
33       * @param b a Book
34       */
35      public PrintPreviewDialog(Book b)
36      {
37         layoutUI(b);
38      }
39
40      /**
41       * Lays out the UI of the dialog.
42       * @param book the book to be previewed
43       */
44      public void layoutUI(Book book)
45      {
46         setSize(DEFAULT_WIDTH, DEFAULT_HEIGHT);
47
48         canvas = new PrintPreviewCanvas(book);
49         add(canvas, BorderLayout.CENTER);
50
51         var buttonPanel = new JPanel();
52
53         var nextButton = new JButton("Next");
54         buttonPanel.add(nextButton);
55         nextButton.addActionListener(event -> canvas.flipPage(1));
56
57         var previousButton = new JButton("Previous");
58         buttonPanel.add(previousButton);
59         previousButton.addActionListener(event -> canvas.flipPage(-1));
60
61         var closeButton = new JButton("Close");
62         buttonPanel.add(closeButton);
63         closeButton.addActionListener(event -> setVisible(false));
64
65         add(buttonPanel, BorderLayout.SOUTH);
66      }
67  }
```

程序清单 11-28 book/PrintPreviewCanvas.java

```java
1  package book;
```

```java
2
3   import java.awt.*;
4   import java.awt.geom.*;
5   import java.awt.print.*;
6   import javax.swing.*;
7
8   /**
9    * The canvas for displaying the print preview.
10   */
11  class PrintPreviewCanvas extends JComponent
12  {
13     private Book book;
14     private int currentPage;
15
16     /**
17      * Constructs a print preview canvas.
18      * @param b the book to be previewed
19      */
20     public PrintPreviewCanvas(Book b)
21     {
22        book = b;
23        currentPage = 0;
24     }
25
26     public void paintComponent(Graphics g)
27     {
28        var g2 = (Graphics2D) g;
29        PageFormat pageFormat = book.getPageFormat(currentPage);
30
31        double xoff; // x offset of page start in window
32        double yoff; // y offset of page start in window
33        double scale; // scale factor to fit page in window
34        double px = pageFormat.getWidth();
35        double py = pageFormat.getHeight();
36        double sx = getWidth() - 1;
37        double sy = getHeight() - 1;
38        if (px / py < sx / sy) // center horizontally
39        {
40           scale = sy / py;
41           xoff = 0.5 * (sx - scale * px);
42           yoff = 0;
43        }
44        else
45           // center vertically
46        {
47           scale = sx / px;
48           xoff = 0;
49           yoff = 0.5 * (sy - scale * py);
50        }
51        g2.translate((float) xoff, (float) yoff);
52        g2.scale((float) scale, (float) scale);
53
54        // draw page outline (ignoring margins)
55        var page = new Rectangle2D.Double(0, 0, px, py);
```

```
56        g2.setPaint(Color.white);
57        g2.fill(page);
58        g2.setPaint(Color.black);
59        g2.draw(page);
60
61        Printable printable = book.getPrintable(currentPage);
62        try
63        {
64           printable.print(g2, pageFormat, currentPage);
65        }
66        catch (PrinterException e)
67        {
68           g2.draw(new Line2D.Double(0, 0, px, py));
69           g2.draw(new Line2D.Double(px, 0, 0, py));
70        }
71     }
72
73     /**
74      * Flip the book by the given number of pages.
75      * @param by the number of pages to flip by. Negative values flip backwards.
76      */
77     public void flipPage(int by)
78     {
79        int newPage = currentPage + by;
80        if (0 <= newPage && newPage < book.getNumberOfPages())
81        {
82           currentPage = newPage;
83           repaint();
84        }
85     }
86  }
```

11.5.3 打印服务程序

到目前为止，我们已经介绍了如何打印 2D 图形。然而，Java SE 1.4 中的打印 API 提供了更大的灵活性。该 API 定义了大量的数据类型，并且可以让你找到能够打印这些数据类型的打印服务程序。这些类型有：

- GIF、JPEG 或者 PNG 格式的图像。
- 纯文本、HTML、PostScript 或者 PDF 格式的文档。
- 原始的打印机代码数据。
- 实现了 Printable、Pageable 或 RenderableImage 的某个类的对象。

数据本身可以存放在一个字节源或字符源中，比如一个输入流、一个 URL 或者一个数组中。文档风格（document flavor）描述了一个数据源和一个数据类型的组合。DocFlavor 类为不同的数据源定义了许多内部类，每一个内部类都定义了指定风格的常量。例如，常量

DocFlavor.INPUT_STREAM.GIF

描述了从输入流中读入一个 GIF 格式的图像。表 11-4 中列出了数据源和数据类型的各种组合。

表 11-4 打印服务的文档风格

数据源	数据类型	MIME 类型
INPUT_STREAM	GIF	image/gif
URL	JPEG	image/jpeg
BYTE_ARRAY	PNG	image/png
	POSTSCRIPT	application/postscript
	PDF	application/pdf
	TEXT_HTML_HOST	text/html（使用主机编码）
	TEXT_HTML_US_ASCII	text/html; charset=us-ascii
	TEXT_HTML_UTF_8	text/html; charset=utf-8
	TEXT_HTML_UTF_16	text/html; charset=utf-16
	TEXT_HTML_UTF_16LE	text/html; charset=utf-16le（小尾数法）
	TEXT_HTML_UTF_16BE	text/html; charset=utf-16be（大尾数法）
	TEXT_PLAIN_HOST	text/plain（使用主机编码）
	TEXT_PLAIN_US_ASCII	text/plain; charset=us-ascii
	TEXT_PLAIN_UTF_8	text/plain; charset=utf-8
	TEXT_PLAIN_UTF_16	text/plain; charset=utf-16
	TEXT_PLAIN_UTF_16LE	text/plain; charset=utf-16le（小尾数法）
	TEXT_PLAIN_UTF_16BE	text/plain; charset=utf-16be（大尾数法）
	PCL	application/vnd.hp-PCL（惠普公司打印机控制语言）
	AUTOSENSE	application/octet-stream（原始打印数据）
READER	TEXT_HTML	text/html; charset=utf-16
STRING	TEXT_PLAIN	text/plain; charset=utf-16
CHAR_ARRAY		
SERVICE_FORMATTED	PRINTABLE	无
	PAGEABLE	无
	RENDERABLE_IMAGE	无

假设我们想打印一个位于文件中的 GIF 格式的图像。首先，确认是否有能够处理该打印任务的打印服务程序。PrintServiceLookup 类的静态 lookupPrintServices 方法返回一个能够处理给定文档风格的 PrintService 对象的数组。

```
DocFlavor flavor = DocFlavor.INPUT_STREAM.GIF;
PrintService[] services = PrintServiceLookup.lookupPrintServices(flavor, null);
```

当 lookupPrintServices 方法的第二个参数值为 null 时，表示我们不想通过设定打印机属性来限制对文档的搜索。我们在下一节中介绍打印机的属性。

如果对打印服务程序的查找返回的数组带有多个元素的话，那就需要从打印服务程序列表中选择所需的打印服务程序。通过调用 PrintService 类的 getName 方法，可以获得打印机的名称，然后让用户进行选择。

接着，从该打印服务获取一个文档打印作业：

```
DocPrintJob job = services[i].createPrintJob();
```

如果要执行打印操作，需要一个实现了 Doc 接口的类的对象。Java 为此提供了一个 Simple-Doc 类。SimpleDoc 类的构造器必须包含数据源对象、文档风格和一个可选的属性集。例如，

```
var in = new FileInputStream(fileName);
var doc = new SimpleDoc(in, flavor, null);
```

最后，就可以执行打印输出了。

```
job.print(doc, null);
```

与前面一样，null 参数可以被一个属性集取代。

请注意，这个打印进程和上一节的打印进程之间有很大的差异。这里不需要用户通过打印对话框来进行交互式操作。例如，可以实现一个服务器端的打印机制，这样，用户就可以通过 Web 表单提交打印作业了。

API *javax.print.PrintServiceLookup* 1.4

- PrintService[] lookupPrintServices(DocFlavor flavor, AttributeSet attributes)
  查找能够处理给定文档风格和属性的打印服务程序。
  参数：flavor   文档风格
        attributes   需要的打印属性，如果不考虑打印属性的话，其值应该为 null

API *javax.print.PrintService* 1.4

- DocPrintJob createPrintJob()
  为了打印实现了 Doc 接口（如 SimpleDoc）的对象而创建一个打印作业。

API *javax.print.DocPrintJob* 1.4

- void print(Doc doc, PrintRequestAttributeSet attributes)
  打印带有给定属性的给定文档。
  参数：doc   要打印的 Doc
        attributes   需要的打印属性，如果不需要任何打印属性的话，其值为 null

API *javax.print.SimpleDoc* 1.4

- SimpleDoc(Object data, DocFlavor flavor, DocAttributeSet attributes)
  构建一个能够用 DocPrintJob 打印的 SimpleDoc 对象。
  参数：data   带有打印数据的对象，比如一个输入流或者一个 Printable
        flavor   打印数据的文档风格
        attributes   文档属性，如果不需要文档属性，其值为 null

11.5.4 流打印服务程序

打印服务程序将打印数据发送给打印机。流打印服务程序产生同样的打印数据，但是并不把数据发送给打印机，而是发给流。这么做的目的也许是为了延迟打印或者因为打印数据

格式可以由其他程序来进行解释。尤其是，如果打印数据格式是 PostScript 时，那么可将打印数据保存到一个文件中，因为有许多程序都能够处理 PostScript 文件。Java 平台引入了一个流打印服务程序，它能够从图像和 2D 图形中产生 PostScript 输出。可以在任何系统中使用这种服务程序，即使这些系统中没有本地打印机，也可以使用该服务程序。

枚举流打印服务程序要比定位普通的打印服务程序复杂一些。既需要打印对象的 DocFlavor 又需要流输出的 MIME 类型，接着获得一个 StreamPrintServiceFactory 类型的数组，如下所示：

```
DocFlavor flavor = DocFlavor.SERVICE_FORMATTED.PRINTABLE;
String mimeType = "application/postscript";
StreamPrintServiceFactory[] factories
    = StreamPrintServiceFactory.lookupStreamPrintServiceFactories(flavor, mimeType);
```

StreamPrintServiceFactory 类没有任何方法能够帮助我们区分不同的 factory，所以我们只提取 factories[0]。我们调用带有输出流参数的 getPrintService 方法来获得一个 StreamPrintService 对象。

```
var out = new FileOutputStream(fileName);
StreamPrintService service = factories[0].getPrintService(out);
```

StreamPrintService 类是 PrintService 的子类。如果要产生一个打印输出，只要按照上一节介绍的步骤进行操作即可。

API javax.print.StreamPrintServiceFactory 1.4

- StreamPrintServiceFactory[] lookupStreamPrintServiceFactories(DocFlavor flavor, String mimeType)

  查找所需的流打印服务程序工厂，它能够打印给定文档风格，并且产生一个给定 MIME 类型的输出流。

- StreamPrintService getPrintService(OutputStream out)

  获得一个打印服务程序，以便将打印输出发送到指定的输出流中。

程序清单 11-29 展示了如何使用流打印服务程序将 Java 2D 的形状打印到 PostScript 文件中。你可以用任何生成 Java 2D 形状的代码替换其中的样例绘图代码，然后将这些形状转换为 PostScript。然后，通过使用外部工具，你可以很容易地将得到的结果转换成 PDF 或 EPS。（遗憾的是，Java 不支持直接打印成 PDF）。

> **注释**：在这个示例中，我们在 Graphics2D 对象上调用了绘制 Java 2D 形状的 draw 方法。如果想要绘制一个构件的表层（例如表格或树），那么使用下面的代码：
>
> ```
> private static int IMAGE_WIDTH = component.getWidth();
> private static int IMAGE_HEIGHT = component.getHeight();
> public static void draw(Graphics2D g2) { component.paint(g2); }
> ```

程序清单 11-29 printService/PrintServiceTest.java

```
1 package printService;
2
```

```java
 3  import java.awt.*;
 4  import java.awt.font.*;
 5  import java.awt.geom.*;
 6  import java.awt.print.*;
 7  import java.io.*;
 8  import javax.print.*;
 9  import javax.print.attribute.*;
10
11  /**
12   * This program demonstrates the use of stream print services. The program prints
13   * Java 2D shapes to a PostScript file. If you don't supply a file name on the command
14   * line, the output is saved to out.ps.
15   * @version 1.0 2018-06-01
16   * @author Cay Horstmann
17   */
18  public class PrintServiceTest
19  {
20     // Set your image dimensions here
21     private static int IMAGE_WIDTH = 300;
22     private static int IMAGE_HEIGHT = 300;
23
24     public static void draw(Graphics2D g2)
25     {
26        // Your drawing instructions go here
27        FontRenderContext context = g2.getFontRenderContext();
28        var f = new Font("Serif", Font.PLAIN, 72);
29        var clipShape = new GeneralPath();
30
31        var layout = new TextLayout("Hello", f, context);
32        AffineTransform transform = AffineTransform.getTranslateInstance(0, 72);
33        Shape outline = layout.getOutline(transform);
34        clipShape.append(outline, false);
35
36        layout = new TextLayout("World", f, context);
37        transform = AffineTransform.getTranslateInstance(0, 144);
38        outline = layout.getOutline(transform);
39        clipShape.append(outline, false);
40
41        g2.draw(clipShape);
42        g2.clip(clipShape);
43
44        final int NLINES = 50;
45        var p = new Point2D.Double(0, 0);
46        for (int i = 0; i < NLINES; i++)
47        {
48           double x = (2 * IMAGE_WIDTH * i) / NLINES;
49           double y = (2 * IMAGE_HEIGHT * (NLINES - 1 - i)) / NLINES;
50           var q = new Point2D.Double(x, y);
51           g2.draw(new Line2D.Double(p, q));
52        }
53     }
54
55     public static void main(String[] args) throws IOException, PrintException
56     {
```

```
57      String fileName = args.length > 0 ? args[0] : "out.ps";
58      DocFlavor flavor = DocFlavor.SERVICE_FORMATTED.PRINTABLE;
59      var mimeType = "application/postscript";
60      StreamPrintServiceFactory[] factories
61         = StreamPrintServiceFactory.lookupStreamPrintServiceFactories(flavor, mimeType);
62      var out = new FileOutputStream(fileName);
63      if (factories.length > 0)
64      {
65         PrintService service = factories[0].getPrintService(out);
66         var doc = new SimpleDoc(new Printable()
67            {
68               public int print(Graphics g, PageFormat pf, int page)
69               {
70                  if (page >= 1) return Printable.NO_SUCH_PAGE;
71                  else
72                  {
73                     double sf1 = pf.getImageableWidth() / (IMAGE_WIDTH + 1);
74                     double sf2 = pf.getImageableHeight() / (IMAGE_HEIGHT + 1);
75                     double s = Math.min(sf1, sf2);
76                     var g2 = (Graphics2D) g;
77                     g2.translate((pf.getWidth() - pf.getImageableWidth()) / 2,
78                        (pf.getHeight() - pf.getImageableHeight()) / 2);
79                     g2.scale(s, s);
80
81                     draw(g2);
82                     return Printable.PAGE_EXISTS;
83                  }
84               }
85            }, flavor, null);
86         DocPrintJob job = service.createPrintJob();
87         var attributes = new HashPrintRequestAttributeSet();
88         job.print(doc, attributes);
89      }
90      else
91         System.out.println("No factories for " + mimeType);
92   }
93 }
```

11.5.5 打印属性

打印服务程序 API 包含了一组复杂的接口和类，用以设定不同种类的属性。重要的属性共有四组，前两组属性用于设定对打印机的访问请求。

- 打印请求属性（Print request attribute）为一个打印作业中的所有 doc 对象请求特定的打印属性，例如，双面打印或者纸张的大小。
- Doc 属性（Doc attribute）是仅作用在单个 doc 对象上的请求属性。

另外两组属性包含关于打印机和作业状态的信息。

- 打印服务属性（Print service attribute）提供了关于打印服务程序的信息，比如打印机的种类和型号，或者打印机当前是否接受打印作业。
- 打印作业属性（Print job attribute）提供了关于某个特定打印作业状态的信息，比如该打印作业是否已经完成。

如果要描述各种不同的打印属性，可以使用带有如下子接口的 Attribute 接口。

```
PrintRequestAttribute
DocAttribute
PrintServiceAttribute
PrintJobAttribute
SupportedValuesAttribute
```

各个属性类都实现了上面的一个或几个接口。例如，Copies 类的对象描述了一个打印输出的拷贝数量，该类就实现了 PrintRequestAttribute 和 PrintJobAttribute 两个接口。显然，一个打印请求可以包含一个需要多个拷贝的请求。反过来，打印作业的某个属性可能表示的是实际上打印出来的拷贝数量。这个拷贝数量可能很小，也许是因为打印机的限制或者是因为打印机的纸张已经用完了。

SupportedValuesAttribute 接口表示某个属性值反映的不是实际的打印请求或状态数据，而是某个服务程序的能力。例如，实现了 SupportedValuesAttribute 接口的 CopiesSupported 类，该类的对象可以用来描述某个打印机能够支持 1～99 份拷贝的打印输出。

图 11-65 显示了属性分层结构的类图。

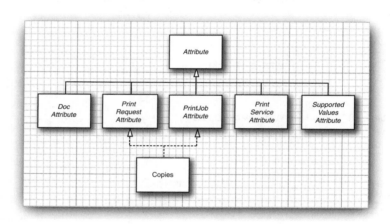

图 11-65　属性分层结构图

除了为各个属性定义的接口和类以外，打印服务程序 API 还为属性集定义了接口和类。父接口 AttributeSet 有四个子接口：

```
PrintRequestAttributeSet
DocAttributeSet
PrintServiceAttributeSet
PrintJobAttributeSet
```

对于每个这样的接口，都有一个实现类，因此会产生下面 5 个类：

```
HashAttributeSet
HashPrintRequestAttributeSet
HashDocAttributeSet
HashPrintServiceAttributeSet
HashPrintJobAttributeSet
```

图 11-66 显示了属性集分层结构的类图。

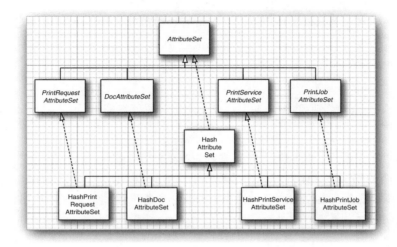

图 11-66　属性集的分层结构

例如，可以用如下的方式构建一个打印请求属性集。

var attributes = new HashPrintRequestAttributeSet();

当构建完属性集后，就不用担心使用 Hash 前缀的问题了。

为什么要配有所有这些接口呢？因为，它们使得"检查属性是否被正确使用"成为可能。例如，DocAttributeSet 只接受实现了 DocAttribute 接口的对象，添加其他属性的任何尝试都会导致运行期错误的产生。

属性集是一个特殊的映射表，其键是 Class 类型的，而值是一个实现了 Attribute 接口的类。例如，如果要插入一个对象

new Copies(10)

到属性集中，那么它的键就是 Class 对象 Copies.class。该键被称为属性的类别。Attribute 接口声明了下面这样一个方法：

Class getCategory()

该方法就可以返回属性的类别。Copies 类定义了用以返回 Copies.class 对象的方法。但是，属性的类别和属性的类没有必要是相同的。

当将一个属性添加到属性集中时，属性的类别就会被自动地获取。你只需添加该属性的值：

attributes.add(new Copies(10));

如果后来添加了一个具有相同类别的另一个属性，那么新属性就会覆盖第一个属性。如果要检索一个属性，需要使用它的类别作为键，例如，

AttributeSet attributes = job.getAttributes();
var copies = (Copies) attribute.get(Copies.class);

最后，属性是按照它们拥有的值来进行组织的。Copies 属性能够拥有任何整数值。Copies 类继承了 IntegerSyntax 类，该类负责处理所有带有整数值的属性。getValue 方法将返回属性的整数值，例如，

```
int n = copies.getValue();
```

下面这些类：

```
TextSyntax
DateTimeSyntax
URISyntax
```

用于封装一个字符串、日期与时间，或者 URI（通用资源标识符）。

最后要说明的是，许多属性都能够接受数量有限的值。例如，PrintQuality 属性有三个设置值：draft（草稿质量）、normal（正常质量）和 high（高质量），它们用三个常量来表示：

```
PrintQuality.DRAFT
PrintQuality.NORMAL
PrintQuality.HIGH
```

拥有有限数量值的属性类继承了 EnumSyntax 类，该类提供了许多便利的方法，用来以类型安全的方式设置这些枚举。当使用这样的属性时，不必担心该机制，只需要将带有名字的值添加给属性集即可：

```
attributes.add(PrintQuality.HIGH);
```

下面的代码说明了如何来检查一个属性的值：

```
if (attributes.get(PrintQuality.class) == PrintQuality.HIGH)
    ...
```

表 11-5 列出了各个打印属性。表中的第二列列出了属性类的超类（例如，Copies 属性的 IntegerSyntax 类）或者是具有一组有限值属性的枚举值。最后四列表示该属性是否实现了 DocAttribute (DA)、PrintJobAttribute (PJA)、PrintRequestAttribute (PRA) 和 PrintServiceAttribute (PSA) 几个接口。

表 11-5 打印属性一览表

属性	超类或枚举常量	DA	PJA	PRA	PSA
Chromaticity	MONOCHROME, COLOR		√	√	
ColorSupported	SUPPORTED, NOT_SUPPORTED				√
Compression	COMPRESS, DEFLATE, GZIP, NONE	√			
Copies	IntegerSyntax			√	√
DateTimeAtCompleted	DateTimeSyntax		√		
DateTimeAtCreation	DateTimeSyntax		√		
DateTimeAtProcessing	DateTimeSyntax		√		
Destination	URISyntax			√	√
DocumentName	TextSyntax	√			

（续）

属性	超类或枚举常量	DA	PJA	PRA	PSA
Fidelity	FIDELITY_TRUE, FIDELITY_FALSE		√	√	
Finishings	NONE, STAPLE, EDGE_STITCH, BIND, SADDLE_STITCH, COVER, ...	√	√	√	
JobHoldUntil	DateTimeSyntax		√		
JobImpressions	IntegerSyntax		√	√	
JobImpressionsCompleted	IntegerSyntax		√		
JobKOctets	IntegerSyntax		√	√	
JobKOctetsProcessed	IntegerSyntax		√		
JobMediaSheets	IntegerSyntax		√	√	
JobMediaSheetsCompleted	IntegerSyntax		√		
JobMessageFromOperator	TextSyntax		√		
JobName	TextSyntax		√	√	
JobOriginatingUserName	TextSyntax		√		
JobPriority	IntegerSyntax		√	√	
JobSheets	STANDARD, NONE		√	√	
JobState	ABORTED, CANCELED, COMPLETED, PENDING, PENDING_HELD, PROCESSING, PROCESSING_STOPPED		√		
JobStateReason	ABORTED_BY_SYSTEM, DOCUMENT_FORMAT_ERROR, 其他				
JobStateReasons	HashSet		√		
MediaName	ISO_A4_WHITE, ISO_A4_TRANSPARENT, NA_LETTER_WHITE, NA_LETTER_TRANSPARENT	√	√	√	
MediaSize	ISO.A0-ISO.A10, ISO.B0-ISO.B10, ISO.C0-ISO.C10, NA.LETTER, NA.LEGAL, 各种其他纸张和信封尺寸				
MediaSizeName	ISO_A0-ISO_A10, ISO_B0-ISO_B10, ISO_C0-ISO_C10, NA_LETTER, NA_LEGAL, 各种其他纸张和信封尺寸名称	√	√	√	
MediaTray	TOP, MIDDLE, BOTTOM, SIDE, ENVELOPE, LARGE_CAPACITY, MAIN, MANUAL	√	√	√	
MultipleDocumentHandling	SINGLE_DOCUMENT, SINGLE_DOCUMENT_NEW_SHEET, SEPARATE_DOCUMENTS_COLLATED_COPIES, SEPARATE_DOCUMENTS_UNCOLLATED_COPIES		√	√	
NumberOfDocuments	IntegerSyntax		√		
NumberOfInterveningJobs	IntegerSyntax		√		
NumberUp	IntegerSyntax	√	√	√	
OrientationRequested	PORTRAIT, LANDSCAPE, REVERSE_PORTRAIT, REVERSE_LANDSCAPE	√	√	√	
OutputDeviceAssigned	TextSyntax		√		
PageRanges	SetOfInteger	√	√	√	
PagesPerMinute	IntegerSyntax				√
PagesPerMinuteColor	IntegerSyntax				√
PDLOverrideSupported	ATTEMPTED, NOT_ATTEMPTED				√
PresentationDirection	TORIGHT_TOBOTTOM, TORIGHT_TOTOP, TOBOTTOM_TORIGHT, TOBOTTOM_TOLEFT, TOLEFT_TOBOTTOM, TOLEFT_TOTOP, TOTOP_TORIGHT, TOTOP_TOLEFT		√	√	
PrinterInfo	TextSyntax				√

(续)

属性	超类或枚举常量	DA	PJA	PRA	PSA
PrinterIsAcceptingJobs	ACCEPTING_JOBS, NOT_ACCEPTING_JOBS				√
PrinterLocation	TextSyntax				√
PrinterMakeAndModel	TextSyntax				√
PrinterMessageFromOperator	TextSyntax				√
PrinterMoreInfo	URISyntax				√
PrinterMoreInfoManufacturer	URISyntax				√
PrinterName	TextSyntax				√
PrinterResolution	ResolutionSyntax	√	√	√	
PrinterState	PROCESSING, IDLE, STOPPED, UNKNOWN				√
PrinterStateReason	COVER_OPEN, FUSER_OVER_TEMP, MEDIA_JAM, 其他				
PrinterStateReasons	HashMap				
PrinterURI	URISyntax				√
PrintQuality	DRAFT, NORMAL, HIGH	√	√	√	
QueuedJobCount	IntegerSyntax				√
ReferenceUriSchemesSupported	FILE, FTP, GOPHER, HTTP, HTTPS, NEWS, NNTP, WAIS				
RequestingUserName	TextSyntax			√	
Severity	ERROR, REPORT, WARNING				
SheetCollate	COLLATED, UNCOLLATED	√	√	√	
Sides	ONE_SIDED, DUPLEX (= TWO_SIDED_LONG_EDGE), TUMBLE (= TWO_SIDED_SHORT_EDGE)	√	√	√	

📝 **注释**：可以看到，属性的数量很多，其中许多属性都是专用的。大多数属性都来源于因特网打印协议 1.1 版（RFC 2911）。

📝 **注释**：打印 API 的早期版本引入了 JobAttributes 和 PageAttributes 类，其目的与本节所介绍的打印属性类似。这些类现在已经弃用了。

API *javax.print.attribute.Attribute* 1.4

- Class getCategory()

  获取该属性的类别。

- String getName()

  获取该属性的名字。

API *javax.print.attribute.AttributeSet* 1.4

- boolean add(Attribute attr)

  向属性集中添加一个属性。如果集中有另一个属性和此属性有相同的类别，那么集中的属性被新添加的属性所取代。如果由于添加属性的操作改变了属性集，则返回 true。

- Attribute get(Class category)

  检索带有指定属性类别键的属性，如果该属性不存在，则返回 null。
- boolean remove(Attribute attr)
- boolean remove(Class category)

  从属性集中删除给定属性，或者删除具有指定类别的属性。如果由于这个操作改变了属性集，则返回 true。
- Attribute[] toArray()

  返回一个带有该属性集中所有属性的数组。

API *javax.print.PrintService* 1.4

- PrintServiceAttributeSet getAttributes()

  获取打印服务程序的属性。

API *javax.print.DocPrintJob* 1.4

- PrintJobAttributeSet getAttributes()

  获取打印作业的属性。

现在，我们来到了本章的尾声，这长长的一章涵盖了高级 Swing 和 AWT 特性。在最后一章，我们将转而研究 Java 编程的另一个完全不同的方面：在同一台机器上与用其他编程语言编写的"本地"代码交互。

第 12 章 本地方法

- ▲ 从 Java 程序中调用 C 函数
- ▲ 数值参数与返回值
- ▲ 字符串参数
- ▲ 访问域
- ▲ 编码签名
- ▲ 调用 Java 方法
- ▲ 访问数组元素
- ▲ 错误处理
- ▲ 使用调用 API
- ▲ 完整的示例：访问 Windows 注册表

原则上说，"100% 纯 Java"的解决方案是非常好的，但有时你也会想要编写或使用其他语言的代码（这种代码通常称为本地代码）。

特别是在 Java 的早期阶段，许多人都认为使用 C 或 C++ 来加速 Java 应用中关键部分是个好主意。但是，实际上，这基本上是徒劳的。1996 年 JavaOne 会议上有一个演讲很明确地说明了这一点，来自 Sun Microsystems 的密码库的实现者报告说他们的加密函数的纯 Java 平台实现已臻化境。他们的代码确实没有已有的 C 实现快，但是事实证明这无关紧要。Java 平台实现比网络 I/O 要快得多，而后者是真正的瓶颈。

当然，求助于本地代码是有缺陷的。如果应用的某个部分是用其他语言编写的，那么就必须为需要支持的每个平台都提供一个单独的本地类库。用 C 或 C++ 编写的代码没有对通过使用无效指针所造成的内存覆写提供任何保护。编写本地代码很容易破坏你的程序，并感染操作系统。

因此，我们建议只有在必需的时候才使用本地代码。特别是在以下三种情况下，也许可以使用本地代码：

- 你的应用需要访问的系统特性和设备通过 Java 平台是无法实现的。
- 你已经有了大量的测试过和调试过的用另一种语言编写的代码，并且知道如何将其导出到所有的目标平台上。
- 通过基准测试，你发现所编写的 Java 代码比用其他语言编写的等价代码要慢得多。

Java 平台有一个用于和本地 C 代码进行互操作的 API，称为 Java 本地接口（JNI）。我们将在本章讨论 JNI 编程。

> **C++ 注释**：你可以使用 C++ 代替 C 来编写本地方法。这样会有一些好处：类型检查会更严格一些，访问 JNI 函数会更便捷一些。然而，JNI 并不支持 Java 类和 C++ 类之间的任何映射机制。

12.1 从 Java 程序中调用 C 函数

假设你有一个 C 函数，它能为你实现某个功能，因为某种原因，你不想费事使用 Java 编程语言重新实现它。为了方便说明问题，我们从一个很简单的打印问候语的 C 函数入手。

Java 编程语言使用关键字 native 表示本地方法，而且很显然，你还需要在类中放置一个方法。其结果显示在程序清单 12-1 中。

关键字 native 提醒编译器该方法将在外部定义。当然，本地方法不包含任何 Java 编程语言编写的代码，而且方法头后面直接跟着一个表示终结的分号。因此，本地方法声明看上去和抽象方法声明类似。

程序清单 12-1 helloNative/HelloNative.java

```
1  /**
2   * @version 1.11 2007-10-26
3   * @author Cay Horstmann
4   */
5  class HelloNative
6  {
7     public static native void greeting();
8  }
```

> **注释：** 与前一章一样，为了保持样例的简单性，我们在这里也不使用包。

在这个特定示例中，本地方法也被声明为 static。本地方法既可以是静态的也可以是非静态的，使用静态方法是因为我们此刻还不想处理参数传递。

你实际上可以编译这个类，但是在程序中使用它时，虚拟机就会告诉你它不知道如何找到 greeting，它会报告一个 UnsatisfiedLinkError 异常。为了实现本地代码，需要编写一个相应的 C 函数，你必须完全按照 Java 虚拟机预期的那样来命名这个函数。其规则是：

1. 使用完整的 Java 方法名，比如：HelloNative.greeting。如果该类属于某个包，那么在前面添加包名，比如：com.horstmann.HelloNative.greeting。

2. 用下划线替换掉所有的句号，并加上 Java_ 前缀，例如，Java_HelloNative_greeting 或 Java_com_horstmann_HelloNative_greeting。

3. 如果类名含有非 ASCII 字母或数字，如：'_'，'$' 或是大于 '\u007F' 的 Unicode 字符，用 _0xxxx 来替代它们，xxxx 是该字符的 Unicode 值的 4 个十六进制数序列。

> **注释：** 如果你重载了本地方法，也就是说，你用相同的名字提供了多个本地方法，那么你必须在名称后附加两个下划线，后面再加上已编码的参数类型。在本章后面，我们将描述参数类型的编码方法。例如，如果你有一个本地方法 greeting 和另一个本地方法 greeting(int repeat)，那么，第一个称为 Java_HelloNative_greeting__，第二个称为 Java_HelloNative_greeting__I。

实际上，没人会手工完成这些操作。相反，你应该用 -h 标志运行 javac，并提供头文件放置的目录：

```
javac -h . HelloNative.java
```

这条命令在当前目录中创建了一个名为 HelloNative.h 的头文件，正如程序清单 12-2 所示。

程序清单 12-2　helloNative/HelloNative.h

```c
1  /* DO NOT EDIT THIS FILE - it is machine generated */
2  #include <jni.h>
3  /* Header for class HelloNative */
4
5  #ifndef _Included_HelloNative
6  #define _Included_HelloNative
7  #ifdef __cplusplus
8  extern "C" {
9  #endif
10 /*
11  * Class:     HelloNative
12  * Method:    greeting
13  * Signature: ()V
14  */
15 JNIEXPORT void JNICALL Java_HelloNative_greeting
16   (JNIEnv *, jclass);
17
18 #ifdef __cplusplus
19 }
20 #endif
21 #endif
```

如你所见，这个文件包含了函数 Java_HelloNative_greeting 的声明（宏 JNIEXPORT 和 JNICALL 是在头文件 jni.h 中定义的，它们为那些来自动态装载库的导出函数标明了依赖于编译器的说明符）。

现在，需要将函数原型从头文件中复制到源文件中，并且给出函数的实现代码，如程序清单 12-3 所示。

程序清单 12-3　helloNative/HelloNative.c

```c
1  /*
2     @version 1.10 1997-07-01
3     @author Cay Horstmann
4  */
5
6  #include "HelloNative.h"
7  #include <stdio.h>
8
9  JNIEXPORT void JNICALL Java_HelloNative_greeting(JNIEnv* env, jclass cl)
10 {
11    printf("Hello Native World!\n");
12 }
```

在这个简单的函数中，我们忽略了 env 和 cl 参数。后面你会看到它们的用处。

> **C++ 注释**：你可以使用 C++ 实现本地方法。然而，那样你必须将实现本地方法的函数声明为 extern"C"（这可以阻止 C++ 编译器混编方法名）。例如：
>
> ```
> extern "C"
> JNIEXPORT void JNICALL Java_HelloNative_greeting(JNIEnv* env, jclass cl)
> {
>     cout << "Hello, Native World!" << endl;
> }
> ```

将本地 C 代码编译到一个动态装载库中，具体方法依赖于编译器。

例如，Linux 下的 Gnu C 编译器，使用如下命令：

```
gcc -fPIC -I jdk/include -I jdk/include/linux -shared -o libHelloNative.so HelloNative.c
```

用 Windows 下的微软编译器，命令是：

```
cl -I jdk\include -I jdk\include\win32 -LD HelloNative.c -FeHelloNative.dll
```

这里 *jdk* 是含有 JDK 的目录。

> **提示**：如果你要从命令 shell 中使用微软的编译器，首先要运行批处理文件 vcvars32.bat 或 vcvarsall.bat。这个批处理文件设置了编译器需要的路径和环境变量。你可以在目录 c:\Program Files\Microsoft Visual Studio .14.0\Common7\Tools，或类似位置找到该文件，细节请查看 Visual Studio 的文档。

也可以使用可从 http://www.cygwin.com 处免费获取的 Cygwin 编程环境。它包含了 GNU C 编译器和 Windows 下的 UNIX 风格编程的库。使用 Cygwin 时，用以下命令：

```
gcc -mno-cygwin -D __int64="long long" -I jdk/include/ -I jdk/include/win32 \
    -shared -Wl,--add-stdcall-alias -o HelloNative.dll HelloNative.c
```

整个命令应该键入在同一行中。

> **注释**：Windows 版本的头文件 jni_md.h 含有如下类型声明：
>
> ```
> typedef __int64 jlong;
> ```
>
> 它是专门用于微软编译器的。如果你使用的是 GNU 编译器，那么你可能需要编辑这个文件，例如：
>
> ```
> #ifdef __GNUC__
>     typedef long long jlong;
> #else
>     typedef __int64 jlong;
> #endif
> ```
>
> 或者，如编译器调用的示例那样，使用 -D __int64="long long" 进行编译。

最后，我们要在程序中添加一个对 System.loadLibrary 方法的调用。为了确保虚拟机在第一次使用该类之前就会装载这个库，需要使用静态初始化代码块，如程序清单 12-4 所示。

程序清单 12-4　helloNative/HelloNativeTest.java

```java
/**
 * @version 1.11 2007-10-26
 * @author Cay Horstmann
 */
class HelloNativeTest
{
   public static void main(String[] args)
   {
      HelloNative.greeting();
   }

   static
   {
      System.loadLibrary("HelloNative");
   }
}
```

图 12-1 给出了对本地代码处理的总结。

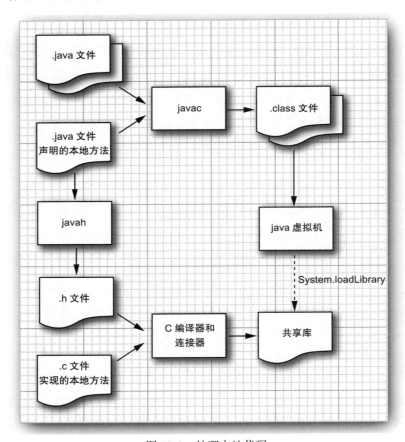

图 12-1　处理本地代码

如果编译并运行该程序，终端窗口会显示消息"Hello, Native World!"。

> **注释**：如果运行在 Linux 下，必须把当前目录添加到库路径中。实现方式可以是通过设置 LD_LIBRARY_PATH 环境变量：
>
>   export LD_LIBRARY_PATH=.:$LD_LIBRARY_PATH
>
> 或者是设置 java.library.path 系统属性：
>
>   java -Djava.library.path=. HelloNativeTest

当然，这个消息本身并不会给人留下深刻印象。然而，如果你记得这个信息是由 C 的 printf 命令产生而不是由任何 Java 编程语言代码产生的话，你就会明白我们已经在连接两种语言上走出了第一步。

总之，遵循下面的步骤就可以将一个本地方法链接到 Java 程序中：

1. 在 Java 类中声明一个本地方法。
2. 运行 javah 以获得包含该方法的 C 声明的头文件。
3. 用 C 实现该本地方法。
4. 将代码置于共享类库中。
5. 在 Java 程序中加载该类库。

API java.lang.System 1.0

- void loadLibrary(String libname)

  装载指定名字的库，该库位于库搜索路径中。定位该库的确切方法依赖于操作系统。

> **注释**：一些本地代码的共享库必须先运行初始化代码。你可以把初始化代码放到 JNI_OnLoad 方法中。类似地，如果你提供该方法，当虚拟机关闭时，将会调用 JNI_OnUnload 方法。它们的原型是：
>
>   jint JNI_OnLoad(JavaVM* vm, void* reserved);
>   void JNI_OnUnload(JavaVM* vm, void* reserved);
>
> JNI_OnLoad 方法要返回它所需的虚拟机的最低版本，例如：JNI_VERSION_1_2。

12.2 数值参数与返回值

当在 C 和 Java 之间传递数字时，应该知道它们彼此之间的对应类型。例如，C 也有 int 和 long 的数据类型，但是它们的实现却是取决于平台的。在一些平台上，int 类型是 16 位的，在另外一些平台上是 32 位的。然而，在 Java 平台上 int 类型总是 32 位的整数。基于这个原因，Java 本地接口定义了 jint、jlong 等类型。

表 12-1 显示了 Java 数据类型和 C 数据类型的对应关系。

表 12-1　Java 数据类型和 C 数据类型

Java 编程语言	C 编程语言	字节	Java 编程语言	C 编程语言	字节
boolean	jboolean	1	int	jint	4
byte	jbyte	1	long	jlong	8
char	jchar	2	float	jfloat	4
short	jshort	2	double	jdouble	8

在头文件 jni.h 中，这些类型被 typedef 语句声明为在目标平台上等价的类型。该头文件还定义了常量 JNI_FALSE = 0 和 JNI_TRUE = 1。

直到 Java 5.0，Java 才有了与 C 语言的 printf 函数相类似的方法。在下面的示例中，我们假设你依然坚持使用古老版本的 JDK，并且决定通过调用本地方法中的 C 的 printf 函数来实现同样的功能。

程序清单 12-5 给出了一个名为 Printf1 的类，它使用本地方法来打印给定域宽度和精度的浮点数。

程序清单 12-5　printf1/Printf1.java

```
 1  /**
 2   * @version 1.10 1997-07-01
 3   * @author Cay Horstmann
 4   */
 5  class Printf1
 6  {
 7     public static native int print(int width, int precision, double x);
 8
 9     static
10     {
11        System.loadLibrary("Printf1");
12     }
13  }
```

注意，用 C 实现该方法时，所有的 int 和 double 参数都要转换成 jint 和 jdouble，如程序清单 12-6 所示。

程序清单 12-6　printf1/Printf1.c

```
 1  /**
 2     @version 1.10 1997-07-01
 3     @author Cay Horstmann
 4  */
 5
 6  #include "Printf1.h"
 7  #include <stdio.h>
 8
 9  JNIEXPORT jint JNICALL Java_Printf1_print(JNIEnv* env, jclass cl,
10        jint width, jint precision, jdouble x)
11  {
12     char fmt[30];
```

```
13     jint ret;
14     sprintf(fmt, "%%%d.%df", width, precision);
15     ret = printf(fmt, x);
16     fflush(stdout);
17     return ret;
18  }
```

该函数只是装配了变量 fmt 中的格式字符串 "%w.pf"，然后调用 printf 函数，接着返回打印出的字符的个数。

程序清单 12-7 给出了验证 Printf1 类的测试程序。

程序清单 12-7　printf1/Printf1Test.java

```
1  /**
2   * @version 1.10 1997-07-01
3   * @author Cay Horstmann
4   */
5  class Printf1Test
6  {
7     public static void main(String[] args)
8     {
9        int count = Printf1.print(8, 4, 3.14);
10       count += Printf1.print(8, 4, count);
11       System.out.println();
12       for (int i = 0; i < count; i++)
13          System.out.print("-");
14       System.out.println();
15    }
16 }
```

12.3　字符串参数

接着，我们要考虑怎样把字符串传入、传出本地方法。字符串在这两种语言中很不一样，Java 编程语言中的字符串是 UTF-16 编码点的序列，而 C 的字符串则是以 null 结尾的字节序列。JNI 有两组操作字符串的函数，一组把 Java 字符串转换成 "modified UTF-8" 字节序列，另一组将它们转换成 UTF-16 数值的数组，也就是说转换成 jchar 数组。（UTF-8、"modified UTF-8" 和 UTF-16 格式都已经在第 2 章中讨论过了，请回忆一下，"modified UTF-8" 编码保持 ASCII 字符不变，但是其他所有 Unicode 字符被编码为多字节序列。）

> **注释**：标准 UTF-8 编码和 "modified UTF-8" 编码的差别仅在于编码大于 0xFFFF 的增补字符。在标准 UTF-8 编码中，这些字符编码为 4 字节序列；然而，在 modified UTF-8 编码中，这些字符首先被编码为一对 UTF-16 编码的 "替代品"，然后再对每个替代品用 UTF-8 编码，总共产生 6 字节编码。这有点笨拙，但这是个由历史原因造成的意外，编写 Java 虚拟机规范的时候 Unicode 还局限在 16 位。

如果你的 C 代码已经使用了 Unicode，那么你可以使用第二组转换函数。另一方面，如果你的字符串都仅限于使用 ASCII 字符，那么就可以使用"modified UTF-8"转换函数。

带有字符串参数的本地方法实际上都要接受一个 jstring 类型的值，而带有字符串参数返回值的本地方法必须返回一个 jstring 类型的值。JNI 函数将读入并构造出这些 jstring 对象。例如，NewStringUTF 函数会从包含 ASCII 字符的字符数组，或者是更一般的"modified UTF-8"编码的字节序列中，创建一个新的 jstring 对象。

JNI 函数有一个有些古怪的调用惯例。下面是对 NewStringUTF 函数的一个调用：

```
JNIEXPORT jstring JNICALL Java_HelloNative_getGreeting(JNIEnv* env, jclass cl)
{
    jstring jstr;
    char greeting[] = "Hello, Native World\n";
    jstr = (*env)->NewStringUTF(env, greeting);
    return jstr;
}
```

注释：本章中的所有代码都是 C 代码，除了指明为别的代码。

所有对 JNI 函数的调用都使用到了 env 指针，该指针是每一个本地方法的第一个参数。env 指针是指向函数指针表的指针（参见图 12-2）。所以，你必须在每个 JNI 调用前面加上 (*env)->，以便解析对函数指针的引用。而且，env 是每个 JNI 函数的第一个参数。

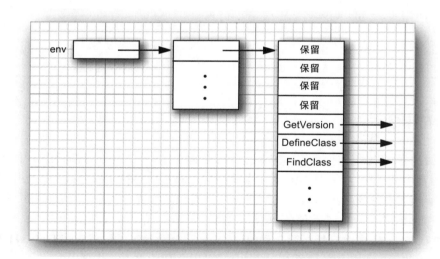

图 12-2　env 指针

C++ 注释：C++ 中对 JNI 函数的访问要简单一些。JNIEnv 类的 C++ 版本有一个内联成员函数，它负责帮你查找函数指针。例如，你可以这样调用 NewStringUTF 函数：

```
jstr = env->NewStringUTF(greeting);
```

注意，这里删除了该调用的参数列表里的 JNIEnv 指针。

NewStringUTF 函数可以用来构造一个新的 jstring，而读取现有 jstring 对象的内容，需要使用 GetStringUTFChars 函数。该函数返回指向描述字符串的"modified UTF-8"字符的 const jbyte* 指针。注意，具体的虚拟机可以为其内部的字符串表示方法自由地选择编码机制。所以，你可以得到实际的 Java 字符串的字符指针。因为 Java 字符串是不可变的，所以慎重处理 const 就显得非常重要，不要试图将数据写到该字符数组中。另一方面，如果虚拟机使用 UTF-16 或 UTF-32 字符作为其内部字符串的表示，那么该函数会分配一个新的内存块来存储等价的"modified UTF-8"编码字符。

虚拟机必须知道你何时使用完字符串，这样它就能进行垃圾回收（垃圾回收器是在一个独立线程中运行的，它能够中断本地方法的执行）。基于这个原因，你必须调用 ReleaseStringUTFChars 函数。

另外，可以通过调用 GetStringRegion 或 GetStringUTFRegion 方法来提供你自己的缓存，以存放字符串的字符。

最后 GetStringUTFLength 函数返回字符串的"modified UTF-8"编码所需的字符个数。

> **注释**：你可以在 http://docs.oracle.com/javase/7/docs/technotes/guides/jni 处找到 JNI API。

API 从 C 代码访问 Java 字符串

- jstring NewStringUTF(JNIEnv* env, const char bytes[])

  根据以全 0 字节结尾的"modified UTF-8"字节序列，返回一个新的 Java 字符串对象，或者当字符串无法构建时，返回 NULL。

- jsize GetStringUTFLength(JNIEnv* env, jstring string)

  返回进行 UTF-8 编码所需的字节个数（作为终止符的全 0 字节不计入内）。

- const jbyte* GetStringUTFChars(JNIEnv* env, jstring string, jboolean* isCopy)

  返回指向字符串的"modified UTF-8"编码的指针，或者当不能构建字符数组时返回 NULL。直到 ReleaseStringUTFChars 函数调用前，该指针一直有效。isCopy 指向一个 jboolean，如果进行了复制，则填入 JNI_TRUE，否则填入 JNI_FALSE。

- void ReleaseStringUTFChars(JNIEnv* env, jstring string, const jbyte bytes[])

  通知虚拟机本地代码不再需要通过 bytes（GetStringUTFChars 返回的指针）访问 Java 字符串。

- void GetStringRegion(JNIEnv *env, jstring string, jsize start, jsize length, jchar *buffer)

  将一个 UTF-16 双字节序列从字符串复制到用户提供的尺寸至少大于 2×length 的缓存中。

- void GetStringUTFRegion(JNIEnv *env, jstring string, jsize start, jsize length, jbyte *buffer)

  将一个"modified UTF-8"字符序列从字符串复制到用户提供的缓存中。为了存放要复制的字节，该缓存必须足够长。最坏情况下，要复制 3×length 个字节。

- jstring NewString(JNIEnv* env, const jchar chars[], jsize length)

根据 Unicode 字符串返回一个新的 Java 字符串对象，或者在不能构建时返回 NULL。
- jsize GetStringLength(JNIEnv* env, jstring string)

  返回字符串中字符的个数。
- const jchar* GetStringChars(JNIEnv* env, jstring string, jboolean* isCopy)

  返回指向字符串的 Unicode 编码的指针，或者当不能构建字符数组时返回 NULL。直到 ReleaseStringChars 函数调用前，该指针一直有效。isCopy 要么为 NULL；要么在进行了复制时，指向用 JNI_TRUE 填充的 jboolean，否则指向用 JNI_FALSE 填充的 jboolean。
- void ReleaseStringChars(JNIEnv* env, jstring string, const jchar chars[])

  通知虚拟机本地代码不再需要通过 chars（GetStringChars 返回的指针）访问 Java 字符串。

让我们使用这些函数来编写一个调用 C 函数 sprintf 的类，我们要像程序清单 12-8 所示那样调用这个函数。

程序清单 12-8   printf2/Printf2Test.java

```
 1  /**
 2   * @version 1.10 1997-07-01
 3   * @author Cay Horstmann
 4   */
 5  class Printf2Test
 6  {
 7     public static void main(String[] args)
 8     {
 9        double price = 44.95;
10        double tax = 7.75;
11        double amountDue = price * (1 + tax / 100);
12
13        String s = Printf2.sprint("Amount due = %8.2f", amountDue);
14        System.out.println(s);
15     }
16  }
```

程序清单 12-9 给出了带有本地 sprint 方法的类。

因此，格式化浮点数的 C 函数原型如下：

```
JNIEXPORT jstring JNICALL Java_Printf2_sprint(JNIEnv* env, jclass cl,
    jstring format, jdouble x)
```

程序清单 12-9   printf2/Printf2.java

```
 1  /**
 2   * @version 1.10 1997-07-01
 3   * @author Cay Horstmann
 4   */
 5  class Printf2
 6  {
 7     public static native String sprint(String format, double x);
 8
 9     static
10     {
```

```
11      System.loadLibrary("Printf2");
12   }
13 }
```

程序清单 12-10 给出了 C 的实现代码。注意，我们通过调用 GetStringUTFChars 来读取格式参数，通过调用 NewStringUTF 来产生返回值，通过调用 ReleaseStringUTFChars 来通知虚拟机不再需要访问该字符串。

程序清单 12-10 printf2/Printf2.c

```
1  /**
2     @version 1.10 1997-07-01
3     @author Cay Horstmann
4  */
5
6  #include "Printf2.h"
7  #include <string.h>
8  #include <stdlib.h>
9  #include <float.h>
10
11 /**
12    @param format a string containing a printf format specifier
13    (such as "%8.2f"). Substrings "%%" are skipped.
14    @return a pointer to the format specifier (skipping the '%')
15    or NULL if there wasn't a unique format specifier
16 */
17 char* find_format(const char format[])
18 {
19    char* p;
20    char* q;
21
22    p = strchr(format, '%');
23    while (p != NULL && *(p + 1) == '%') /* skip %% */
24       p = strchr(p + 2, '%');
25    if (p == NULL) return NULL;
26    /* now check that % is unique */
27    p++;
28    q = strchr(p, '%');
29    while (q != NULL && *(q + 1) == '%') /* skip %% */
30       q = strchr(q + 2, '%');
31    if (q != NULL) return NULL; /* % not unique */
32    q = p + strspn(p, " -0+#"); /* skip past flags */
33    q += strspn(q, "0123456789"); /* skip past field width */
34    if (*q == '.') { q++; q += strspn(q, "0123456789"); }
35    /* skip past precision */
36    if (strchr("eEfFgG", *q) == NULL) return NULL;
37    /* not a floating-point format */
38    return p;
39 }
40
41 JNIEXPORT jstring JNICALL Java_Printf2_sprint(JNIEnv* env, jclass cl,
42    jstring format, jdouble x)
43 {
```

```c
44      const char* cformat;
45      char* fmt;
46      jstring ret;
47
48      cformat = (*env)->GetStringUTFChars(env, format, NULL);
49      fmt = find_format(cformat);
50      if (fmt == NULL)
51         ret = format;
52      else
53      {
54         char* cret;
55         int width = atoi(fmt);
56         if (width == 0) width = DBL_DIG + 10;
57         cret = (char*) malloc(strlen(cformat) + width);
58         sprintf(cret, cformat, x);
59         ret = (*env)->NewStringUTF(env, cret);
60         free(cret);
61      }
62      (*env)->ReleaseStringUTFChars(env, format, cformat);
63      return ret;
64   }
```

在本函数中,我们选择简化错误处理。如果打印浮点数的格式代码不是 %w.pc 形式的(其中 c 是 e、E、f、g 或 G 中的一个),那么我们将不对数字进行格式化。后面我们会介绍如何让本地方法抛出异常。

12.4 访问域

目前为止你看到的所有本地方法都是带有数字或字符串参数的静态方法。下面,我们考虑在对象上进行操作的本地方法。作为一个练习,我们用本地方法实现卷 I 第 4 章中的 Employee 类的一个方法。通常情况下并不需要这么做,但是这里演示了当你需要的时候可以怎样从本地方法访问对象域。

12.4.1 访问实例域

为了了解怎样从本地方法访问实例域,我们用 Java 重新实现了 raiseSalary 方法。其代码很简单:

```java
public void raiseSalary(double byPercent)
{
   salary *= 1 + byPercent / 100;
}
```

让我们重写代码,使其成为一个本地方法。与此前的本地方法不同,它并不是一个静态方法。运行 javac-h 给出以下原型:

```
JNIEXPORT void JNICALL Java_Employee_raiseSalary(JNIEnv *, jobject, jdouble);
```

注意,第二个参数不再是 jclass 类型而是 jobject 类型。实际上,它和 this 引用等价。静态方法得到的是类的引用,而非静态方法得到的是对隐式的 this 参数对象的引用。

现在，我们访问隐式参数的 salary 域。在 Java1.0 中"原生的"Java 到 C 的绑定中，这很简单，程序员可以直接访问对象数据域。然而，直接访问要求虚拟机暴露它们的内部数据布局。基于这个原因，JNI 要求程序员通过调用特殊的 JNI 函数来获取和设置数据的值。

在我们的例子里，要使用 GetdoubleField 和 SetDoubleField 函数，因为 salary 是 double 类型的。对于其他类型，可以使用的函数有：GetIntField/SetIntField、GetObjectField/SetObjectField 等等。其通用语法是：

```
x = (*env)->GetXxxField(env, this_obj, fieldID);
(*env)->SetXxxField(env, this_obj, fieldID, x);
```

这里，fieldID 是一个特殊类型 jfieldID 的值，jfieldID 标识结构中的一个域，而 Xxx 代表 Java 数据类型 (Object、Boolean、Byte 或其他)。为了获得 fieldID，必须先获得一个表示类的值，有两种方法可以实现此目的。GetObjectClass 函数可以返回任意对象的类。例如：

```
jclass class_Employee = (*env)->GetObjectClass(env, this_obj);
```

FindClass 函数可以让你以字符串形式来指定类名（有点奇怪的是，要以 / 代替句号作为包名之间的分隔符）。

```
jclass class_String = (*env)->FindClass(env, "java/lang/String");
```

之后，可以使用 GetFieldID 函数来获得 fieldID。必须提供域的名字、它的签名以及它的类型的编码。例如，下面是从 salary 域得到域 ID 的代码：

```
jfieldID id_salary = (*env)->GetFieldID(env, class_Employee, "salary", "D");
```

字符串 "D" 表示类型是 double。你将在下一节中学习到编码签名的全部规则。

你可能会认为访问数据域相当令人费解。JNI 的设计者不想把数据域直接暴露在外，所以他们不得不提供获取和设置数据域值的函数。为了使这些函数的开销最小化，从域名计算域 ID (代价最大的一个步骤) 被分解出来作为单独的一步操作。也就是说，如果你反复地获取和设置一个特定的域，你计算域标识符的开销就只有一次。

让我们把各部分汇总起来，下面的代码以本地方法形式重新实现了 raiseSalary 方法。

```
JNIEXPORT void JNICALL Java_Employee_raiseSalary(JNIEnv* env, jobject this_obj,
    jdouble byPercent)
{
   /* get the class */
   jclass class_Employee = (*env)->GetObjectClass(env, this_obj);

   /* get the field ID */
   jfieldID id_salary = (*env)->GetFieldID(env, class_Employee, "salary", "D");

   /* get the field value */
   jdouble salary = (*env)->GetDoubleField(env, this_obj, id_salary);

   salary *= 1 + byPercent / 100;

   /* set the field value */
   (*env)->SetDoubleField(env, this_obj, id_salary, salary);
}
```

> **警告**：类引用只在本地方法返回之前有效。因此，不能在你的代码中缓存 GetObject-Class 的返回值。不要将类引用保存下来以供以后的方法调用重复使用。必须在每次执行本地方法时都调用 GetObjectClass。如果你无法忍受这一点，必须调用 NewGlobalRef 来锁定该引用：
>
> ```
> static jclass class_X = 0;
> static jfieldID id_a;
> . . .
> if (class_X == 0)
> {
>    jclass cx = (*env)->GetObjectClass(env, obj);
>    class_X = (*env)->NewGlobalRef(env, cx);
>    id_a = (*env)->GetFieldID(env, cls, "a", ". . .");
> }
> ```
>
> 现在，你可以在后面的调用中使用类引用和域 ID 了。当你结束对类的使用时，务必调用：
>
> ```
> (*env)->DeleteGlobalRef(env, class_X);
> ```

程序清单 12-11 和程序清单 12-12 给出了测试程序和 Employee 类的 Java 代码。程序清单 12-13 包含了本地 raiseSalary 方法的 C 代码。

程序清单 12-11　employee/EmployeeTest.java

```java
1  /**
2   * @version 1.11 2018-05-01
3   * @author Cay Horstmann
4   */
5
6  public class EmployeeTest
7  {
8     public static void main(String[] args)
9     {
10        var staff = new Employee[3];
11
12        staff[0] = new Employee("Harry Hacker", 35000);
13        staff[1] = new Employee("Carl Cracker", 75000);
14        staff[2] = new Employee("Tony Tester", 38000);
15
16        for (Employee e : staff)
17           e.raiseSalary(5);
18        for (Employee e : staff)
19           e.print();
20     }
21  }
```

程序清单 12-12　employee/Employee.java

```java
1  /**
2   * @version 1.10 1999-11-13
3   * @author Cay Horstmann
4   */
5
```

```java
 6  public class Employee
 7  {
 8     private String name;
 9     private double salary;
10
11     public native void raiseSalary(double byPercent);
12
13     public Employee(String n, double s)
14     {
15        name = n;
16        salary = s;
17     }
18
19     public void print()
20     {
21        System.out.println(name + " " + salary);
22     }
23
24     static
25     {
26        System.loadLibrary("Employee");
27     }
28  }
```

程序清单 12-13 employee/Employee.c

```c
 1  /**
 2     @version 1.10 1999-11-13
 3     @author Cay Horstmann
 4  */
 5
 6  #include "Employee.h"
 7
 8  #include <stdio.h>
 9
10  JNIEXPORT void JNICALL Java_Employee_raiseSalary(
11        JNIEnv* env, jobject this_obj, jdouble byPercent)
12  {
13     /* get the class */
14     jclass class_Employee = (*env)->GetObjectClass(env, this_obj);
15
16     /* get the field ID */
17     jfieldID id_salary = (*env)->GetFieldID(env, class_Employee, "salary", "D");
18
19     /* get the field value */
20     jdouble salary = (*env)->GetDoubleField(env, this_obj, id_salary);
21
22     salary *= 1 + byPercent / 100;
23
24     /* set the field value */
25     (*env)->SetDoubleField(env, this_obj, id_salary, salary);
26  }
27
```

12.4.2 访问静态域

访问静态域和访问非静态域类似，要使用 GetStaticFieldID 和 GetStaticXxxField/ SetStatic-XxxField 函数。它们几乎与非静态的情形一样，只有两个区别：
- 由于没有对象，所以必须使用 FindClass 代替 GetObjectClass 来获得类引用。
- 访问域时，要提供类而非实例对象。

例如，下面给出的是怎样得到 System.out 的引用的代码：

```
/* get the class */
jclass class_System = (*env)->FindClass(env, "java/lang/System");

/* get the field ID */
jfieldID id_out = (*env)->GetStaticFieldID(env, class_System, "out",
    "Ljava/io/PrintStream;");

/* get the field value */
jobject obj_out = (*env)->GetStaticObjectField(env, class_System, id_out);
```

> **API** 访问实例域
>
> - jfieldID GetFieldID(JNIEnv *env, jclass cl, const char name[], const char fieldSignature[])
>
>   返回类中一个域的标识符。
>
> - *Xxx* Get*Xxx*Field(JNIEnv *env, jobject obj, jfieldID id)
>
>   返回域的值。域类型 *Xxx* 是 Object、Boolean、Byte、Char、Short、Int、Long、Float 或 Double 之一。
>
> - void Set*Xxx*Field(JNIEnv *env, jobject obj, jfieldID id, *Xxx* value)
>
>   把某个域设置为一个新值。域类型 *Xxx* 是 Object、Boolean、Byte、Char、Short、Int、Long、Float 或 Double 之一。
>
> - jfieldID GetStaticFieldID(JNIEnv *env, jclass cl, const char name[], const char fieldSignature[])
>
>   返回某类型的一个静态域的标识符。
>
> - *Xxx* GetStatic*Xxx*Field(JNIEnv *env, jclass cl, jfieldID id)
>
>   返回某静态域的值。域类型 *Xxx* 是 Object、Boolean、Byte、Char、Short、Int、Long、Float 或 Double 之一。
>
> - void SetStatic*Xxx*Field(JNIEnv *env, jclass cl, jfieldID id, *Xxx* value)
>
>   把某个静态域设置为一个新值。域类型 *Xxx* 是 Object、Boolean、Byte、Char、Short、Int、Long、Float 或 Double 之一。

12.5 编码签名

为了访问实例域和调用用 Java 编程语言定义的方法，你必须学习将数据类型的名称和方法签名进行"混编"的规则（方法签名描述了参数和该方法返回值的类型）。下面是编码方案：

B	byte
C	char
D	double
F	float
I	int
J	long
L*classname*;	类的类型
S	short
V	void
Z	boolean

为了描述数组类型，要使用 [。例如，一个字符串数组如下：

[Ljava/lang/String;

一个 float[][] 可以描述为：

[[F

要建立一个方法的完整签名，需要把括号内的参数类型都列出来，然后列出返回值类型。例如，一个接收两个整型参数并返回一个整数的方法编码为：

(II)I

12.3 节中的 Sprint 方法有下面的混编签名：

(Ljava/lang/String;D)Ljava/lang/String;

也就是说，该方法接收一个 String 和一个 double，返回值是一个 String。

注意，在 L 表达式结尾处的分号是类型表达式的终止符，而不是参数之间的分隔符。例如，构造器：

Employee(java.lang.String, double, java.util.Date)

具有如下签名：

"(Ljava/lang/String;DLjava/util/Date;)V"

注意，在 D 和 Ljava/util/Date; 之间没有分隔符。另外要注意在这个编码方案中，必须用 / 代替 . 来分隔包和类名。结尾的 V 表示返回类型为 void。即使对 Java 的构造器没有指定返回类型，也需要将 V 添加到虚拟机签名中。

> **提示**：可以使用带有选项 -s 的 javap 命令来从类文件中产生方法签名。例如，运行
>
>   javap -s -private Employee
>
> 可以得到以下显示所有域和方法的输出：
>
>   Compiled from "Employee.java"
>   public class Employee extends java.lang.Object{
>   private java.lang.String name;
>   Signature: Ljava/lang/String;
>   private double salary;
>   Signature: D
>   public Employee(java.lang.String, double);

```
      Signature: (Ljava/lang/String;D)V
    public native void raiseSalary(double);
      Signature: (D)V
    public void print();
      Signature: ()V
    static {};
      Signature: ()V
    }
```

> **注释：** 没有任何理由强迫程序员使用这种混编方案来描述签名。本地调用机制的设计者可以非常容易地编写一个函数来读取 Java 编程语言风格的签名，比如 void (int,java.lang.String)，并且将它们编码为他们喜欢的某种内部表示法。再者，使用混编签名使你能够分享接近虚拟机的编程奥秘。

12.6 调用 Java 方法

当然，Java 编程语言的函数可以调用 C 函数，这正是本地方法要做的。我们能不能换一种方式呢？为什么我们要这么做？答案是，本地方法常常需要从传递给它的对象那里得到某种服务。我们首先介绍非静态方法如何进行这种操作，然后介绍静态方法如何进行这种操作。

12.6.1 实例方法

作为从本地代码调用 Java 方法的一个例子，我们先增强 Printf 类，给它增加一个与 C 函数 fprintf 类似的方法。也就是说，它能够在任意 PrintWriter 对象上打印一个字符串。下面是用 Java 编写的该方法的定义：

```
class Printf3
{
   public native static void fprint(PrintWriter out, String s, double x);
   ...
}
```

我们首先把要打印的字符串组装成一个 String 对象 str，就像我们在 sprint 方法中已经实现的那样。然后，我们从实现本地方法的 C 函数中调用 PrintWriter 类的 print 方法。

使用如下函数调用，可以从 C 中调用任何 Java 方法：

(*env)->Call*Xxx*Method(env, *implicit parameter*, *methodID*, *explicit parameters*)

根据方法的返回类型，用 Void、Int、Object 等来替换 *Xxx*。就像需要一个 fieldID 来访问某个对象的一个域一样，还需要一个方法的 ID 来调用方法。可以通过调用 JNI 函数 GetMethodID，并且提供该类、方法的名字和方法签名来获得方法 ID。

在我们的例子中，我们想要获得 PrintWriter 类的 print 方法的 ID。PrintWriter 类有几个名为 print 的重载方法。基于这个原因，还必须提供一个字符串，描述想要使用的特定函数的参数和返回值。例如，我们想要使用 void print(java.lang.String)，正如前一节讲到的那样，我们必须把签名"混编"为字符串 "(Ljava/lang/String;)V"。

下面是进行方法调用的完整代码：

```
/* get the class of the implicit parameter */
class_PrintWriter = (*env)->GetObjectClass(env, out);

/* get the method ID */
id_print = (*env)->GetMethodID(env, class_PrintWriter, "print", "(Ljava/lang/String;)V");

/* call the method */
(*env)->CallVoidMethod(env, out, id_print, str);
```

程序清单 12-14 和程序清单 12-15 给出了测试程序和 Printf3 类的 Java 代码。程序清单 12-16 包含了本地 fprintf 方法的 C 代码。

> **注释**：数值型的方法 ID 和域 ID 在概念上和反射 API 中的 Method 和 Field 对象相似。可以使用以下函数在两者间进行转换：
>
> ```
> jobject ToReflectedMethod(JNIEnv* env, jclass class, jmethodID methodID);
>     // returns Method object
> methodID FromReflectedMethod(JNIEnv* env, jobject method);
> jobject ToReflectedField(JNIEnv* env, jclass class, jfieldID fieldID);
>     // returns Field object
> fieldID FromReflectedField(JNIEnv* env, jobject field);
> ```

程序清单 12-14 printf3/Printf3Test.java

```java
 1  import java.io.*;
 2
 3  /**
 4   * @version 1.11 2018-05-01
 5   * @author Cay Horstmann
 6   */
 7  class Printf3Test
 8  {
 9     public static void main(String[] args)
10     {
11        double price = 44.95;
12        double tax = 7.75;
13        double amountDue = price * (1 + tax / 100);
14        var out = new PrintWriter(System.out);
15        Printf3.fprint(out, "Amount due = %8.2f\n", amountDue);
16        out.flush();
17     }
18  }
```

程序清单 12-15 printf3/Printf3.java

```java
 1  import java.io.*;
 2
 3  /**
 4   * @version 1.10 1997-07-01
```

```
5    * @author Cay Horstmann
6    */
7   class Printf3
8   {
9      public static native void fprint(PrintWriter out, String format, double x);
10
11     static
12     {
13        System.loadLibrary("Printf3");
14     }
15  }
```

程序清单 12-16　printf3/Printf3.c

```
1   /**
2      @version 1.10 1997-07-01
3      @author Cay Horstmann
4   */
5
6   #include "Printf3.h"
7   #include <string.h>
8   #include <stdlib.h>
9   #include <float.h>
10
11  /**
12     @param format a string containing a printf format specifier
13     (such as "%8.2f"). Substrings "%%" are skipped.
14     @return a pointer to the format specifier (skipping the '%')
15     or NULL if there wasn't a unique format specifier
16  */
17  char* find_format(const char format[])
18  {
19     char* p;
20     char* q;
21
22     p = strchr(format, '%');
23     while (p != NULL && *(p + 1) == '%') /* skip %% */
24        p = strchr(p + 2, '%');
25     if (p == NULL) return NULL;
26     /* now check that % is unique */
27     p++;
28     q = strchr(p, '%');
29     while (q != NULL && *(q + 1) == '%') /* skip %% */
30        q = strchr(q + 2, '%');
31     if (q != NULL) return NULL; /* % not unique */
32     q = p + strspn(p, " -0+#"); /* skip past flags */
33     q += strspn(q, "0123456789"); /* skip past field width */
34     if (*q == '.') { q++; q += strspn(q, "0123456789"); }
35        /* skip past precision */
36     if (strchr("eEfFgG", *q) == NULL) return NULL;
37        /* not a floating-point format */
38     return p;
39  }
40
```

```
41   JNIEXPORT void JNICALL Java_Printf3_fprint(JNIEnv* env, jclass cl,
42        jobject out, jstring format, jdouble x)
43   {
44       const char* cformat;
45       char* fmt;
46       jstring str;
47       jclass class_PrintWriter;
48       jmethodID id_print;
49
50       cformat = (*env)->GetStringUTFChars(env, format, NULL);
51       fmt = find_format(cformat);
52       if (fmt == NULL)
53          str = format;
54       else
55       {
56          char* cstr;
57          int width = atoi(fmt);
58          if (width == 0) width = DBL_DIG + 10;
59          cstr = (char*) malloc(strlen(cformat) + width);
60          sprintf(cstr, cformat, x);
61          str = (*env)->NewStringUTF(env, cstr);
62          free(cstr);
63       }
64       (*env)->ReleaseStringUTFChars(env, format, cformat);
65
66       /* now call ps.print(str) */
67
68       /* get the class */
69       class_PrintWriter = (*env)->GetObjectClass(env, out);
70
71       /* get the method ID */
72       id_print = (*env)->GetMethodID(env, class_PrintWriter, "print", "(Ljava/lang/String;)V");
73
74       /* call the method */
75       (*env)->CallVoidMethod(env, out, id_print, str);
76   }
```

12.6.2　静态方法

从本地方法调用静态方法与调用非静态方法类似。两者的差别是：

- 要用 GetStaticMethodID 和 CallStatic*Xxx*Method 函数。
- 当调用方法时，要提供类对象，而不是隐式的参数对象。

作为一个例子，让我们从本地方法调用以下静态方法：

System.getProperty("java.class.path")

这个调用的返回值是给出了当前类路径的字符串。

首先，我们必须找到要用的类。因为我们没有 System 类的对象可供使用，所以我们使用 FindClass 而非 GetObjectClass：

jclass class_System = (*env)->FindClass(env, "java/lang/System");

接着，我们需要静态 getProperty 方法的 ID。该方法的编码签名是：

"(Ljava/lang/String;)Ljava/lang/String;"

既然参数和返回值都是字符串。因此，我们这样获取方法 ID：

```
jmethodID id_getProperty = (*env)->GetStaticMethodID(env, class_System, "getProperty",
    "(Ljava/lang/String;)Ljava/lang/String;");
```

最后，我们进行调用。注意，类对象被传递给了 CallStaticObjectMethod 函数。

```
jobject obj_ret = (*env)->CallStaticObjectMethod(env, class_System, id_getProperty,
    (*env)->NewStringUTF(env, "java.class.path"));
```

该方法的返回值是 jobject 类型的。如果我们想要把它当作字符串操作，必须把它转型为 jstring：

```
jstring str_ret = (jstring) obj_ret;
```

> **C++ 注释**：在 C 中，jstring 和 jclass 类型同后面将要介绍的数组类型一样，都是与 jobject 等价的类型。因此，在 C 语言中，前面例子中的转型并不是严格必需的。但是在 C++ 中，这些类型被定义为指向拥有正确继承层次关系的"哑类"的指针。例如，将一个 jstring 不经过转型便赋给 jobject 在 C++ 中是合法的，但是将 jobject 赋给 jstring 必须先转型。

12.6.3 构造器

本地方法可以通过调用构造器来创建新的 Java 对象。可以调用 NewObject 函数来调用构造器。

```
jobject obj_new = (*env)->NewObject(env, class, methodID, construction parameters);
```

可以通过指定方法名为 "<init>"，并指定构造器（返回值为 void）的编码签名，从 GetMethodID 函数中获取该调用必需的方法 ID。例如，下面是本地方法创建 FileOutputStream 对象的情形：

```
const char[] fileName = "...";
jstring str_fileName = (*env)->NewStringUTF(env, fileName);
jclass class_FileOutputStream = (*env)->FindClass(env, "java/io/FileOutputStream");
jmethodID id_FileOutputStream
    = (*env)->GetMethodID(env, class_FileOutputStream, "<init>", "(Ljava/lang/String;)V");
jobject obj_stream
    = (*env)->NewObject(env, class_FileOutputStream, id_FileOutputStream, str_fileName);
```

注意，构造器的签名接受一个 java.lang.String 类型的参数，返回类型为 void。

12.6.4 另一种方法调用

有若干种 JNI 函数的变体都可以从本地代码调用 Java 方法。它们没有我们已经讨论过的那些函数那么重要，但偶尔也会很有用。

CallNonvirtual*Xxx*Method 函数接收一个隐式参数、一个方法 ID、一个类对象（必须对应于隐式参数的超类）和一个显式参数。这个函数将调用指定的类中的指定版本的方法，而不使用常规的动态调度机制。

所有调用函数都有后缀 "A" 和 "V" 的版本，用于接收数组中或 va_list 中的显式参数（就像在 C 头文件 stdarg.h 中所定义的那样）。

> **API** 执行 Java 方法

- jmethodID GetMethodID(JNIEnv *env, jclass cl, const char name[], const char methodSignature[])

  返回类中某个方法的标识符。

- *Xxx* Call*Xxx*Method(JNIEnv *env, jobject obj, jmethodID id, args)
- *Xxx* Call*Xxx*MethodA(JNIEnv *env, jobject obj, jmethodID id, jvalue args[])
- *Xxx* Call*Xxx*MethodV(JNIEnv *env, jobject obj, jmethodID id, va_list args)

  调用一个方法。返回类型 *Xxx* 是 Object、Boolean、Byte、Char、Short、Int、Long、Float 或 Double 之一。第一个函数有可变数量参数，只要把方法参数附加到方法 ID 之后即可。第二个函数接受 jvalue 数组中的方法参数，其中 jvalue 是一个联合体，定义如下：

  ```
  typedef union jvalue
  {
    jboolean z;
    jbyte b;
    jchar c;
    jshort s;
    jint i;
    jlong j;
    jfloat f;
    jdouble d;
    jobject l;
  } jvalue;
  ```

  第三个函数接收 C 头文件 stdarg.h 中定义的 va_list 中的方法参数。

- *Xxx* CallNonvirtual*Xxx*Method(JNIEnv *env, jobject obj, jclass cl, jmethodID id, args)
- *Xxx* CallNonvirtual*Xxx*MethodA(JNIEnv *env, jobject obj, jclass cl, jmethodID id, jvalue args[])
- *Xxx* CallNonvirtual*Xxx*MethodV(JNIEnv *env, jobject obj, jclass cl, jmethodID id, va_list args)

  调用一个方法，并绕过动态调度。返回类型 *Xxx* 是 Object、Boolean、Byte、Char、Short、Int、Long、Float 或 Double 之一。第一个函数有可变数量参数，只要把方法参数附加到方法 ID 之后即可。第二个函数接受 jvalue 数组中的方法参数。第三个函数接受 C 头文件 stdarg.h 中定义的 va_list 中的方法参数。

- jmethodID GetStaticMethodID(JNIEnv *env, jclass cl, const char name[], const char methodSignature[])

  返回类的某个静态方法的标识符。

- *Xxx* CallStatic*Xxx*Method(JNIEnv *env, jclass cl, jmethodID id, args)
- *Xxx* CallStatic*Xxx*MethodA(JNIEnv *env, jclass cl, jmethodID id, jvalue args[])

- *Xxx* CallStatic*Xxx*MethodV(JNIEnv *env, jclass cl, jmethodID id, va_list args)

  调用一个静态方法。返回类型 *Xxx* 是 Object、Boolean、Byte、Char、Short、Int、Long、Float 或 Double 之一。第一个函数有可变数量参数，只要把方法参数附加到方法 ID 之后即可。第二个函数接受 jvalue 数组中的方法参数。第三个函数接受 C 头文件 stdarg.h 中定义的 va_list 中的方法参数。

- jobject NewObject(JNIEnv *env, jclass cl, jmethodID id, args)
- jobject NewObjectA(JNIEnv *env, jclass cl, jmethodID id, jvalue args[])
- jobject NewObjectV(JNIEnv *env, jclass cl, jmethodID id, va_list args)

  调用构造器。函数 ID 从带有函数名为 "<init>" 和返回类型为 void 的 GetMethodID 获取。第一个函数有可变数量参数，只要把方法参数附加到方法 ID 之后即可。第二个函数接收 jvalue 数组中的方法参数。第三个函数接收 C 头文件 stdarg.h 中定义的 va_list 中的方法参数。

12.7 访问数组元素

Java 编程语言的所有数组类型都有相对应的 C 语言类型，见表 12-2。

表 12-2 Java 数组类型和 C 数组类型之间的对应关系

Java 数组类型	C 数组类型	Java 数组类型	C 数组类型
boolean[]	jbooleanArray	long[]	jlongArray
byte[]	jbyteArray	float[]	jfloatArray
char[]	jcharArray	double[]	jdoubleArray
int[]	jintArray	Object[]	jobjectArray
short[]	jshortArray		

> **C++ 注释**：在 C 中，所有这些数组类型实际上都是 jobject 的同义类型。然而，在 C++ 中它们被安排在如图 12-3 所示的继承层次结构中。jarray 类型表示一个泛型数组。

GetArrayLength 函数返回数组的长度。

```
jarray array = ...;
jsize length = (*env)->GetArrayLength(env, array);
```

怎样访问数组元素取决于数组中存储的是对象还是基本类型的数据（如 bool、char 或数值类型）。可以通过 GetObjectArrayElement 和 SetObjectArrayElement 方法访问对象数组的元素。

```
jobjectArray array = ...;
int i, j;
jobject x = (*env)->GetObjectArrayElement(env, array, i);
(*env)->SetObjectArrayElement(env, array, j, x);
```

这个方法虽然简单，但是效率明显低下，当想要直接访问数组元素，特别是在进行向量或矩阵计算时更是如此。

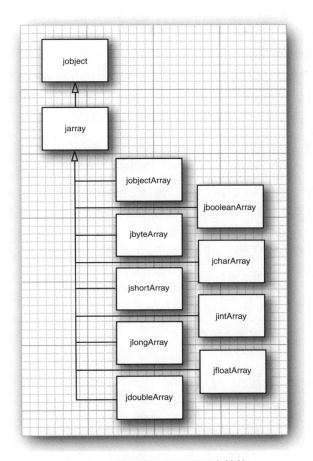

图 12-3　数组类型的继承层次结构

Get*Xxx*ArrayElements 函数返回一个指向数组起始元素的 C 指针。与普通的字符串一样，当不再需要该指针时，必须记得要调用 Release*Xxx*ArrayElements 函数通知虚拟机。这里，类型 *Xxx* 必须是基本类型，也就是说，不能是 Object。这样就可以直接读写数组元素了。另一方面，由于指针可能会指向一个副本，只有调用相应的 Release*Xxx*- ArrayElements 函数时，所做的改变才能保证在源数组里得到反映。

> **注释**：通过把一个指向 jboolean 变量的指针作为第三个参数传递给 Get*Xxx*Array-Elements 方法，就可以发现一个数组是否是副本了。如果是副本，则该变量被 JNI_TRUE 填充。如果对这个信息不感兴趣，传一个空指针即可。

下面是对 double 类型数组中的所有元素乘以一个常量的示例代码。我们获取一个 Java 数组的 C 指针 a，并用 a[i] 访问各个元素。

```
jdoubleArray array_a = . . .;
double scaleFactor = . . .;
double* a = (*env)->GetDoubleArrayElements(env, array_a, NULL);
for (i = 0; i < (*env)->GetArrayLength(env, array_a); i++)
    a[i] = a[i] * scaleFactor;
(*env)->ReleaseDoubleArrayElements(env, array_a, a, 0);
```

虚拟机是否确实需要对数组进行拷贝取决于它是如何分配数组和如何进行垃圾回收的。有些"拷贝"型的垃圾回收器例行地移动对象，并更新对象引用。该策略与将数组锁定在特定位置是不兼容的，因为回收器不能更新本地代码中的指针值。

> 📖 **注释**：Oracle 的 JVM 实现中，boolean 数组是用打包的 32 位字数组表示的。GetBooleanArrayElements 方法能将它们复制到拆包的 jboolean 值的数组中。

如果要访问一个大数组的多个元素，可以用 GetXxxArrayRegion 和 SetXxxArray-Region 方法，它能把一定范围内的元素从 Java 数组复制到 C 数组中或从 C 数组复制到 Java 数组中。

可以用 New*Xxx*Array 函数在本地方法中创建新的 Java 数组。要创建新的对象数组，需要指定长度、数组元素的类型和所有元素的初始值（典型的是 NULL）。下面是一个例子。

```
jclass class_Employee = (*env)->FindClass(env, "Employee");
jobjectArray array_e = (*env)->NewObjectArray(env, 100, class_Employee, NULL);
```

基本类型的数组要简单一些。只需提供数组长度。

```
jdoubleArray array_d = (*env)->NewDoubleArray(env, 100);
```

该数组被 0 填充。

> 📖 **注释**：下面的方法用来操作"直接缓存"：
> ```
> jobject NewDirectByteBuffer(JNIEnv* env, void* address, jlong capacity)
> void* GetDirectBufferAddress(JNIEnv* env, jobject buf)
> jlong GetDirectBufferCapacity(JNIEnv* env, jobject buf)
> ```
> java.nio 包中使用了直接缓存来支持更高效的输入输出操作，并尽可能减少本地和 Java 数组之间的复制操作。

API 操作 Java 数组

- jsize GetArrayLength(JNIEnv *env, jarray array)

  返回数组中的元素个数。

- jobject GetObjectArrayElement(JNIEnv *env, jobjectArray array, jsize index)

  返回数组元素的值。

- void SetObjectArrayElement(JNIEnv *env, jobjectArray array, jsize index, jobject value)

  将数组元素设为新值。

- *Xxx** Get*Xxx*ArrayElements(JNIEnv *env, jarray array, jboolean* isCopy)

  产生一个指向 Java 数组元素的 C 指针。域类型 *Xxx* 是 Boolean、Byte、Char、Short、Int、Long、Float 或 Double 之一。指针不再使用时，该指针必须传递给 Release*Xxx*ArrayElements。

iscopy 可能是 NULL，或者在进行了复制时，指向用 JNI_TRUE 填充的 jboolean；否则，指向用 JNI_FALSE 填充的 jboolean。
- void Release*Xxx*ArrayElements(JNIEnv *env, jarray array, *Xxx* elems[], jint mode)

  通知虚拟机通过 GetXxxArrayElements 获得的一个指针已经不再需要了。Mode 是 0（更新数组元素后释放 elems 缓存）、JNI_COMMIT（更新数组元素后不释放 elems 缓存）或 JNI_ABORT（不更新数组元素便释放 elems 缓存）之一。
- void Get*Xxx*ArrayRegion(JNIEnv *env, jarray array, jint start, jint length, *Xxx* elems[])

  将 Java 数组的元素复制到 C 数组中。域类型 *Xxx* 是 Boolean、Byte、Char、Short、Int、Long、Float 或 Double 之一。
- void Set*Xxx*ArrayRegion(JNIEnv *env, jarray array, jint start, jint length, *Xxx* elems[])

  将 C 数组的元素复制到 Java 数组中。域类型 *Xxx* 是 Boolean、Byte、Char、Short、Int、Long、Float 或 Double 之一。

12.8 错误处理

在 Java 编程语言中，使用本地方法对于程序来说是要冒很大的安全风险的。C 的运行期系统对数组越界错误、不良指针造成的间接错误等不提供任何防护。所以，对于本地方法的程序员来说，处理所有的出错条件以保持 Java 平台的完整性显得格外重要。尤其是，当本地方法诊断出一个它无法解决的问题时，那么它应该将此问题报告给 Java 虚拟机。然后，在这种情况下，很自然地会抛出一个异常。然而，C 语言没有异常，必须调用 Throw 或 ThrowNew 函数来创建一个新的异常对象。当本地方法退出时，Java 虚拟机就会抛出该异常。

要使用 Throw 函数，需要调用 NewObject 来创建一个 Throwable 子类的对象。例如，下面我们分配了一个 EOFException 对象，然后将它抛出。

```
jclass class_EOFException = (*env)->FindClass(env, "java/io/EOFException");
jmethodID id_EOFException = (*env)->GetMethodID(env, class_EOFException, "<init>", "()V");
   /* ID of no-argument constructor */
jthrowable obj_exc = (*env)->NewObject(env, class_EOFException, id_EOFException);
(*env)->Throw(env, obj_exc);
```

通常调用 ThrowNew 会更加方便，因为只需提供一个类和一个"modified UTF-8"字节序列，该函数就会构建一个异常对象。

```
(*env)->ThrowNew(env, (*env)->FindClass(env, "java/io/EOFException"),
   "Unexpected end of file");
```

Throw 和 ThrowNew 都只是发布异常，它们不会中断本地方法的控制流。只有当该方法返回时，Java 虚拟机才会抛出异常。所以，每一个对 Throw 和 ThrowNew 的调用语句之后总是紧跟着 return 语句。

> **C++ 注释**：如果用 C++ 实现本地方法，那么就无法用 C++ 代码抛出 Java 异常。在 C++ 绑定中，是可以实现一个在 C++ 异常和 Java 异常之间的转换的。然而，到目前

为止还没有实现这个功能。需要在本地方法中使用 Throw 或 ThrowNew 函数来抛出 Java 异常，并且要确保你的本地方法不抛出 C++ 异常。

通常，本地代码不需要考虑捕获 Java 异常。但是，当本地方法调用 Java 方法时，该方法可能会抛出异常。而且，一些 JNI 函数也会抛出异常。例如，如果索引越界，SetObjectArrayElement 方法会抛出一个 ArrayIndexOutOfBoundsException 异常，如果所存储的对象的类不是数组元素类的子类，该方法会抛出一个 ArrayStoreException 异常。在这类情况下，本地方法应该调用 ExceptionOccurred 方法来确认是否有异常抛出。如果没有任何异常等待处理，则下面的调用：

```
jthrowable obj_exc = (*env)->ExceptionOccurred(env);
```

将返回 NULL。否则，返回一个当前异常对象的引用。如果只要检查是否有异常抛出，而不需要获得异常对象的引用，那么应使用：

```
jboolean occurred = (*env)->ExceptionCheck(env);
```

通常，有异常出现时，本地方法应该直接返回。那样，虚拟机就会将该异常传送给 Java 代码。但是，本地方法也可以分析异常对象，确定它是否能够处理该异常。如果能够处理，那么必须调用下面的函数来关闭该异常：

```
(*env)->ExceptionClear(env);
```

在我们的例子中，我们实现了 fprint 本地方法，这是基于该方法适合编写为本地方法的假设而实现的。下面是我们抛出的异常：

- 如果格式字符串是 NULL，则抛出 NullPointerException 异常。
- 如果格式字符串不含适合打印 double 所需的 % 说明符，则抛出 IllegalArgument-Exception 异常。
- 如果调用 malloc 失败，则抛出 OutOfMemoryError 异常。

最后，为了说明本地方法调用 Java 方法时，怎样检查异常，我们将一个字符串发送给数据流，一次一个字符，并且在每次调用 Java 方法后调用 ExceptionOccurred。程序清单 12-17 给出了本地方法的代码，程序清单 12-18 展示了含有本地方法的类的定义。注意，在调用 PrintWriter.print 出现异常时，本地方法并不会立即终止执行，它会首先释放 cstr 缓存。当本地方法返回时，虚拟机再次抛出异常。程序清单 12-19 的测试程序说明了当格式字符串无效时，本地方法是如何抛出异常的。

程序清单 12-17 printf4/Printf4.c

```
1  /**
2     @version 1.10 1997-07-01
3     @author Cay Horstmann
4  */
5
6  #include "Printf4.h"
7  #include <string.h>
```

```
8   #include <stdlib.h>
9   #include <float.h>
10
11  /**
12      @param format a string containing a printf format specifier
13      (such as "%8.2f"). Substrings "%%" are skipped.
14      @return a pointer to the format specifier (skipping the '%')
15      or NULL if there wasn't a unique format specifier
16  */
17  char* find_format(const char format[])
18  {
19      char* p;
20      char* q;
21
22      p = strchr(format, '%');
23      while (p != NULL && *(p + 1) == '%') /* skip %% */
24          p = strchr(p + 2, '%');
25      if (p == NULL) return NULL;
26      /* now check that % is unique */
27      p++;
28      q = strchr(p, '%');
29      while (q != NULL && *(q + 1) == '%') /* skip %% */
30          q = strchr(q + 2, '%');
31      if (q != NULL) return NULL; /* % not unique */
32      q = p + strspn(p, " -0+#"); /* skip past flags */
33      q += strspn(q, "0123456789"); /* skip past field width */
34      if (*q == '.') { q++; q += strspn(q, "0123456789"); }
35          /* skip past precision */
36      if (strchr("eEfFgG", *q) == NULL) return NULL;
37          /* not a floating-point format */
38      return p;
39  }
40
41  JNIEXPORT void JNICALL Java_Printf4_fprint(JNIEnv* env, jclass cl,
42          jobject out, jstring format, jdouble x)
43  {
44      const char* cformat;
45      char* fmt;
46      jclass class_PrintWriter;
47      jmethodID id_print;
48      char* cstr;
49      int width;
50      int i;
51
52      if (format == NULL)
53      {
54          (*env)->ThrowNew(env,
55              (*env)->FindClass(env,
56                  "java/lang/NullPointerException"),
57              "Printf4.fprint: format is null");
58          return;
59      }
60
61      cformat = (*env)->GetStringUTFChars(env, format, NULL);
```

```c
 62        fmt = find_format(cformat);
 63
 64        if (fmt == NULL)
 65        {
 66           (*env)->ThrowNew(env,
 67              (*env)->FindClass(env,
 68                 "java/lang/IllegalArgumentException"),
 69                 "Printf4.fprint: format is invalid");
 70           return;
 71        }
 72
 73        width = atoi(fmt);
 74        if (width == 0) width = DBL_DIG + 10;
 75        cstr = (char*)malloc(strlen(cformat) + width);
 76
 77        if (cstr == NULL)
 78        {
 79           (*env)->ThrowNew(env,
 80              (*env)->FindClass(env, "java/lang/OutOfMemoryError"),
 81              "Printf4.fprint: malloc failed");
 82           return;
 83        }
 84
 85        sprintf(cstr, cformat, x);
 86
 87        (*env)->ReleaseStringUTFChars(env, format, cformat);
 88
 89        /* now call ps.print(str) */
 90
 91        /* get the class */
 92        class_PrintWriter = (*env)->GetObjectClass(env, out);
 93
 94        /* get the method ID */
 95        id_print = (*env)->GetMethodID(env, class_PrintWriter, "print", "(C)V");
 96
 97        /* call the method */
 98        for (i = 0; cstr[i] != 0 && !(*env)->ExceptionOccurred(env); i++)
 99           (*env)->CallVoidMethod(env, out, id_print, cstr[i]);
100
101        free(cstr);
102     }
```

程序清单 12-18   printf4/Printf4.java

```java
 1  import java.io.*;
 2
 3  /**
 4   * @version 1.10 1997-07-01
 5   * @author Cay Horstmann
 6   */
 7  class Printf4
 8  {
 9     public static native void fprint(PrintWriter ps, String format, double x);
10
```

```java
11    static
12    {
13        System.loadLibrary("Printf4");
14    }
15 }
```

程序清单 12-19　printf4/Printf4Test.java

```java
 1 import java.io.*;
 2
 3 /**
 4  * @version 1.11 2018-05-01
 5  * @author Cay Horstmann
 6  */
 7 class Printf4Test
 8 {
 9     public static void main(String[] args)
10     {
11         double price = 44.95;
12         double tax = 7.75;
13         double amountDue = price * (1 + tax / 100);
14         var out = new PrintWriter(System.out);
15         /* This call will throw an exception--note the %% */
16         Printf4.fprint(out, "Amount due = %%8.2f\n", amountDue);
17         out.flush();
18     }
19 }
```

> **API** 处理 Java 异常
>
> - jint Throw(JNIEnv *env, jthrowable obj)
>
>   准备一个在本地代码退出时抛出的异常。成功时返回 0，失败时返回一个负值。
>
> - jint ThrowNew(JNIEnv *env, jclass cl, const char msg[])
>
>   准备一个在本地代码退出时抛出的类型为 cl 的异常。成功时返回 0，失败时返回一个负值。msg 是表示异常对象的 String 构造参数的"modified UTF-8"字节序列
>
> - jthrowable ExceptionOccurred(JNIEnv *env)
>
>   如果有异常挂起，则返回该异常对象，否则返回 NULL。
>
> - jboolean ExceptionCheck(JNIEnv *env)
>
>   如果有异常挂起，则返回 true。
>
> - void ExceptionClear(JNIEnv *env)
>
>   清除挂起的异常。

12.9　使用调用 API

到现在为止，我们主要讨论的都是进行了一些 C 调用的用 Java 编程语言编写的程序，

这大概是因为 C 的运行速度更快一些，或者允许访问一些 Java 平台无法访问的功能。假设在相反的情况下，你有一个 C 或者 C++ 的程序，并且想要调用一些 Java 代码。调用 API（invocation API）使你能够把 Java 虚拟机嵌入到 C 或者 C++ 程序中。下面是初始化虚拟机所需的基本代码。

```
JavaVMOption options[1];
JavaVMInitArgs vm_args;
JavaVM *jvm;
JNIEnv *env;

options[0].optionString = "-Djava.class.path=.";
memset(&vm_args, 0, sizeof(vm_args));
vm_args.version = JNI_VERSION_1_2;
vm_args.nOptions = 1;
vm_args.options = options;

JNI_CreateJavaVM(&jvm, (void**) &env, &vm_args);
```

对 JNI_CreateJavaVM 的调用将创建虚拟机，并且使指针 jvm 指向虚拟机，使指针 env 指向执行环境。

可以给虚拟机提供任意数目的选项，这只需增加选项数组的大小和 vm_args.nOptions 的值。例如，

```
options[i].optionString = "-Djava.compiler=NONE";
```

可以钝化即时编译器。

> ✓ 提示：当你陷入麻烦导致程序崩溃，从而不能初始化 JVM 或者不能装载你的类时，请打开 JNI 调试模式。设置一个选项如下：
>
> ```
> options[i].optionString = "-verbose:jni";
> ```
>
> 你会看到一系列说明 JVM 初始化进程的消息。如果看不到你装载的类，请检查你的路径和类路径的设置。

一旦设置完虚拟机，就可以如前面小节介绍的那样调用 Java 方法了。只要按常规方法使用 env 指针即可。

只有在调用 API 中的其他函数时，才需要 jvm 指针。目前，只有四个这样的函数。最重要的一个是终止虚拟机的函数：

```
(*jvm)->DestroyJavaVM(jvm);
```

遗憾的是，在 Windows 下，动态链接到 jre/bin/client/jvm.dll 中的 JNI_CreateJavaVM 函数变得非常困难，因为 Vista 改变了链接规则，而 Oracle 的类库仍旧依赖于旧版本的 C 运行时类库。我们的示例程序通过手工加载该类库解决了这个问题，这种方式与 Java 程序所使用的方式一样，请参阅 JDK 中的 src.jar 文件里的 launcher/java_md.c 文件。

程序清单 12-20 的 C 程序设置了虚拟机，然后调用了 Welcome 类的 main 方法，这个类在卷 I 第 2 章中讨论过了（在开始启用测试程序之前，务必编译 Welcome.java 文件）。

程序清单 12-20　invocation/InvocationTest.c

```c
/**
   @version 1.20 2007-10-26
   @author Cay Horstmann
*/

#include <jni.h>
#include <stdlib.h>

#ifdef _WINDOWS

#include <windows.h>
static HINSTANCE loadJVMLibrary(void);
typedef jint (JNICALL *CreateJavaVM_t)(JavaVM **, void **, JavaVMInitArgs *);

#endif

int main()
{
   JavaVMOption options[2];
   JavaVMInitArgs vm_args;
   JavaVM *jvm;
   JNIEnv *env;
   long status;

   jclass class_Welcome;
   jclass class_String;
   jobjectArray args;
   jmethodID id_main;

#ifdef _WINDOWS
   HINSTANCE hjvmlib;
   CreateJavaVM_t createJavaVM;
#endif

   options[0].optionString = "-Djava.class.path=.";

   memset(&vm_args, 0, sizeof(vm_args));
   vm_args.version = JNI_VERSION_1_2;
   vm_args.nOptions = 1;
   vm_args.options = options;

#ifdef _WINDOWS
   hjvmlib = loadJVMLibrary();
   createJavaVM = (CreateJavaVM_t) GetProcAddress(hjvmlib, "JNI_CreateJavaVM");
   status = (*createJavaVM)(&jvm, (void **) &env, &vm_args);
#else
   status = JNI_CreateJavaVM(&jvm, (void **) &env, &vm_args);
#endif

   if (status == JNI_ERR)
   {
      fprintf(stderr, "Error creating VM\n");
```

```c
53        return 1;
54    }
55
56    class_Welcome = (*env)->FindClass(env, "Welcome");
57    id_main = (*env)->GetStaticMethodID(env, class_Welcome, "main", "([Ljava/lang/String;)V");
58
59    class_String = (*env)->FindClass(env, "java/lang/String");
60    args = (*env)->NewObjectArray(env, 0, class_String, NULL);
61    (*env)->CallStaticVoidMethod(env, class_Welcome, id_main, args);
62
63    (*jvm)->DestroyJavaVM(jvm);
64
65    return 0;
66 }
67
68 #ifdef _WINDOWS
69
70 static int GetStringFromRegistry(HKEY key, const char *name, char *buf, jint bufsize)
71 {
72    DWORD type, size;
73
74    return RegQueryValueEx(key, name, 0, &type, 0, &size) == 0
75        && type == REG_SZ
76        && size < (unsigned int) bufsize
77        && RegQueryValueEx(key, name, 0, 0, buf, &size) == 0;
78 }
79
80 static void GetPublicJREHome(char *buf, jint bufsize)
81 {
82    HKEY key, subkey;
83    char version[MAX_PATH];
84
85    /* Find the current version of the JRE */
86    char *JRE_KEY = "Software\\JavaSoft\\Java Runtime Environment";
87    if (RegOpenKeyEx(HKEY_LOCAL_MACHINE, JRE_KEY, 0, KEY_READ, &key) != 0)
88    {
89        fprintf(stderr, "Error opening registry key '%s'\n", JRE_KEY);
90        exit(1);
91    }
92
93    if (!GetStringFromRegistry(key, "CurrentVersion", version, sizeof(version)))
94    {
95        fprintf(stderr, "Failed reading value of registry key:\n\t%s\\CurrentVersion\n",
96            JRE_KEY);
97        RegCloseKey(key);
98        exit(1);
99    }
100
101    /* Find directory where the current version is installed. */
102    if (RegOpenKeyEx(key, version, 0, KEY_READ, &subkey) != 0)
103    {
104        fprintf(stderr, "Error opening registry key '%s\\%s'\n", JRE_KEY, version);
105        RegCloseKey(key);
106        exit(1);
```

```
107        }
108
109        if (!GetStringFromRegistry(subkey, "JavaHome", buf, bufsize))
110        {
111           fprintf(stderr, "Failed reading value of registry key:\n\t%s\\%s\\JavaHome\n",
112              JRE_KEY, version);
113           RegCloseKey(key);
114           RegCloseKey(subkey);
115           exit(1);
116        }
117
118        RegCloseKey(key);
119        RegCloseKey(subkey);
120  }
121
122  static HINSTANCE loadJVMLibrary(void)
123  {
124        HINSTANCE h1, h2;
125        char msvcdll[MAX_PATH];
126        char javadll[MAX_PATH];
127        GetPublicJREHome(msvcdll, MAX_PATH);
128        strcpy(javadll, msvcdll);
129        strncat(msvcdll, "\\bin\\msvcr71.dll", MAX_PATH - strlen(msvcdll));
130        msvcdll[MAX_PATH - 1] = '\0';
131        strncat(javadll, "\\bin\\client\\jvm.dll", MAX_PATH - strlen(javadll));
132        javadll[MAX_PATH - 1] = '\0';
133
134        h1 = LoadLibrary(msvcdll);
135        if (h1 == NULL)
136        {
137           fprintf(stderr, "Can't load library msvcr71.dll\n");
138           exit(1);
139        }
140
141        h2 = LoadLibrary(javadll);
142        if (h2 == NULL)
143        {
144           fprintf(stderr, "Can't load library jvm.dll\n");
145           exit(1);
146        }
147        return h2;
148  }
149
150  #endif
```

要在 Linux 下编译该程序，请用：

```
gcc -I jdk/include -I jdk/include/linux -o InvocationTest \
   -L jdk/jre/lib/i386/client -ljvm InvocationTest.c
```

在 Windows 下用微软的 C 编译器时，请用下面的命令行：

```
cl -D_WINDOWS -I jdk\include -I jdk\include\win32 InvocationTest.c \
   jdk\lib\jvm.lib advapi32.lib
```

需要确保 INCLUDE 和 LIB 环境变量包含了 Windows API 头文件和库文件的路径。

用 Cygwin 时，用下面的语句进行编译：

```
gcc -D_WINDOWS -mno-cygwin -I jdk\include -I jdk\include\win32 -D__int64="long long" \
    -I c:\cygwin\usr\include\w32api -o InvocationTest
```

在 Linux/UNIX 下运行该程序之前，需要确保 LD_LIBRARY_PATH 包含了共享类库的目录。例如，如果使用 Linux 上的 bash 命令行，则需要执行下面的命令：

```
export LD_LIBRARY_PATH=jdk/jre/lib/i386/client:$LD_LIBRARY_PATH
```

> **API** 调用 API 函数
>
> - jint JNI_CreateJavaVM(JavaVM** p_jvm, void** p_env, JavaVMInitArgs* vm_args)
>
>   初始化 Java 虚拟机。如果成功，则返回 0，否则返回 JNI_ERR。
>
> - jint DestroyJavaVM(JavaVM* jvm)
>
>   销毁虚拟机。如果成功，则返回 0，否则返回一个负值。该函数必须通过一个虚拟机指针调用。例如，(*jvm)->DestroyJavaVM(jvm)。

12.10 完整的示例：访问 Windows 注册表

在本节中，我们介绍一个完整的可运行的例子，涵盖了我们在本章讨论的所有内容：使用带有字符串、数组和对象的本地方法，构造器调用和错误处理。我们将展示如何用 Java 平台包装器来包装普通的基于 C 的 API 子集，用于进行 Windows 注册表操作。当然，由于 Windows 的具体特性，使用 Windows 注册表的程序天生就不可移植。基于这个原因，标准的 Java 库不支持注册表，所以使用本地方法访问注册表是有意义的。

12.10.1 Windows 注册表概述

Windows 注册表是一个存放 Windows 操作系统和应用程序的配置信息的数据仓库。它提供了对系统和应用程序参数的单点管理和备份。其不足的方面是，注册表的错误也是单点的。如果你弄乱了注册表，你的电脑就会出故障，甚至无法启动。

我们不建议你使用注册表来存储 Java 程序的配置参数。Java 配置 API（preferences API）是一个更好的解决方案（更多信息请参见卷 I 第 10 章）。我们使用注册表只是为了说明怎样把重要的本地 API 包装成 Java 类。

检查注册表的主要工具是注册表编辑器。由于可能存在幼稚而狂热的用户，所以 Windows 没有配备任何图标来启动注册表编辑器。你必须启动 DOS shell（或打开"开始"→"运行"对话框）然后键入 regedit。图 12-4 给出了一个运行中的注册表编辑器。

左边是树形结构排列的注册表键。请注意，每个键都以 HKEY 节点开始，如：

```
HKEY_CLASSES_ROOT
HKEY_CURRENT_USER
HKEY_LOCAL_MACHINE
...
```

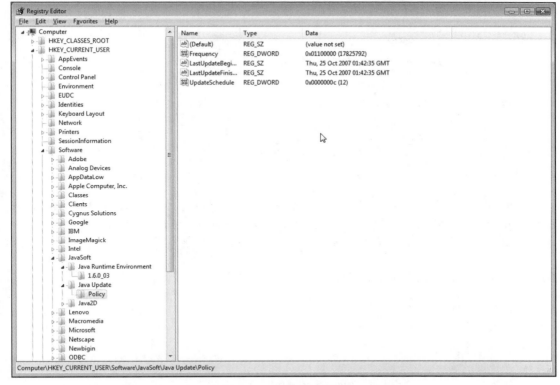

图 12-4 注册表编辑器

右边是与特定键关联的名 / 值对。例如，如果你安装了 Java 11，那么键：

HKEY_LOCAL_MACHINE\Software\JavaSoft\Java Runtime Environment

就包含下面这样的名值对：

CurrentVersion="11.0_10"

在本例中，值是字符串。值也可以是整数或字节数组。

12.10.2 访问注册表的 Java 平台接口

我们创建了一个从 Java 代码访问注册表的简单接口，然后用本地代码实现了这个接口。我们的接口只允许几个注册表操作，为了保持较小的代码规模，我们省略了其他重要的操作，如：添加、删除和枚举注册表键（添加剩余的这些注册表 API 函数是很容易的）。

即使使用我们提供的受限的子集，你也可以：

- 枚举某个键中存储的所有名字。
- 读出用某个名字存储的值。
- 设置用某个名字存储的值。

下面是封装注册表键的 Java 类：

```java
public class Win32RegKey
{
    public Win32RegKey(int theRoot, String thePath) { . . . }
    public Enumeration names() { . . . }
    public native Object getValue(String name);
    public native void setValue(String name, Object value);

    public static final int HKEY_CLASSES_ROOT = 0x80000000;
    public static final int HKEY_CURRENT_USER = 0x80000001;
    public static final int HKEY_LOCAL_MACHINE = 0x80000002;
    . . .
}
```

names 方法返回与该键存放在一起的所有名字的一个枚举，你可以用你熟悉的 hasMore-Elements/nextElement 方法获取它们。getValue 方法返回一个对象，该对象可以是字符串、Integer 对象或字节数组。setValue 方法的 value 参数也必须是上述三种类型之一。

12.10.3　以本地方法实现注册表访问函数

我们需要实现三个操作：
- 获取某个键的值。
- 设置某个键的值。
- 迭代键的名字。

在本章中，你基本上已经看到了所有必需的工具，如 Java 字符串和数组到 C 的字符串和数组的转换，还了解了如何在出错时抛出异常。

有两个问题使得这些本地方法比之前的例子更加复杂。getValue 和 setValue 方法处理的是 Object 类型，它可以是 String、Integer 或 byte[ ] 之一。枚举对象需要用来存放连续的对 hasMoreElements 和 nextElement 的调用之间的状态。

让我们先看一下 getValue 方法，该方法（见程序清单 12-22）经历了以下几个步骤：

1. 打开注册表键。为了读取它们的值，注册表 API 要求这些键是开放的。
2. 查询与名字关联的值的类型和大小。
3. 把数据读到缓存。
4. 如果类型是 REG_SZ（字符串），调用 NewStringUTF，用该值来创建一个新的字符串。
5. 如果类型是 REG_DWORD（32 位整数），调用 Integer 构造器。
6. 如果类型是 REG_BINARY，调用 NewByteArray 来创建一个新的字节数组，并调用 SetByteArrayRegion，把值数据复制到该字节数组中。
7. 如果不是以上类型或调用 API 函数时出现错误，那就抛出异常，并小心地释放到此为止所获得的所有资源。
8. 关闭键，并返回创建的对象（String、Integer 或 byte[]）。

如你所见，这个例子很好地说明了怎样产生不同类型的 Java 对象。

在本地方法中，处理泛化的返回类型并不困难，jstring、jobject 或 jarray 引用都可以直接作为一个 jobject 返回。但是，setValue 方法接受的是一个对 Object 的引用，并且，为了把该 Object 保存为字符串、整数或字节数组，必须确定该 Object 的确切类型。我们可以通过查询 value 对象的类，找出对 java.lang.String、java.lang.Integer 和 byte[] 的引用，将其与 IsAssignableFrom 函数进行比较，从而确定它的确切类型。

如果 class1 和 class2 是两个类引用，那么调用：

```
(*env)->IsAssignableFrom(env, class1, class2)
```

当 class1 和 class2 是同一个类或 class1 是 class2 的子类时，返回 JNI_TRUE。在这两种情况下，class1 对象的引用都可以转型到 class2。例如，当：

```
(*env)->IsAssignableFrom(env, (*env)->GetObjectClass(env, value),
    (*env)->FindClass(env, "[B"))
```

为 true 时，那么我们就知道该值是一个字节数组。

下面是对 setValue 方法中的步骤的概述：

1. 打开注册表键以便写入。
2. 找出要写入的值的类型。
3. 如果类型是 String，调用 GetStringUTFChars 获取一个指向这些字符的指针。
4. 如果类型是 Integer，调用 intValue 方法获取该包装器对象中存储的整数。
5. 如果类型是 byte[]，调用 GetByteArrayElements 获取指向这些字节的指针。
6. 把数据和长度传递给注册表。
7. 关闭键。
8. 如果类型是 String 或 byte[]，那么还要释放指向数据的指针。

最后，我们介绍枚举键的本地方法。这些方法属于 Win32RegKeyNameEnumeration 类（参见程序清单 12-21）。当枚举过程开始时，我们必须打开键。在枚举过程中，我们必须保持该键的句柄。也就是说，该键的句柄必须与枚举对象存放在一起。键的句柄是 DWORD 类型的，它是一个 32 位数，所以可以存放在一个 Java 的整数中。它被存放在枚举类的 hkey 域中，当枚举开始时，SetIntField 初始化该域，而后续的调用用 GetIntField 来读取其值。

程序清单 12-21　win32reg/Win32RegKey.java

```
 1  import java.util.*;
 2
 3  /**
 4   * A Win32RegKey object can be used to get and set values of a registry key in the Windows
 5   * registry.
 6   * @version 1.00 1997-07-01
 7   * @author Cay Horstmann
 8   */
 9  public class Win32RegKey
10  {
11      public static final int HKEY_CLASSES_ROOT = 0x80000000;
12      public static final int HKEY_CURRENT_USER = 0x80000001;
```

```java
13    public static final int HKEY_LOCAL_MACHINE = 0x80000002;
14    public static final int HKEY_USERS = 0x80000003;
15    public static final int HKEY_CURRENT_CONFIG = 0x80000005;
16    public static final int HKEY_DYN_DATA = 0x80000006;
17
18    private int root;
19    private String path;
20
21    /**
22     * Gets the value of a registry entry.
23     * @param name the entry name
24     * @return the associated value
25     */
26    public native Object getValue(String name);
27
28    /**
29     * Sets the value of a registry entry.
30     * @param name the entry name
31     * @param value the new value
32     */
33    public native void setValue(String name, Object value);
34
35    /**
36     * Construct a registry key object.
37     * @param theRoot one of HKEY_CLASSES_ROOT, HKEY_CURRENT_USER, HKEY_LOCAL_MACHINE,
38     * HKEY_USERS, HKEY_CURRENT_CONFIG, HKEY_DYN_DATA
39     * @param thePath the registry key path
40     */
41    public Win32RegKey(int theRoot, String thePath)
42    {
43       root = theRoot;
44       path = thePath;
45    }
46
47    /**
48     * Enumerates all names of registry entries under the path that this object describes.
49     * @return an enumeration listing all entry names
50     */
51    public Enumeration<String> names()
52    {
53       return new Win32RegKeyNameEnumeration(root, path);
54    }
55
56    static
57    {
58       System.loadLibrary("Win32RegKey");
59    }
60 }
61
62 class Win32RegKeyNameEnumeration implements Enumeration<String>
63 {
64    public native String nextElement();
65    public native boolean hasMoreElements();
66    private int root;
```

```
67     private String path;
68     private int index = -1;
69     private int hkey = 0;
70     private int maxsize;
71     private int count;
72
73     Win32RegKeyNameEnumeration(int theRoot, String thePath)
74     {
75        root = theRoot;
76        path = thePath;
77     }
78  }
79
80  class Win32RegKeyException extends RuntimeException
81  {
82     public Win32RegKeyException()
83     {
84     }
85
86     public Win32RegKeyException(String why)
87     {
88        super(why);
89     }
90  }
```

在这个例子里,我们用枚举对象存放了另外三个数据项。当枚举一开始,我们可以从注册表中查询到名/值对的个数和最长名字的长度,我们需要这些信息,因此我们分配 C 字符数组以保存这些名字。这些值存放在枚举对象的 count 和 maxsize 域中。最后,index 域被初始化为 –1,表示枚举的开始。一旦其他实例域被初始化,index 域就被置为 0,在完成每个枚举步骤之后,都会进行递增。

让我们简要介绍一下支持枚举的本地方法。hasMoreElements 方法很简单:

1. 获取 index 和 count 域。

2. 如果 index 是 -1,调用 startNameEnumeration 函数打开键,查询数量和最大长度,初始化 hkey、count、maxsize 和 index 域。

3. 如果 index 小于 count,则返回 JNI_TRUE,否则返回 JNI_FALSE。

nextElement 方法要复杂一些。

1. 获取 index 和 count 域。

2. 如果 index 是 -1,调用 startNameEnumeration 函数打开键,查询数量和最大长度,初始化 hkey、count、maxsize 和 index 域。

3. 如果 index 等于 count,抛出一个 NoSuchElementException 异常。

4. 从注册表中读入下一个名字。

5. 递增 index。

6. 如果 index 等于 count,则关闭键。

在编译之前,记得在 Win32RegKey 和 Win32RegKeyNameEnumeration 上都要运行 javac-h。微软编

译器的完整命令行如下：

cl -I *jdk*\include -I *jdk*\include\win32 -LD Win32RegKey.c advapi32.lib -FeWin32RegKey.dll

Cygwin 系统上，请使用：

gcc -mno-cygwin -D __int64="long long" -I *jdk*\include -I *jdk*\include\win32 \
 -I c:\cygwin\usr\include\w32api -shared -Wl,--add-stdcall-alias -o Win32RegKey.dll
Win32RegKey.c

因为注册表 API 是针对 Windows 的，所以这个程序不能在其他操作系统上运行。

程序清单 12-23 给出了测试我们新的注册表函数的程序。我们在键中添加了三个名值对：一个字符串、一个整数和一个字节数组。

HKEY_CURRENT_USER\Software\JavaSoft\Java Runtime Environment

然后，我们枚举该键的所有名字并获取它们的值。该程序应该打印如下信息：

```
Default user=Harry Hacker
Lucky number=13
Small primes=2 3 5 7 11 13
```

虽然在该键中添加这些名值对不会有什么危害，但是在运行该程序后，你可能还是想使用注册表编辑器去移除它们。

程序清单 12-22　win32reg/Win32RegKey.c

```
1  /**
2     @version 1.00 1997-07-01
3     @author Cay Horstmann
4  */
5
6  #include "Win32RegKey.h"
7  #include "Win32RegKeyNameEnumeration.h"
8  #include <string.h>
9  #include <stdlib.h>
10 #include <windows.h>
11
12 JNIEXPORT jobject JNICALL Java_Win32RegKey_getValue(
13    JNIEnv* env, jobject this_obj, jobject name)
14 {
15    const char* cname;
16    jstring path;
17    const char* cpath;
18    HKEY hkey;
19    DWORD type;
20    DWORD size;
21    jclass this_class;
22    jfieldID id_root;
23    jfieldID id_path;
24    HKEY root;
25    jobject ret;
26    char* cret;
27
28    /* get the class */
```

```c
29      this_class = (*env)->GetObjectClass(env, this_obj);
30
31      /* get the field IDs */
32      id_root = (*env)->GetFieldID(env, this_class, "root", "I");
33      id_path = (*env)->GetFieldID(env, this_class, "path", "Ljava/lang/String;");
34
35      /* get the fields */
36      root = (HKEY) (*env)->GetIntField(env, this_obj, id_root);
37      path = (jstring)(*env)->GetObjectField(env, this_obj, id_path);
38      cpath = (*env)->GetStringUTFChars(env, path, NULL);
39
40      /* open the registry key */
41      if (RegOpenKeyEx(root, cpath, 0, KEY_READ, &hkey) != ERROR_SUCCESS)
42      {
43         (*env)->ThrowNew(env, (*env)->FindClass(env, "Win32RegKeyException"),
44             "Open key failed");
45         (*env)->ReleaseStringUTFChars(env, path, cpath);
46         return NULL;
47      }
48
49      (*env)->ReleaseStringUTFChars(env, path, cpath);
50      cname = (*env)->GetStringUTFChars(env, name, NULL);
51
52      /* find the type and size of the value */
53      if (RegQueryValueEx(hkey, cname, NULL, &type, NULL, &size) != ERROR_SUCCESS)
54      {
55         (*env)->ThrowNew(env, (*env)->FindClass(env, "Win32RegKeyException"),
56             "Query value key failed");
57         RegCloseKey(hkey);
58         (*env)->ReleaseStringUTFChars(env, name, cname);
59         return NULL;
60      }
61
62      /* get memory to hold the value */
63      cret = (char*)malloc(size);
64
65      /* read the value */
66      if (RegQueryValueEx(hkey, cname, NULL, &type, cret, &size) != ERROR_SUCCESS)
67      {
68         (*env)->ThrowNew(env, (*env)->FindClass(env, "Win32RegKeyException"),
69             "Query value key failed");
70         free(cret);
71         RegCloseKey(hkey);
72         (*env)->ReleaseStringUTFChars(env, name, cname);
73         return NULL;
74      }
75
76      /* depending on the type, store the value in a string,
77         integer, or byte array */
78      if (type == REG_SZ)
79      {
80         ret = (*env)->NewStringUTF(env, cret);
81      }
82      else if (type == REG_DWORD)
```

```c
83      {
84          jclass class_Integer = (*env)->FindClass(env, "java/lang/Integer");
85          /* get the method ID of the constructor */
86          jmethodID id_Integer = (*env)->GetMethodID(env, class_Integer, "<init>", "(I)V");
87          int value = *(int*) cret;
88          /* invoke the constructor */
89          ret = (*env)->NewObject(env, class_Integer, id_Integer, value);
90      }
91      else if (type == REG_BINARY)
92      {
93          ret = (*env)->NewByteArray(env, size);
94          (*env)->SetByteArrayRegion(env, (jarray) ret, 0, size, cret);
95      }
96      else
97      {
98          (*env)->ThrowNew(env, (*env)->FindClass(env, "Win32RegKeyException"),
99                  "Unsupported value type");
100         ret = NULL;
101     }
102
103     free(cret);
104     RegCloseKey(hkey);
105     (*env)->ReleaseStringUTFChars(env, name, cname);
106
107     return ret;
108 }
109
110 JNIEXPORT void JNICALL Java_Win32RegKey_setValue(JNIEnv* env, jobject this_obj,
111         jstring name, jobject value)
112 {
113     const char* cname;
114     jstring path;
115     const char* cpath;
116     HKEY hkey;
117     DWORD type;
118     DWORD size;
119     jclass this_class;
120     jclass class_value;
121     jclass class_Integer;
122     jfieldID id_root;
123     jfieldID id_path;
124     HKEY root;
125     const char* cvalue;
126     int ivalue;
127
128     /* get the class */
129     this_class = (*env)->GetObjectClass(env, this_obj);
130
131     /* get the field IDs */
132     id_root = (*env)->GetFieldID(env, this_class, "root", "I");
133     id_path = (*env)->GetFieldID(env, this_class, "path", "Ljava/lang/String;");
134
135     /* get the fields */
136     root = (HKEY)(*env)->GetIntField(env, this_obj, id_root);
```

```c
137        path = (jstring)(*env)->GetObjectField(env, this_obj, id_path);
138        cpath = (*env)->GetStringUTFChars(env, path, NULL);
139
140        /* open the registry key */
141        if (RegOpenKeyEx(root, cpath, 0, KEY_WRITE, &hkey) != ERROR_SUCCESS)
142        {
143           (*env)->ThrowNew(env, (*env)->FindClass(env, "Win32RegKeyException"),
144                "Open key failed");
145           (*env)->ReleaseStringUTFChars(env, path, cpath);
146           return;
147        }
148
149        (*env)->ReleaseStringUTFChars(env, path, cpath);
150        cname = (*env)->GetStringUTFChars(env, name, NULL);
151
152        class_value = (*env)->GetObjectClass(env, value);
153        class_Integer = (*env)->FindClass(env, "java/lang/Integer");
154        /* determine the type of the value object */
155        if ((*env)->IsAssignableFrom(env, class_value, (*env)->FindClass(env, "java/lang/String")))
156        {
157           /* it is a string--get a pointer to the characters */
158           cvalue = (*env)->GetStringUTFChars(env, (jstring) value, NULL);
159           type = REG_SZ;
160           size = (*env)->GetStringLength(env, (jstring) value) + 1;
161        }
162        else if ((*env)->IsAssignableFrom(env, class_value, class_Integer))
163        {
164           /* it is an integer--call intValue to get the value */
165           jmethodID id_intValue = (*env)->GetMethodID(env, class_Integer, "intValue", "()I");
166           ivalue = (*env)->CallIntMethod(env, value, id_intValue);
167           type = REG_DWORD;
168           cvalue = (char*)&ivalue;
169           size = 4;
170        }
171        else if ((*env)->IsAssignableFrom(env, class_value, (*env)->FindClass(env, "[B")))
172        {
173           /* it is a byte array--get a pointer to the bytes */
174           type = REG_BINARY;
175           cvalue = (char*)(*env)->GetByteArrayElements(env, (jarray) value, NULL);
176           size = (*env)->GetArrayLength(env, (jarray) value);
177        }
178        else
179        {
180           /* we don't know how to handle this type */
181           (*env)->ThrowNew(env, (*env)->FindClass(env, "Win32RegKeyException"),
182                "Unsupported value type");
183           RegCloseKey(hkey);
184           (*env)->ReleaseStringUTFChars(env, name, cname);
185           return;
186        }
187
188        /* set the value */
189        if (RegSetValueEx(hkey, cname, 0, type, cvalue, size) != ERROR_SUCCESS)
190        {
```

```c
191         (*env)->ThrowNew(env, (*env)->FindClass(env, "Win32RegKeyException"),
192             "Set value failed");
193      }
194
195      RegCloseKey(hkey);
196      (*env)->ReleaseStringUTFChars(env, name, cname);
197
198      /* if the value was a string or byte array, release the pointer */
199      if (type == REG_SZ)
200      {
201         (*env)->ReleaseStringUTFChars(env, (jstring) value, cvalue);
202      }
203      else if (type == REG_BINARY)
204      {
205         (*env)->ReleaseByteArrayElements(env, (jarray) value, (jbyte*) cvalue, 0);
206      }
207   }
208
209   /* helper function to start enumeration of names */
210   static int startNameEnumeration(JNIEnv* env, jobject this_obj, jclass this_class)
211   {
212      jfieldID id_index;
213      jfieldID id_count;
214      jfieldID id_root;
215      jfieldID id_path;
216      jfieldID id_hkey;
217      jfieldID id_maxsize;
218
219      HKEY root;
220      jstring path;
221      const char* cpath;
222      HKEY hkey;
223      DWORD maxsize = 0;
224      DWORD count = 0;
225
226      /* get the field IDs */
227      id_root = (*env)->GetFieldID(env, this_class, "root", "I");
228      id_path = (*env)->GetFieldID(env, this_class, "path", "Ljava/lang/String;");
229      id_hkey = (*env)->GetFieldID(env, this_class, "hkey", "I");
230      id_maxsize = (*env)->GetFieldID(env, this_class, "maxsize", "I");
231      id_index = (*env)->GetFieldID(env, this_class, "index", "I");
232      id_count = (*env)->GetFieldID(env, this_class, "count", "I");
233
234      /* get the field values */
235      root = (HKEY)(*env)->GetIntField(env, this_obj, id_root);
236      path = (jstring)(*env)->GetObjectField(env, this_obj, id_path);
237      cpath = (*env)->GetStringUTFChars(env, path, NULL);
238
239      /* open the registry key */
240      if (RegOpenKeyEx(root, cpath, 0, KEY_READ, &hkey) != ERROR_SUCCESS)
241      {
242         (*env)->ThrowNew(env, (*env)->FindClass(env, "Win32RegKeyException"),
243             "Open key failed");
244         (*env)->ReleaseStringUTFChars(env, path, cpath);
```

```c
245        return -1;
246     }
247     (*env)->ReleaseStringUTFChars(env, path, cpath);
248
249     /* query count and max length of names */
250     if (RegQueryInfoKey(hkey, NULL, NULL, NULL, NULL, NULL, NULL, &count, &maxsize,
251            NULL, NULL, NULL) != ERROR_SUCCESS)
252     {
253        (*env)->ThrowNew(env, (*env)->FindClass(env, "Win32RegKeyException"),
254             "Query info key failed");
255        RegCloseKey(hkey);
256        return -1;
257     }
258
259     /* set the field values */
260     (*env)->SetIntField(env, this_obj, id_hkey, (DWORD) hkey);
261     (*env)->SetIntField(env, this_obj, id_maxsize, maxsize + 1);
262     (*env)->SetIntField(env, this_obj, id_index, 0);
263     (*env)->SetIntField(env, this_obj, id_count, count);
264     return count;
265 }
266
267 JNIEXPORT jboolean JNICALL Java_Win32RegKeyNameEnumeration_hasMoreElements(JNIEnv* env,
268        jobject this_obj)
269 {
270     jclass this_class;
271     jfieldID id_index;
272     jfieldID id_count;
273     int index;
274     int count;
275     /* get the class */
276     this_class = (*env)->GetObjectClass(env, this_obj);
277
278     /* get the field IDs */
279     id_index = (*env)->GetFieldID(env, this_class, "index", "I");
280     id_count = (*env)->GetFieldID(env, this_class, "count", "I");
281
282     index = (*env)->GetIntField(env, this_obj, id_index);
283     if (index == -1) /* first time */
284     {
285        count = startNameEnumeration(env, this_obj, this_class);
286        index = 0;
287     }
288     else
289        count = (*env)->GetIntField(env, this_obj, id_count);
290     return index < count;
291 }
292
293 JNIEXPORT jobject JNICALL Java_Win32RegKeyNameEnumeration_nextElement(JNIEnv* env,
294        jobject this_obj)
295 {
296     jclass this_class;
297     jfieldID id_index;
298     jfieldID id_hkey;
```

```c
299     jfieldID id_count;
300     jfieldID id_maxsize;
301
302     HKEY hkey;
303     int index;
304     int count;
305     DWORD maxsize;
306
307     char* cret;
308     jstring ret;
309
310     /* get the class */
311     this_class = (*env)->GetObjectClass(env, this_obj);
312
313     /* get the field IDs */
314     id_index = (*env)->GetFieldID(env, this_class, "index", "I");
315     id_count = (*env)->GetFieldID(env, this_class, "count", "I");
316     id_hkey = (*env)->GetFieldID(env, this_class, "hkey", "I");
317     id_maxsize = (*env)->GetFieldID(env, this_class, "maxsize", "I");
318
319     index = (*env)->GetIntField(env, this_obj, id_index);
320     if (index == -1) /* first time */
321     {
322         count = startNameEnumeration(env, this_obj, this_class);
323         index = 0;
324     }
325     else
326         count = (*env)->GetIntField(env, this_obj, id_count);
327
328     if (index >= count) /* already at end */
329     {
330         (*env)->ThrowNew(env, (*env)->FindClass(env, "java/util/NoSuchElementException"),
331             "past end of enumeration");
332         return NULL;
333     }
334
335     maxsize = (*env)->GetIntField(env, this_obj, id_maxsize);
336     hkey = (HKEY)(*env)->GetIntField(env, this_obj, id_hkey);
337     cret = (char*)malloc(maxsize);
338
339     /* find the next name */
340     if (RegEnumValue(hkey, index, cret, &maxsize, NULL, NULL, NULL, NULL) != ERROR_SUCCESS)
341     {
342         (*env)->ThrowNew(env, (*env)->FindClass(env, "Win32RegKeyException"),
343             "Enum value failed");
344         free(cret);
345         RegCloseKey(hkey);
346         (*env)->SetIntField(env, this_obj, id_index, count);
347         return NULL;
348     }
349
350     ret = (*env)->NewStringUTF(env, cret);
351     free(cret);
352
```

```
353            /* increment index */
354            index++;
355            (*env)->SetIntField(env, this_obj, id_index, index);
356
357            if (index == count) /* at end */
358            {
359               RegCloseKey(hkey);
360            }
361
362            return ret;
363         }
```

程序清单 12-23 win32reg/Win32RegKeyTest.java

```java
 1   import java.util.*;
 2
 3   /**
 4      @version 1.03 2018-05-01
 5      @author Cay Horstmann
 6   */
 7   public class Win32RegKeyTest
 8   {
 9      public static void main(String[] args)
10      {
11         var key = new Win32RegKey(
12            Win32RegKey.HKEY_CURRENT_USER, "Software\\JavaSoft\\Java Runtime Environment");
13
14         key.setValue("Default user", "Harry Hacker");
15         key.setValue("Lucky number", new Integer(13));
16         key.setValue("Small primes", new byte[] { 2, 3, 5, 7, 11 });
17
18         Enumeration<String> e = key.names();
19
20         while (e.hasMoreElements())
21         {
22            String name = e.nextElement();
23            System.out.print(name + "=");
24
25            Object value = key.getValue(name);
26
27            if (value instanceof byte[])
28               for (byte b : (byte[]) value) System.out.print((b & 0xFF) + " ");
29            else
30               System.out.print(value);
31
32            System.out.println();
33         }
34      }
35   }
```

API 类型质询函数

- jboolean IsAssignableFrom(JNIEnv *env, jclass cl1, jclass cl2)

如果第一个类的对象可以赋给第二个类的对象，则返回 JNI_TRUE，否则返回 JNI_FALSE。这个函数可以测试：两个类是否相同，cl1 是否是 cl2 的子类，cl2 是否表示一个由 cl1 或它的一个超类实现的接口。

- jclass GetSuperclass(JNIEnv *env, jclass cl)

    返回某个类的超类。如果 cl 表示 Object 类或一个接口，则返回 NULL。

一路走来，大家已经学习了许多高级 API，现在，终于结束了。我们从每位 Java 程序员都应该了解的主题开始，即：流、XML、网络、数据库和国际化，又用了非常技术性的几章结尾，即安全、注解处理、高级图形化编程和本地方法。我们希望你能够真正享受这个旅程，掌握这些涉及领域广泛的 Java API，并能够将这些新知识应用到你的项目中。

推荐阅读

Effective Java中文版（原书第3版）

作者：[美]约书亚·布洛克（Joshua Bloch） ISBN：978-7-111-61272-8 定价：119.00元

Java之父James Gosling鼎力推荐、Jolt获奖作品全新升级，针对Java 7、8、9全面更新，Java程序员必备参考书

包含大量完整的示例代码和透彻的技术分析，通过90条经验法则，探索新的设计模式和语言习惯用法，帮助读者更加有效地使用Java编程语言及其基本类库

推荐阅读